Forest Ecosystems

THIRD EDITION

Forest Ecosystems

Analysis at Multiple Scales

THIRD EDITION

Richard H. Waring

DEPARTMENT OF FOREST SCIENCE
OREGON STATE UNIVERSITY
CORVALLIS, OREGON

Steven W. Running

DEPARTMENT OF ECOSYSTEM AND CONSERVATION SCIENCES
NUMERICAL TERRADYNAMIC SIMULATION GROUP
UNIVERSITY OF MONTANA
MISSOULA, MONTANA

AMSTERDAM • BOSTON • HEIDELBERG • LONDON
NEW YORK • OXFORD • PARIS • SAN DIEGO
SAN FRANCISCO • SINGAPORE • SYDNEY • TOKYO

Academic Press is an imprint of Elsevier

ELSEVIER

Elsevier Academic Press
30 Corporate Drive, Suite 400, Burlington, MA 01803, USA
525 B Street, Suite 1900, San Diego, California 92101-4495, USA
84 Theobald's Road, London WC1X 8RR, UK

This book is printed on acid-free paper.

Cover Image–Gross primary production for June 2–10, 2003, draped over a digital elevation map of the complex 50,000 km^2 forested landscape of western Montana. The calculations for this image combined 250 m resolution spectral data from MODIS of the fraction of absorbed PAR with 1 km daily surface meteorology (Running *et al.*, 2004).

Library of Congress Cataloging-in-Publication Data

Waring, Richard H.
   Forest ecosystems analysis at multiple scales / Richard H. Waring, Steven W. Running. – 3rd ed.
     p. cm.
   Includes bibliographical references and index.
   ISBN-13: 978-0-12-370605-8 (pbk. : alk. paper)   1. Forest ecology.   2. Forest management.   I. Running, S. W.   II. Title.
   QH541.5.F6W34 2007
   577.3–dc22   2007017124

British Library Cataloguing in Publication Data
A catalogue record for this book is available from the British Library

ISBN: 978-0-12-370605-8

For all information on all Elsevier Academic Press publications
visit our Web site at www.books.elsevier.com

Printed and bound by CPI Group (UK) Ltd, Croydon, CR0 4YY

Transferred to Digital Print 2012

R.H.W. dedicates this book to his wife, Doris Carlson Waring,
his partner in love and life for 50 years.

S.W.R. dedicates this book to his wife, Constance C. Running,
who has found fabric art to be a good antidote to
her husband's preoccupation with his science.

CONTENTS

SECTION II. Introduction to Temporal Scaling

5. Temporal Changes in Forest Structure and Function

6. Susceptibility and Response of Forests to Disturbance

SECTION III. Introduction to Spatial Scaling and Spatial/Temporal Modeling

7. Spatial Scaling Methods for Landscape and Regional Ecosystem Analysis

Color plates appear between pages 220–221; 260–261; 268–269; and at the back of the book.

PREFACE TO THE THIRD EDITION

With notification that the second edition of *Forest Ecosystems* was out of print, we considered whether a complete revision of the text was warranted. We recognize that the modeling of forest ecosystem responses has increased significantly in the last decade to include biodiversity and climatic limitations, but the underlying principles presented in the second edition appear to remain sound. At the same time, the network of sites that continuously monitor seasonal and interannual variation in CO_2 and water vapor exchange has grown. By combining data from many sites, new generalities have emerged that should be shared. Also, we wished to illustrate that forest disturbance and recovery can be more accurately documented than was previously possible. Finally, with the publication of the 2007 reports by the Intergovernmental Panel on Climate Change, we recognize the pressing need to provide background material to policy makers charged with designing policies that reduce carbon emissions and perpetuate healthy forests in an unstable climate with new mixtures of species.

In our decision to publish a third edition of the textbook, we have corrected errors in the previous edition, updated color plates, and added Chapter 10 that focuses on new information. Specifically, we document in the added chapter how climatic change has already affected forests, offer insights gained from an expanded network of eddy-flux sites, and provide evidence of improvements in remote sensing technology. To keep up with the expanded role that remote sensing and modeling will play in predicting and monitoring the effects of future policies, we provide a web site reference to replace the compact disc available with the previous edition.

R. H. W.
S. W. R.

PREFACE TO THE SECOND EDITION

The first edition of *Forest Ecosystems: Concepts and Management*, with my colleague William Schlesinger of Duke University, was published in 1985. At that time, most of the information on forest ecosystems consisted of mass balance analyses conducted on stands or small watersheds for periods up to one year. Few simulation models were available, and those that could be tested were largely restricted to predictions of streamflow. Today, new methods and new models provide a much wider basis for extrapolation, in space as well as time. In 1991 and again in 1997, William Schlesinger demonstrated his unique abilities to synthesize and expand our understanding of terrestrial and aquatic ecosystems by publishing *Biogeochemistry: An Analysis of Global Change.*

The opportunity to expand the scope of analysis of forest ecosystems was clear. Such an expansion, however, required new techniques and experience beyond those possessed by either author of the first edition. With my colleague's support, I sought a new coauthor with experience that extended to the global scale. The person with the most noteworthy experience in scaling the analysis of forest ecosystems was Steven Running at the University of Montana. To my pleasure, he agreed to join me in the endeavor of writing a major revision of the first edition that emphasized quantitative modeling and extrapolations across large spatial and time scales.

A broadened perspective of management is essential today. The pressing issues include regional and global analyses of biodiversity, changes in climatic cycles, implications of wide-scale pollution, and the possibility of fire, floods, insect out-breaks, and other major disturbances that extend beyond the limits of political boundaries. Organizing the principles and providing examples for expanding the horizon of ecosystem analyses were the challenges in writing the new edition. From our own experience in teaching graduate and undergraduate courses, we recognize the difficulty in presenting material that crosses many fields, but the success of expanded integration and problem analysis lies in acquiring new methods and concepts. To that end we have made an attempt to define terms and to explain concepts in a variety of ways by providing equations, graphs, and tabular examples.

Many critical facets of ecosystem behavior, as well as future changes in the environment, remain unknown. Perhaps the best we can do is to distinguish those processes that have a firm basis for analysis from those that require more research. The search for principles that scale, matched with appropriate methods, will be required, regardless of the quest. We hope that students, faculty, research scientists, and managers of natural resources will gain confidence in their abilities to predict and to monitor the implications of various changes. We encourage concern about the long-term implications of policies, in the hopes that the alternatives considered will sustain ecosystem processes to which many organisms contribute, and on which all life depends.

R. H. Waring

PREFACE TO THE FIRST EDITION

Our challenge and reasons for writing this book are to share an emerging insight that there are key linkages between the processes that operate in forests. We emphasize forests in this book because we know them best and because their long life permits us to evaluate the effects of periodic disturbance more readily. Our examples, most often, are drawn from simple cases in which principles are more easily seen and explained. We believe, however, the principles apply widely, as we show in predicting transpiration and other hydrologic properties for a variety of forests in differing climatic settings.

In many cases, scientists cannot accurately predict the effects of acid rain, fire suppression, short-rotation timber harvest, or increasing carbon dioxide levels in the atmosphere on forests or other ecosystems. We believe a diagnostic approach linking a variety of processes is warranted and that with carefully designed experiments the mysteries will unravel.

We have striven to provide a cosmopolitan flavor to the book. Because most experimental work has been focused on rather simple systems, our examples are drawn mainly from temperate and boreal forests. However, the same processes operate in more complex forests, as references denote. The book is written for upper-level students with some background in general ecology, inorganic chemistry, physics, and plant physiology. We hope that specialists will see new implications to their work and be encouraged to develop integrative experiments. Managers of forest resources (wood products, wildlife, and water) will find explanations for some of their observations and predictions of the effects of various management policies.

We owe a debt to earlier studies of ecosystems, particularly those sponsored by the National Science Foundation in the 1970s as part of the International Biological Program, which established a group of five major ecosystem programs in the United States, in addition to earlier work at Hubbard Brook in New Hampshire. For almost a decade, balance sheets were constructed describing how carbon, water, and minerals are stored or transported in a variety of forest, grassland, desert, tundra, and aquatic systems.

Much of the basic information has been published in books and other periodicals. A summary of the international program with listings of data from all woodland sites appeared in 1981, edited by D. E. Reichle.[1] Regional efforts at synthesis have also been made for the other biomes. These references, as well as the open literature, provide a description of how forest systems operate.

Few of the research programs, however, involved experiments that evaluated linkages between major processes. The influence of fire, erosion, wind storms, and epidemic outbreaks of insects or disease organisms could not be rigorously evaluated until a benchmark

---

[1] Reichle, D. E. (1981). "Dynamic Properties of Forest Ecosystmes." Int. Biol. Programme No. 23, Cambridge Univ. Press, London and New York.

for rates of normal processes had been established. The foundation was laid for critical experiments that could test hypotheses involving how and why ecosystems respond to periodic disturbances of various kinds. We propose that integrated experiments based on ecosystem-level insight can provide answers to managers. Whether this is the case, as we emphasize in interpreting the probability of disturbance in forests, awaits future tests.

R. H. W.
W. H. S.

ACKNOWLEDGMENTS

I GRATEFULLY acknowledge helpful reviews by Michael Ryan, Kevin O'Hara, Michael Unsworth, Beverly Law, Barbara Bond, Hank Margolis, and John Marshall on early drafts of one or more of the first six chapters. Kate Lajtha, Dan Binkley, and Kermit Cromack, Jr. generously offered valuable advice and references for the chapter on mineral cycling. I owe a special debt to Joe Landsberg and Pam Matson, who not only reviewed and edited much of the first six chapters but suggested ways to present major sections in a more logical manner. Ron Neilson and Larry Band provided comprehensive reviews of Chapters 7–9, and Eva Falge redrafted figures from an original publication and consulted on analyses presented from eddy-flux sites in Chapter 10, for which I am most appreciative.

In completing the final manuscript, I leaned heavily on the scientifice advice, editorial assistance, and enduring friendship of Joe Landsberg, who served as my recent host while I was on sabbatical leave in Australia. In addition, I thank John Finnigan, Head of the Centre for Environmental Mechanics, CSIRO, for permitting me access to all CSIRO research facilities in Canberra, A.C.T. I took special advantage of Graham Smith, who introduced me to the marvelous computer network and software that CSIRO provides visiting scientists, and to Greg Heath, who carefully redrafted a number of complicated figures. Support for my sabbatical year in Australia was obtained through a CSIRO McMaster Fellowship and a grant received from the Forestry and Forest Products Research and Development Corporation. In addition, I received salary support from the College of Forestry, Oregon State University, and a NASA grant (NAGW-4436).

Finally, to Doris, my wife, I promise not to write another book before I retire, and recognize that my luxury to pursue this and other scientific endeavors for the last 40 years rests on your love and support, for which I am truly appreciative.

R. H. W.

SEVERAL COLLEAGUES influenced my thinking on how to address regional ecosystem analysis. Ramakrishna Nemani, my invaluable associate since 1981, has been an integral contributor to and coauthor of the ideas and implementations of Chapters 7–9. David L. Peterson and Larry Band introduced me to remote sensing and landscape analysis techniques that were developed into the RHESSys regional modeling package. John Aber, Pam Matson, and Peter Vitousek were instrumental in the conceptual development of FOREST-BGC. I shared many intense discussions with Dave Schimel, Chris Field, and Tom Gower on how to generalize the representation of ecosystem processes at larger scales. Although many agencies have funded the research featured in this book, the National Aeronautics and Space Administration deserves special thanks. Diane Wickland, Bob Murphy, and Tony Janetos have sponsored both of our research programs consistently, beginning in the early 1980s when remote sensing and ecological disciplines seemed to have little in common. In addition, I thank Dan Fagre, Glacier National Park, for funding all work related to the GNP. My Numerical Terradynamic Simulation Group has had a stream of excellent students who have continued to advance the development of principles that apply to regional analyses, including Joe Coughlan, George Riggs, Rob Kremer, Ronni Korol, E. Ray Hunt, Kevin Ryan, Daolan Zheng, Lars Pierce, Joe White, Peter Thornton, John Kimball, Kathy Hibbard, Galina Churkina, and Mike White. A number of them built new simulations and images specifically for Chapters 7–9 of this textbook. Joe Glassy and Saxon Holbrook provided an advanced computing environment for the laboratory.

Finally, Connie, Trina, and Emily have endured many nights with a husband and father on the road, rather than at home helping them.

S. W. R.

COMPANION WEB SITE INFORMATION

The companion Web site for this book can be found at:
   www.books.elsevier.com/companions/9780123706058
This site contains a link to the authors' site which contains modeling software, tutorials, and video clips.

CHAPTER 1

Forest Ecosystem Analysis at Multiple Time and Space Scales

I. INTRODUCTION

Forests currently cover about 40% of Earth's ice-free land surface ($52.4 \times 10^6 \, \text{km}^2$), a loss of $10 \times 10^6 \, \text{km}^2$ from that estimated were it not for the presence of humans (see Chapter 9). Although a large fraction of forestland has been converted to agricultural and urban uses, we remain dependent on that remaining for the production of paper products, lumber, and fuelwood. In addition to wood products, forested lands produce freshwater from mountain watersheds, cleanse the air of many pollutants, offer habitat for wildlife and domestic grazing animals, and provide recreational opportunity. With projected increases in human population and rising standards of living, the importance of the world's remaining forests will likely continue to increase, and, along with it, the challenge to manage and sustain this critical resource.

Humans affect forests at many scales. In individual stands, our activities influence the composition, cover, age, and density of the vegetation. At the scale of landscapes, we alter the kinds of stands present and their spatial arrangement, which influences the movement of wind, water, animals, and soils. At the regional level, we introduce by-products into the air that may fertilize or kill forests. At the global scale, our consumption of fossil fuels has increased atmospheric carbon dioxide levels and possibly changed the way that carbon is distributed in vegetation, soils, and the atmosphere, with implications on global climate. The worldwide demand for forest products has stimulated not only the transfer of processed wood products from one country to another, but also the introduction of nonnative tree species, along with associated pests, that threaten native forests and fauna. While the management of forested lands is becoming increasingly important, it is also becoming more contentious because less land is available for an increasing range of demands. Pressure to extract more resources from a dwindling base is leading to a number of challenging questions. Is it possible to maintain wildlife habitat and timber production on the same land unit, and still retain the land's hydrologic integrity? Where should forested land be preserved for aesthetic values, and where should it be managed for maximum wood production? How can an entire watershed be managed so that the availability of water to distant agricultural fields and cities is assured?

This book does not provide specific answers to these management questions, as each situation is unique. Rather, it offers a framework for analyses and introduces a set of tools that together provide a quantitative basis for judging the implications of a wide variety of management decisions on the natural resource base, viewed at broader spatial scales and longer time dimensions than was previously possible. It is our supposition that if we are to be successful stewards of forests we must find a way to integrate what is known into predictive models and apply new methods to validate or invalidate the predictions of these models over Earth's broad surface. We believe that advances in modeling provide such a basis for the analysis of forest ecosystems at multiple scales and strive to illustrate the underlying principles and their application.

One of the major concessions in scaling that we are required to accept is the need to reduce the amount of detail to a minimum. This requirement has the advantage of reducing the cost and complexity of analyses, but it demands insights into which ecosystem properties are critical and then determining how they may be condensed into integrative indices and monitored at progressively larger scales. By modeling ecosystem behavior at different scales we gain confidence in the appropriateness of key variables, when those variables should best be monitored, and the extent to which the analyses apply generally.

This book is structured to start the analysis of forest ecosystems at the level of individual stands and gradually expand the time and space scales. In doing this, we have incorporated throughout the text many of the principles presented in the U.S. Ecological Society of America report on "the scientific basis for ecosystem management" (Christensen *et al.,* 1996). We emphasize quantifying our present understanding of ecosystem operation with soundly based, tested ecological models, but we also identify some important gaps in research. When covering the breadth of topics needed for multiscale analysis, we are unable to review all topics comprehensively but provide over a thousand references to original sources. Although we incorporate how human activities and forested ecosystems interact, we do not advocate specific management policies. We believe, however, that sounder decisions are possible by projecting the implications of various management policies at a variety of scales when models rest on common underlying biophysical and ecological principles.

II. THE SCIENTIFIC DOMAIN OF FOREST ECOSYSTEM ANALYSIS

A forest ecosystem includes the living organisms of the forest, and it extends vertically upward into the atmospheric layer enveloping forest canopies and downward to the lowest soil layers affected by roots and biotic processes. Ecosystem analysis is a mix of biogeo-chemistry, ecophysiology, and micrometeorology that emphasizes "the circulation, trans-formation, and accumulation of energy and matter through the medium of living things and their activities" (Evans, 1956). For example, rather than concentrating on the growth of individual trees, the ecosystem ecologist often expresses forest growth as net primary production in units of kilograms per hectare per year. Ecosystem ecology is less concerned with species diversity than with the contribution that any complex of species makes to the water, carbon, energy, and nutrient transfer on the landscape.

Ecosystem studies consider not only the flux of energy and materials through a forest, but also the transformations that occur within the forest. These transformations are an index of the role of biota in the behavior of the system. Forest ecosystems are open systems in the sense that they exchange energy and materials with other systems, includ-ing adjacent forests, aquatic ecosystems, and the atmosphere. The exchange is essential for the continued persistence of the ecosystem. A forest ecosystem is never in complete equilibrium, a term appropriate only to closed systems in the laboratory. An excellent primer on ecosystem analysis terminology and principles is a textbook by Aber and Melillo (1991).

Although we are studying forest ecosystems across multiple time and space scales, we initiate our analysis at a forest stand, a scale where most of our measurements and under-standing originated (Burke and Lauenroth, 1993). A hierarchical structure is common to all of science in that a reference level of interest is first defined where patterns are observed and described. A causal explanation is sought for these patterns at a finer resolution of detail, while their implications and broader interactions become apparent at a level above (Passioura, 1979; O'Neill *et al.*, 1986). This book is based on a hierarchical structure of studying ecosystems. For example, in evaluating net primary production, we explore details of photosynthesis, respiration, and carbon allocation at the canopy level in the first section of the book to understand the causal mechanisms and their controls. In Section II, we follow stand development through time, evaluating how net primary production changes. In Section III, the effects of photosynthesis combined with other ecosystem properties are shown to interact across the landscape, modifying the local climate, the flow of rivers, and the seasonal variation in regional atmospheric CO_2.

An initial step in ecosystem analysis is to measure the amount of material stored in different components of the system, for example, the carbon stored in stem biomass, water stored in the snowpack, and nutrients stored in the soil. In systems terminology, these are the *state variables* that can be directly measured at any given time. Innumerable studies have been published measuring the current state of forest ecosystems. Frequently, however, the rates of change of these system states, or *flows of material,* are of greatest interest. What is the rate of snowmelt, stem biomass accumulation, or nutrient leaching in a par-ticular system? These questions require study of the processes controlling energy and matter transfer, a much more difficult undertaking. In these process studies, we wish to identify the *cause–effect relationships* controlling system activity, which is often called a mechanistic approach. This identification of system states and multiple cause–effect rela-tionships that operate in a forest ecosystem to regulate material flows can be quantified and organized with an ecosystem simulation model. This type of model becomes the start-ing point of our space/time scaling of ecosystem principles.

III. THE SPACE/TIME DOMAIN OF ECOSYSTEM ANALYSIS

A. Seasonal Dynamics Operating in Individual Forest Stands

We begin multiscale analysis of forest ecosystems with the *stand* as our reference level, which includes the vegetation and surrounding physical environment, linked together through a variety of biological, chemical, and physical processes. Most scientific understanding of ecosystem processes has been gained by direct field measurements and experiments on small study plots usually <1 ha (10,000 m^2) over a period from a few days to at most a few years (Levin, 1992; Karieva and Andersen, 1988). From an ecological scaling point of view, we like to refer to these studies as the *stand/seasonal* level of analysis (Fig. 1.1). Such studies are designed to clarify the ecological processes and controls on the forest without regard to the spatial heterogeneity of the surrounding landscape, or the temporal changes that forests have undergone or will undergo in future years. In the first section of the book (Chapters 2–4) we review forest ecosystem processes and the mechanisms controlling their activity.

Unfortunately, many of the major ecological concerns facing natural resource managers and policy makers are not tractable at the spatial and time scales within the stand/seasonal domain where we have the most direct insights. Forest managers are generally responsible for decisions that affect large and diverse areas, and policy makers must visualize *future* conditions and ecosystem responses that may result from policies made *today*. Consequently, to make forest ecosystem analyses more relevant it is essential that stand/seasonal understanding of ecosystem processes be extrapolated in space and forward in time. Attempts to execute direct studies over large regions (Sellers *et al.,* 1995) or for long periods (Magnuson, 1990) cost tens of millions of dollars, so are rarely attempted. We must search for an alternative means by which knowledge gained at the stand/seasonal level can be expressed in a quantitative way to serve as a platform for extrapolation to larger space and time scales. A set of conceptually linked computer simulation models offer a valid alternative to large-scale ecosystem experiments if they can represent the mechanisms coupling biogeochemical processes in a realistic way yet not require an exorbitant amount of data that are difficult or impossible to acquire.

B. Role of Models in Ecosystem Analysis

Models have been an integral tool of ecosystem analysis since the earliest days of systems ecology (Odum, 1983). Ecosystems are too complex to describe by a few equations; current ecosystem models have hundreds of equations which present interactions in non-continuous and nonlinear ways. Furthermore, these models provide the organizational basis for interpreting ecosystem behavior. Swartzman (1979) identified six primary objectives for ecosystem simulation models: (1) to replicate system behavior under normal conditions by comparison with field data, (2) to further understand system behavior, (3) to organize and utilize information from field and laboratory studies, (4) to pinpoint areas for future field research, (5) to generalize the model beyond a single site, and (6) to investigate effects of manipulations or major disturbances on the ecosystem over a wide range of conditions. Active ecosystem modeling programs pursue all of these objectives, although relevance to land management is attained only in objectives 5 and 6.

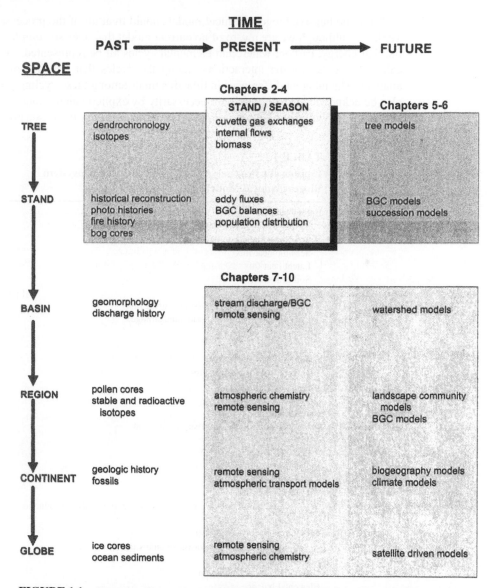

FIGURE 1.1.   Examples of measurement techniques available for forest ecosystem analysis at different time and space scales. Temporal analysis of past ecosystem activity is possible from quasi-permanent records obtained by tissue or elemental analysis, such as tree rings, isotopic ratios, and pollen records from ice and bog cores. Spatial analysis beyond the stand level requires some type of remote sensing technology, and temporal analysis into the future requires some form of modeling. The scales and measurements depicted in the shaded boxes are the subject of this book. The stand/seasonal demarcation represents the space/time scale where most ecological understanding has been attained.

A comprehensive biogeochemical model should treat all of the processes presented in Table 1.1, although we are aware of no current model that does so completely. It is essential that energy, carbon, water, and elemental cycles all be represented, even if simplistically. It is precisely the interactions among the cycles that are the core of ecosystem analysis. The inherent differences in time dynamics among these cycling processes should also be acknowledged, although not necessarily by explicit calculations. Leaf energy balances change within minutes, system gas fluxes change diurnally, tissue growth and carbon

TABLE 1.1

Component Processess of a Comprehensive Ecosystem Biogeochemical Model

Energy balance
   Short-wave radiation balance (incoming—outgoing)
   Long-wave radiation balance (incoming—outgoing)
   Sensible heat flux
   Latent heat flux
   Soil heat flux
Water balance
   Precipitation partitioning (snow versus rain)
   Canopy and litter interception and storage
   Soil surface infiltration
   Soil water content
   Subrooting zone outflow
   Hill slope hydrologic routing
   Evaporation
   Transpiration
Carbon balance
   Photosynthesis, gross primary production
   Maintenance respiration
   Growth respiration
   Photosynthate storage
   Net primary production
   Carbon allocation
     Leaves, stem/branches, roots, defensive compounds, reproduction
   Phenological timing
     Canopy growth/senescence
     Litterfall of leaves, turnover of stems and roots
   Decomposition
   Net ecosystem production
Elemental balance
   Sources (atmosphere, rock weathering, biological fixation)
   Soil solution transformation
   Immobilization, nitrification, denitrification
   Mineralization
   Root uptake
   Tissue storage
   Internal recycling
   Volatilization
   Leaching
   Export through harvesting and erosion

allocation dynamics are observable at weekly to monthly intervals, whereas nutrient mobilization may be measurable seasonally. Different forest ecosystem models have time steps ranging from an hour to a year, and newer models contain sections that represent processes at different time steps.

Of equal importance is that each process is treated with approximately the same level of detail. A model that computes photosynthesis of each age class of needles but fails to couple the nitrogen cycle to photosynthetic capacity is not balanced. Most forest biogeochemical models suffer some deficiencies in balance because they began as single process models and only later, often in much less detail, added other processes critical to ecosystem operation. Beyond some of these basic properties of good ecosystem modeling, every model differs depending on the specific objectives pursued. Some ecosystem models optimize energy partitioning as part of a climate model, whereas others focus on forest productivity, hydrology, or elemental cycles.

Ecosystems, because of their dynamic and interconnected properties, cannot be subjected to classic experimentation where one variable at a time is modified (Rastetter, 1996). Computer simulation models of ecosystem behavior offer a valuable experimental alternative because they allow multivariant interactions to be traced and analyzed. With simulated experiments, the accuracy with which different variables need to be measured can also be estimated. Such ecosystem models establish mathematical relationships in a simple but increasingly mechanistic way, to clarify causal connections and integrate system operation. On this basis, models can *predict responses to new conditions* that do not yet exist. For example, computer simulation models can predict how stream discharge may respond to harvesting in a watershed and identify possible flood problems before any logging commences. Computer simulation models have been the primary means for evaluating potential responses of natural ecosystems to future climate changes. For example, the Vegetation/ Ecosystem Modeling and Analysis Program (VEMAP) has simulated ecosystem biogeographical and biogeochemical responses to climatic change for the entire continental United States (VEMAP, 1995). This project relied almost exclusively on computer simulations to provide a reasonable estimate of future conditions.

1. Example of a Scalable Ecological Model

Because a primary theme of this book is scaling in space and time, we will frequently use a model originally designed for multiscale applications, the FOREST-Bio-Geo-Chemical simulation model (FOREST-BGC) to explore ecosystem interactions. FOREST-BGC originated as a stand-level model of forest biogeochemical cycles, in effect, a model quantifying our understanding of the mechanistic processes of energy and mass fluxes in the stand/season space/time domain. Other forest ecosystem models are also available, and results from these will be illustrated (see reviews by Ågren *et al.,* 1991; Tiktak and van Grinsven, 1995; Ryan *et al.,* 1996a; Thornley and Cannell, 1996). FOREST-BGC is a process-level simulation model that calculates the cycling of carbon, water, and nitrogen through forest ecosystems (Fig. 1.2; Running and Coughlan, 1988; Running and Gower, 1991). FOREST-BGC calculates most of the important ecosystem processes covered in this book. The model has both daily and annual time steps, recognizing the substantial differences in the response of different ecosystem processes. Hydrologic and canopy gas

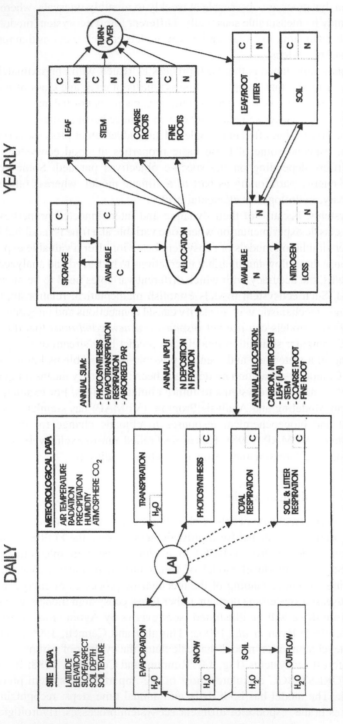

FIGURE 1.2. Compartment flow diagram for the FOREST-BGC ecosystem simulation model. This diagram illustrates the state variables of carbon, water, and nitrogen, the critical mass flow linkages, the combined daily and annual time resolution, and the daily meteorological data required for executing the model. The major variables and underlying principles associated with the model were developed specifically for application at multiple time and space scales, and for compatibility with remote-sensed definition of key ecosystem properties.

exchange processes are calculated daily, while ecosystem carbon and nitrogen processes are computed annually.

The model structure of FOREST-BGC is based on some key simplifying assumptions that have proved particularly valuable in extending ecosystems analyses to regional levels (Chapters 7–9). These simplifying assumptions give us examples of how much ecological detail must be sacrificed in an ecosystem model to attempt regional scaling. Trees are not defined individually as in a typical forest stand growth model, only the cycling of carbon, water, and nitrogen is expressed in mass units. Species also are not explicitly identified, although species-specific physiological characteristics can be represented. Geometric complexities of different tree canopies are reduced to a simple quantification of the sum of all leaf layers as leaf area index (LAI). Treatments of internal tree physiology such as carbohydrate and water transport are minimized. Belowground root system and soil processes are treated in a simplified way to reduce requirements for difficult to obtain field data. Finally, FOREST-BGC was designed to require only standard meteorological data, namely, daily maximum–minimum temperature and precipitation, so that the model might be applied beyond those few sites with sophisticated instrumentation. Many of the other models presented in this book make different assumptions of model structure that are valid for other objectives but do impede spatial applications.

Even more simplifying assumptions are incorporated in a Light Use Efficiency (LUE) model of forest productivity we present in Chapter 3 (Landsberg and Waring, 1997). This model estimates forest productivity simply by quantifying the photosynthetically active radiation absorbed by a forest canopy, and then places some climatological and physiological restrictions on the conversion of that solar energy into above- and belowground biomass, thus bypassing all of the complexity of quantifying carbon, water, and nitrogen cycles found in FOREST-BGC. The LUE models can appropriately be operated with satellite data only, and hence they provide a global estimate of primary productivity, the ultimate spatial scaling exercise (see Chapters 9 and 10).

2. Importance of Model Validation

Because the initial stand/seasonal ecosystem model is a vehicle central to the success of the space/time scaling to follow, high confidence in the validity of this first type of simulation model is essential. How well does the model capture the structure, controls, and dynamics of a real forest ecosystem? Obviously that question can never be answered for all forests and circumstances, but by testing some of the key assumptions, and finding good agreement between predicted and measured values for a wide range of systems, a basis is provided for judging the model's soundness. Rykiel (1996) formalized the steps required in ecological model validation as beginning with evaluation of the theoretical basis of the model, then testing the operational integrity of the computer code, and finally making comparisons of model output with measured data relevant to the intended domain and purpose of the model.

Some parts of FOREST-BGC have undergone considerable testing, particularly the hydrologic and carbon cycle components (Running, 1984a,b; Knight *et al.*, 1985; McLeod and Running, 1988; Nemani and Running, 1989a; Hunt *et al.*, 1991; Korol *et al.*, 1991; Band *et al.*, 1993; Running, 1994; White and Running, 1994; Korol *et al.*, 1995). It is difficult, however, to validate every aspect of an ecosystem model. We recommend a group of variables that can be accurately measured in the field and reflect a range of ecosystem interactions linked to carbon, water, and nutrient cycles. These include

- snowmelt
- soil moisture depletion
- predawn leaf water potential
- leaf area index
- net primary production
- stem biomass
- leaf litterfall
- leaf nitrogen content
- leaf litter nitrogen content

Throughout this book we will present results of studies that confirm the ability of forest ecosystem models to predict these variables, often within the accuracy that can be obtained with direct field measurements. Confidence in ecosystem models is best developed at the stand level first, where a wealth of data exists. Tests of model predictions at larger space and time scales are much more difficult, although new technologies are providing a means of validation at these scales also.

IV. TIME AND SPACE SCALING FROM THE STAND/SEASONAL LEVEL

Once we have gained some confidence in our understanding of how major processes interact in a single stand seasonally, we can begin to extrapolate principles to larger space and time scales required for land management. Making these extrapolations, however, requires a much wider array of tools, including multiple models that demand different input data. FOREST-BGC evolved first as a stand/season model, but, because of our interest in space and time extrapolation, a family of conceptually related models was produced (see Sections II and III). The logical connections among this family of models allow the validation activity from one model to improve our confidence in a subsequent model designed for larger spatial scales, Because FOREST-BGC has undergone validation tests at the stand level for many years, we have higher confidence in simulations extrapolated forward in time, or spatially to a region, despite the validation not being at the same scale as the current model implementation.

A. Scaling in Time

Section II of this book (Chapters 5 and 6) concentrates on temporal scaling of our stand-level understanding. Looking back in time allows us to develop an appreciation and understanding of the temporal dynamics previously experienced by an ecosystem. We can quantify the ecological history of a forest through analysis of tree ring growth increments, fire scars, tissue stable isotope composition, and a variety of other methods including written and photographic records (Fig. 1.1). While we can measure current activity of ecosystem processes with gas exchange cuvettes, water budgets, and biomass analysis, to project into the future we must rely on computer models, which justifies care in constructing them and testing them as widely as possible (Dale and Rauscher, 1994).

1. Stable Isotopes

Stable isotope analyses provide one of the few consistent and available methods for *temporal scaling backward in time*. In this book we make an effort to illustrate opportunities for greater use of stable isotope analyses in nearly every chapter. Many elements are composed of multiple isotopes differing slightly in atomic mass because of extra neutrons but possessing nearly identical chemical activities (Table 1.2; Schimel, 1993). These elements with higher atomic mass do not decay radioactively, so are known as stable isotopes. Kinetic reactions, such as diffusion, and biochemical reactions, such as photosynthesis, slightly discriminate against these heavier isotopes, causing subtle variations in the concentrations of the stable isotope relative to the standard isotope for each element (Fig. 1.3). For example, although the standard carbon in any biological tissue is ^{12}C, some fraction less than 1% will be ^{13}C, depending on the kinetic and biochemical reactions that occurred to synthesize a carbon-bearing molecule such as cellulose. Carbon also exists in an unstable radioactive form as ^{14}C, whose half-life is 5715 years. In biological material, the decay of ^{14}C since its incorporation provides a means of dating the age.

The natural abundances of stable isotopes found in rocks, plants, animals, atmosphere, and water bear witness to their origin. Improvements in the sensitivity of mass spectrometers have allowed detection of isotopic ratios that previously could not be done, and have opened an entire new array of retrospective ecosystem analysis. As ecosystem studies have expanded in scope to assess changes over landscapes and geologic time scales, analyses of multiple isotopes of C, N, S, H, O, Sr, and Pb have provided integrated, long-term records on the frequency and extent of drought, documented shifts in the diets and distribution of animals, and provided a quantitative record of the amount of anthropogenic pollutants deposited on forests, fields, and water bodies throughout the world.

When analyses are extended across ecosystems that include urban areas, our increasing dependence on fossil fuels becomes clear through changes recorded in the isotopic composition of $^{13}CO_2$ in the atmosphere and in the annual rings of trees. Organic matter contains hydrogen and oxygen derived from water which shows through its isotopic composition the temperature at which precipitation fell, so that climatic changes also can be assessed through isotopic analyses. Rocks, and soils derived from them, differ in the isotopic composition of sulfur, strontium, and lead. These differences are reflected in the

TABLE 1.2

Percent Abundance of Stable Isotopes Commonly Employed in Ecosystem Studies[a]

Carbon	Oxygen	Hydrogen	Nitrogen	Sulfur	Strontium	Potassium	Magnesium
^{12}C 98.89	^{16}O 99.763	H 99.9844	^{14}N 99.64	^{32}S 95.02	^{84}Sr 0.56	^{39}K 93.08	^{24}Mg 78.8
^{13}C 1.11	^{17}O 0.0375	D 0.0156	^{15}N 0.36	^{33}S 0.75	^{86}Sr 9.86	^{40}K 0.0119	^{25}Mg 10.15
	^{18}O 0.1995			^{34}S 4.21	^{87}Sr 7.00	^{41}K 6.91	^{26}Mg 11.06
				^{36}S 0.02	^{88}Sr 82.58		

[a]Adapted from Schimel (1993).

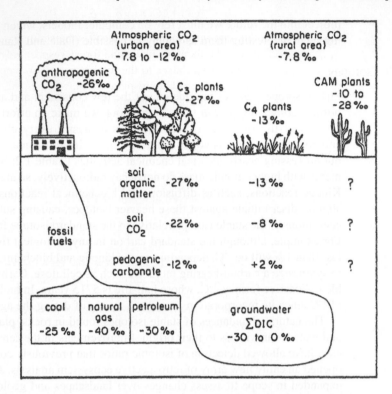

FIGURE 1.3.   The concentration of the stable isotope carbon-13 expressed as a reduction in ^{13}C relative to a background reference (standard delta ^{13}C = 0, with units of parts per thousand, or $\delta‰$) shows that trees, tropical grasses, and desert cacti produce biomass with distinctly different isotopic composition from that of CO_2 in the atmosphere as a result of differences in photosynthetic pathways. Further isotopic discrimination occurs as organic matter is decomposed or converted to various fossil fuels. Combustion of fossil fuels and burning of biomass cause a reduction in the ^{13}C/^{12}C ratio of CO_2 in the atmosphere which, through analysis of trapped gases in dated ice cores, provides a quantitative record of anthropogenic activities for many centuries into the past. Similar isotopic fractionations occur with other elements, which provide ecologists quantitative estimates of the diets and home ranges of animals, the rates of nitrogen fixation and mineral weathering, and the amounts of nitrogen and sulfur deposited as pollutants. (From Boutton, 1991.)

tissue composition of plants and the bones and other tissues of all animal life, which provides a basis for locating the origin of elephant tusks and the ranges of other animals in geologically variable landscapes. Stable isotope analyses also play a role at regional and global scales in verifying model assumptions concerning the movement of pollutants and the relative importance of terrestrial and oceanic production on the global carbon balance.

2. Range of Ecosystem Time Dynamics

Many ecosystem processes occur continuously, although the process rates are controlled by prevailing conditions. One can consider litter decomposition to occur almost continuously except under extreme conditions of temperature and aridity. Litter decomposition can be described based on cause–effect relationships between the temperature and hydration of the substrate tissues and observed rates of litter decay. Other processes occur in predictable cycles; daily photosynthesis begins every morning at sunrise, or seasonally with canopy leaf emergence every spring. Because of the high level of regularity and predictability, these process activities can be treated as a straightforward *deterministic* relationship in a simulation model. Other ecosystem processes, however, are initiated by discrete events that can be highly unpredictable in time, originate well beyond the boundaries of the ecosystem, yet produce major consequences. Major disturbances such as forest fires and hurricanes are good examples, and these are often treated as random events with *stochastic* models. A model is stochastic if even only one process is described mathematically as a random event in the system.

Considering the time domain of tree life cycles, what we find in the historical development of most forests is a progression of phases following establishment. Without disturbance the progression is fairly predictable, but disturbance is a part of forest history, although its occurrence is not necessarily random. Disturbance that leads to change in structure often occurs when competition for resources becomes intense and the ability of trees to withstand fire, insect, or disease outbreaks approaches a minimum. At such time, a disturbance is likely to open the forest canopy and improve conditions for surviving trees and their replacements.

To project a forest ecosystem response forward in time, we use models that simulate forest development and vegetation dynamics (Shugart *et al.,* 1992). Such models are able to predict the expected changes in biogeochemical cycling, and the progressive life cycle of species through their recruitment, growth, death, and ultimate replacement, following specified kinds and intensities of disturbance. The models, to accommodate current conditions, must also account for introductions and extinctions of species of plants, animals, and pathogens that alter normal competitive relations. In Chapter 5, examples of models predicting stand development and vegetation dynamic are presented.

In Chapter 6, stress in forest development is specifically linked to disturbance. Experiments and observations are reported which quantify changes in resource availability and subsequent system response to fire, insects, and disease. A general measure of ecosystem carbon stress is the growth efficiency by which sunlight is captured through photosynthesis and transformed into stem growth. Other indices related to nutrient and water availability provide explanations to observed changes in growth efficiency. These *stress indices* can be predicted with ecosystem process models such as FOREST-BGC, or with more simplified LUE models, and thus serve as a type of validation. In addition, some indices derived from dendrochronology and stable isotope analyses provide retrospective information about disturbances that occurred over the life span of the trees.

B. Scaling in Space

In Section III, we focus on spatial scaling. Chapter 7 presents a hierarchical framework for integrating the tools of remote sensing, geographic information systems (GIS), topographic climatic extrapolation, and ecosystem modeling to analyze current ecosystem activity over large areas and to project ecosystem responses into the future. The Regional

Ecosystem Simulation System, RESSys, is introduced, and some of the component models for local climate extrapolation, topographic partitioning, and hydrologic routing are discussed (Running *et al.,* 1989; Band *et al.,* 1993). The necessity for using a forest ecosystem model with carefully simplified input variables becomes evident, for how could one obtain data of leaf age distribution or root lengths for simulations over a 100,000-ha forest?

The integration of remote sensing, GIS, and computer modeling has proved essential for regional scale ecosystem analysis, and we describe some of the latest examples in Chapters 8 and 9. The most advanced regional analyses drive sophisticated ecosystem process models forward in both time and space, providing simulations of vegetation dynamics and biogeochemistry covering whole continents for hundreds of years into the future (VEMAP, 1995). Validation of these large-scale computer model projections is becoming one of the biggest challenges in ecology. It is from these space/time integrated modeling tools, however, that ecology is able to offer a major advance in resource analysis. An important decision in the spatial representation of landscapes is whether the smallest recognizable elements (or cells) are considered connected or unconnected in the horizontal dimension. The question asked is, Does the ecosystem activity in one cell influence the activity in adjacent cells? Depending on the answer to this seemingly mundane question, the tools and theory for landscape representation are substantially different, as illustrated in Chapters 8 and 9. For example, if decomposition can be considered spatially independent of the surrounding mosaic of forest ecosystems, it greatly simplifies the subsequent spatial analyses.

On the other hand, processes that control the spread of fire, outbreaks of insects and diseases, or the transport of water and sediment in streams require an explicit spatial understanding of surrounding ecosystems, which greatly complicates the analyses. The discipline of landscape ecology has evolved numerous methods to quantify and interpret spatial connectiveness. These techniques attempt to describe the pattern of the landscape by defining the spatial arrangements of different ecosystem types, and by recording changes in the sizes and shapes of those recently disturbed. We introduce techniques for evaluating landscape structure, heterogeneity, vegetation redistribution, and disturbance processes in Chapters 7 and 8. However, because the focus of this book is primarily on ecosystem structure and function rather than pattern and interpretation, we leave the detailed treatment of these geographic topics to others (Forman, 1995; Turner and Gardner, 1991).

V. MANAGEMENT APPLICATIONS OF ECOSYSTEM ANALYSIS

Management applications of ecosystem analysis commonly encompass large areas, which imposes a requirement that the types and accuracy of data match the available sources. Ecosystem analysis can provide through model simulations some estimates of important variables that are difficult to measure directly. For example, using hydrologic equilibrium theory, one can infer a balance that is commonly established among climatic properties, soil water holding capacity, and the maximum leaf area that forests will support. It is a seeming contradiction that these rather sophisticated ecosystem models and analytic tools are particularly valuable in data-poor areas. A handful of key measurements, some acquired by satellite and synthesized with a model, can allow an inference of ecosystem activity that would be nearly impossible to acquire through standard ground surveys. The first requirement in preparing for regional scale assessments is to construct a coordinated, geographically specific information base that includes the most important system attributes such as weather data, satellite imagery of the mosaic of vegetation and soils, snowpack depth, streamflow, and location of wildlife populations. Most established land management agencies have acquired a tremendous amount of these kinds of data, but they are often not available in a consistent, geographically referenced format. The second requirement is to maintain the array of ecosystem and environmental data in an immediately accessible form. Finally, ecological process models are needed that use the archived data sets and real-time information to project both near and long-term ecosystem responses. We will illustrate how these data sets are developed, archived, and utilized in models developed specifically for projecting regional ecosystem responses to changing conditions, including both natural disturbances and those associated with management policies.

Many of the concepts and analyses presented in this book come from studies conducted across the Oregon transect (Fig. 1.4). The 250-km transect provides a climatic gradient for the distribution of temperate evergreen forests that encompasses the full range in productivity represented by this biome (Waring and Franklin, 1979). Because of this unique compression of ecological diversity and the commercial value of the Pacific Northwestern

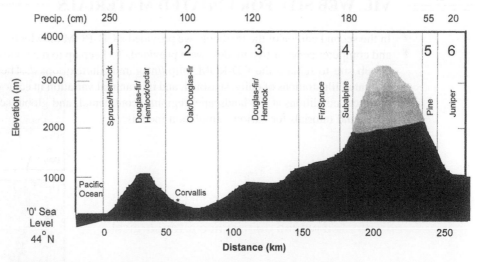

FIGURE 1.4. Graphic depiction of the elevational and climatic gradient across the Oregon transect, the location of many studies cited in this book. The transect begins in the temperate rain forests along the Pacific Ocean and ends in juniper woodland only 250 km to the east after crossing two mountain ranges. The transect encompasses virtually the global range of structure and productivity of evergreen coniferous forests. A long history of forest ecosystem research makes this region one of the most studied and best understood forest landscapes on Earth (Waring and Franklin, 1979; Gholz, 1982; Edmonds, 1982; Peterson and Waring, 1994). Competing socioeconomic values for timber production, wildlife habitat, anadromous fish populations, hydroelectric power production, and recreation also make this region ideal for testing the soundness of multiresource ecosystem management principles.

forests, a long and rich history of forest research is available from which to draw examples (Edmonds, 1982; Peterson and Waring, 1994). This text will highlight the many innovations in forest ecosystem research that have come from studies in this region.

VI. RELATED TEXTBOOKS

This text would not be possible without the historical development of forest ecosystem analysis. Important books on forest ecosystems were published under the auspices of the International Biological Program (Reichle, 1981; Edmonds, 1982) and precursors to that program (Bormann and Likens, 1979). More recent texts related to forest ecosystem analysis include *Forest Stand Dynamics,* emphasizing species-specific stand development and individual species (Oliver and Larson, 1996). *Terrestrial Ecosystems* by Aber and Melillo (1991) has a particularly good chapter on ecosystem analysis principles. Principles of landscape analysis which emphasize the quantification and interpretation of landscape patterns are well treated in publications by Formann (1995) and Turner and Gardner (1991). Ecophysiology of forests are covered by Smith and Hinckley (1995a,b). *Biogeochemistry* at the global scale is the topic of a valuable book by Schlesinger (1991, 1997). Two other texts on forest ecosystems have been published that contain additional information and offer somewhat different perspectives (Perry, 1994; Landsberg and Gower, 1997).

VII. WEB SITE FOR UPDATED MATERIALS

In the second edition of the textbook, we provided a CD-ROM on which additional images and computer code for two models were provided. To keep up to date, we have established a web site to replace the CD-ROM: http://ntsg.umt.edu/textbooks/. There you will find animated illustrations of daily, seasonal, and interannual variation in ecological and meteorological conditions at the landscape, regional, continental, and global scales, along with codes and tutorials for various simulation models.

SECTION I

Introduction to Analysis of Seasonal Cycles of Water, Carbon, and Minerals through Forest Stands

IN SECTION I, we provide a detailed analysis of the daily and seasonal cycling of water, carbon, and minerals in undisturbed forest ecosystems. This stand/seasonal level of analysis allows us to introduce methods, basic principles, and equations that support extrapolations to broader time (Section II) and spatial (Section III) scales. The analyses also serve as the foundation for the most sophisticated kinds of ecosystem simulation models, those that attempt to address the interactions of simultaneous changes in climate and the availability of critical resources on selected ecosystem processes.

CHAPTER 2

Water Cycle

I. INTRODUCTION

The hydrologic cycle is an important feature of all ecosystems, and particularly forests, which generally grow in climates where precipitation provides more water than the vegetation can use or soils can store. The excess water contributes to stream flow, which provides for irrigation and urban needs far from the source of precipitation. Vegetation is a major factor in the hydrologic cycle, as shown in Fig. 2.1. Before precipitation reaches the soil, water is intercepted and evaporated from the surface of vegetation and the litter layer. The rate at which water infiltrates into the soil, runs off the surface, or percolates through to the water table is affected by the density and depth of root channels and organic residue incorporated into the soil.

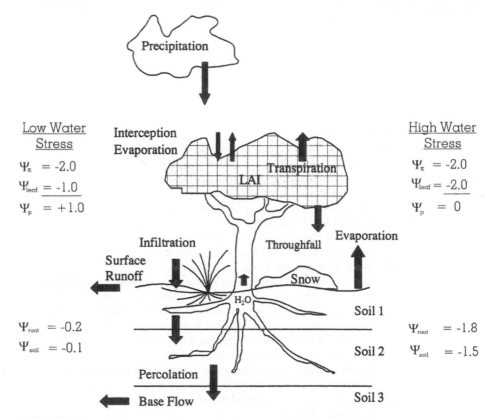

Vegetation Water Balance

Low Water Stress

$\Psi_\pi = -2.0$

$\Psi_{leaf} = -1.0$

$\Psi_p = +1.0$

$\Psi_{root} = -0.2$

$\Psi_{soil} = -0.1$

High Water Stress

$\Psi_\pi = -2.0$

$\Psi_{leaf} = -2.0$

$\Psi_p = 0$

$\Psi_{root} = -1.8$

$\Psi_{soil} = -1.5$

FIGURE 2.1.   Stand water balance. Precipitation which falls as rain, snow, or fog is intercepted on vegetation; the amount intercepted depends to some extent on leaf area index (LAI). Some precipitation reaches the ground surface by throughfall and stemflow, wets the litter, runs off, or infiltrates into the soil where it may percolate to the water table. Some water is held against gravitational forces within the soil matrix; this is available for extraction by plant roots. Water taken up by plants may be stored temporarily within the stem, branches, and foliage, but most is transpired quite rapidly from the leaves into the atmosphere. Evaporation of water also occurs from plant and soil surfaces when they are wet. Example components of water potential for a tree under low and high stress are given. Note how leaves under high stress can have a pressure potential of 0.0 MPa causing wilting, and that the xylem pressure potential gradient from root to leaves is −0.8 MPa for the low stress example but only −0.2 MPa under high stress, quantifying reduced water flow rates that occur under high stress. See Section III,A for details of plant water potential. (After Neilson, 1995.)

Trees differ from other types of vegetation in their vertical height, which has important consequences for local climates because heat and water vapor transfer processes are strongly influenced by vegetation roughness, which is proportional to height. Trees also extend their roots deeper than many other types of vegetation and thus tap additional sources of water (Canadell *et al.*, 1996). The amount of water stored in the rooting zone often determines whether a stand of trees can continue to grow during a drought or whether mortality rates are likely to increase in association with outbreaks of disease and insects (Chapter 6).

Forest hydrologists are particularly interested in the fraction of precipitation that enters streams during storms, a component that may vary from <10 to >50% of annual streamflow and causes the majority of flooding, erosion, and transfer of pollutants to downstream areas (Johnson and Lindberg, 1992). By continuously measuring precipitation and streamflow in gauged watersheds, hydrologists have obtained empirical relations to answer many practical questions. Long-term studies at gauged watersheds, moreover, have provided baseline data for developing and testing a host of increasingly detailed process models that couple water, carbon, and nutrient cycles to changes in vegetation, climate, soils, and transfer of chemicals from the atmosphere (Bormann and Likens, 1979; Johnson and Lindberg, 1992). These process models, which will be featured in Chapter 8, have the advantage of more general applicability than empirical data but require more information than is commonly available.

In our search for general principles that apply across a range of scales, we emphasize the basic physics involved in water movement through the soil–plant–atmosphere continuum and indicate how specific measurements of climate, soils, and plant structural and physiological properties interact to affect the hydrologic cycle. A few basic equations are presented, supplemented with graphic presentations of important functional relationships. More formal mathematical treatments are provided in other texts (Monteith and Unsworth, 1990; Landsberg and Gower, 1997).

In this chapter we limit discussion to the hydrologic balance of forest stands over a period of 1 year. Using a simple energy balance approach, we first examine the rates of water loss by evaporation or transpiration from forests compared with other types of vegetation. We then consider the factors controlling interception and evaporation. Water vapor is lost from dry canopies by transpiration, which involves the limitations imposed by leaf stomata. A major section concerns the uptake of water from soil and its flow through stems and branches to leaves. From this analysis, we discover how the accuracy of ecosystem model predictions of seasonal water balances can be significantly improved by monitoring a few physiological properties of the vegetation, along with a select number of environmental variables. Consideration of water storage in soils and snow, together with surface runoff, infiltration, and percolation, completes the coverage of all basic processes. The major processes are coupled into a generalized hydrologic model that places structural limits on the kind of forests that can develop. These stand-level hydrologic models serve, with some modifications, as a basis for expanded analysis in time and space (Chapters 5, 8, and 9).

II. HEAT AND WATER VAPOR TRANSFER FROM VEGETATION

A. Energy Balance

The total flux of water vapor from an ecosystem can be estimated from the energy balance, determined by measuring the radiant energy absorbed and retained by the system (net radiation) and the mean gradients of temperature and humidity between the canopy upward to the atmosphere (Monteith and Unsworth, 1990). The radiation balance includes both short-wave (0.3–4 μm) and long-wave (4–80 μm) components of radiation. About 95% of solar radiation is short-wave, which contains about equal proportions of visible light (0.4–0.7 μm; the photosynthetically active component of radiation, PAR) and near-infrared radiation (NIR). Short-wave radiation incident on vegetation or other surfaces may be reflected or absorbed; the absorbed radiation may heat the surface and be transformed into sensible heat, or it may evaporate water (latent heat). All surfaces above absolute zero emit long-wave radiation at a rate proportional to the fourth power of the temperature (Stefan–Boltzmann law). The total radiation incident on any surface is the sum of (1) direct short-wave radiation from the sun; (2) diffuse short-wave radiation from the sky; (3) reflected short-wave from nearby surfaces; (4) long-wave radiation from atmospheric emission; and (5) long-wave emitted from nearby surfaces (Fig. 2.2). Net radiation (R_n) is quantified according to the following equation:

$$R_n = (1 + \alpha)I_s - \varepsilon_L \, \sigma T^4 \text{ (surface)} + \sigma T^4 \text{ (sky)} \qquad (2.1)$$

where α is the albedo or reflectivity of the surface as a fraction of intercepted incident short-wave radiation (I_s); ε_L is emissivity compared to a perfect black body, with ε_L for most soils and vegetation being between 0.9 and 0.98; σ is the Stefan–Boltzmann constant ($5.67 \times 10^{-8}\,\text{W m}^{-2}\,\text{K}^{-4}$); and T is Kelvin temperature in reference to absolute zero ($10°C$ = 283 K).

Depending on the reflective properties of leaves, the net radiation above dense forests is typically about 80–90% of incident short-wave radiation (Landsberg and Gower, 1997). If soils or snow intercept a major part of incoming solar radiation, more precise calculations are required, based on the fractions of short- and long-wave radiation that penetrate through the canopy. Reflectivities (*albedos*) for a wide range of surfaces are presented in Table 2.1.

The surface energy balance may be written

$$R_n + G = H + \lambda E \qquad (2.2)$$

where R_n is the net radiant energy, in Joules (J) m^{-2} s^{-1} or Watts (W) m^{-2}; G represents soil and stem storage of heat (into the system during the day, out at night); H is the sensible heat flux into the atmosphere (W m^{-2}); and λE represents the latent heat flux (W m^{-2}), the energy used to evaporate water, where λ is the heat of vaporization of water (2454 kJ kg^{-1} at 20°C) and E is the evaporation rate (kg m^{-2} s^{-1}).

The available energy, represented by the difference in net radiation and soil heat flux, is dissipated almost entirely as either sensible or latent heat, with a small but important component going into metabolic activity (photosynthesis and respiration, Chapter 3). On a 24-hr basis, the storage term (G) is very small, but seasonally it is important because it may change in sign and cause soil to warm or to cool to considerable depth. The ratio of sensible to latent heat losses ($H/\lambda E$), termed the *Bowen ratio* (β), is an important ecosystem index in relation to how it varies diurnally and seasonally. If G in Eq. (2.2) is small, it may be neglected so that $R_n \approx \lambda E + H$. By definition, if $\beta = H/\lambda E$, then $\beta \lambda E = H$. By substituting for H in Eq. (2.2), $R_n = \beta \lambda E + \lambda E = \lambda E(1 + \beta)$, and hence we can predict latent heat loss with the Bowen ratio and net radiation:

$$\lambda E = R_n/(1 + \beta). \qquad (2.3)$$

Short vegetation, such as a well-watered pasture, maintains a fairly constant Bowen ratio throughout the day, with both latent and sensible heat transfer essentially tracking net radiation. The Bowen ratio above a well-watered forest, on the other hand, exhibits

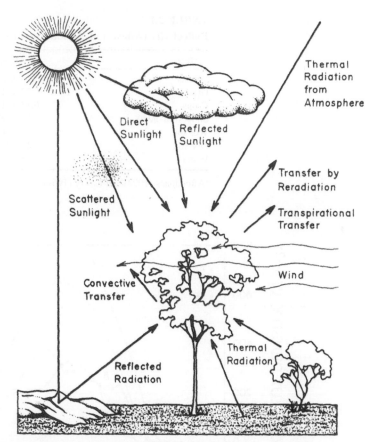

FIGURE 2.2.   Energy exchange between vegetation and the environment involves a number of processes. Solar radiation reaches plant canopies as direct, scattered, and reflected sunlight, all of which contain some short-wave components important for photosynthesis. On partly cloudy days, reflection from clouds can increase incident short-wave radiation at the ground surface by as much as 30%. On clear days, less than 10% of the short-wave radiation is scattered by the atmosphere; on overcast days, incident short-wave radiation is reduced and diffuse, casting no shadows. Plant and other surfaces absorb and reflect short-wave and long-wave radiation, and they emit thermal radiation as a function of their absolute (Kelvin) temperature. The bulk of the heat load on plants is reradiated; evaporative cooling by transpiration and heat transfer by convection and wind (advection) remove the rest. Some heat is stored temporarily in the soil and plant tissue, which is later reradiated. (After Gates, 1980.)

considerable diurnal variation. For example, in Fig. 2.3, in the early morning (0600 hr), β ≤ 0.01 associated with the rapid evaporation of dew; at 1000, with dry surfaces, β is near 1.0, which indicates that sensible and latent heat transfer are equal at about $250\,W\,m^{-2}$, with the latter equivalent to about $0.36\,mm\,hr^{-1}$, whereas by 1400, after the peak solar radiation, β increases to nearly 1.9 as a result of physiological limitations on transpiration (Price and Black, 1990). For more insight into interpretation of the diurnal variation in H and λE, we need information on whether the canopy is wet or dry and other properties that affect exchange rates.

TABLE 2.1
Reflectivity (Albedo) of Various Surfaces[a]

Surface	Albedo
Forests	0.05–0.18
Grass	0.22–0.28
Crops	0.15–0.26
Snow (old–new)	0.75–0.95
Wet soil	0.09 ± 0.04
Dry soil	0.19 ± 0.06
Water	0.05 to >0.20

[a]After Jones (1992) and Lowry (1969).

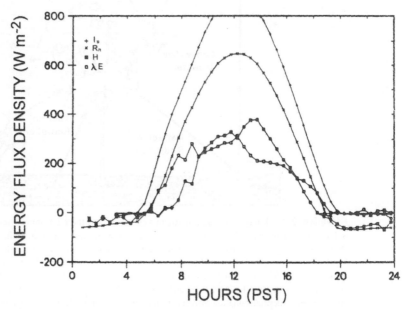

FIGURE 2.3.   Ecosystem energy exchange balance for a Douglas-fir stand in British Columbia for a clear day in July. Variation in incoming short-wave radiation (I_s), net radiation (R_n), heat storage (G), and fluxes of sensible heat (H) and latent heat (λE) are presented for a 24-hr period. The Bowen ratio ($H/\lambda E$) was small in early morning as dew evaporated freely from leaf surfaces. From 0800 to 1200 transpiration dissipated nearly as much energy as was lost through sensible heat transfer (Bowen ratio approximately 1.0). During midafternoon, partial closure of leaf stomata reduced λE and increased the Bowen ratio to nearly 2.0. (Modified from *Agricultural and Forest Meteorology*, Volume 50, D. T. Price and T. A. Black, "Effects of short-term variation in weather on diurnal canopy CO_2 flux and evaporation of a juvenile Douglas-fir stand," pp. 139–158, 1990, with kind permission of Elsevier Science–NL, Sara Burgerhartstraat 25, 1055 KV Amsterdam, The Netherlands.)

B. Evaporation from Wet Surfaces

In this section we provide background on the historical development of equations describing evaporation, borrowing from portions of an excellent review paper written by Kelliher and Scotter (1992). Evaporation involves the diffusion of water vapor molecules away from wet surfaces and is proportional to the difference between the saturated vapor pressure of water (determined by water temperature) and the vapor pressure of the air (a function of air temperature and humidity). The amount of water vapor held in a saturated atmosphere increases exponentially with temperature (Fig. 2.4); the degree of saturation is often expressed as the *relative humidity*. The *dew point* is the temperature at which the water vapor pressure equals the saturated vapor pressure; below that temperature, condensation occurs. Where water vapor pressure cannot be measured directly, it is often approximated from a correlation with minimum night temperatures (Chapter 7). The *water vapor pressure deficit* (*D*) is an important term in many models of evaporation and represents the difference between the vapor pressure at saturation and the actual vapor pressure, determined by the amount of water vapor in the air (Jones, 1992).

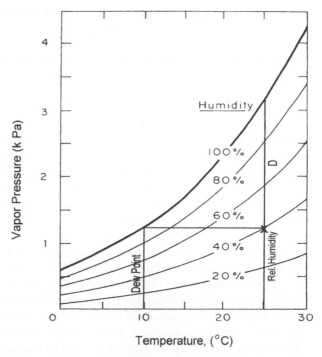

FIGURE 2.4.   The amount of water vapor that may be held in the air increases exponentially with temperature. The saturated vapor pressure (e_{sat}) between 0° and 45°C when expressed in kilopascals (kPa) is approximated as a function of air temperature (*T*) as $e_{sat} = 0.61078 \exp[(17.269T)/(237 + T)]$. The water vapor deficit of the air (*D*) is the difference between the saturated value and that actually held at a given temperature. In this example, air at 25°C with a relative humidity of 40% must be cooled to 10°C to reach the dew point. The vapor pressure deficit (*D*) is the difference between e_{sat} at 25°C (3.2 kPa) and that at the dew point (1.2 kPa), or 2.0 kPa.

Only if the air is absolutely still does molecular diffusion alone control the rates of evaporation. Most of the time, the air circulates over forests and other natural surfaces causing turbulence that greatly enhances the potential for evaporation. Immediately adjacent to a solid surface the air is still, but away from that surface wind movement causes drag, just as moving water does on objects in a stream. As the momentum increases, eddies and currents are formed. Turbulence is increased if the underlying surface is rough, as is the case for forests compared to the surface of a lawn. Dalton in 1801 first recognized the importance of including turbulence in estimating evaporation (cited in Kelliher and Scotter, 1992). Turbulence is expressed in an aerodynamic transfer equation, generally expressed as

$$E = (e_w - e_a)\, f(u) \tag{2.4}$$

where E is evaporation rate, e_w is saturated vapor pressure of water at the surface temperature, e_a is vapor pressure of the air, and f is an empirical function that describes the aerodynamic exchange property of the surface as a function of air circulation (turbulence) associated with wind speed (u).

Evaporation, as we have shown in presenting the Bowen ratio, is a component of the energy balance. In 1948, Penman combined the energy balance with Dalton's equation to calculate evaporation rates without needing to know the surface temperature. In its modern form the Penman equation can be written (Monteith and Unsworth, 1990)

$$E = E_{eq} + Dg_b/\zeta. \tag{2.5}$$

In Eq. (2.5), E_{eq} is the evaporation rate ($m\,s^{-1}$) obtained in equilibrium with an extensive, homogeneous wet surface via the energy balance (McNaughton, 1976) and is given by

$$E_{eq} = \frac{\varepsilon}{\rho\lambda(\varepsilon+1)} R_n \tag{2.6}$$

where ρ is the density of air and λ is the latent heat of vaporization, defined earlier, both properties of water in air and only weakly temperature dependent; ε is the change of latent heat relative to the change of sensible heat of saturated air which is 1.26 at 10°C and increases exponentially with temperature (T): $\varepsilon = 0.7185\, e^{0.0544T}$. Other terms in Eq. (2.5) are the air saturation deficit D (Pa) and g_b, the boundary-layer conductance for water vapor ($m\,s^{-1}$), which approaches zero under calm conditions, increases with wind speed, and depends on the roughness of the evaporating surface (Fig. 2.5). Zeta (ζ), the final term in Eq. (2.5), is a combination function so that $\zeta = \rho(\varepsilon + 1)G_vT_k$, where G_v is the gas constant for water vapor ($0.462\,m^3\,kPa\,kg^{-1}\,K^{-1}$) and T_k is air temperature in degrees Kelvin.

The relative importance of E_{eq} and Dg_b/ζ in Eq. (2.5) depends on the relative magnitudes of net radiation and boundary-layer conductance. When the boundary-layer conductance is small, the deep boundary layer largely isolates the wet surface from the effects of the saturation deficit in the air above, and evaporation rates approach the equilibrium rate E_{eq}. Alternatively, the energy for evaporation may be increased substantially if *advection* provides for the horizontal transfer of warmer and drier air into the system by increasing the saturation deficit D (Fig. 2.4).

The rate of evaporation from wet surfaces depends strongly on the boundary-layer conductance (g_b). Representative values of boundary-layer conductance for pasture, crops, and forests, with closed canopies and nominal heights of 0.05, 0.5, and 20 m, respectively,

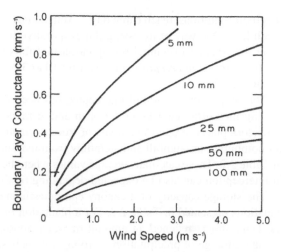

FIGURE 2.5. The aerodynamic or boundary-layer conductance (g_b) is generally high for tall vegetation because of turbulence. With shorter vegetation, however, less turbulence is present, and the width of leaves (shown from 5 to 100 mm) and wind speed become increasingly important in determining g_b. (After Grace, 1981.)

are 0.01, 0.02, and 0.2 m s^{-1}. Differences in what approaches maximum wet canopy evaporation rates for the three types of vegetation can be illustrated using typical daytime values during periods of rainfall, for the variables in Eq. (2.5): 50 W m^{-2} for R_n, 0.1 kPa for D, 10°C for air temperature, and the g_b values given above. These give evaporation rates of 0.05 mm hr^{-1} for the pasture, 0.065 mm hr^{-1} for crops, and 0.3 mm hr^{-1} for the forest. At night, evaporation continues at nearly the same rate for forests (assuming D remains at 0.1 kPa), adding up to a maximum 24-hr water loss of nearly 6 mm day^{-1}. Shorter vegetation is more dependent on the net radiation balance, which is negative at night, so maximum daily rates are unlikely to exceed 1–1.5 mm day^{-1}, unless the vegetation is situated on exposed sites where wind speeds are sufficient to increase the boundary-layer conductance to values similar to that typical for forests (Fig. 2.5).

Evaporation from bare soil, which has a boundary-layer conductance similar to a pasture, may also reach 6 mm day^{-1} on sunny summer days, hut these rates are soon reduced as the surface dries (Kelliher and Scotter, 1992). The litter layer beneath a dense forest, which may store two to three times its dry mass in water (10,000 kg of water per hectare is equivalent to 1 mm), is shielded from direct radiation and wind turbulence so normally contributes less than 5% to daily evaporation (Waring and Schlesinger, 1985). When the canopy is more open, however, turbulent exchange and direct radiation on the understory vegetation and leaf litter increase substantially and losses from these surfaces can account for up to 30% of the daily total. As these surfaces dry, solar radiation begins to heat the air, which results in increasing the air saturation deficit.

C. Interception

In the previous section we established that maximum rates of evaporation from the wet surface of various (homogeneous) types of vegetation and from bare soil could, under specified conditions, reach 6 mm day^{-1}. Actual rates of evaporation, however, are generally much less. One way of assessing the evaporative losses from forest canopies is to determine the fraction of annual precipitation that falls but does not reach the ground. This fraction, termed *interception,* ranges from 10 to 50% of annual precipitation (Kelliher and Scotter, 1992).

The proportion of rainfall intercepted has been determined for a variety of forests by placing measuring devices in the open and at random points beneath the canopy. Other collectors placed around the stems of trees estimate stemflow. A summary of equations for computing throughfall and stemflow from measurements of gross rainfall is presented in Table 2.2. These calculate average values when the canopy is fully developed. Variation in interception can also be expected depending on the intensity and duration of a storm.

The storage capacity of a canopy can be estimated from interception calculated from equations presented in Table 2.2. For a plantation of Scots pine, Rutter (1963) found that the leafy shoots retained an amount of water approximately equal to the dry weight of foliage. Storage of precipitation on foliage, branches, and stems was estimated to be 0.8, 0.3, and 0.25 mm, respectively. Deciduous trees in general store less water on their canopies than conifers, unless the canopy is heavily loaded with epiphytes. If the precipitation is in the form of snow or ice, the water equivalent stored may be more than double that in the liquid state (Zinke, 1967).

Large deviations from these values occur in different seasons. The distribution of storage also changes with stand development because the surface areas of branches (living and dead) and stems increase with stand age up to a point, then decrease (Chapter 5). Epiphytes (mosses, lichens, liverworts) on branches and stems can greatly increase the capacity of those structures to absorb surface water, which explains, in part, why some large trees may exhibit almost no stemflow even during intense storms. Another factor that controls stemflow is the smoothness of the bark surface. Species with smooth surfaces such as beech (*Fagus*) may transport as much as 12% of the precipitation as stemflow, whereas pine normally transfers less than 2% by this route. Stemflow may have special

TABLE 2.2

Summary of Equations for Computing Throughfall and Stemflow for Coniferous and Hardwood Forests from Measurement of Rainfall[a]

Vegetation	Throughfall	Stemflow	Interception per 1 cm of rain
Red pine	$0.87P - 0.04$	$0.02P$	0.15
Loblolly pine	$0.80P - 0.01$	$0.08P - 0.02$	0.15
Shortleaf pine	$0.88P - 0.05$	$0.03P$	0.14
Ponderosa pine	$0.89P - 0.05$	$0.04P - 0.01$	0.13
Eastern white pine	$0.85P - 0.04$	$0.06P - 0.01$	0.14
Pine (average)	$0.86P - 0.04$	$0.05P - 0.01$	0.14
Spruce–fir–hemlock	$0.77P - 0.05$	$0.02P$	0.26
Hardwoods, in leaf	$0.90P - 0.03$	$0.041P - 0.005$	0.10
Hardwood, deciduous	$0.91P - 0.015$	$0.062P - 0.005$	0.05

[a]After Helvey (1971) and Helvey and Patric (1965). Last column presents estimates of interception for 1 cm of precipitation, where Interception = 1 − (Throughfall + Stemflow).

significance in distributing potassium to the area around certain smooth-barked trees, because that nutrient is easily leached from foliage (Gersper and Holowaychuk, 1971; Chapter 4). It may also concentrate water near the roots of some arid forest species, such as *Acacia aneura* (Doley, 1981). Details of this type are significant in explaining the composition of forests, but they cannot be incorporated into generalized models predicting interception.

Most hydrologic models that directly consider exchange of water vapor from vegetation incorporate some estimate of canopy leaf area. Detailed studies show that the capacity of foliage to store water increases in direct proportion to the surface area, although the retention coefficient (kg water per m^2 of surface) varies with leaf dimensions, orientation, and surface smoothness (Fig. 2.6). A reasonable estimate of the surface area of foliage, branches, and stems may be calculated from regressions with stem diameter (Chapter 3); however, because the amount of foliage present may vary by 100% throughout the year, indirect ways of estimating seasonal variation in the canopy surface area, expressed as the

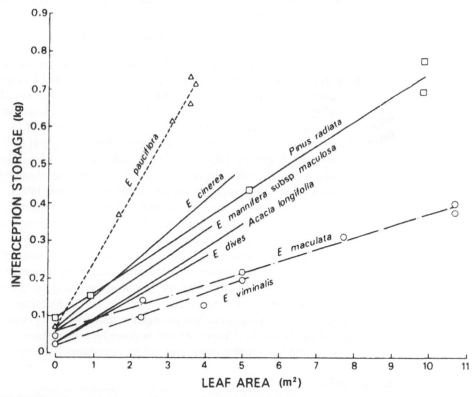

FIGURE 2.6.   Interception storage capacity of six species of eucalyptus varying from 0.032 to 0.178 mm per unit leaf area, depending on surface properties and leaf orientation. *Acacia longifolia* and *Pinus radiata* had intermediate interception capacities. (Modified from *Journal of Hydrology,* Volume 42, A. R. Aston, "Rainfall interception by eight small trees," pp. 383–396, 1979, with kind permission of Elsevier Science–NL, Sara Burgerhartstraat 25, 1055 KV Amsterdam, The Netherlands.)

projected (or silhouette) area of leaves per area of ground surface [leaf area index (LAI)], have been sought.

D. Seasonal Estimation of Leaf Area Index

Leaf area index is defined as the projected area of leaves over a unit of land ($m^2 m^{-2}$), so one unit of LAI is equivalent to $10,000 m^2$ of leaf area per hectare. Sometimes LAI is expressed on the basis of all leaf surfaces. For broadleaf trees, the total is twice the projected area; for needle-leaf trees the conversion is between 2.4 and 2.6 (Waring *et al.*, 1982) and only rarely pi (π), because leaves are seldom perfect cylinders (Grace, 1987).

Seasonal variation in LAI may be estimated by determining a mid-season maximum value and then reducing this in proportion to the amount of leaf litterfall collected periodically throughout the year (Burton *et al.*, 1991). The most common method of estimating seasonal variation in LAI is from measurements of the fraction of visible light transmitted through the canopy to the ground (Pierce and Running, 1988; Gower and Norman, 1991). The *Beer–Lambert law* is often applied to calculate the fraction of light intercepted (I) by increasing layers of leaves:

$$I = 1 - Q_t/Q_o = 1 - e^{(-k)(\text{LAI})} \tag{2.7}$$

where Q_o is the visible radiation on cloudless days incident above the canopy, Q_t is visible radiation transmitted through accumulated layers of leaves, and k is an extinction coefficient (usually between 0.3 and 0.6), which represents the fraction of visible radiation intercepted by a unit area of leaves. The relation is exponential, so that at $k = 0.5$, a LAI of 1, 3, 6, and 9 intercepts, respectively, 39, 78, 95, and 99% of the visible light. The Beer–Lambert law can also be applied to estimate the interception of near-infrared radiation by canopies. Because leaves are relatively transparent to infrared radiation, the short-wave radiation deep in plant canopies is relatively enriched in the near-infrared component, with k values often half those recorded for visible light (Jones, 1992).

In more open canopies, the vertical and horizontal distribution of leaves becomes important because certain configurations alter the normal exponential decrease in the net radiation and wind speed from the top of a canopy downward (Shuttleworth, 1989). *Roughness length* is a term related to boundary-layer conductance that describes how height and other structural features of vegetation modify the generally logarithmic wind profile through canopies and defines the height above the ground surface where wind speeds extrapolate to zero (Monteith and Unsworth, 1990). Roughness length and related expressions of how vegetation affects momentum transfer are important in estimating wet and dry fall of chemicals (Chapter 4) and in describing turbulence created by variations in vegetation and topographic conditions across landscapes (Chapter 8).

E. Modeling Evaporation

Evaporation occurs only from the wet surfaces of foliage, branches, stems, and litter or, when litter is absent, from the surface soil. Detailed models account for the storage and movement of water through ecosystems, estimating evaporation as a function of location of the wetted surfaces, energy loading, and other factors that affect the rate at which water vapor is transferred to the atmosphere (Rutter *et al.,* 1971). Foliage temperature is lower when evaporation occurs and approaches air temperature when completely dry, a fact which permits the fraction of wet/dry foliage in a canopy to be estimated from direct or indirect measurements of the difference between temperatures of the canopy surface and air (Teklehaimanot and Jarvis, 1991).

To apply the Penman equation to partially wet canopies, Rutter (1975) developed a model that keeps continuous account during a storm of the fraction of surfaces that are wet and estimates evaporation exclusively from those surfaces. The position of the wet surfaces is important, as mentioned previously, because both the net radiation and wind speed are generally reduced exponentially from the top of the vegetation downward through layers of leaves (Shuttleworth, 1989). When gaps appear in a forest canopy, large eddies are created that increase turbulent transfer between the top of the canopy and the soil. Turbulent transfer around individual trees increases linearly as the spaces between trees become greater, but the boundary-layer conductance of the whole canopy approaches an asymptote as leaf area index increases (Teklehaimanot and Jarvis, 1991).

Gash (1979) modified Rutter's model so that it could be applied to estimate interception during storms of known average intensities ($mm\,hr^{-1}$) by accounting for evaporation, throughfall, and stemflow. This kind of model can be applied to situations where only daily estimates of precipitation are available, but average rainfall intensities must still be known because 6 mm of light drizzle may be fully evaporated whereas a cloudburst of similar magnitude may result in only 1–2 mm being intercepted.

F. Transpiration from Plant Canopies

When the surfaces of leaves are dry, water loss takes place by transpiration through stomata. Plants can control stomatal opening and thus influence transpiration rates. By including a stomatal conductance term, Dalton's equation can be extended to calculate transpiration from a single leaf, and in 1965 Monteith used this approach to extend Penman's equation so that it could describe transpiration from a plant canopy using a single bulk canopy conductance (G_s) that is the product of the mean stomatal conductance (g_s) and LAI. The so-named Penman–Monteith equation may be written (McNaughton and Jarvis, 1983) as

$$E_T = \Omega E_{eq} + (1 - \Omega)E_{imp} \tag{2.8}$$

where transpiration (E_T) from a dry canopy is related to the evaporation rate (E_{imp}) imposed by the effects of the air saturation deficit D:

$$E_{imp} = D g_s / \rho G_v T_k \tag{2.9}$$

and the coefficient Ω indicating the degree of coupling between the canopy and D is

$$\Omega = \frac{1 + \varepsilon}{1 + \varepsilon + g_b / G_S}. \tag{2.10}$$

The Penman–Monteith equation incorporates and defines the relative importance of the two environmental driving forces on evaporation, namely, net radiation (R_n), and the dryness of the air (D), with two controls: one physical, the boundary-layer conductance

(g_b), and the other physiological, the canopy stomatal conductance (G_s). When the boundary-layer conductance is much less than canopy conductance, Ω approaches 1, and transpiration rates approximate the equilibrium evaporation rate. This tends to be true for short vegetation; however, for tall vegetation, such as forests, the boundary-layer conductance usually exceeds canopy conductance, so that Ω approaches zero. Under such conditions, where other factors do not limit stomatal conductance, the air saturation deficit (D) rather than net radiation largely controls the rate of transpiration.

At LAI < 3, evaporation from the soil surface contributes greatly to the bulk conductance, but at LAI > 3, the canopy stomatal conductance (G_s) plays the dominant role. Körner (1994) published a distillation of leaf stomatal conductance (g_s) values for over 20 vegetation types and 200 species that showed maximum values were similar for most woody plant communities, at about $0.006 \, m \, s^{-1}$. Kelliher *et al.* (1995) expanded the comparison to include maximum whole canopy conductance (G_s) and concluded that forests with LAI > 3.0 have nearly stable maximum G_s values (about three times the maximum g_s, see Table 9.3), because the maximum g_s of individual leaves progressively decreases with additional increments of LAI (Lloyd *et al.*, 1995; Whitehead *et al.*, 1996).

G. Empirical Model of Stomatal Conductance

The Penman–Monteith equation requires estimates of stomatal conductance in order to predict transpiration from canopies. Stomatal conductance is often predicted as a function of environmental factors which directly or indirectly restrict stomata opening below a maximum (g_{max}), so that

$$g_s = g_{max} \, f_1(L) \, f_2(T) \, f_3(D) \, f_4(H_2O) \, f_5(CO_2). \qquad (2.11)$$

The variables $f_1, f_2, \ldots, f_x$ vary from 0 to 1 and are nonlinear functions of light (L), temperature (T), water vapor pressure deficit (D), available soil water (H_2O), ambient CO_2 concentrations, and air pollutants (Chapter 6). Equation (2.11) describes a response surface that may be defined by leaf–gas exchange measurements under controlled conditions or by determining the maximum G_s observed for whole canopies over a wide range of conditions (Jarvis, 1976). Lloyd *et al.* (1995) constructed a series of response surfaces showing interactions between light, leaf temperature, and air vapor pressure deficits on G_s from continuous monitoring of water vapor exchange above a tropical forest (Fig. 2.7a). With these empirical relations, measured canopy stomatal conductance was accurately predicted (Fig. 2.7b). The canopy stomatal responses shown in Fig. 2.7a are similar to those obtained for other types of forests, although the air temperature response is much less pronounced for temperate and boreal forests (Shuttleworth, 1989).

The most striking feature of the canopy stomatal response patterns illustrated in Fig. 2.7 is the limitation that the foliage to air vapor pressure gradient exerts as radiation and temperature increase. Between values for D of 0.5 to 3.0 kPa, canopy stomatal conductance is reduced exponentially. It is primarily because of this reduction in G_s associated with increasing D that transpiration rates from forests, even on clear days, rarely exceed $0.5 \, mm \, hr^{-1}$ or $6 \, mm \, day^{-1}$ (Kelliher *et al.*, 1995). If stomata were not to close in response to D, transpiration could theoretically be expected to exceed $30 \, mm \, day^{-1}$ in some situations.

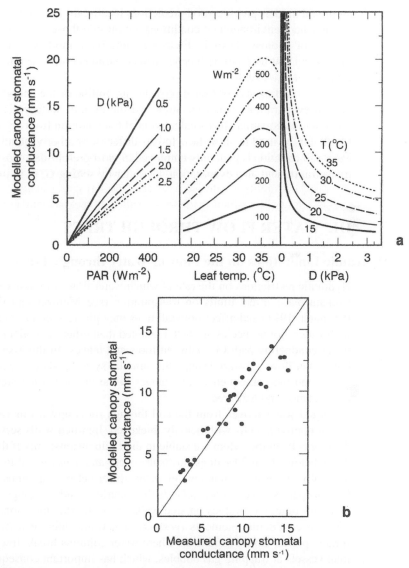

FIGURE 2.7.   (a) Empirical data collected above an Amazonian rain forest illustrate that canopy stomatal conductance (G_s) varies with meteorological conditions. in general, G_s increases with photosynthetically active radiation (PAR) and air temperature (T) and decreases with vapor pressure gradient between foliage and the air (D). (b) Combined into a model, predicted canopy stomatal conductance closely matched that measured. (After Lloyd *et al.*, 1995.)

The empirical approach to modeling stomatal response has been widely applied in comparative studies to document differences among species in their sensitivity to various environmental factors. This type of information is useful in predicting changes in forest composition if supplemented with additional information regarding differential growth allocations and sensitivity to various types of disturbances (Chapters 5 and 6). Although empirical models provide good predictions locally, they should not be extrapolated to situations where environmental conditions are outside those studied. In Chapter 3, where the process of photosynthesis is discussed, more basic models of stomatal response will be introduced that incorporate biochemical constraints (von Caemmerer and Farquhar, 1981; Leuning *et al.,* 1995).

The preceding sections have provided an outline of the energy balance principles that govern evaporation from vegetation and the ground surface. The Penman–Monteith or combination equation allows calculation of transpiration from dry canopies; Eqs. (2.8) and (2.10) can be used to show how forests differ from other vegetation types. The Penman–Monteith equation is generally employed in most present-day ecosystem models to predict transpiration and also evaporation, the latter by setting G_s to infinity.

III. WATER FLOW THROUGH TREES

A. Plant Hydraulic Limitations on the Flow of Water through Trees

Hydraulic restrictions on the rate at which water flows through stem wood, branches, and stomata place direct limits on transpiration (see Meinzer and Grantz, 1991; Mott and Parkhurst, 1991) and affect how tall trees may grow, even in environments well supplied with water. Some species are better adapted than others in maintaining the integrity of the water conducting pathway between roots and leaves. In this section, we briefly describe how water flows upward through the vascular system, where the major resistances to flow occur, and the consequences of changes in the internal water status of trees on stomatal conductance and leaf area.

Liquid water moves from the soil through roots upward in vascular tissue (*xylem*) in which nonliving cells with heavily thickened, lignified walls serve as conduits. The conducting cells in the xylem are stable in their dimensions; only if the relative water content drops below about 20% does shrinkage occur and cause wood to split (Siau, 1971). Anatomically, the water conducting elements differ between gymnosperms and angiosperms. Gymnosperms have short *tracheids* that interlock and exchange water through bordered pits with functional valves on side walls. Angiosperms have longer and wider and more efficient conducting elements (*vessels*), which are interconnected at their ends by perforated plates. Under conditions where water columns break, tracheids are more efficient than vessels at trapping gas bubbles, which has important consequences in relation to the restoration of water to empty (*cavitated*) xylem elements.

Only the sapwood actually conducts water through stems and branches (Fig. 2.8). Heartwood, which forms internally from sapwood, has a majority of cells filled with gas or impermeable metabolic products. Water columns in the capillaries of sapwood have great cohesiveness as a result of strong surface tensions associated with the way that water molecules are hydrogen bonded (Jones, 1992). The strong cohesion of water molecules to one another allows them to be pulled upward through the stem to wet surfaces of leaf cell walls from which water is transpired through stomata, or more slowly through the leaf cuticle.

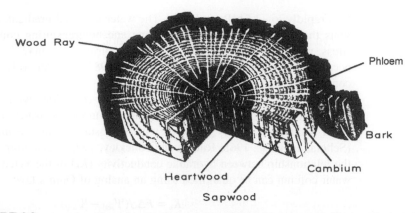

FIGURE 2.8.   A section of an oak stem shows various anatomical components. The outer bark protects the cambium where cell divisions occur, producing wood inward and phloem outward. Water and nutrients are conducted from the roots to the leaves via conducting elements in the sapwood. Large wood rays transverse the sapwood and inner heartwood. The ray cells are alive in the sapwood but no longer function in the heartwood. (From Raven *et al.*, 1981.)

The movement of water through trees is dependent on the difference in the energy state of water from point to point in the system (Fig. 2.1). The energy state of water is usually expressed as the potential (Ψ) relative to pure, free water with units of pressure (megapascals), equivalent to a force per unit area ($1\,\text{MPa} = 1.02 \times 10^5\,\text{kg}\,\text{m}^{-2}$). Flow is always in the direction of more negative Ψ. Water potential in plants may be separated into four components:

$$\Psi = \Psi_p + \Psi_s + \Psi_m + \Psi_g \tag{2.12}$$

where Ψ_p, Ψ_s, Ψ_m, and Ψ_g, respectively, are related to pressure, solute, matric, and gravitational forces. The pressure component (Ψ_p) represents the difference in hydrostatic pressure from one cell to another, and it is the only component which can be positive or negative. The solute potential (Ψ_s) represents the contribution of dissolved sugars and salts. The matric potential (Ψ_m) results from small negative forces at the surface of solids such as cell walls or soil particles. The gravitational component (Ψ_g) increases with height above ground at $0.01\,\text{MPa}\,\text{m}^{-1}$.

As molecules of water are vaporized and diffuse from leaves, tension in the sapwood water columns increases. Living cells possess semipermeable membranes and can accumulate solutes which lower the water potential within cells, increasing the potential gradient and causing water to be extracted from the vascular system. As a result of accumulating solutes and by changes in cell wall permeability, living cells remain relatively turgid through most droughts. The high concentration of solutes in the thin layer of *phloem* located between the cell-dividing *cambium* and bark (Fig. 2.8) can create a reverse flow from the top of a tree downward toward growing roots or other organs where sugar and other metabolites become concentrated.

In nontranspiring small trees, with water readily available to roots, the xylem potential of leaves is about $-0.2\,\text{MPa}$. Transpiration causes tension in the water columns, and Ψ

falls rapidly to $-1.5\,\text{MPa}$ or below. The water potential gradient, $\Delta\Psi$, in this case, represents the difference between transpiring and nontranspiring conditions. Under drought stress, predawn water potentials in nontranspiring trees may fall well below $-1.5\,\text{MPa}$, occasionally to $-8\,\text{MPa}$ (Tyree and Dixon, 1986), but $\Delta\Psi$ values are small under such conditions because stomata remain closed.

Predawn Ψ status, which usually represents conditions with zero transpiration, is an important reference base for interpreting and modeling water flow through trees. Leaf water potential is usually measured with a pressure chamber using twigs or single leaves (Scholander *et al.*, 1965; Ritchie and Hinckley, 1975). Assuming a steady-state situation, the relationship between hydraulic conductivity (K_h) of the xylem and the tension on the water column can be quantified using an analog of Ohm's law:

$$K_h = F\Delta z/(\Psi_{soil} - \Psi_{leaf}) \qquad (2.13)$$

where F is the flux of water ($\text{kg\,s}^{-1}$) and ($\Psi_{soil} - \Psi_{leaf}$) is the water potential gradient ($\Delta\Psi$) between the leaf and soil, per unit distance (Δz, m) the water travels. Ψ_{leaf} measured under nontranspiring conditions is often substituted for Ψ_{soil}. The hydraulic conductivity (K_h, $\text{kg\,m}^{-1}\,\text{s}^{-1}\,\text{MPa}^{-1}$) is independent of the diameter of the stem at any particular point. Specific conductivity (K_s, $\text{kg\,m}^{-1}\,\text{s}^{-1}\,\text{MPa}^{-1}$) of a stem (or branch) is K_h normalized by the area of conducting tissue (m^2) in the cross section and is a measure of porosity. Leaf-specific conductivity (LSC, $\text{kg\,m}^{-1}\,\text{s}^{-1}\,\text{MPa}^{-1}$) is K_h normalized by the leaf area (m^2) distal to the stem and is the best general reference for comparing the efficiency of one species with another in conducting water to transpiring surfaces (Tyree and Ewers, 1991).

Wood permeability may be determined in the laboratory by measuring the flow of water under constant pressure through samples of specified length and cross-sectional area. From such measurements, it has been demonstrated that the outer zone of sapwood may have a permeability as much as 10 times that of the inner zone (Comstock, 1965). Also, progressing up the stem to the base of the live crown, sapwood permeability tends to increase. The significance of these trends is that hydraulic conductivity (K_h), represented by the product of the average permeability and sapwood area, remains constant from the base of the stem to the live crown, although the total cross-sectional area of sapwood may be reduced by more than threefold (Shelburne and Hedden, 1996).

From such analyses of wood permeability, we find that species well adapted to their environment exhibit significantly higher wood permeability and conductances than less well adapted species, or those situated in an unfavorable competitive position relative to neighboring individuals (Sellin, 1993; Shelburne and Hedden, 1996). Once trees reach maximum height, the total resistance to water flow, when sapwood is saturated, tends to approach a common value (Pothier *et al.*, 1989). Much effort has gone into analyzing anatomical differences in wood structure, with wood density an obviously important variable, but the dominant control on stem conductance appears to be the fraction of sapwood that remains in a saturated state (Waring and Running, 1978; Shelburne and Hedden, 1996).

Hydraulic properties of various structural components can also be determined in the field or laboratory by sequentially cutting off portions of a tree under water and providing the remaining system free access to a monitored water reservoir. From such analyses, the branches and twigs have been found to offer the highest resistance to flow, and, being at the end of the vascular system, they are more likely to suffer permanent damage from cavitation than the main stem (Zimmermann, 1978; Tyree *et al.,* 1994).

B. Seasonal Variation in Water Content and Hydraulic Conductivity

In transpiring plants, the water stored in tissues is not in steady state, as assumed by Ohm's law. The tissues in a plant may be considered as a number of alternative sources of water linked in parallel with each other and the soil. As water is lost through transpiration, the flows out of storage at any one time and their relative phasing depend on the resistances between the stores and the xylem, the capacity of the stores, and the relationship between Ψ_{leaf} and tissue moisture characteristics (see Landsberg, 1986a). Initially, water will move out of storage from the leaves; however, this source is small (Whitehead and Jarvis, 1981), and the potential will quickly drop, shifting the main source of supply progressively lower down the plant. The extraction of water from branches and stems may contribute up to 1–2 mm day^{-1}, and it explains why measurements of sap flow made at the base of trees, by introducing metal probes and measuring the velocity of a heat pulse upward, show that flow there often lags 1–2 hr behind measured rates of transpiration (Fig. 2.9; see Granier *et al.,* 1996, for comparison of four related methods). Extraction of water from the sapwood appears to account for the majority of water supplied to buffer the difference between the recorded flow in the stem and losses from the canopy (Waring *et al.,* 1980; Tognetti *et al.,* 1996).

As a consequence of emptying the larger diameter xylem elements first, the hydraulic conductivity of stem and branches is reduced exponentially as the relative water content of sapwood decreases. This in turn causes stomatal closure at progressively lower vapor pressure deficits (Waring and Running, 1978; Sperry and Pockman, 1993; Lo Gullo *et al.,* 1995). Wet temperate species begin to experience cavitation at moderate water potentials between −0.8 and −1.5 MPa, as documented by the recording of acoustic emissions from tree stems (Sandford and Grace, 1985; Sperry *et al.,* 1988; Sperry and Pockman,

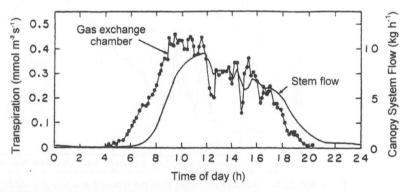

FIGURE 2.9.   Water uptake, tracked by measuring the velocity of a heat pulse up the stem, may lag 20–30% behind cuvette measurements of water transpired through leaf stomata during the day. Recharge of the reservoirs in sapwood and other tissues occurs as water uptake continues through the night. (After Schulze *et al.,* 1987. © 1987 American Institute of Biological Sciences.)

1993; Jackson *et al.*, 1995). Species adapted to harsher environments, such as *Juniperus,* do not experience cavitation until water potentials fall below −3.5 MPa and still maintain some functional xylem at potentials below −8.8 MPa (Tyree and Dixon, 1986).

One practical way of assessing whether major seasonal changes in hydraulic conductance occur that could affect predictions of stomatal conductance (g_s) is to extract small wood core samples of known volume from near the base of trees and determine the relative water content (RWC) by measuring wood density and moisture content, assuming that cell wall constituents have a constant density of 1530 kg m^{-3} (Siau, 1971). From such an analysis, an old-growth Douglas-fir forest subjected to a long period of summer drought was shown to experience major seasonal changes in RWC, and recovery to a fully saturated state required many months following the commencement of precipitation (Waring and Running, 1978; Fig. 2.10).

Conifers with a high proportion of sapwood in their stems (such as *Pinus*) or species that accumulate large biomass (such as Douglas-fir) may store more than 200 Mg H$_2$O ha^{-1}, equivalent to 20 mm in sapwood (Waring and Running, 1978; Waring *et al.*, 1979). The mechanisms by which cavitated cells in xylem refill are not fully understood. Mobilization of sugars in roots on some angiosperms can create root pressure and force water into previously cavitated xylem (Sperry *et al.*, 1988). For tall conifers, other processes must

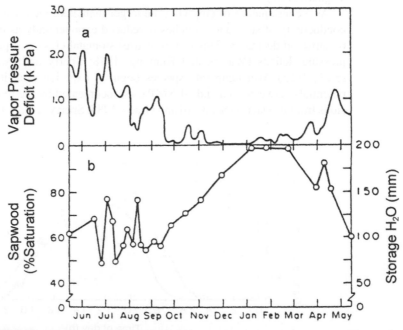

FIGURE 2.10.   Withdrawal of water from the sapwood of a 500-year-old Douglas-fir forest growing in the Pacific Northwest begins in March at the start of the growing season, and stored water reaches its lowest level during summer drought periods. Full recharge of the sapwood occurs in the winter when rainfall is high. Over the course of the year, the sapwood water reservoir was depleted by 125 mm and its relative water content reduced from 100 to 50%. (From Waring and Running, 1978.)

operate, some of which appear dependent on dew or periods of high rainfall to minimize internal water potential gradients (Sobrado *et al.,* 1992; Edwards *et al.,* 1994; Boucher *et al.,* 1995; Magnani and Borghetti, 1995).

Trees adapted to harsh environments, where freezing and drought commonly occur, support less leaf area per unit of sapwood area (measured usually at 1.3 m and extrapolated, with knowledge of stem taper, to the base of the live crown) than species or varieties adapted to less stressful conditions. For example, ponderosa pine and Douglas-fir native to Oregon support almost twice the leaf area per unit of sapwood compared with related forms growing in the more continental state of Montana (Table 2.3). Adjustments between leaf area and sapwood area demonstrate a structural property of trees that helps maintain functional xylem and results in similar values of specific leaf conductivity.

TABLE 2.3

Ratio of Projected Leaf Area to Sapwood Cross-sectional Area at Breast Height for Selected Tree Species

Tree species	Leaf area/sapwood area, ratio, m^2/m^2	Reference
Conifers		
Abies balsamea	7100	Coyea and Margolis (1992)
Ahies lasiocarpa	7500	Kaufmann and Troendle (1981)
Abies procera	2700	Grier and Waring (1974)
Picea engelmannii	3500	Waring *et al.* (1982)
Pinus contorta	1500	Waring *et al.* (1982)
Pinus ponderosa (Montana)	1400	Gower *et al.* (1993)
Pinus ponderosa (Oregon)	2500	Waring *et al.* (1982)
Pinus sylvestris (England)	800	Mencuccini and Grace (1995)
Pinus sylvestris (Scotland)	1700	Mencuccini and Grace (1995)
Pseudotsuga menziesii var. *glauca* (Colorado)	2500	Snell and Brown (1978)
Pseudotsuga menziesii var. *menziessi* (W. Oregon)	5400	Waring *et al.* (1982)
Tsuga heterophylla	4600	Waring *et al.* (1982)
Tsuga mertensiana	1600	Waring *et al.* (1982)
Hardwoods		
Acer macrophyllum	2100	Waring *et al.* (1982)
Castanopsis chrysophylla	4600	Waring *et al.* (1982)
Eucalyptus regnans	3100	Vertessy *et al.* (1995)
Nothofagus solanderi (subalpine)	700	Benecke and Nordmeyer (1982)
Nothofagus solanderi (montane)	1200	Benecke and Nordmeyer (1982)
Populus tremuloides	1000	Kaufmann and Troendle (1981)
Quercus alba	4000	Rogers and Hinckley (1979)
Tectona grandis	6500	Whitehead *et al.* (1981)

C. Carbon Isotope Discrimination of Hydraulic Limitations on Stomatal Conductance

Carbon dioxide in the air diffuses into leaves through stomata and is incorporated through photosynthesis into biomass (Chapter 3). In the atmosphere, two stable isotopic forms, carbon-13 (^{13}C) and carbon-12 (^{12}C), exist in a ratio of about $1:84$ in today's fossil fuel-enriched atmosphere. During photosynthesis, plants selectively favor the lighter and more abundant form (^{12}C) over the heavier isotope, but the degree of discrimination is dependent to a large extent on the stomatal conductance. Discrimination (Δ) is defined as the difference in relative molar abundance of $^{13}C/^{12}C$ in the atmosphere compared to that in photosynthetic products. Because the normalized ratios of heavy to lighter forms of stable isotopes are so small, isotopic composition is conventionally expressed as parts per thousand (‰) in reference to deviations (δ) from the abundance ratio in a nonvarying reference standard, which, in the case of carbon, is a carbonate rock, from a Pee Dee Belemnite formation in South Carolina with a $^{13}C/^{12}C$ ratio of 0.01124. Deviations are calculated as

$$\delta^{13}C‰ = \{[(^{13}C/^{12}C)_{sample}/(^{13}C/^{12}C)_{standard}] - 1\}1000. \qquad (2.14)$$

As CO_2 diffuses into leaves through stomata, ^{13}C moves more slowly than ^{12}C and is discriminated against by 4.4‰. An additional discrimination of 30‰ against ^{13}C occurs in carbon fixation. When stomatal conductance is reduced, photosynthesis continues, and the internal concentration of CO_2 falls so that a greater proportion of the intercellular $^{13}CO_2$ reacts with the carboxylating enzyme. As a result, an indirect measure of the restrictions on stomatal conductance may be obtained by analyzing, with a mass spectrometer, the carbon isotopic composition of leaves or wood samples.

In New Zealand, the importance of branch length as a major factor limiting stomatal conductance was confirmed by comparing the isotopic composition of foliage situated at the ends of pine branches of varying length but at similar height above the ground (Fig. 2.11). In a less maritime climate than New Zealand, where the relative water content in branch wood does not remain near saturation, $\delta^{13}C$ values measured in foliage at the end of branches still reflect the leaf-specific conductivity (Panek, 1996). As trees approach maximum height, hydraulic limitations increase, which reduces g_{max} from values typically obtained on younger trees (Yoder *et al.*, 1994). The $\delta^{13}C$ analysis of foliage or wood samples quantifies the extent to which age and inherently stressful site conditions need to be taken into account in modeling transpiration and photosynthesis.

D. Soil Limitations on the Flow of Water through Plants and Stomatal Conductance

All the water transpired by vegetation is taken up by roots from the soil. The amount of water stored in soils and available to roots is among the most difficult but critical measurements to obtain. Forest soils are generally not as uniform as agricultural fields, and rooting depths of trees may vary from 0.1 to >10 m. In this section we describe how estimates of soil water storage are obtained, what limitations are placed on water extraction by roots, and what implications these limitations have on restricting stomatal conductance.

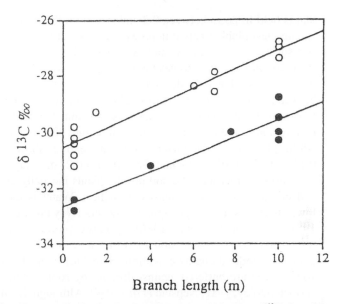

FIGURE 2.11.   The normalized ratio of carbon-13 to carbon-12 isotopes ($\delta^{13}C$) from foliage of *Pinus radiata* becomes less negative, which indicates relatively more enrichment in the heavier isotope carbon-13, with increasing branch length. Branches on more exposed aspects (○) show proportionally more enrichment of carbon-13 than those less exposed to direct solar radiation (●). The $\delta^{13}C$ isotopic ratios reflect time-integrated differences in hydraulic conductivity of branches. As stomata are forced to close, the internal concentration of CO_2 decreases, increasing the proportion of the heavier isotope (^{13}C), which is normally discriminated against by photosynthetic enzymes, that becomes incorporated in photosynthetic products. (After Waring and Silvester, 1994.)

Water uptake by roots depends on the absorbing surface area, membrane permeability, and the $\Delta\Psi$ between roots and soil (Landsberg and Fowkes, 1978; Landsberg and McMurtrie, 1984). The soil water potential (Ψ_{soil}) is the sum of gravitational, pressure, solute, and matric potentials. In most forest soils, salt concentrations are low, and thus the matric potential is the most important component. As soils dry, the energy required to extract water increases quickly as the water most freely available in large pores is depleted first.

The volumetric water content θ is the fraction of soil volume consisting of water and is calculated as

$$\theta = (m_{water}/m_{soil})\rho_{soil}/\rho_{water} \tag{2.15}$$

where m_{water}/m_{soil} is the mass of water per unit mass of soil, and ρ_{soil}/ρ_{water} is the ratio of the soil bulk density to the density of water ($1000\,kg\,m^{-3}$). The volume of water in the rooting zone would be equivalent to the product of θ and the rooting depth. Soil porosity is the ratio of the volume of pore space per unit volume of soil; when $\theta = 1.0$, all pore spaces are filled and the soil is saturated. In most cases, soil porosity is between 0.3 and 0.6, depending on the bulk density, texture, and content of organic matter and rock.

Soil water can be separated into a number of functional categories: a rapidly draining component associated with pores >50 μm in diameter, a fraction available to plant roots that is held in pores between 0.2 and 50 μm (at Ψ_{soil} between about -0.01 and -1.5 MPa), and an unavailable fraction in pores $<0.2\,\mu$m that is only lost through evaporation near the soil surface (Fig. 2.12). A sandy soil with a bulk density of 1200 kg m^{-3} is shown (at the intercept on the x axis) to have a porosity only slightly less than loam or clay soils (0.45–0.48) with similar bulk densities. Sandy soils, however, have large and uniform pores between particles, compared to smaller and more variable pore-size distributions characteristic of most finer textured soils. As a result, the moisture-retention curves in relation to soil water potentials differ with soil texture. The maximum storage of water available (θ) for plants in a cubic meter of soil with textures shown in Fig. 2.12 is as follows: sand, $0.12 - 0.05 = 0.07$ m^3; loam, $0.38 - 0.14 = 0.24$ m^3; clay, $0.46 - 0.31 = 0.15$ m^3.

Following recharge of a soil profile, plants normally extract water first from the upper soil horizons where root density is highest. Water is then progressively extracted from lower horizons, unless precipitation occurs (Nnyamah and Black, 1977; Hinckley *et al.*, 1981; Talsma and Gardner, 1986). Species exploit the soil to differing extents. Shallowly rooted trees or shrubs exhibit more negative water potentials than more deeply rooted associates under droughty conditions (Hinckley *et al.*, 1983). In both arid and tropical wet zones, trees are buffered against drought by root systems that extend >10 m below the surface (Doley, 1981; Nepstad *et al.*, 1994). Although detailed models exist for calculating

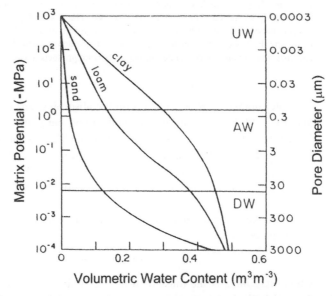

FIGURE 2.12. Water storage capacity of soils varies with texture, reflecting pore size distribution. Water in pores below 0.2 μm diameter is unavailable to roots (UW), being held at soil matrix potentials at or below -1.5 MPa. Available water (AW) is held in pores in the range from 0.2 to 50 μm, equivalent to matrix potentials between about -0.01 and -1.5 MPa. As pore size increases above 50 μm, capillary forces are unable to hold water, and it drains rapidly (DW zone in the graph). Soils of similar density (here about 1200 kg m^{-3}) and porosity (θ between 0.43 and 0.48) vary in the volume of available water per cubic meter of soil from $\theta = 0.07$ m^3 in the sand to 0.24 m^3 in the loam. (After Ulrich *et al.*, in *Dynamic Properties of Forest Ecosystems*, 1981, courtesy of Cambridge University Press.)

water uptake from different soil horizons as a function of soil properties and root density (see Landsberg, 1986a), data requirements are too high for inclusion in most ecosystem models. These models generally include only an estimate of maximum rooting depth and assume that roots are equally distributed or, alternatively, concentrated mostly in the upper soil horizon.

Depending on soil texture and root distribution, between 45 and 75% of the available water can usually be extracted from the rooting zone before transpiration is reduced below maximum rates and plants exhibit a significant reduction in predawn Ψ (Fahey and Young, 1984; Dunin *et al.,* 1985; Fig. 2.13a). Small additions of water following a single storm may permit a full recovery to minimum predawn Ψ while most of the soil profile remains dry. Generally, the maximum opening of stomata during the day decreases exponentially as predawn Ψ falls (Running, 1976; Schulze, 1986; Reich and Hinckley, 1989; Ni and Pallardy, 1991; Fig. 2.13b).

Water uptake by roots is affected by soil temperature and oxygen levels, as well as by limited availability of water in dry soils. Low root temperatures limit water uptake because the viscosity of water near freezing is approximately twice that at 25°C (Nobel, 1991). Plants adapted to warm or to cold climates differ in their root permeabilities as a result of varying lipid–protein ratios in root membranes. As a result, water uptake by the roots of a subalpine species such as *Pinus contorta* shows sensitivity to temperatures only below 5°C (Running and Reid, 1980), whereas for a maritime species such as *Pinus radiata,* water uptake is inhibited until soil temperatures exceed 15°C (Kaufmann, 1977). In the southern hemisphere, where an ancient subtropical tree flora has evolved and now survives in a temperate climate, root resistance to water uptake at temperatures below 10°C appear to limit stomatal conductance more than for typical temperate forest species (Hawkins and Sweet, 1989; Waring and Winner, 1996).

As soils drain, the oxygen levels increase as more void spaces become air filled. At saturation, soils have no unfilled void spaces, and oxygen in the water is rapidly depleted by respiration of roots and microorganisms. Some species are resistant to flooding, such as *Fraxinus pennsylvanica, Nyssa aquatica,* and *Taxodium distichum,* and continue to grow roots when inundated (Hook *et al.,* 1970; Dickson and Broyer, 1972; Keeley, 1979). The ability to grow under such conditions indicates the passage of oxygen from above the water level, aerating the rooting zone. Upland plants subjected to flooding rarely exhibit water stress, probably because their stomata remain closed (Pereira and Kozlowski, 1977; Kozlowski and Pallardy, 1979) as a result of an altered hormone balance (Rinne *et al.,* 1992; Jackson, 1994). Compacted soils, whether wet or dry, offer limited air spaces for oxygen to diffuse to roots. As the bulk density of soils increases, average pore size decreases. As a result, roots are unable to penetrate soils with soil bulk densities much above $1500\,kg\,m^{-3}$ (Heilman, 1981; Tworkoski *et al.,* 1983).

E. Indirect Approaches to Defining Water Available to Plant Roots

The water available in the soil rooting zone (θ, mm) is one of the most difficult variables to quantify. Given good estimates of daily transpiration, soil water storage capacities may be indirectly estimated by adjusting the value of available θ to about 30% of the storage capacity, which corresponds to when predawn Ψ first begins to fall below its maximum

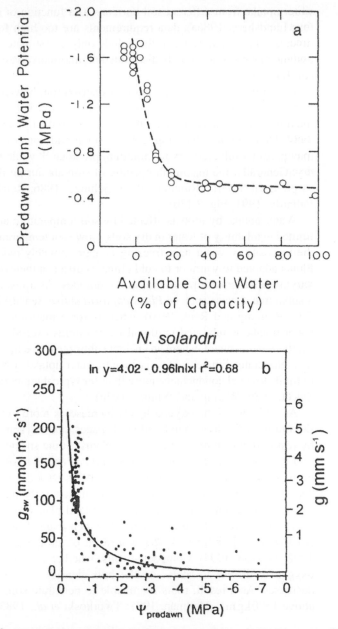

FIGURE 2.13.   (a) Until more than three-quarters of the available water held between −0.01 and −1.5 MPa is depleted from the rooting zone, a pine stand exhibits a constant predawn water potential of −0.5 MPa. As the remaining available water is withdrawn from the soil, predawn Ψ falls rapidly to below −1.5 MPa. (After "Water potential in red pine: Soil moisture, evapotranspiration, crown position" by E. Sucoff, *Ecology,* 1972, **53,** 681–686. Copyright © 1972 by the Ecological Society of America. Reprinted by permission.) (b) Maximum daily leaf stomatal conductance (g_s in two alternative units) for *Nothofagus solandri,* native to New Zealand, shows an exponential decrease as predawn water potential (Ψ$_{predawn}$) falls. (After Sun *et al.,* 1995.)

(Fig. 2.13a). When roots can tap groundwater or seeps, transpiration will not be limited by the water holding properties of the soil. Other methods are required to assess whether trees have access to such sources, although measurement of predawn Ψ will provide indirect evidence by showing values are maintained near maximum independent of the calculated soil–water balance.

Stable isotopes of oxygen ($^{18}O/^{16}O$) and hydrogen ($^{2}H/^{1}H$ or D/H) in the sap stream and cellulose of annual growth rings provide a means of distinguishing the extent to which plants tap groundwater sources (Ehleringer and Dawson, 1992). Isotopic differences between groundwater and surface sources result from not only atmospheric circulation patterns that distribute precipitation but also isotopic fractionations taking place during evaporation and condensation. The source of most atmospheric water vapor is tropical oceans. As water evaporates, the heavier isotopes are discriminated against so that water vapor, in relation to seawater (the standard), has a more negative $\delta^{18}O$ and δD. As the vapor moves into higher latitudes and cools, rain or snow is condensed from the air mass. Condensation is a fractionating process that removes the heavier isotopes and leaves the lighter behind as vapor. Thus, as atmospheric water vapor moves up in latitude or in elevation, its isotopic signatures become progressively lighter, which is reflected in the composition of the precipitation and results in a good correlation with mean annual temperature (Fig. 2.14).

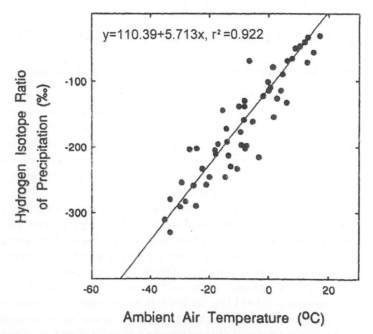

FIGURE 2.14. The hydrogen isotopic ratio in precipitation decreases linearly in association with ambient air temperature, as does that of $\delta^{18}O$, which is functionally related: $\delta^{18}O = (\delta D/8) - 10‰$. (After Ehleringer and Dawson, 1992.)

Precipitation that falls during winter months, when temperatures are low, is less likely to be evaporated and therefore contributes most to the groundwater (or seepage water derived from higher elevations). During the summer, if plant roots tap the water table, the water transferred from the deepest plant roots may diffuse out from other roots occupying very dry surface soils. In arid environments shallow-rooted plants may thus gain access to water far below their rooting depth (Fig. 2.15a). Even in areas where summer precipitation is common, plants that tap deep sources of water, including the groundwater, show much more negative δD values in their xylem sap and wood cellulose than those that draw on only surface horizons and therefore mirror the isotopic composition of summer precipitation (White *et al.,* 1985; Fig. 2.15b).

In summary, the upper limit to flow of liquid water through trees is set by structural features: the amount of leaves, tree height, branch length, xylem anatomy, and root development. Low soil temperatures, flooding, drought, and other factors further limit flow. In trees, sapwood and deep roots provide buffers against water deficits, but as wood and soil reservoirs are exhausted, reductions in leaf area and plant (and soil) hydraulic conductances occur. Measurements of predawn plant water potential and wood relative water contents provide functional measures of drought and seasonal limitations on estimates of maximum stomatal conductance. The stable isotopic composition of carbon, found in leaves and wood, provides an integrated signal of hydraulic limitations on water uptake and flow through stems and branches. Stable isotopes in water, when analyzed in the xylem sap and wood cellulose, provide a way of determining the relative contributions of subsoil and groundwater to transpiration.

IV. WATER STORAGE AND LOSSES FROM SNOW

A. Interception, Accumulation, and Energy Exchange Processes

To complete the story of surface storage and losses of water, we must consider water in its frozen state as snow or ice. As mentioned previously, snow accumulates on the canopy in proportion to leaf area index, but most is shed or falls directly to the soil surface where it may accumulate to great depths. Fresh snow contains between 6 and 35% water by volume. The density of a snowpack increases to a maximum when snowmelt begins, at which time the water content per cubic meter is uniform and the temperature throughout the pack is isothermal at 0°C.

With the exception of freezing of liquid water within the pack and conduction of heat from the ground, all components of the energy balance can be treated as within a thin surface layer of snow (Anderson, 1968). The surface temperature and albedo of snow (Table 2.1) are two key properties needed to calculate energy exchange and to determine whether snow is converted directly to water vapor (*sublimation*) or melts. When warm, moist air condenses on the cold surface of snow, heat as well as moisture is added to the snowpack ($2501 \, kJ \, kg^{-1}$ at 0°C). If the gradient is reversed, moisture and heat are lost. If the snow surface is below 0°C, the latent heat of sublimation required to turn snow directly to water vapor ($355 \, kJ \, kg^{-1}$) must be added. Latent heat transfer can be important in areas where warm, moist, turbulent air is advected over snowpacks, because 1 mm of condensate can produce about 7.5 mm of melt (Anderson, 1968).

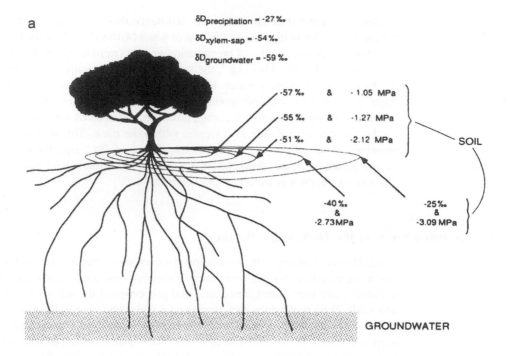

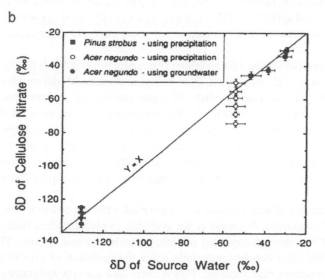

FIGURE 2.15. (a) The hydrogen isotopic composition of xylem sap (δD‰) indicates that an isolated tree (*Acer saccharum*) obtains >90% of its water during a summer drought from snowmelt-enriched groundwater. Some water leaks from the tree's roots into the upper soil horizon (0–35 cm), which explains the decreasing water potential gradient (MPa) measured up to 5 m from the tree. (b) Cellulose nitrate δD compares well with that of source water for white pine (*Pinus strobus*) growing in New York State (■) and for shallow- (○) and deep-rooted (●) boxelder (*Acer negundo*) growing near a stream in Utah. (After Dawson, 1993.)

Rain is warmer than the snowpack, and hence there is a transfer of heat from rain to the snowpack. Knowing the specific heat of water ($4.18\,\text{kJ}\,\text{kg}^{-1}\,\text{K}^{-1}$), the heat added is easily calculated from the amount of precipitation and air (rain) temperature. Because the latent heat of fusion is only $334\,\text{kJ}\,\text{kg}^{-1}$, the temperature of the snowpack is not much affected by the process of freezing water. For example, 4 mm of rain at 20°C must be added to a "ripe" snowpack (at 0°C and maximum water-holding capacity) to produce 1 mm of melt. When the snowpack is below freezing, however, rain adds its small heat content and gives up the latent heat of fusion as it freezes within the pack. This may quickly bring a cold snowpack to a uniform (isothermal) temperature of 0°C, ripe for melting. When melt occurs, water percolates down where it may refreeze. Melt is held until the water-holding capacity of the pack is exceeded ($350\,\text{kg}\,\text{m}^{-3}$).

B. Modeling Snow in the Hydrologic Balance

Coughlan and Running (1997a) developed a computer simulation model of snow accumulation and melt that requires a minimum of meteorological information (daily precipitation, maximum and minimum temperature) and properties of the site (elevation, slope, aspect, and LAI). With these data, they generated a water and energy balance to calculate daily snowfall, evaporation and sublimation, storage water in the pack, and melt at 10 sites scattered across the western United States. Soils were assumed to be at saturation to simplify calculations further (see Section V). The model accounts for changes in albedo as snow ages and applies the Beer–Lambert law to calculate short-wave radiation penetration through canopies to permit calculation of snow melt and other processes beneath a forest canopy.

Snow pillows situated on weighing lysimeters provided independent estimates of snow water content and snow melt over a 3-year period. Model predictions were in close agreement with measured values of water content of the snowpack (Fig. 2.16a) and spring melt for most sites (Fig. 2.16b). Sensitivity analyses confirmed the general importance of LAI in influencing snow accumulation and melt. Topographic variables became particularly important at higher elevation sites in early spring when sunny weather predominates and large differences in radiation occur between north and south aspects. In early summer, when solar zenith angles are higher, aspect has much less influence on melt.

The amount of water stored in a snowpack may exceed that stored in soils by some orders of magnitude, but most is not available to plants. When melt occurs, the majority of water flows through saturated soil into groundwater and streams. The presence of snow may, on the other hand, significantly delay the initiation of growth by keeping soil and root temperatures near freezing. Forest cover has a major influence on snow hydrology.

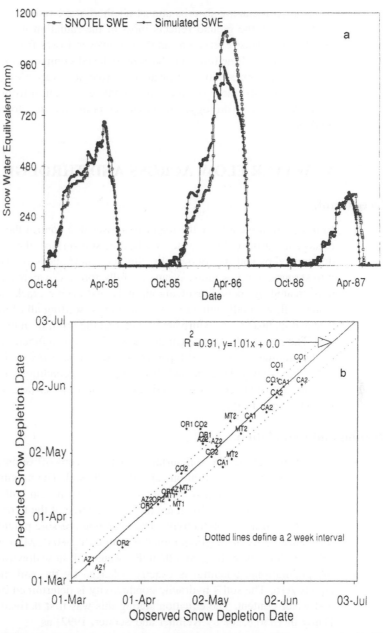

FIGURE 2.16. (a) Changes in snow water content measured on a snow pillow compared well with FOREST-BGC modeled daily predictions over 3 years. (b) The dates of complete snowmelt were predicted by a snow hydrology model over a 3-year period at 10 locations in the western United States (represented by abbreviations of the states). In general, model predictions were in good agreement with observed values. (After *Landscape Ecology,* Volume 12, J. C. Coughlan and S. W. Running, "Regional ecosystem simulation: A general model for simulating snow accumulation and melt in mountainous terrain," pp. 119–136, 1997, with kind permission of Elsevier Science–NL, Sara Burgerhartstraat 25, 1055 KV Amsterdam, The Netherlands.)

An evergreen canopy absorbs almost all short-wave radiation, while a deciduous canopy with snow on the ground reflects most of the radiation upward. Interception losses are highest from dense evergreen canopies; however, losses from the snowpack on the ground are minimal, and melting may be long delayed compared to forests with more open or deciduous canopies. No additional meteorological variables or structural description of canopies, beyond those required to estimate evaporation from wet surfaces, are required to model snow hydrology, an important point for developing generalized ecosystem models.

V. WATER FLOW ACROSS AND THROUGH SOIL

A. Surface Runoff

In this section we review the major processes that control the movement of water after it reaches the ground. Once water reaches the soil surface, the rate and direction of flow are determined by soil and topography. Surface runoff may sweep litter and pollutants directly into streams, but this rarely occurs on fully forested watersheds. In forests, natural channels created by roots, burrowing animals, and drying cracks are important in permitting water to flow rapidly into streams without first wetting all of the soil. The kind of vegetation and its history of disturbance also affect the infiltration properties of surface litter and soil. The rate at which throughfall reaches the soil is a function of the kind of litter as well as the amount. Certain soil types are known to be *hydrophobic* when dry and to absorb water very slowly (DeBano and Rice, 1973); these conditions are often enhanced following fire through transfer of dissolved residues into the surface soil (Shahlaee *et al.,* 1991; Everett *et al.,* 1995).

B. Infiltration and Percolation

Once water reaches the soil surface, whether it runs off or enters the soil depends on the rate of delivery and infiltration capacity of the soil. The maximum infiltration capacity of soils is largely determined by pore size distribution, and bulk density. The two determinants of soil water movement are the driving force of the hydraulic potential gradient and the soil's hydraulic conductivity. The hydraulic pressure potential gradient ($\Delta\Psi/\Delta z$) is the difference between the gravitational potential (Ψ_g, $+0.01\,\mathrm{MPa\,m^{-1}}$) and the matric potential (Ψ_m). As long as Ψ_g is greater than Ψ_m, water will flow downward through the soil. When the gravitational potential is exactly balanced by the soil matric potential, water flow equals zero. The soil's hydraulic conductivity is determined by the pore size distribution and water content. The equation stating this was first derived for saturated materials by Darcy in 1856 (cited by Kelliher and Scotter, 1992) as

$$F = -K_{sat}\,(\Delta\Psi/\Delta z) \tag{2.16}$$

where F is the volume flux of water through unit cross-sectional area per unit time in the direction of the lower potential, and z is the distance. Soil hydraulic conductivity (K) falls rapidly as θ decreases, in the same way as the hydraulic conductivity of sapwood. The unsaturated hydraulic conductivity of soil is very much a function of soil texture and pore size distribution, which can be described reasonably accurately with empirical equations that are a function of the clay fraction and the ratio θ/θ_{sat} (Clapp and Hornberger, 1978). Alternatively, K can be expressed in reference to changes in Ψ_p (Fig. 2.17). For a pure sand, at saturation, K_{sat}, the maximum sustainable infiltration rate, is above 10,000 mm day^{-1}, whereas for a silt loam with very uniform pore sizes, K_{sat} is only 0.05 mm day^{-1}. A sandy loam with a mixture of large and small pores has an intermediate K_{sat}, around 300 mm day^{-1}. Most undisturbed forest soils have saturated conductivities in their surface horizons above 20 mm hr^{-1} or 500 mm day^{-1}. Approaching the other extreme, minimum flows of less than 1 mm day^{-1} can be expected on all soils once the potential gradient falls below −0.01 MPa (Fig. 2.17).

Saturated and unsaturated hydraulic conductivity are usually estimated from laboratory analyses on undisturbed cores collected at different soil depths (Vepraskas *et al.*, 1991). When soils are variable, many core samples are required. In such cases, field determination of hydraulic conductivity, made on soils as they are wetted and drained, is often more reliable (Luxmoore, 1983; Talsma and Gardner, 1986; Whitehead and Kelliher, 1991).

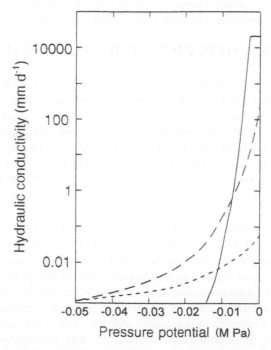

FIGURE 2.17. The hydraulic conductivity of soils decreases rapidly when the volumetric water content drops below saturated, here represented when the pressure potential equals zero. Sandy soils (solid line) exhibit at 0 pressure potential $K_{sat} > 10,000$ mm day^{-1}. Fine clays or extremely uniform silty soils (- - -) can make a subsoil nearly impervious, with $K_{sat} < 0.05$ mm day^{-1}. A sandy loam soil (– – –) has a relatively high K_{sat} of 300 mm day^{-1} and the capacity to contribute slowly to subsurface and streamflow as the pressure potential decreases. Below a pressure potential of −0.01 MPa, however, the hydraulic conductivity for most soils falls below 1 mm day^{-1}. (After Kelliher and Scotter, 1992.)

Drainage characteristics change within soil profiles so the flow of water, even in saturated zones, is not uniform. This has led to development of very complicated models, but most are based on some variation of Darcy's law (Golden, 1980; Prevedello *et al.,* 1991).

Deleterious effects of forest practices on surface soil permeability, where most compaction occurs, can be easily measured. Compaction from logging, road building, or heavy grazing can substantially increase the bulk density of surface soils and reduce infiltration rates. As a result of such activities, surface runoff and erosion may be increased by orders of magnitude in steep mountainous areas, particularly those with monsoon climate (Riley, 1984; Gilmour *et al.,* 1987; Kamaruzaman *et al.,* 1987; Malmer and Grip, 1990). If measurements are made of soil hydraulic properties throughout the surface and subsoil, unstable slopes which contain impervious horizons that foster lateral water flow can be recognized before roads and culverts are in place.

In summary, soil texture and soil depth both play an important role in determining the rate at which water flows into streams. Shallow, coarse textured soils store little water and drain rapidly, whereas deep, fine textured soils have the capacity to store large amounts of water and release the excess slowly into groundwater reservoirs and streams. Abrupt transitions in soil texture or bulk density within the profile alter hydraulic properties to cause abrupt changes in lateral flow. As a result, slopes may become unstable and subject to slumping and mass failure when saturated.

VI. COUPLED WATER BALANCE MODELS

We can now assemble the key components of the hydrologic system: precipitation is the input; snow, soil, and sapwood are storage components; and evapotranspiration and runoff are outputs (Fig. 2.18). A variety of forest ecosystem models exist which share a common structure and apply many of the formulas presented in this chapter (Ågren *et al.,* 1991; Whitehead and Kelliher, 1991; Aber and Federer, 1992; McMurtrie, 1993; Williams *et al.,* 1996). These models predict vapor losses in the form of transpiration and evaporation (and sublimation in the case of FOREST-BGC). They all calculate a daily water balance for the soil, but rarely for tree stems (except see Williams *et al.,* 1996). Excess water is routed into seepage or runoff.

Models can be run from daily weather data but are improved when precipitation, solar radiation, temperature, humidity, and wind speed can be provided hourly. In addition to general site descriptors (latitude, elevation, slope, and aspect), knowledge of seasonal changes in LAI is a prerequisite for all integrated ecosystem models. Likewise, some estimates of rooting depth and the water storage and drainage characteristics of soils are essential. Additional structural information is needed if water stored within the sapwood of vegetation is considered, and if the hydrologic responses of separate strata of vegetation are of interest (e.g., height, LAI, leaf dimensions, and rooting depth). The stomatal response of different species to atmospheric humidity deficits and to soil temperature limitations can further improve model results (Körner, 1985; Kelliher *et al.,* 1993; Schulze *et al.,* 1994a).

The most convincing tests of these forest hydrologic models are provided by comparing predicted against observed snowpack dynamics (Fig. 2.16), canopy evapotranspiration (Fig. 2.19a), and seasonal patterns in soil water depletion (Fig. 2.19b). Most models apply

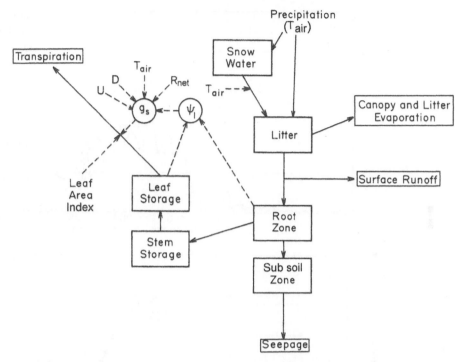

FIGURE 2.18.   General structure of a forest water balance model that accounts for precipitation entering a snowpack, litter, soil surface, and subsoil horizons. Water is eventually lost from the system through transpiration, evaporation, runoff, or seepage. Within trees, water may be stored temporarily in the sapwood of stems and branches, and in leaves. Evaporation from wet canopies depends on the driving variables (R_n), wind speed (U). and vapor pressure deficit (D) and structural variables associated with height and the leaf area index (LAI) that affect boundary-layer conductance. The calculation of transpiration requires an additional parameter, stomatal conductance (g_s), which reflects all the hydraulic resistances in the path between roots and leaves. Values of leaf water potential (Ψ_l) may be simulated and compared against measurements. A daily estimate of change in each storage compartment represents the finest resolution predicted by most coupled ecosystem models. (Modified from *Oecologia*, "Physiological control of water flux in conifers: A computer simulation model," S. W. Running, R. H. Waring, and R. A. Rydell, Volume 18, p. 11, Fig. 5, 1975, © 1975 by Springer-Verlag; and from Running, 1984a.)

the simple "big-leaf" Penman–Monteith equation and derive estimates of radiation and humidity with climatological models rather than from direct measurement (see Chapter 7 for details). In situations where rooting depth is unknown or difficult to establish, hydrologic models such as FOREST-BGC which predict predawn plant Ψ offer a realistic check on estimates of rooting depth and the amount of water extracted during long periods of drought (Running, 1994).

Modeling the annual water balance with a range of LAI values helps set realistic limits on hydrologic flows. Figure 2.20 shows that where precipitation is high and evenly distributed seasonally (e.g., Jacksonville, Florida, and Knoxville, Tennessee), FOREST-BGC predicts that outflow will drop from 80 to 20 cm as LAI increases from 3 to 9; consequently, streamflow records imply that reasonable maximum LAI values do not exceed 6.

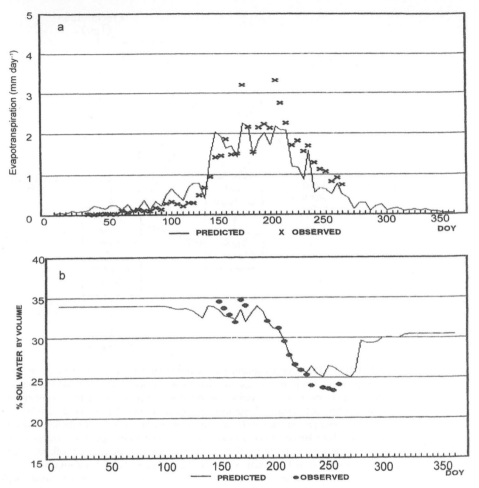

FIGURE 2.19. (a) The FOREST-BGC simulation model (Running and Coughlan, 1988) predicted daily evapotranspiration (continuous line) from an aspen forest that generally agreed with micrometeorological measurements (x) made throughout most days of a year (DOY). (b) Model predictions of soil water depletion also matched those determined with a neutron probe. (After Kimball *et al.,* 1997a.)

In more arid climates (Missoula, Montana; Fairbanks, Alaska; and Tucson, Arizona), no outflow is predicted at the lowest LAI of 3.0; values of 9 are likewise unrealistic in these climates. Direct observations of maximum leaf area index for Missoula, Montana, confirmed that values there were less than 3.0 (Running and Coughlan, 1988). Additionally, in the cold and arid Montana climate, snowpack is more important than soil depth for sustaining forest LAI. At sites with intermediate climates, like Seattle, Washington, and Madison Wisconsin, stream discharge is inversely proportional to LAI, and the distinction between evergreen and deciduous forest cover becomes increasingly important.

Grier and Running (1977) showed that the LAI of mature coniferous forests across the steep environmental gradient associated with the Oregon transect at 44°N (Fig. 1.4) was

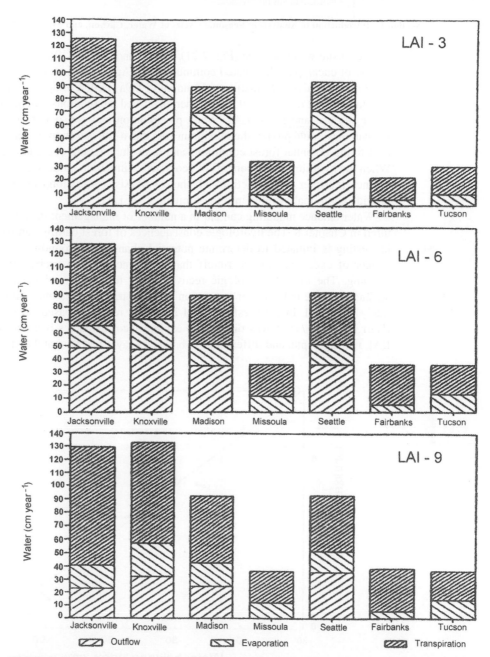

FIGURE 2.20. A sensitivity analysis with FOREST-BGC performed with weather data gathered at different cities throughout the United States indicates that increasing the leaf area index (LAI) from 3 to 9 would affect outflow and other hydrologic losses most on sites which receive >80cm year^{-1} of precipitation. In more arid environments (Missoula, Montana; Fairbanks, Alaska; and Tucson, Arizona), no outflow is predicted at an LAI of 3. (Reprinted from *Ecological Modelling*, Volume 41, S. W. Running and J. C. Coughlan, "A general model of forest ecosystem processes for regional applications. I. Hydrologic balance, canopy gas exchange and primary production processes," pp. 125–154, 1988, with kind permission of Elsevier Science–NL, Sara Burgerhartstraat 25, 1055 KV Amsterdam, The Netherlands.)

related to site water balance (Fig. 2.21). Specht and Specht (1989) showed a similar sensitivity of eucalyptus-dominated communities in Australia to aridity gradients. Following this type of analysis, Nemani and Running (1989a) built a simplified hydrological equilibrium concept into a model that further quantified the relations among precipitation, soil water storage, and forest LAI. When precipitation and soil water holding capacity are known, they could predict the maximum LAI that a site can support, an important concept in defining potential forest conditions and in evaluating site degradation at larger scales. When the hydrologic components of ecosystem simulation models are linked with carbon balance (Chapter 3), predictions of photosynthesis, respiration, and growth allocations provide additional insights into the limitations on LAI.

Water balance modeling can aid in a number of management decisions, as will be featured in Chapter 8. The hydrologic consequences of forest cutting can be explored before harvesting is initiated to determine potential changes in groundwater recharge and the danger of excessive surface runoff that could increase the probability of downstream flooding. The rate of hydrologic recovery after forest harvesting can be simulated in advance and the influence of vegetation on evapotranspiration and soil compaction separately evaluated. In early estimates of hydraulic recovery based on gauged watersheds (Harr *et al.*, 1979), it was difficult to explain inconsistencies because measurements of LAI, rooting depth, and differences in boundary layer and stomatal conductance were not

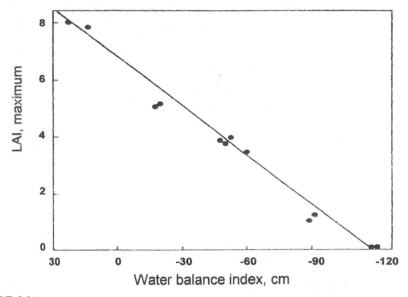

FIGURE 2.21.   In the Pacific Northwest, climate and topography vary considerably. As conditions change, so do the species that dominate the forests, as shown in the transect across western Oregon (see Fig. 1.4). The maximum observed leaf area index of the forests ranges from less than 1 to 8 across the transect, correlated with a summer water balance computed by adding soil water storage to measured growing season precipitation and then subtracting open-pan evaporation (data from Grier and Running, 1977; Gholz, 1982). The LAI estimates were converted from all surfaces by dividing by 2.5 and values were resealed by 0.5, based on corrections published by Marshall and Waring (1986) indicating overestimates in the original analyses in 1977.

available. As a result, summer flows could return to normal within a decade while peak flows during storms might remain high for many years, depending on the rate of recovery from soil compaction and reestablishment of conifer LAI. Simulations indicate that 20–30 years should be adequate on most sites in the U.S. Pacific Northwest to permit a return to maximum LAI, and observations substantiate this prediction (Turner and Long, 1975).

VII. SUMMARY

In this chapter we identified the major components of a hydrologic model that can be linked to carbon and nutrient cycling models. Five meteorological variables drive the hydrologic model: solar radiation, temperature, vapor pressure deficits, precipitation, and wind speed. The soil and snowpack are major sources of temporary water storage. The leafy canopy and surface litter present the main surfaces from which water is transpired or evaporated. The vertical height, seasonal variation in LAI, and rooting depth are important properties of vegetation that affect water movement through ecosystems. Seasonal limitations on the source and flow of water from roots through stems and branches can be assessed by monitoring the relative water content of sapwood and its δD, the $\delta^{13}C$ of foliage, and the predawn Ψ of twigs. By incorporating additional properties of the soil, hydrologic models have wide application. Such models provide insights into the implications of various policies that alter forest composition and structure.

CHAPTER 3

Carbon Cycle

I. INTRODUCTION

Carbon is a constituent of all terrestrial life. Carbon begins its cycle through forest ecosystems when plants assimilate atmospheric CO_2 through photosynthesis into reduced sugars (Fig. 3.1). Usually about half the gross photosynthetic products produced (GPP) are expended by plants in autotrophic respiration (R_a) for the synthesis and maintenance of living cells, releasing CO_2 back into the atmosphere. The remaining carbon products (GPP − R_a) go into net primary production (NPP): foliage, branches, stems, roots, and plant reproductive organs. As plants shed leaves and roots, or are killed, the dead organic matter forms detritus, a substrate that supports animals and microbes, which through their heterotrophic metabolism (R_h) release CO_2 back into the atmosphere. On an annual basis, undisturbed forest ecosystems generally show a small net gain in carbon exchange with the atmosphere. This represents net ecosystem production (NEP). The ecosystem may lose carbon if photosynthesis is suddenly reduced or when organic materials are removed as a result of disturbance (Chapter 6). Soil humus represents the major accumulation of carbon

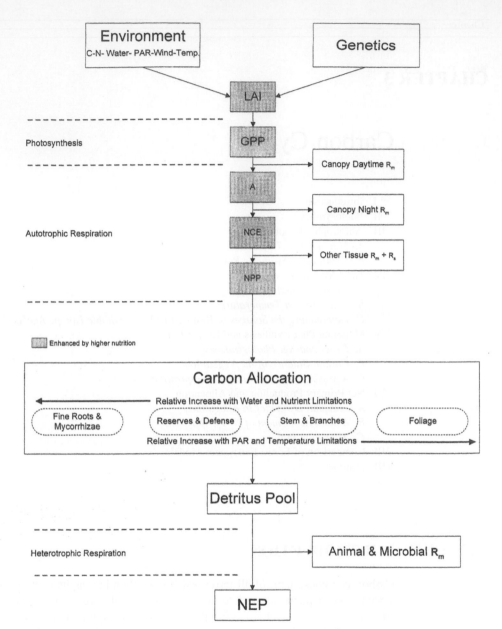

FIGURE 3.1. Carbon balance models that are coupled to water and nutrient cycling operate by predicting carbon uptake and losses through a series of processes, starting with photosynthesis and the absorption of solar radiation by leaves. Gross primary production (GPP) is further limited by other environmental variables affecting canopy stomatal conductance. Deducting foliar maintenance respiration during the daylight hours provides an estimate of net assimilation (A). Including canopy respiration at night yields an estimate of daily net canopy exchange (NCE) for a 24-hr period. Net primary production (NPP) is calculated by accounting for additional autotrophic losses associated with synthesis (R_s) and maintenance (R_m) throughout each day. NPP is partitioned into various components based on schemes associated with C:N ratios which change with the availability of water and nutrients. Leaf and fine-root turnover are the major contributors to litter on a seasonal basis, but all biomass components eventually enter the detrital pool. The annual turnover of leaves and roots is correlated with seasonal variation in LAI, specific leaf area, and nitrogen content. Decomposition of litter and release of CO_2 by heterotrophic organisms are functions of substrate quality (C:N ratio), temperature, and moisture conditions. Net ecosystem production (NEP) is calculated as the residual, after deducting heterotrophic respiration (R_h).

in most ecosystems because it remains unoxidized for centuries. It is the most important long-term carbon storage site in ecosystems. We delay discussion on soil humus until Chapter 4.

Figure 3.1 provides a general framework for modeling carbon flow through ecosystems and is the basis for organizing the material presented in this chapter. All of the environmental variables associated with modeling the water cycle (Chapter 2) are closely linked with the carbon cycle. Atmospheric carbon dioxide concentrations and the availability of soil nitrogen (N) must also be considered when modeling photosynthesis, carbon allocation, and respiration. Confidence in the reliability of models has greatly increased with the development of an eddy correlation technique that uses fast-response sensors to record the net exchange of CO_2 and water vapor from forests and other types of terrestrial ecosystems.

As ecosystem scientists, we consider the exchange of carbon into the system through photosynthesis to be a positive flux and respiration to represent a loss to the atmosphere. Atmospheric scientists would consider the signs to be reversed. Net ecosystem exchange (NEE) measured during the daylight hours includes gross photosynthesis (P_g or GPP), photorespiration (R_p), maintenance respiration (R_m), and synthesis (growth) respiration (R_s) of autotrophic plants, as well as heterotrophic respiration (R_h) by animals and microbes:

$$\text{Day NEE} = P_g - R_p - R_m - R_S - R_h. \tag{3.1}$$

At night the photosynthetic terms, P_g and R_p, are absent:

$$\text{Night NEE} = -R_m - R_s - R_h = -R_e \tag{3.2}$$

where R_e is total ecosystem respiration, exclusive of R_p. On a given day, R_e is largely controlled by temperature, which allows us to make adjustments for its rise during daylight periods from values recorded at lower temperatures during the night. Gross ecosystem production (GEP) includes photorespiration, which is usually small, so GEP is often assumed to approximate P_g:

$$\text{GEP} = \text{Day NEE} + \text{Day } R_e = P_g - R_p \approx \text{GPP}. \tag{3.3}$$

To separate the sources of respired CO_2, a series of chambers are often installed and CO_2 effluxes monitored at frequent intervals from soil, stem, branches, and leaves. Alternatively, respiration sources can be identified by monitoring the isotopic composition of carbon ($\delta^{13}C$) and oxygen ($\delta^{18}O$) in CO_2 diffusing into the turbulent transfer steam. The scientific value of full system analyses with eddy correlation techniques has proved immense, particularly when conducted over a series of years (Goulden *et al.,* 1996). Eddy-flux installations therefore serve for testing the underlying assumptions and accuracy of stand-level ecosystem models.

Among species (and genetic varieties), important differences exist in the pattern of carbon allocation. These differences affect competitive relationships (Chapter 5), the susceptibility of trees and other plants to various stresses (Chapter 6), as well as the annual carbon balance of a stand. Foresters have developed good empirical models to predict volume growth of trees and whole stands. These are helpful in supporting general assumptions built into stand-level ecosystem models; stem growth, however, is highly dependent on the fraction of NPP allocated to foliage versus roots, and so is difficult to predict as environmental conditions change seasonally and from site to site. Different theories that provide a basis for modeling carbon allocation are presented in this chapter. We will identify general principles governing the way the environment affects carbon allocation seasonally and over the course of a year, and apply these principles in later chapters.

II. PHOTOSYNTHESIS

Photosynthesis is the process by which plants convert atmospheric CO_2 to carbon products. Photosynthesis takes place within cells containing chloroplasts. Chloroplasts contain chlorophyll and other pigments that absorb sunlight. Energy from the sun causes electrons to become excited and water molecules to be split into hydrogen and oxygen:

$$CO_2 + 2H_2O + \text{light energy} \rightarrow CO_2 + 4H^+ + O_2 \rightarrow CH_2O + H_2O + O_2. \qquad (3.4)$$

The hydrogen joins with carbon from CO_2 to produce simple three- or four-carbon products which ultimately are synthesized into larger molecules that are incorporated into biomass or consumed in metabolic processes. Photosynthesis is restricted by both physical and biochemical processes and involves some reactions that require light and others that can take place in the dark. At the leaf surface, stomata limit the diffusion of carbon dioxide into intercellular spaces. Inside leaves, CO_2 must dissolve in water and pass through cell walls to reach the sites where chemical reactions take place within chloroplasts.

Net photosynthesis has three separate, potentially limiting components: (a) light reactions, in which radiant energy is absorbed and used to generate high-energy compounds (ATP and NADPH); (b) dark reactions, which include the biochemical reduction of CO_2 to sugars using the high-energy compounds previously generated; and (c) the rate at which CO_2 in ambient air is supplied to the site of reduction in the chloroplast.

In the light reactions, radiation absorbed by chlorophyll causes excitation of electrons which are transferred down a chain of specialized pigment molecules to reaction centers where high-energy compounds are formed, water is split, and O_2 released [Eq. (3.4)]. The initial part of the light reaction is only limited by the irradiance and amount of chlorophyll present. The rate of electron transfer is sensitive to temperature but independent of CO_2 concentrations. In the dark reactions, C_3 plants use the enzyme ribulose-bisphosphate carboxylase–oxygenase (Rubisco) for the primary fixation of CO_2. In light, photorespiration also occurs in the process of generating the substrate ribulose bisphosphate (RuBP), particularly when the ratio of O_2 to CO_2 increases within the chloroplast. Additional high-energy compounds are required to create six-carbon sugars, and, as a result, additional CO_2 is respired. Dark reactions are CO_2 and temperature limited, and also dependent on sufficient nitrogen and other substrate being available to synthesize the Rubisco enzyme. The rate at which CO_2 can be supplied to chloroplasts is limited by the CO_2 partial pressure and by stomatal conductance, which limits diffusion of CO_2 to 0.625 times that of water vapor, because of the difference in molecular mass.

Farquhar *et al.* (1980) developed a set of basic equations that incorporate the limiting processes to net assimilation of CO_2. Specifically, the equations consider limitations by enzymes (A_v), by electron transport (A_j), and by stomatal conductance (g_{CO_2}). Enzyme-limited assimilation (A_v) is defined as

$$A_v = [V_{max}(c_i - G^*)/[K_c(1 + pO_2/K_0) + c_i]] - R_c. \tag{3.5}$$

where V_{max} is the maximum carboxylation rate when the enzyme is saturated, pO_2 is the ambient partial pressure of oxygen, K_c and K_0 are Michaelis–Menten constants for carboxylation and oxygenation by Rubisco, respectively; c_i is the partial pressure of CO_2 in the chloroplast, G^* is the CO_2 compensation partial pressure in the absence of dark respiration, and R_c is the dark respiration by the leaf. K_c, K_0, and V_{max} are temperature sensitive and are adjusted from a base rate at 25°C (298.2 K) with Arrhenius-type relationships (with K_x representing all three variables):

$$K_x = K_{x,25} \exp[(E_x/298.2R)(1 - 298.2/T_c)] \tag{3.6}$$

where T_c is the canopy temperature (K), R is the universal gas constant ($8.314\,J\,mol^{-1}\,K^{-1}$), and E_x refer to the approximate activation energies for K_c, K_0, and V_{max}.

When electron transport limits photosynthesis (A_j):

$$A_j = (J/4)[(c_i - G^*)/(c_i + 2G^*)] - R_c \tag{3.7}$$

where J is the potential rate of electron transport, which is related to the maximum light-saturated rate of electron transport and the absorbed irradiance. The sensitivities of A_j and R_c to temperature are also considered in equations similar to Eq. (3.6).

Finally, after accounting for effects of PAR and temperature on photosynthesis, stomatal conductance to CO_2 (g_{CO_2}) is calculated, assuming that c_i (partial pressure of CO_2 within the chloroplasts) is proportional to ambient CO_2 partial pressure (c_a):

$$c_i = c_a - (A/g_{CO_2}) \tag{3.8}$$

A semiempirical model developed by Ball *et al.* (1987) described stomatal conductance for CO_2 diffusion (g_{CO_2}) as:

$$g_{CO_2} = g_0 + (g_1AhP)/(c_a - G) \tag{3.9}$$

where g_0 is the minimum stomatal conductance and g_1 is an empirical coefficient which represents the composite sensitivity of conductance to assimilation, CO_2, humidity, and temperature, h is the relative humidity at the leaf surface, P is the atmospheric pressure, G is the CO_2 compensation point, and c_a is the ambient partial pressure of CO_2 at the leaf surface. Equation (3.9) has been modified to express relative humidity as vapor pressure deficit with reference to leaf temperature (Leuning, 1995). Dewar (1995) suggested a more mechanistic form of the equation which reflects stomatal guard cell response and has the potential to incorporate restrictions on water uptake that affect stomatal conductance (Chapter 2).

Many additional equations are required to solve all the unknowns in such fundamental models, but Eqs. (3.5) to (3.9) include all the major variables. They allow prediction of the consequence of rising levels of CO_2 in the atmosphere, while also identifying potential limitations associated with the availability of nitrogen, light, temperature, and all the additional variables affecting stomatal conductance discussed in Chapter 2. This basic model has been applied in temperate deciduous (Baldocchi and Harley, 1995), temperate evergreen (Thornley and Cannell, 1996), tropical (Lloyd *et al.*, 1995), Mediterranean (Valentini *et al.*, 1991), and a variety of boreal forests (Wang and Polglase, 1995) and will serve as a reference against which we will compare more empirical models. For a critical review of the basic equations and modeling assumptions, see Leuning *et al.* (1995).

Because of the interactions of light, CO_2, and the rate at which RuBP can be regenerated, net photosynthesis always shows an asymptotic curve with increasing irradiance (Fig. 3.2). With reference to photosynthesis, where light energy in different wavelengths is involved, irradiance is expressed in units of moles (mol) per photon of light, where 1 mol is equivalent to the atomic mass of carbon (12) or molecular mass of CO_2 (44) fixed and $4.6\,\mu mol$ photons $m^{-2}s^{-1} = 1\,J\,m^{-2}s^{-1}$ or $1\,W\,m^{-2}$ of absorbed PAR (APAR). Below some minimum irradiance level, net carbon uptake (assimilation, A) is negative in reference to the leaf, as foliar respiration exceeds photosynthesis. As irradiance increases, a compensation point is reached where the uptake of CO_2 through photosynthesis is exactly balanced by losses through respiration. Above the *light compensation point,* uptake increases linearly until the availability of RuBP or CO_2 limits the process. Increasing ambient CO_2 increases the maximum net photosynthetic rates (A_{max}) but not the linear part of the curve, because at low irradiance photochemistry, not CO_2, limits the process.

The apparent quantum efficiency (α) is the slope (A/PAR) of the linear part of the photosynthesis–irradiance curve, that is, the rate of increase in assimilation with irradiance at levels below those at which CO_2 has an effect (e.g., no photorespiration occurs). The apparent quantum efficiency is relatively conservative, with an average of ~$0.03\,mol\,CO_2/$ mol photons of APAR (equivalent to $1.65\,g\,C\,MJ^{-1}$ PAR absorbed) throughout most forest canopies (and far below the theoretical maximum of $0.08\,mol\,CO_2$/mol photons of APAR), but it changes, as one might expect, as chlorophyll levels fluctuate (Jones, 1992; Waring *et al.*, 1995a).

The maximum rate of assimilation, on the other hand, decreases from the top to the bottom of a forest canopy (Fig. 3.3a). A_{max} is generally higher in the leaves of deciduous tree species than in evergreens (Ceulemans and Saugier, 1991), and the longer that leaves live, the more their A_{max} is reduced (Reich *et al.*, 1995a). One of the underlying reasons for these relationships is that thicker leaves associated with evergreens offer a more restricted diffusion pathway to CO_2 than deciduous leaves (Robinson, 1994). A second reason is that evergreen foliage tends to hold less nitrogen per unit area than deciduous leaves (Field and Mooney, 1986; Ellsworth and Reich, 1993; Reich and Walters, 1994). Because some of the nitrogen in leaves is in forms other than photosynthetic structures, total nitrogen content is not always directly related to photosynthetic capacity. Kull and Jarvis (1995) showed theoretically how measurements made on foliage collected from the upper part of a canopy might allow prediction of the amount of nitrogen bound in photosynthetic machinery throughout the entire canopy. Field investigations in boreal forests, however, do not fully support this interpretation (Dang *et al.*, 1997).

Both canopy leaf nitrogen and canopy leaf mass tend to increase with absorbed PAR (Kull and Jarvis, 1995). Because highly illuminated foliage contains more cell layers than shaded leaves, *specific leaf mass* (SLM, $kg\,m^{-2}$) and A_{max} generally decrease in parallel

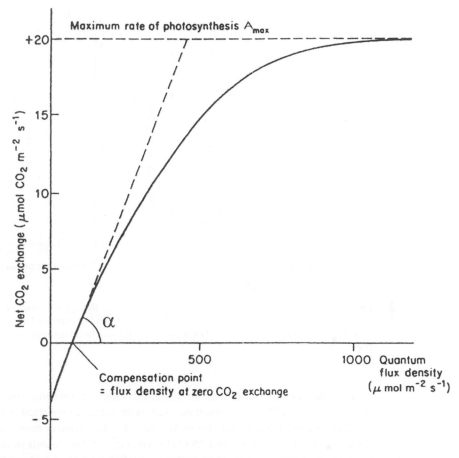

FIGURE 3.2. Representative net photosynthetic response (*A*) to increasing irradiance. The slope (α) at low irradiances, denoted by the dotted line, represents the apparent quantum efficiency. Light compensation occurs where net CO_2 exchange is zero, and gross photosynthesis equals dark respiration. Net photosynthesis continues to increase asymptotically with irradiance until maximum rates (A_{max}) are achieved.

through a canopy (Fig. 3.3b). Specific leaf mass, or its reciprocal specific leaf area ($m^2 kg^{-1}$), is an essential component of allocation models, and it tends to change in predictable ways not only with the light environment, but also with the availability of nitrogen and general harshness of the environment (Specht and Specht, 1989; Pierce *et al.*, 1994). The actual concentration of chlorophyll pigments per unit of leaf surface area, however, offers a more accurate measure of quantum efficiency and photosynthetic capacity than total nitrogen or specific leaf mass (Waring *et al.*, 1995a). The amount of chlorophyll per square meter of foliage varies considerably among tree species and with season (Escarré *et al.*, 1984). When expressed as chlorophyll per unit of ground surface area, which integrates the entire canopy, values may be quite similar for evergreen and deciduous forests (Cannell, 1989; Reich *et al.*, 1995b; Dang *et al.*, 1997).

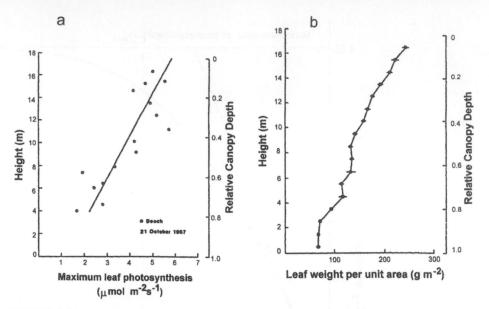

FIGURE 3.3.   (a) Photosynthetic capacity (A_{max}) decreased through a *Nothofagus* canopy in parallel with changes in leaf weight per unit area. (b) Leaf weight per unit area in a pure stand of *Nothofagus* growing in New Zealand decreased from the top to the bottom of the canopy as a function of light intercepted. (From Hollinger, 1989.)

The optimum temperature range for photosynthesis varies with species but is commonly between 15° and 25°C for temperate trees, with extremes reported between about 10° and 35°C (Kozlowski and Keller, 1966; Mooney, 1972). Tropical trees show higher optimum temperatures, between 30° and 35°C (Lloyd *et al.*, 1995). A shift in temperature optimum may occur if other factors change the intercellular level of CO_2 or the efficiency of the photosynthetic machinery. In general, increased internal CO_2 concentrations allow the temperature optimum to be shifted upward by reducing photorespiration. At temperatures above 40°C, gross photosynthesis decreases abruptly because of changes in chloroplast and enzyme activity (Berry and Downton, 1982). Protein denatures at 55°C (Levitt, 1980). As a result of these factors, the net carbon uptake by leaves usually increases gradually to an optimum temperature and then decreases more abruptly as the maximum temperature limit is approached. The broadness of the temperature optimum, together with its ability to shift with season (Strain *et al.*, 1976; Slatyer and Morrow, 1977), reduces its significance for modeling gross photosynthesis, except perhaps where species may have become geographically isolated and survive, if not thrive, under a changed climate from that for which they originally evolved (Waring and Winner, 1996).

Because trees are often adapted to a wide range of seasonal temperature variations, the frequency and duration of extremes are more critical than mean temperature in limiting photosynthesis (Van *et al.*, 1994; Perkins and Adams, 1995). For example, a 3-hr exposure of seedlings of *Pinus sylvestris* to temperatures between −5° and −12°C reduced quantum efficiency from 10 to over 80% (Strand and Öquist, 1985). Recovery may be delayed for weeks or months with repeated exposure to frost, particularly if leaves are exposed to high irradiance (Strand and Öquist, 1985). Different mechanisms may operate, but increased respiration that follows exposure to low temperatures suggests that repair of damaged membranes is involved (see review by Hällgren *et al.*, 1991). Short exposure to high temperatures likewise may be injurious and may alter competitive relationships among species (Bassow *et al.*, 1994).

Water is essential to all living cells, so any reduction in its availability might be expected to affect photosynthesis as well as many other processes. In Chapter 2, we discussed the water relations of forests. Here, only the effects of water limitations on photosynthesis and the related process of photorespiration will be covered. The availability of water in leaf tissue is not directly dependent on the water content. Changes in cell wall elasticity, in membrane permeability, and in concentration of solutes in cells counterbalance the effects of decreasing water content (Edwards and Dixon, 1995). Diurnal variations of 5–10% in leaf water content relative to saturation often have no direct effect on photosynthesis (Hanson and Hitz, 1982). Eventually, if leaf tissue continues to lose water, stomata are forced to close, and the rate of CO_2 diffusion into the leaf is reduced or halted. Under extended drought, some of the Rubisco enzyme will be broken down, reducing the biochemical capacity of the photosynthetic system. Concentrations of chlorophyll and other pigments important in photochemical reactions are also reduced (Farquhar and Sharkey, 1982; Jones, 1992). The main effects of water limitations on photosynthesis are through reduction in stomatal conductance, as described in Chapter 2.

III. AUTOTROPHIC RESPIRATION

Autotrophic respiration (R_a) involves the oxidation of organic substances to CO_2 and water, with the production of ATP and reducing power (NADPH):

$$O_2 + CH_2O \rightarrow CO_2 + H_2O. \tag{3.10}$$

Total autotrophic respiration consists of two major components associated with the metabolic energy expended in the synthesis of new tissue and in the maintenance of living tissue already synthesized.

A. Maintenance Respiration

Maintenance respiration, the basal rate of metabolism, includes the energy expended on ion uptake and transfer within plants. Repair of injured tissue may greatly increase the basal rates of metabolism. Because trees accumulate a large amount of conducting and storage tissues as they age, the observed decrease in relative growth rate associated with age has often been assumed to reflect increasing maintenance costs (Whittaker, 1975; Waring and Schlesinger, 1985). Most of the conducting tissue in trees, however, is sapwood, which contains relatively few living cells (Fig. 3.4; Table 3.1). Enzymatic activity, associated with N concentration in living tissue, is also much lower in the sapwood than in leaves (Amthor, 1984; Ryan, 1991a, 1995). Leaf tissue, with 2% N, might respire about 0.45% of their weight daily at 10°C and 0.9% at 20°C. In comparison, an equal weight of sapwood metabolizes at less than one-tenth that of foliage. As a result, the annual

Xylem cells

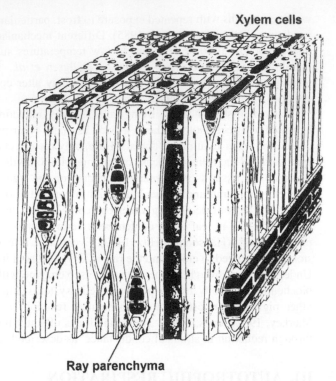

Ray parenchyma

FIGURE 3.4.   Most of the cells in the sapwood of trees are dead, which permits efficient transport of water. Traversing the wood, however, are bands of living parenchyma cells (darkened) which store carbohydrates. Diagram shows a redwood (*Sequoia sempervirens*). (From Weier *et al.,* 1974, by copyright permission of John Wiley & Sons, Inc.)

maintenance cost of sapwood in large trees is generally less than 10% of annual GPP (Fig. 3.5).

In modeling maintenance respiration (R_m), an exponential increase with temperature is normally observed within the range of biological activity, as described by the formula

$$R_m(T) = R_0 Q_{10}^{[(T-T_0)/10]} \qquad (3.11)$$

where R_0 is the basal respiration rate at $T_0 = 0°C$ (or other reference temperature). The parameter Q_{10} is the *respiration quotient* and represents the change in the rate of respiration for a $10°C$ change in temperature (T). The Q_{10} is relatively conservative, usually between about 2.0 and 2.3. Under field conditions, where daily temperature variation is large, the nonlinearity in the response of maintenance respiration should be taken into account. This has been accomplished by fitting a sine function to minimum–maximum temperature differences and integrating the response not only daily, but throughout the year (Hagihara and Hozumi, 1991; Ryan, 1991a). As mentioned previously, R_0 is highly variable, depending on the protein (or N) content in the tissue (Jarvis and Leverenz, 1983). Seasonal changes in LAI and fine-root mass, together with possible variation in the ratio of active to inactive enzymes, make it difficult to estimate carbon expenditures annually

TABLE 3.1

Percentage of Living Parenchyma Cells in the Sapwood of Representative Hardwoods and Conifers Native to the United States[a]

Species	%	Range
Hardwoods		
Populus tremuloides	9.6	4.4
Betula alleghaniensis	10.7	0.9
Fagus grandifolia	20.4	5.3
Quercus alba	27.9	—
Liriodendron tulipifera	14.2	2.5
Robinia pseudoacacia	20.9	3.1
Acer saccharum	17.9	5.2
Tilia americana	6.0	3.8
Conifers		
Pinus taeda	7.6	1.6
Larix occidentalis	10.0	1.1
Picea engelmannii	5.9	2.5
Pseudotsuga menziesii	7.3	2.1
Tsuga canadensis	5.9	0.7
Abies balsamea	5.6	2.3
Sequoia sempervirens	7.8	2.5
Taxodium distichum	6.6	2.6

[a]After Panshin *et al.* (1964).

to much better than ±25%, but this is a great improvement over earlier efforts and provides an important advance in constructing tree and stand carbon balances (Ryan, 1991b, 1996b).

Some special concerns should be mentioned in relation to measuring respiration with chambers. When chambers are placed over thin bark surfaces that have photosynthetic capacity, the respiration measured may overestimate (by up to 100%) the net exchange from such surfaces (Linder and Troeng, 1981). Also, when determining root respiration, care must be taken because maintenance respiration is reduced exponentially as CO_2 concentrations increase; thus, rates in the soil, where CO_2 concentrations may exceed 5000 ppm, are only one-tenth of those recorded at atmospheric concentrations (Qi *et al.*, 1994; Burton *et al.*, 1997).

B. Growth and Synthesis Respiration

Growth requires the metabolism of more resources than can be found in the final product. Rates of *synthesis respiration* (R_s) for various tissues differ, depending on the biochemical pathways involved. The production of 1 g of lipid would require 3.02 g of glucose, whereas 1 g of lignin, protein, or sugar polymer might require, respectively, 1.90, 2.35, and 1.18 g of glucose (Penning de Vries, 1975). More recently, empirical relations have been derived to estimate the total cost of construction based on correlations with the heat of combustion, ash, and organic N content of tissue and the biochemical constituents (McDermitt and

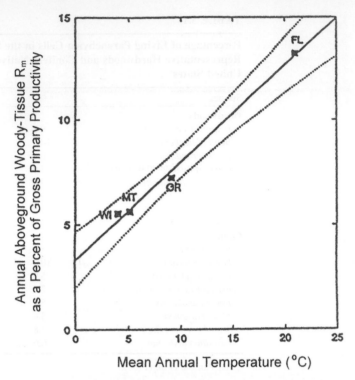

FIGURE 3.5.   Maintenance respiration of aboveground woody tissue in evergreen trees growing in Wisconsin (WI), Montana (MT), Oregon (OR), and Florida (FL) was usually <10% of estimated gross primary production (GPP) but increased linearly with mean annual temperature ($y = 2.76 + 0.48x$, $r^2 = 0.99$). (From *Oecologia*, "Stem maintenance respiration of four conifers in contrasting climates," M. G. Ryan, S. T. Gower, R. M. Hubbard, R. H. Waring, H. L. Gholz, W. P. Cropper, and S. W. Running, Volume 101, p. 138, Fig. 4, 1995, © 1995 by Springer-Verlag.)

Loomis, 1981; Vertregt and Penning de Vries, 1987; Williams *et al.,* 1987; Griffin, 1994). In Table 3.2, the cost of producing various tissues and organs, based on their biochemical composition, is presented. The total construction cost for a gram of pine shoot is equivalent to 1.57 g glucose (Chung and Barnes, 1977). Assuming foliage is 50% C, 1.28 g C in glucose would produce 1 g C in foliage. For large trees, synthesis costs are usually assumed to be about 25% of the carbon incorporated into new tissues (Ryan, 1991a,b). For smaller trees, particularly deciduous species with high concentrations of proteins in foliage and other organs, synthesis costs may rise to 35% of the carbon sequestered in biomass.

To assess the contribution of synthesis respiration to total plant respiration, monitoring must include periods in both active and dormant seasons. Periods of rapid growth can be monitored by recording the expansion in diameter of calibrated bands placed around branches, stems, and large-diameter roots, and through measuring stem elongation (Dougherty *et al.,* 1994). Dormancy and annual growth increments should be confirmed by anatomical inspection (Gregory, 1971; Emmingham, 1977) so that respiration rates measured throughout the growing season can be properly interpreted (Ryan *et al.,* 1995).

TABLE 3.2.
Estimated Construction Costs of Various Organs and Tissues[a]

Type of tree	Organ or tissue	Total construction cost (g C required/g C in product)
Pine	Needles	1.28
	Branches	1.21
	Bark	1.30
	Roots	1.20
Eucalyptus	Phloem	1.18
	Cambium	1.05
	Sapwood	1.11
	Heartwood	1.14

[a]After Chung and Barnes (1977). Glucose equivalents were converted by recognizing that glucose is 40% C and by assuming that the dry weight of all products contained 50% C.

The determination of root growth and maintenance is particularly difficult because of the heterogeneity of roots in size, distribution, and degree of colonization with symbiotic organisms (mycorrhizal fungi and nitrogen-fixing bacteria). In most field programs where attempts are made to separate the sources of respiration, chambers or CO_2 traps are placed over the litter surface and an estimate of total CO_2 efflux is obtained. When such data are available throughout the year, they provide some constraints on the estimate of below-ground carbon allocation to roots and the contribution of heterotrophic respiration.

Raich and Nadelhoffer (1989) provided an indirect approach to estimate belowground carbon allocation by assuming that, over a year, leaf litter and root turnover would add an amount of carbon approximately equal to that respired from the soil, and that there would be no net change in the soil carbon pool. Under these conditions, soil respiration minus leaf litterfall approximately equals root production plus root respiration. With soil respiration and litterfall data assembled from a range of forests from boreal to tropical, Raich and Nadelhoffer presented a linear relationship between carbon in leaf litterfall and the total carbon allocation to small-diameter roots (Fig. 3.6). Ryan *et al.* (1996b) extended the approach to ecosystems with a significant annual increment of coarse roots by assuming that total belowground carbon allocation was equal to the difference between annual soil respiration and aboveground litterfall plus coarse root increment. The carbon allocated to NPP_{roots} is often assumed to be about half the total. The general application of this approach requires a measurement of annual litterfall or knowledge of the fraction of the total foliage mass that turns over annually, taking into account a loss of mass (usually 10–25%) per unit area before leaves are shed (Chapter 4).

IV. HETEROTROPHIC RESPIRATION

The detritus produced by autotrophic plants serves as food or substrate for heterotrophic organisms which respire CO_2 or methane (CH_4). In estimating carbon balances of ecosystems, the rate that litter (including large woody components) decomposes (above- and belowground) is important to quantify. In Chapter 4, we discuss how the activities of micro- and macroorganisms respond to changes in the size, biochemical composition, and

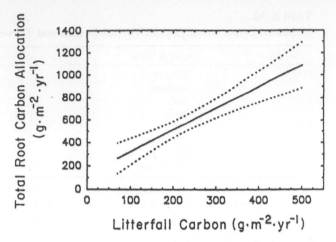

FIGURE 3.6.   Predicted rates of total belowground carbon allocation in forest ecosystems are linearly related to aboveground litterfall ($y = 130 + 1.92x$, $r^2 = 0.52$). Belowground carbon allocation was calculated as the difference between soil respiration and litterfall for individual forests where the soil carbon pool was assumed to be in approximate steady state over a year. (From Raich and Nadelhoffer, 1989.)

physical environment associated with different substrates, with the goal of estimating CO_2 evolution from the breakdown and decomposition of all forms of detritus. In this section, we focus on leaf and fine-root detritus because these components turn over rapidly and are therefore likely to contribute the most to seasonal fluctuations in heterotrophic respiration.

When heterotrophic respiration is monitored by enclosing samples of fresh leaf or fine-root litter within small-mesh nylon bags, the mass loss per unit time can be measured, and, because organic matter is approximately 50% carbon, the CO_2 evolved can be calculated. Alternatively, under laboratory conditions, litter samples may be placed in chambers with controls on moisture and temperature and CO_2 efflux directly monitored. From a combination of studies, three major variables have been identified as limiting heterotrophic respiration: substrate quality, relative water content, and temperature.

A. Substrate Quality

Substrate quality is correlated with the energy that microorganisms must expend in processing detritus. Sugars, starch, fats, and proteins are relatively reduced forms of carbon and easily metabolized, whereas tannins, cellulose, and lignin are more oxidized forms and less efficiently metabolized. A crude index of substrate quality is its nitrogen content per gram C, usually expressed as a C:N ratio, with high ratios indicative of material that will be processed very slowly. Conifer wood at 0.1% N has a C:N ratio of 500:1, whereas fresh hardwood leaves with 2% N have a C:N ratio of 25:1. Over time, C:N ratios in decomposing substrates are reduced as microbes process the carbon, and nitrogen accumulates in their biomass. Microbial biomass has a C:N range from about 4 to over 20, with bacteria having <10 and fungi >10 (Sparling and Williams, 1986; Schimel *et al.,*

1989; Martikainen and Palojarvi, 1990; Lavelle *et al.,* 1993). Fungi have some advantage over bacteria in forest ecosystems because of their higher C:N ratios. Unless environmental conditions become unfavorable and cause microbes to die, little nitrogen will be released from the substrate until its C:N ratio is reduced to below that of the organisms involved. In reality, the simple C:N ratio is not a very precise index; a better index is derived from the ratio of lignin to nitrogen (Melillo *et al.,* 1982), or other biochemical constituents which reflect the free energy extractable from a given mass of substrate.

B. Moisture and Temperature

The metabolic activity of microorganisms also varies with moisture and temperature of surface litter and soil. In surface litter, fairly stable metabolic activity is usually observed until the substrate is nearly dry (Fig. 3.7). In soils, however, microbial activity appears to be reduced more linearly with decreasing gravimetric water content and to show a log–linear decrease with falling Ψ_{soil} (Orchard *et al.,* 1992). The temperature response is usually exponential but highly variable. Note in Fig. 3.7 that over the range from 4° to 35°C that the Q_{10} for microbial activity on jarrah leaf litter is 1.6 whereas the Q_{10} on karri substrate is 2.3. The base reference temperature also varies depending on the climate in which specific microorganisms have evolved. In general, decomposition rates are extremely low below 0°C, increase rapidly between 10° and 30°C, and decrease at temperatures >40°C (Ågren *et al.,* 1991). These nonlinear responses to moisture and temperature provide, together with lignin content or related indices of substrate quality, the foundation for modeling heterotrophic respiration and decomposition in forest and other terrestrial ecosystems (Jansson and Berg, 1985; O'Connell, 1990; Ågren *et al.,* 1991).

C. Determining the Sources of Respired CO_2 from Stable Isotope Analyses

The isotopic composition of carbon and oxygen in CO_2 differs significantly, depending on whether it is derived from leaves, stems, roots, or decaying organic matter. This allows us to discriminate the relative contribution of different sources of CO_2 in soil, air, and leaves of forest canopies (Keeling, 1958; Sternberg, 1989; Flanagan *et al.,* 1997a). In the biochemical synthesis of various plant constituents from sugars, some isotopic fractionation occurs. For example, lignin is depleted in ^{13}C compared to whole cellulose (Benner *et al.,* 1987). It is also likely that some isotopic fractionation occurs during microbial respiration, leading to a ^{13}C enrichment in microbial carbon compared to plant-derived carbon in soils (Macko and Estep, 1984).

During the process of photosynthesis, a portion of the CO_2 that enters the leaf and equilibrates with chloroplast water is not fixed and diffuses back out of the leaf with an altered oxygen isotopic ratio. The oxygen in CO_2 evolved from the soil has exchanged with soil water and isotopically matches that source. The isotopic composition of water in the xylem does not change from that in the soil, until transpiration from the leaf takes place. During transpiration, [18]O is discriminated against, so that leaf water is enriched and the vapor released into the atmosphere is depleted. The CO_2 metabolically respired by plants is also depleted in ^{13}C compared to the atmosphere because, during photosynthesis,

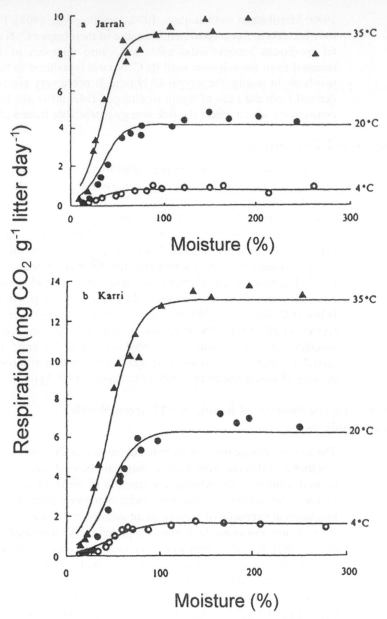

FIGURE 3.7.   Laboratory incubation of foliage from (a) jarrah (*Eucalyptus marginata*) and (b) karri (*Eucalyptus diversicolor*) show that microbial respiration (R_h) increases exponentially with temperature between 4° and 35°C, and is relatively insensitive to changes in litter moisture content until below about 100% or near saturation (not shown). Jarrah litter represents a less favorable substrate for decomposition than that of karri. Separate equations for the two species that combined both temperature and moisture responses accounted for 93 to 94% of the observed variation in measured respiration. (Reprinted from *Soil Biology and Biochemistry*, Volume 22, A. M. O'Connell, "Microbial decomposition (respiration) of litter in eucalypt forests of southwestern Australia: An empirical model based on laboratory incubations," pp. 153–160, Copyright 1990, with kind permission of Elsevier Science Ltd, The Boulevard, Langford Lane, Kidlington OX5 1GB, UK.)

^{13}C is discriminated against during diffusion through stomata and also by Rubisco. As a result of these differences in isotopic fractionation, which are now well established and can be quantitatively modeled, it has become possible to separate the major sources of CO_2 evolved diurnally and seasonally from forest ecosystems (Buchmann et al., 1997; Flanagan et al., 1997a,b).

Sternberg (1989) developed a two-ended gas-mixing model to describe the relationship between the concentration and isotope ratio of CO_2 within a forest canopy. He assumed that four different fluxes influence the forest atmosphere: photosynthetic CO_2 uptake (F_p), turbulent flux of CO_2 out of the forest (F_a), respiration (F_r), and turbulent flux of CO_2 into the forest (F_{fa}). The relationship is described by

$$\delta_f = ([CO_2]_a/[CO_2]_f)(\delta_a - \delta_r)(1 - P) + \delta_r + P\Delta_a \qquad (3.12)$$

where $[CO_2]$ is the concentration of CO_2 and δ is the stable isotope ratio of CO_2; the subscripts a and f represent bulk atmosphere and forest, respectively; P is the uptake of carbon dioxide by photosynthesis relative to the total loss of CO_2 from the forest $[P = F_p/(F_p + F_{fa})]$; δ_r is the isotopic ratio of CO_2 respired by plants and soil; and Δ_a is the isotopic discrimination that occurs during photosynthetic gas exchange.

Under conditions where removal of carbon dioxide from the forest by photosynthesis is small relative to turbulent mixing, P tends toward zero and Eq. (3.12) is reduced, as presented by Keeling (1958), to

$$\delta_f = ([CO_2]_a/[CO_2]_f)(\delta_a - \delta_r) + \delta_r. \qquad (3.13)$$

From Eq. (3.13), it can be seen that a plot of $1/[CO_2]_f$ against δ_f gives a straight-line relationship with slope $[CO_2]_a(\delta_a - \delta_r)$, and intercept δ_r. This relationship can be applied to estimate the isotopic composition of CO_2 respired by plants and soil if data are available from at least two heights in the forest canopy.

These kinds of isotopic analyses have been applied to a number of undisturbed forests throughout the growing season and during dormant periods when evergreen forests still hold leaves but deciduous forests do not. During the growing season, in a wide variety of evergreen and deciduous types, more than 70% of the respired carbon appears derived from autotrophic respiration (Flanagan et al., 1997b). Because of turbulent conditions present during the daytime, less than 5% of respired carbon appears to be reincorporated through photosynthesis, even in dense tropical forests (Buchmann et al., 1997). Seasonally, the efflux of CO_2 from evergreen and deciduous forests shows different isotopic patterns as a result of the absence of any leaf photosynthesis by deciduous forests during the dormant season. These isotopic analyses are particularly valuable when combined with continuous eddy-flux and chamber measures of CO_2 and water vapor exchange.

V. MODELING PHOTOSYNTHESIS AND RESPIRATION

A. Gross and Net Photosynthesis

With an understanding of the underlying processes by which water vapor and carbon dioxide are exchanged from forest ecosystems, there is an opportunity to compare model predictions against eddy-flux measurements and to learn what simplifications can be made if the time scale of interest is expanded. As an example, we note analyses of measurements made at Harvard Forest in Massachusetts. The forest is composed of 75% deciduous hardwoods and 25% evergreen conifers. At the forest, eddy-flux data have been acquired continuously over a number of years (Wofsy *et al.*, 1993; Goulden *et al.*, 1996).

Williams *et al.* (1996) developed a fine-scale, soil–plant–atmosphere canopy model which incorporates the Farquhar model of leaf-level net and gross photosynthesis (Farquhar and Sharkey, 1982; Farquhar *et al.*, 1989) and the Penman–Monteith equation to determine leaf-level transpiration and evaporation (Monteith and Unsworth, 1990). The two process models were linked by a model of stomatal conductance (g_s) that optimized daily C gain per unit of leaf N, with limitations imposed by canopy water storage and soil-to-canopy water transport. The model assumed that the maximum carboxylation capacity (V_{cmax}) and maximum electron transport rate (J_{max}) were proportional to foliar N concentration, which changed seasonally with leaf development and senescence (Harley *el al.*, 1992; Waring *et al.*, 1995a).

The unique feature of the model lies in its treatment of stomatal opening, which explicitly couples water flow from soil to atmosphere with C fixation. The rate at which water can be supplied to the canopy is restricted by plant hydraulics and soil water availability. This rate ultimately limits transpiration because stomata close at a threshold minimum leaf water potential to prevent irreversible xylem cavitation. Because plant canopies also use water that has accumulated and been stored during periods of low transpiration (at night and when rain falls), the model optimizes carbon uptake by extracting some of the reserve of water in the morning, thus delaying the onset of stomatal closure associated with rising air vapor pressure deficits later in the day. As the canopy grows taller, hydraulic limitations on GPP are increased, but branch length was not considered.

To drive the model, meteorological data, including the fraction of direct to diffuse radiation, we required at 30-minute intervals to permit accurate estimates of irradiance through 10 canopy layers and the calculation of leaf-air vapor pressure deficits. Independent estimates of CO_2 efflux from stems and soil surface were available to allow predictions of net ecosystem exchange hourly and to model separately GPP and assimilation. Using this model, for which the required structural and environmental data were available, both hourly CO_2 exchange rate ($r^2 = 0.86$) and latent energy flux ($r^2 = 0.87$) were strongly correlated with independent whole-forest measurements obtained with eddy-flux measurements (Williams *et al.*, 1996).

Sensitivity analyses showed that major simplifications could be made in the canopy photosynthesis model, particularly if the time scale could be extended from hours to days. In forests, as we might expect, atmospheric turbulence is sufficient to allow wind speed to be set at a constant $2\,\mathrm{m\,s^{-1}}$. The large differences in A_{max} associated with crown position were severely damped because the upper, more exposed part of the canopy was the first to suffer hydraulic restrictions that caused stomatal closure. As a result, photosynthesis was often constrained by gas diffusion before reaching light saturation.

When solar radiation was integrated over periods of a full day, the importance of distinguishing diffuse from direct solar radiation also became less important. Longer integration times, however, reduce the precision of estimates of canopy interception and could affect the accuracy of soil water balance calculations, as discussed in Chapter 2. Full-day time integration, however, significantly reduced data requirements, while still providing good agreement ($r^2 > 0.9$) with GPP estimated from hourly data (Williams et al., 1996).

Jarvis and Leverenz (1983) also found, in dense Sitka spruce plantations in Scotland, that modeling of canopy photosynthesis could be simplified, without introducing much error, by assuming a linear rather than curvilinear relation with incident PAR. In more open woodlands, however, a fair proportion of the canopy may become light saturated during the day (Baldocchi and Harley, 1995). Over periods of a week to a month, however, we can assume that even forests with relatively low LAI will tend to show a linear increase in photosynthesis with absorbed PAR (Wang et al., 1992a; Wang and Polgase, 1995). Of course other factors reduce potential gross photosynthesis below that predicted from light absorption alone. Landsberg (1986a) suggested that potential gross photosynthesis (GPP) be reduced on the basis of three restrictive environmental constraints to obtain an estimate of actual GPP for integration periods up to a month:

$$GPP = \alpha \, APAR \, f(H_2O) \, f(D) \, f(T) \qquad (3.14)$$

where α is maximum quantum efficiency, APAR is absorbed PAR by the canopy, and modifying factors $f(i)$, which range from 1 to 0, describe reductions related to soil water deficits (H_2O), low temperature effects (T), and vapor pressure deficits of the air (D). The values of $f(H_2O)$ is reduced to zero as root zone soil water content drops below a critical point; $f(D)$ decreases linearly or nonlinearly with increasing daytime values of D. The temperature modifer $f(T)$ remains at zero for a day when temperatures drop below freezing. In essence, this calculation defines the fraction of APAR that can be effectively utilized by photosynthesis when stomata are at least partially open. McMurtrie et al. (1994) applied this type of analysis to 10 pine forests growing in strikingly different environments and discovered that, when deductions were made for the fraction of PAR intercepted during periods when stomata constrained CO_2 diffusion, the resulting calculation of quantum efficiency (α) approached a constant $\sim 1.8 \, g \, C \, MJ^{-1}$ APAR.

When GPP was calculated using the simplified relation described by Eq. (3.14) with a maximum quantum efficiency during the summer of $1.65 \, g \, C \, MJ^{-1}$ APAR for the stand at Harvard Forest, predictions of GPP integrated monthly showed good agreement with GPP derived from day and night eddy-flux measurements (Fig. 3.8). Seasonal differences in canopy LAI were pronounced, and reductions in leaf chlorophyll concentrations also contributed to reducing quantum efficiency and canopy photosynthesis in the autumn (Waring et al., 1995a).

In regard to our interest in scaling, we see that there are trade-offs between achieving accurate estimates of the diurnal variation in canopy photosynthesis by strata and obtaining acceptable daily or monthly estimates of photosynthesis for the entire canopy. The finer resolution models serve as excellent standards for reference and allow estimates of individual tree growth (Chapter 5), but they require much additional meteorological data and

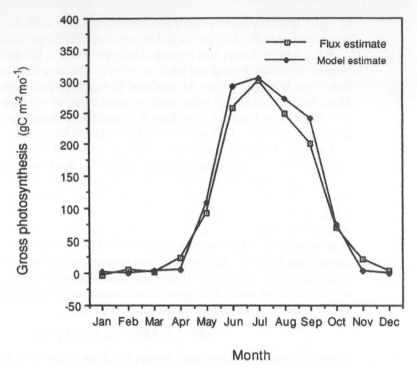

FIGURE 3.8. Monthly integrated estimates of gross photosynthesis made with a stomatal constrained quantum-efficiency model (α depended on seasonal changes in chlorophyll content of foliage but for the growing season was a constant $1.65\,\text{g}\,\text{C}\,\text{MJ}^{-1}$ APAR) agreed well with integrated monthly values acquired from continuous flux measurements made throughout an entire year ($r^2 = 0.97$). (From Waring *et al.*, 1995a.)

detailed description of the absorption of light through the canopy. By extending the time scale of analysis from days to months there is little lost in our ability to predict whole-canopy photosynthesis. Moreover, at monthly time steps, satellite imagery is generally available to record variation in leaf area index and nitrogen status, which effectively represent the fraction of PAR that can be absorbed by all vegetation present (Sellers *et al.*, 1992a,b; Prince and Goward, 1995; Dang *et al.*, 1997; Chapters 7–9).

B. Carbon Balance of the Vegetation

If we have fairly reliable models of GPP, then they should compare well with annual estimates of the sum of carbon required for growth and autotrophic respiration. Williams *et al.* (1997) applied their daily integrated model of GPP developed at Harvard Forest to a wide range of forests in Oregon, arrayed along a 250-km transect at 44° N latitude (Table 3.3). Data on canopy LAI, foliar biomass, and N content were available from Runyon *et al.* (1994) and Matson *et al.* (1994). Predawn water potentials reached levels that might significantly constrain stomatal conductance only on sites 2, 5, and 6 (Runyon *et al.*, 1994). GPP was not measured at any of the sites but was estimated with an annual component carbon budget:

$$\text{GPP} = \text{NPP}_A + \text{NPP}_B + R_{SA} + R_{SB} + R_{Msap} + R_{Mfol} + R_{Mroot} \qquad (3.15)$$

TABLE 3.3

Ecosystem Variables and Annual Environmental Variables for Sites across the Oregon Transect[a]

Site	Species	LAI	Mean foliar N (g m⁻²)	Mean annual temp. (°C)	Growing season[b] (Julian dates)	Annual total PAR (MJ m⁻² year⁻¹)	Minimum ψ (MPa)	Average maximum canopy height (m)
1	*Picea sitchensis/Tsuga heterophylla*	6.4	1.2	10.1	75–320	1934	−0.5	50
1A	*Alnus rubra*	4.3	2.4	10.1	110–275	1934	−0.5	13
2	*Pseudotsuga menziesii/Quercus garryana*	5.3	1.8	11.2	75–280	2267	−1.7	40
3	*Tsuga heterophylla/Pseudotsuga menziesii*	8.6	1.7	10.6	75–305	2259	−0.5	30
4	*Tsuga mertensiana/Abies lasiocarpa/ Picea engelmanii*	1.9	3.0	6.0	160–256	2088	−0.5	20
5	*Pinus ponderosa*	0.9	2.7	7.4	125–275	2735	−1.7	30
6	*Juniperus occidentalis*	0.4	5.8	9.1	125–275	2735	−2.5	10

[a]After Williams *et al.* (1997), developed from Runyon *et al.* (1994) and Matson *et al.* (1994).
[b]Growing season signifies those days when leaves are present and photosynthesis is possible. In some climates the season is abbreviated because of long periods below freezing.

where NPP_A and NPP_B are aboveground and belowground net primary production; R_{SA} and R_{SB} are aboveground and belowground synthesis respiration; and R_{MSap}, R_{Mfol}, and R_{Mroot} are, respectively, sapwood, foliage, and fine-roots maintenance respiration. NPP_A includes new foliage production and branch and stem growth. Runyon *et al.* (1994) provided estimates of these quantities from equations related to annual increases in stem diameter determined from extracted wood cores. Belowground net primary production (NPP_B) was not measured; instead, the relationship based on litterfall (Raich and Nadelhoffer, 1989; Fig. 3.6) was applied. Ryan *et al.* (1995) provided estimates of sapwood maintenance respiration at site 3 (western hemlock, Douglas-fir); the annual value was around 5% of estimated GPP (Fig. 3.5). For the other sites, Williams *et al.* assumed a similar ratio of R_{Msap} to NPP_A (18%). Foliage respiration (R_{Mfol}) was determined from total canopy N and daily and seasonal variations in temperature following the approach developed by Ryan (1991b, 1995).

Daily climatic data from the site meteorological stations were used to run the unmodified Harvard Forest model through the annual cycle to estimate GPP. On days that temperature dropped below −2°C, no C fixation was assumed to occur as a result of frost. LAI in the coniferous stands varied by around 30% each year (Runyon *et al.*, 1994), and its seasonal variation was taken into account. Foliar N concentrations, however, were assumed constant throughout the year, except for the deciduous forest. Predawn water potentials below −1.5 MPa were assumed to cause complete stomatal closure, except for the more drought-resistant *Juniperus occidentalis,* where limits were set at −2.5 MPa. GPP estimates obtained with the simulation model (Williams *et al.*, 1996, 1997) compared well with those calculated with Eq. (3.15) ($r^2 = 0.97$; Fig. 3.9a). The only exception was at site 1, where an old-growth forest exhibited greater hydraulic restrictions of water flow than the model assumed (Chapter 2).

An important insight gained from constructing the carbon balances for the stands distributed across the Oregon transect was that the ratio of NPP/GPP is conservative, averaging 0.46 with a total range from 0.40 to 0.52. Additional comparisons were added from similar carbon balance analyses made at Harvard Forest (Williams *et al.*, 1997), at three pine plantations in Australia (Ryan *et al.*, 1996b), and in a native *Nothofagus* forest in New Zealand (Benecke and Evans, 1987), and all showed that the ratio NPP/GPP remains essentially constant (Fig. 3.9b). The finding that NPP/GPP is a conservative ratio has been reported previously in growth room studies where respiration and photosynthesis have been monitored for short periods on a variety of species exposed to a range of temperatures from 15° to 30°C (Gifford, 1994). The balance between photosynthesis, respiration, and growth could reflect the key role of nitrogen, as will be discussed in later sections. The NPP/GPP ratio is relatively conservative but may be significantly lower in boreal forests (~0.25) than in other biomes (Ryan *et al.*, 1997a).

C. Assessment of Heterotrophic Respiration

Although heterotrophic respiration may be a relatively small component of total ecosystem respiration under undisturbed conditions, it is a critical component and one that could change measurably with disturbance or climatic warming. In a steady-state condition, one

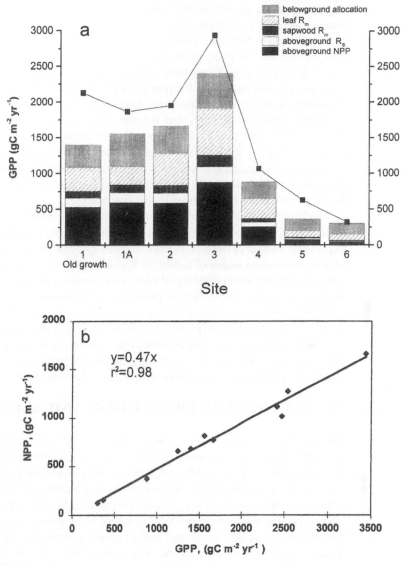

FIGURE 3.9. (a) For seven sites across the Oregon transect, GPP (■) estimated with a daily resolution process model is compared with a component analysis that sums to GPP (stacked bars). The dominant species at each site are given in Table 3.3. (Provided by Mathew Williams, personal communications, with data from Williams *et al.,* 1997.) (b) The ratio NPP/GPP, determined by component carbon balance analyses, is essentially constant for a dozen forests that include seven on the Oregon transect (Williams *et al.,* 1997), three pine plantations in Australia (Ryan *et al.* 1996b), Harvard Forest (Williams *et al.,* 1997), and a *Nothofagus* forest in New Zealand (Benecke and Evans, 1987). Regression of NPP to GPP has been forced through the origin. The slope of the relation is 0.47 with a standard deviation of 0.04 for temperate forests analyzed in Table 3.3. (After Waring *et al.,* 1998.)

might expect on an annual basis that heterotrophic respiration might approach a fixed fraction of total ecosystem respiration and GPP, and certainly total soil respiration is closely related to NPP (see global review paper by Raich and Schlesinger, 1992). Steady-state conditions, however, rarely apply, as demonstrated at Harvard Forest where continuous eddy-flux measurements were made over a 5-year period. When the total respiration $(R_a + R_h)$ was compared as a fraction of GPP it averaged 0.82 but varied from 0.75 to 0.90 (Goulden *et al.,* 1996). With more detailed studies, Goulden *et al.* found, during an extended period of drought, that heterotrophic respiration decreased much more than photosynthesis and autotrophic respiration. Because heterotrophic respiration is largely confined to the surface litter and upper soil horizons that dry quickly, R_h is very responsive to drought, whereas R_a is less sensitive because few perennial roots actually die as a result of drought (Marshall and Waring, 1985) and because fine-root growth continues in the lower soil horizons where water is still available.

Modeling seasonal variation in soil respiration realistically requires the recognition of at least three distinct horizons: surface litter, surface soil, and the deeper rooting zone. We can expect this level of definition, or even finer, to apply in predicting the seasonal availability of soil nutrients to plants (Chapter 4). It is clear why we require seasonal resolution and some underlying understanding of basic processes to provide an alternative to long-term monitoring and to allow, where possible, simplifications that are justified from more detailed spatial and temporal analyses. One promising example of the application of a coupled ecosystem model (FOREST-BGC) is its ability to predict the separate components of carbon acquisition and loss in forests other than those where it was originally developed (Fig. 3.10). In a later section we will compare and contrast the structure and function of other dynamic ecosystem models that generate seasonal estimates of fluxes (C, H_2O, and N).

VI. NET PRIMARY PRODUCTION AND ALLOCATION

The amount of carbon that may be synthesized into new tissues, storage reserves, and protective compounds is determined from the photosynthate remaining after accounting for autotrophic respiration $(R_m + R_s)$. Partitioning of assimilate into various products and the biochemical composition of those products are affected by the relative availability of critical resources (solar energy, water, nitrogen, CO_2, and temperature). From an evolutionary perspective, carbon products synthesized by plants might be expected to increase the chances of an individual tree and its progeny surviving. Species differ, however, in their adaptations and thus in the way carbon and other resources are allocated. For example, some pine species produce serotinous cones that provide viable seeds following a destructive fire. In contrast, redwood and eucalyptus respond to fire by producing epicormic branches from their main stems.

In this section we first describe how trees allocate assimilated carbon seasonally. A set of concepts is introduced that have served as a basis for modeling seasonal shifts in allocation. Simpler analyses are introduced for assessing annual patterns. With some understanding of why allocation patterns shift, we define a number of indices which reflect responses of trees to competition in general as well as to specific stresses. These allocation indices provide a means of interpreting how different tree species and whole forests are likely to respond to changing environmental conditions, seasonally as well as over their life span.

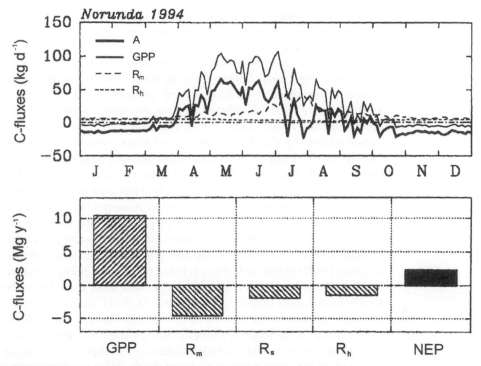

FIGURE 3.10. Net ecosystem CO_2 fluxes were measured continuously with eddy-flux instrumentation at the NOPEX site in central Sweden above a 25-m-tall spruce and pine forest which had a range in LAI between 3 and 6. Simulation of the components of net ecosystem production [assimilation, A; gross primary production, GPP, maintenance respiration, R_m, heterotrophic respiration, R_h, and growth (synthesis) respiration, R_s] were estimated with FOREST-BGC. (Reprinted from *Journal of Hydrology*, E. Cienciala, S. W. Running, A. Lindroth, A. Grelle, and M. G. Ryan, "Analysis of carbon and water fluxes from the NOPEX boreal forest: Comparison of measurements with FOREST-BGC simulations," 1998, with kind permission of Elsevier Science–NL, Sara Burgerhartstraat 25, 1055 KV Amsterdam, The Netherlands.)

A. Seasonal Dynamics in Allocation

In perennial plants, and particularly long-lived trees, there is much seasonal variation in NPP and its allocation (Mooney and Chu, 1974). Generally, we can consider that no growth or storage of carbohydrates occurs until the basal metabolic requirements for all living cells are first met. Storage reserves must be present to cover maintenance costs during the night and other nonphotosynthetic periods. These key limitations on allocation of reserves are usually applied in process models designed to predict seasonal growth patterns (Cannell and Dewar, 1994).

In favorable climates, where growth may be continuous, allocation patterns might be assumed fixed, and growth rates would then be proportional to the pool of carbohydrates available. In reality, the growth of reproductive organs, roots, and shoots is rarely in phase. Internal controls on allocation are hormonal, but they reflect evolutionary adaptations to climatic, edaphic, and biotic pressures. At present, the allocation process is very poorly understood in comparison to photosynthesis, yet it is critically important because a 5% shift in allocation away from fine-root production may permit a 30% increase in foliage mass (Cropper and Gholz, 1994).

A number of schemes have been proposed to model carbon allocation, as summarized by Cannell and Dewar (1994). They postulate that (i) photosynthate is allocated from sources in accordance with assigned sink strengths (Ford and Kiester, 1990; Luxmoore, 1991; Thornley, 1991); (ii) photosynthate is allocated to points where a resource deficiency first occurs (Ewel and Gholz, 1991); (iii) plants maintain a functional balance between carbon fixation by shoots and nutrient and water uptake by roots (Davidson, 1969); (iv) carbohydrates are dispersed in proportion to distance from the site of fixation, and nutrients and water are allocated in the opposite order (Weinstein *et al.,* 1991); and (v) carbon is allocated to optimize net carbon gain or total plant growth rate (Ågren and Ingestad, 1987; Johnson and Thornley, 1987). The various assumptions on allocation are interrelated; however, none deals fully with underlying processes, and some are in direct conflict with one another. Even within one scheme (scheme v), Thornley's model assumes that sink strength is controlled by carbon and nitrogen substrate concentrations, whereas Ågren and Ingestad (1987) define sink strength based on the total nitrogen content in the canopy.

Models that relate root uptake of nitrogen and shoot uptake of carbon have progressed slowly toward more realistic assumptions (Fig. 3.11). Initially, growth was assumed to be a simple function of the amount of carbon and nitrogen available, with the product of carbohydrate and nitrogen resources in foliage and roots determining the relative allocation. More refined models include resistance to transport and demands for resources by intermediate structures between roots and shoots. More mechanistic models consider sugar and amino acid transport separately through phloem and sapwood (Cannell and Dewar, 1994). Root-to-shoot gradients in water potential also influence the rates at which resources are transported through the two vascular systems (Dewar, 1993).

Carbon allocation models apply particularly well to species showing *indeterminate* growth. In many forests, however, trees are dormant in certain seasons, independent of the amount of carbohydrates or nitrogen available. For example, at full leaf, an oak tree may have accumulated sufficient starch reserves to replace its canopy three times (McLaughlin *et al.,* 1980). The seasonal timing of growth (*phenology*) must be predicted or continuously monitored.

Plant phenology has been correlated with a variety of environmental signals. In tropical or desert regions, the onset of a wet season (or sometimes the end of a dry season) may initiate conditions favorable for growth (Borchert, 1973; Reich, 1995). In boreal and temperate regions, day length and temperature are major controlling variables, with different species exhibiting different genetically programmed thresholds. Shorter day lengths, or, more precisely, longer nights, induce dormancy, whereas longer day lengths induce hormonal changes that favor growth. Cessation of growth and senescence of leaves may also be triggered by drought or subfreezing temperatures. Growth in temperate, subalpine, and boreal zones is not initiated until soil temperatures rise at least a few degrees above freezing. A number of empirical models predict budbreak, elongation, and budset on the basis of accumulated daily temperature values (heat sums) above some minimum threshold temperature (Hari and Hakkinen, 1991; Dougherty *et al.,* 1994; Whitehead *et al.,* 1994). Other phenology models integrate soil and air temperature values to predict leaf and stem

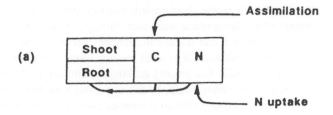

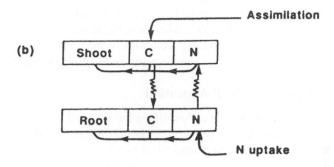

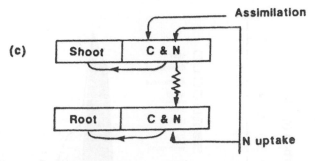

FIGURE 3.11. Progressively more sophisticated allocation models are presented from top to bottom. In (a) partitioning of assimilate is determined by the product of carbon (C) and nitrogen (N) concentrations in roots or shoots. In (b) opposite gradients in C and N are envisioned between shoots and roots with resistances against transport in both directions. In (c), transport of C and N are both considered to be through the phloem, which is the most realistic model and accommodates changes in plant water relations. (From Dewar *et al.*, 1994.)

phenology (Cleary and Waring, 1969). With daily satellite imagery now available, the phenology of vegetation can be assessed by observations at regional and global scales (Chapters 7–9).

With knowledge of phenology and environmental conditions, the pools of available carbon and nitrogen can be allocated into biomass, storage reserves, and losses through respiration. Weinstein *et al.* (1991) generated seasonal estimates of carbon allocation for red spruce by assuming priorities based on proximity to the resource (scheme iv; Fig.

3.11b). In this model, leaf growth had first priority for carbon after maintenance requirements of all living tissue were met. Storage of carbon in leaves had the next priority, followed by growth and storage in branches, stem, coarse roots, and, last, fine roots. Carbon in excess of that needed to meet the maximum growth rate of an organ was passed down the priority chain. The priority ranking for allocation of water and nutrients would, according to this proximity logic, be the opposite, with fine roots ranked first and new foliage ranked last. Because critical resources come from opposite directions, no given organ is likely to grow at its full potential, unless phenology limits all growth elsewhere. Predictions of photosynthate allocation seasonally in *Picea rubens,* based on phenology and the proximity logic (Fig. 3.12), showed reasonable growth patterns and matched measured storage reserves within 10% and total carbon content within 2% at the end of the year (Weinstein *et al.,* 1991). Most seasonal models of carbon allocation require specific knowledge of plant phenology, definition of the limits to growth of various organs, and specification of the size of storage reserves in all major organs.

B. Annual Assessment of NPP Allocation

Annual changes in carbon allocation are much easier to assess and to model than seasonal patterns. Annual primary production represents all carbon sequestered into dry matter during a year and is equivalent to total carbon uptake through photosynthesis minus the loss through autotrophic respiration. In practice, net primary production (NPP) is estimated

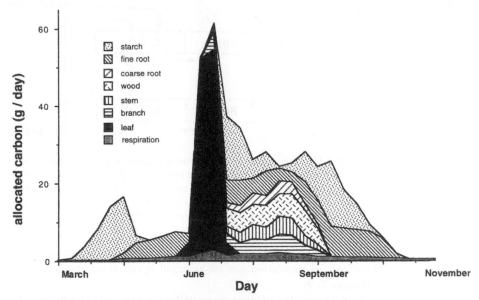

FIGURE 3.12.   Seasonal carbon allocation patterns generated with a simulation model for red spruce (*Picea rubens*) are dependent on phenology and proximity to resources. Starch reserves accumulate when growth is at a minimum compared to photosynthesis. Leaf growth attains priority in June while the production of other structural components peaks thereafter. (After Weinstein *et al.,* 1991.)

by summing the growth of all tissue produced during a year, whether or not the tissue was consumed by herbivores or entered the detrital pool. The equation is

$$NPP = \Delta B + D_B + C_B \tag{3.16}$$

where ΔB is the change in biomass over a period of a year, D_B is detritus produced during the year, and C_B represents consumption of biomass by herbivores during the year.

Consumption by animals is usually a small fraction of total NPP in forests unless there is an outbreak of defoliating insects (Chapter 6). Even if 10% of the foliage is consumed, this represents less than 3% of total NPP in deciduous or evergreen forests. Estimates of foliage consumption are made by deducting the area of sample leaves that are partially consumed (Reichle *et al.*, 1973) relative to comparable foliage on twigs that developed normally, or by measuring insect frass in litterfall traps and calculating the weight of tissue consumed to produce the amount of frass. Animals also eat roots and fruits, but this consumption is ignored in most calculations of NPP in forests.

From destructive analysis of trees, information can be obtained on how growth is distributed. In a particular climatic zone, the annual pattern of carbon allocation to foliage and stem wood shows a general consistency (Fig. 3.13a), except when large crops of seeds are produced or unusual weather conditions prevail (Eis *et al.*, 1965; Pregitzer and Burton, 1991). Proportional increments in biomass of stem wood, leaves, branches, and large-diameter roots are related exponentially to increases in stem diameter (Whittaker and Woodwell, 1968; Kira and Ogawa, 1971; Gholz *et al.*, 1979; Deans, 1981; Pastor *et al.*, 1984; Fig. 3.13b). *Allometric* relations derived from such analyses allow computation of forest biomass and also the estimation of stores of carbon and nutrients in various components as forests grow (Chapter 5). Production is determined by periodic measurement of stem diameter or by extracting wood cores and measuring annual increments. Biomass and production of shrubs and herbs in forest undergrowth are estimated separately, often by obtaining correlations with cover which can be estimated with increasing accuracy using digital imagery acquired from the ground and from satellites (Levine *et al.*, 1994; Law, 1995). Biomass increment is calculated by measuring all trees within a known area or by using variable-plot surveys based on the diameter of trees intercepted by a selected angle. Variable-plot surveys are generally more efficient because only a few trees require measurement at each sampling point (Sukwong *et al.*, 1971).

Soil organic matter consists of fresh litter, partially decomposed material, and humus. The total dead organic matter in an ecosystem is sometimes called "detritus," although we prefer to restrict the term to dead material from which the source can still be recognized. The annual loss of leaves, twigs, flowers, fruits, and bark fragments represents the obvious forms of litterfall in forest ecosystems. Leaf litter typically makes up 70% of the total annual litterfall from above ground that is collected in litter screens (Waring and Schlesinger, 1985). The composition and quantity of litterfall are often variable between years. During insect outbreaks the production of frass may represent more than the annual production of foliage in evergreen forests (Chapter 6). The resulting dead trees disappear much more slowly than leaves from an ecosystem, and in their slow decay they play special roles that will be described in Chapter 5.

In this chapter, we consider the smaller detrital components which consist mainly of leaf litter, twigs, and small-diameter roots that die and accumulate throughout a year on

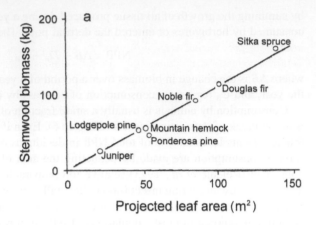

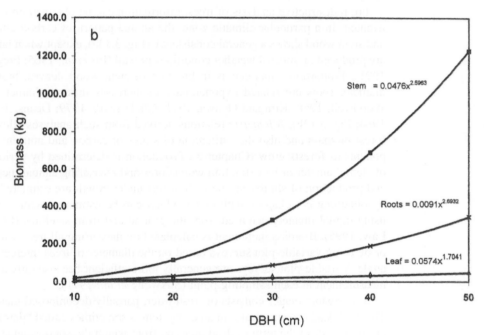

FIGURE 3.13.   (a) For conifers distributed across a steep environmental gradient in Oregon, trees of a comparable diameter, here 20 cm, show an increase in stem wood mass and projected leaf area (derived from leaf area/sapwood area relationships) that reflects a transition from the most extreme dry and cold environments supporting *Juniperus* toward a mild, frost-free, and moist coastal environment where Sitka spruce (*Picea sitchensis*) is restricted. Variation occurs in these allometric relations for species such as *Pinus ponderosa* and Douglas-fir (*Pseudotsuga menziesii*) with wide environmental ranges. (After Waring, 1980.) (b) Allometric relations developed for Douglas-fir growing in western Oregon show that the biomass of stems, roots, and leaves increases exponentially with stem diameter (*x*, cm) at breast height (DBH). Equations were transformed from natural log–linear relationships presented by Gholz *et al.* (1979).

the soil surface or within the soil. Aboveground estimates of litterfall are obtained by installing sets of screens to catch material which is then weighed. The smaller diameter roots (0.5 to 5 mm) which are produced within a year usually die and begin to decompose (Harris *et al.,* 1975; McGinty, 1976; Persson, 1978). The smaller the diameter of a root, the shorter is its life span (Schoettle and Fahey, 1994). Typically, estimates of fine-root production and turnover have been obtained by comparing seasonal changes in the standing crop of live roots extracted from soil cores (Fig. 3.14). A large number of samples is required to reduce the standard error of the estimate to less than 10%, and a bias to overestimation is common (Singh *et al.,* 1984; Kurz and Kimmins, 1987; Schoettle and Fahey, 1994). Nondestructive measures of root growth and turnover have been obtained by analyzing sequential digital images of the same soil surface acquired with miniature video cameras inserted in root periscopes (Olsthoorn and Tiktak, 1991; Hendrick and Pregitzer, 1993; Reid and Bowden, 1995).

As mentioned previously, the total carbon allocated to roots may also be estimated indirectly through a correlation with litterfall and CO_2 efflux from the soil (Raich and Nadelhoffer, 1989; Hanson *et al.,* 1993). Temperature is assumed to have a similar influence on root and microbial activity in the relationship. This assumption may be unwarranted, however, because of interactions with the depletion of surface soil water and because of the ability of fine roots to respire as long as they have starch or other reserves to expend (Marshall and Waring, 1985).

Much additional insight has been gained by a few groups who established plantations of trees in one area and then experimentally manipulated the availability of resources to quantify shifts in carbon allocation (Linder, 1986; Linder *et al.,* 1987). These kinds of experiments have, along with eddy-flux measurements, greatly advanced our abilities to

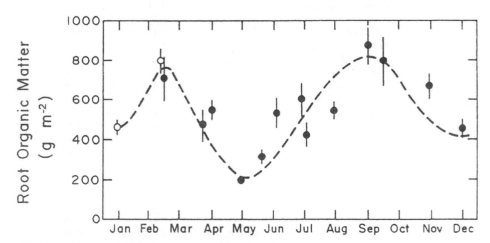

FIGURE 3.14.   Seasonal changes in the standing crop of small-diameter roots in a *Liriodendron* forest were obtained by core samples collected throughout the year. Estimates of production and turnover rates are obtained from differences in peak and nadir biomass. Organic matter production in this case was about 1460 g m^{-2} year^{-1} with an average standing crop of 680 g m^{-2}. (After "Carbon cycling in a mixed deciduous forest floor" by N. T. Edwards and W. F. Harris, *Ecology,* 1977, **58**, 431–437. Copyright © 1977 by the Ecological Society of America. Reprinted by permission.)

generalize and model ecosystem responses. As an example, we cite a long-term study initiated on a *Pinus radiata* plantation in Australia described in more detail by Landsberg (1986b) and Benson *et al.* (1992). Treatments applied to the pine plantation were irrigation, fertilizer without irrigation, and fertilizer plus irrigation with a balanced nutrient solution, along with the untreated "control." Some results of this experiment (Table 3.4) show that significant shifts occurred in allocation and respiration by trees subjected to the different treatments. Irrigation increased aboveground and fine-root growth by about 25%, without increasing leaf area index above that of the untreated control. It was assumed that photosynthetic rate decreased as leaf N concentrations fell from 1.3% in the control to 1.1% with irrigation (Ryan *et al.*, 1996b). Addition of water and nutrients together resulted in a 40% increase in GPP and a 150% increase in foliage production, raising LAI to 4.6. Stem and coarse-root production in the irrigation plus fertilizer treatment increased by a similar amount to that observed with irrigation alone. The analysis indicates that with irrigation and fertilization about 10% of GPP was expended on fine roots and mycorrhizae, whereas in the irrigated treatment the allocation to these components reached 30% of GPP.

Similar results have been observed in other studies (Alexander and Fairly, 1983; Beets and Whitehead, 1996). Although the precision of estimates varied, with less accuracy in the fraction of carbon allocated below ground, the sum of all estimates of growth and

TABLE 3.4

Annual Carbon Budget of *Pinus radiata* Stands in Australia with Different Treatments[a]

Variable	C (Mg ha^{-1} year^{-1})		
	Control	Irrigated	Irrigated and fertilized
Foliage production	0.84	1.13	2.13
Foliage R_s	0.21	0.28	0.53
Foliage R_m	4.00	2.67	6.28
Branch production	0.22	0.22	0.27
Branch R_s	0.05	0.05	0.07
Branch R_m	1.11	1.20	1.87
Stem + bark production	4.93	6.15	10.51
Stem + bark R_s	1.23	1.54	2.63
Stem wood R_m	1.34	1.71	2.70
Aboveground NPP	*(5.99)*	*(7.50)*	*(12.91)*
Coarse-root production	1.19	1.56	2.96
Coarse-root R_s	0.30	0.39	0.74
Coarse-root R_m	0.50	0.64	1.24
Fine-root production	1.85	2.35	1.33
Fine-root R_s	0.46	0.59	0.33
Fine-root R_m	1.48	1.47	1.42
Mycorrhizae + exudates	4.43	3.36	−0.63
Total belowground allocation, B_a	*(10.21)*	*(10.36)*	*(7.39)*
Gross primary production	24.14	25.31	34.38

[a]Respiration is partitioned into components: synthesis (R_s) and maintenance (R_m). Total belowground carbon allocation was estimated as the difference between annual soil respiration and aboveground litterfall plus coarse toot increment. Mycorrhizae or root exudation was estimated as the difference between B_a and root respiration plus production. After Ryan *et al.* (1996b).

autotrophic respiration averaged within 10% of annual simulated gross photosynthesis (Ryan *et al.*, 1996b). If these analyses are correct, fine-root production, including the support of mycorrhizae, might be expected to represent between 10 and 60% of total NPP (Ryan *et al.*, 1996b; Beets and Whitehead, 1996). From similar but more extensive analyses performed on other pine plantations, Beets and Whitehead (1996) demonstrated that by increasing the availability of nitrogen, the annual fraction of NPP allocated to fine roots decreased from 60 to about 10% (Fig. 3.15a), while the fraction allocated to stem wood

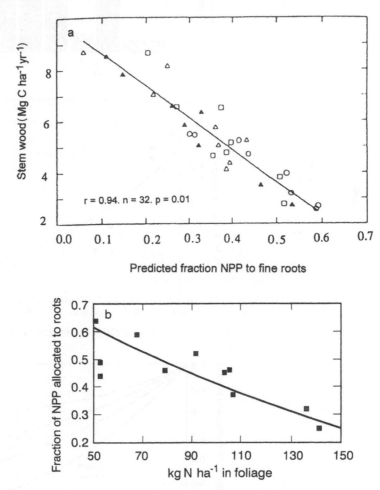

Predicted fraction NPP to fine roots

FIGURE 3.15.   (a) Increasing the availability of nitrogen to *Pinus radiata* plantations in New Zealand did not change the concentration of N in foliage, but it caused a proportional shift in carbon allocation of NPP away from fine roots into stem wood. Root production was estimated from a component analysis as a residual. (From Beets and Whitehead, 1996.) (b) As the total content of nitrogen in the radiata pine canopy increased from 50 to nearly 150 kg N ha^{-1}, the fraction of NPP allocated to all roots showed a decrease from a maximum of about 0.65 to a minimum of about 0.25. Graph was made from a composite of information collected by Beets and Madgwick (1988) and from Beets and Whitehead (1996). (Modified from *Forest Ecology and Management*, Volume 95, J. J. Landsberg and R. H. Waring, "A generalized model of forest productivity using simplified concepts of radiation-use efficiency, carbon balance and partitioning," pp. 209–228, 1997, with kind permission of Elsevier Science–NL, Sara Burgerhartstraat 25, 1055 KV Amsterdam, The Netherlands.)

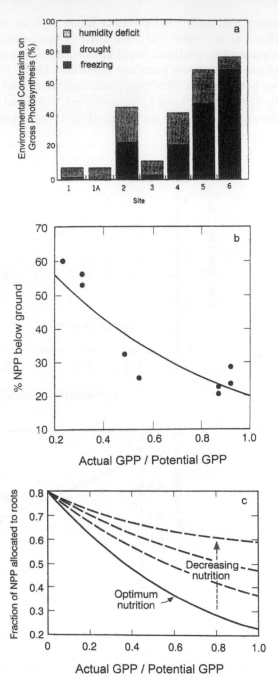

and leaf area increased proportionally. If only nitrogen were a limiting factor, the allocation could be predicted from a correlation between total foliage N content and the proportion of carbon allocated to all sizes of roots, with the remaining NPP going to aboveground growth (Fig. 3.15b). Deficiencies in nitrogen, and to a lesser extent phosphorus and sulfur, favor growth of small-diameter roots over that of shoots, whereas deficiencies in potassium, magnesium, and manganese have the opposite effect (Wikström and Ericsson, 1995). Nutrient imbalances in general require plants to expend more energy and reduce overall growth (Sheriff *et al.*, 1986; Schulze, 1989).

In addition to limiting nutrients, dry soils and restrictive temperatures, as mentioned earlier, may result in an increase in the fraction of NPP allocated below ground. Across the Oregon transect, where a combination of factors limited stomatal conductance (Fig. 3.16a), the fraction of NPP allocated to roots varied over a similar range to that observed in Fig. 3.15a and was correlated with the ratio of actual GPP to potential GPP (Fig. 3.16b).

The relative importance of climatic and edaphic limitations of carbon allocation to roots should change seasonally. As the ratio GPP/potential GPP approaches unity, soil fertility will play an increasingly important role, with the most infertile soils requiring the maximum fraction of NPP allocation under otherwise favorable environmental conditions (Fig. 3.16c).

Landsberg and Waring (1997) applied this reasoning in the construction of a stand growth model with monthly time steps that contained the following simplifying assumptions: (a) potential GPP is a linear function of APAR ($1.8\,g\,C\,MJ^{-1}$), (b) actual GPP reflects additional environmental constraints that limit diffusion of CO_2 through stomata, (c) NPP is 45% of GPP, and (d) the fraction of NPP to roots varies depending on the monthly calculated ratio GPP/potential GPP. The remaining NPP not allocated to roots was proportioned into foliage and stem production on the basis of the ratio of the rate of change in weight of foliage to the rate of change in stem biomass determined with species-specific allometric equations (McMurtrie and Wolf, 1983; McMurtrie and Landsberg, 1992). The model reduced the fraction of NPP allocation to leaves when environmental conditions favored root growth and resulted in a net decrease in LAI when monthly production of leaves was less than a set value for leaf litterfall. Model predictions of annual aboveground stem biomass for two *Pinus radiata* plantations agreed almost 1:1 with measured values

FIGURE 3.16.   (a) The environmental constraints on gross photosynthesis vary across the Oregon transect depending on limitations from humidity deficits, drought, and freezing. The forest types include old-growth Sitka spruce (1) and alder (1A), mixed forests of conifers with deciduous oak (2), fast-growing Douglas-fir and western hemlock on the western slopes of the Cascade Mountains (3), a subalpine forest (4), an open pine forest (5), and juniper woodland (6). (From Runyon *et al.*, 1994.). (b) With increasingly climatic constraints on photosynthesis across the Oregon transect, the ratio of actual GPP/potential GPP falls and, with it, the fraction of NPP allocated belowground annually. (After Runyon *et al.*, 1994.) (c) The ratio actual GPP/potential GPP controls the minimum fraction of NPP allocated to roots in a growth model with monthly resolution. The additional possible constraints of limiting nutrition are incrementally added. In months favorable for photosynthesis, the model shows the greatest effect of nutrition on allocation. As environmental conditions become progressively less favorable, the availability of nutrients has a decreasing effect on allocation. (Modified from *Forest Ecology and Management*, Volume 95, J. J. Landsberg and R. H. Waring, "A generalized model of forest productivity using simplified concepts of radiation-use efficiency, carbon balance and partitioning," pp. 209–228, 1997, with kind permission of Elsevier Science–NL, Sara Burgerhartstraat 25, 1055 KV Amsterdam, The Netherlands.)

acquired annually over periods up to 30 years, when harvesting begins, taking into account natural mortality predicted with a self-thinning rule that will be discussed in Chapter 5.

For scaling principles, the approach outlined above has several lessons. First, simplifying assumptions must be derived from comparative studies across a range of environments and forest conditions. Second, a minimum time step must be identified (e.g., a year is too long, integrated monthly data are near minimum). Third, the reliability of simplified models can often best be tested at longer time steps with data not originally required in the model (e.g., annual increments in stem biomass recorded over decades). Because the Landsberg and Waring model represents a formulation of the Light Use Efficiency model (Table 3.5) that provides for estimates of monthly allocation of NPP above and below ground, it has much wider potential application than other models and warrants comparison with data acquired for other life forms with different allocation patterns (Law and Waring, 1994). Moreover, because of the monthly time steps, the modified LUE model can be easily run with condensed averaged weather data and monthly estimates of the ability of the vegetation to absorb PAR.

C. Allocation Indices

A number of simple allocation indices can be distilled from our discussion on allocation. One is based on the idea that canopies can be assumed to respond to the total PAR absorbed in a linear fashion and that yield can be related directly to APAR under favorable environments. John Monteith first proposed this scheme and verified its application on crops (Monteith, 1977) and wet tropical forests (Monteith, 1972). He assigned the symbol epsilon (ε) to the amount of dry matter or its carbon equivalent produced above ground per unit of the annually absorbed PAR. Epsilon for crops and other vegetation that grow in well-watered and fertile soils usually has a maximum value of ~0.7 g C MJ^{-1} APAR (Monteith, 1977; Jones, 1992; Landsberg *et al.*, 1996). This maximum value, however, is rarely observed in forests, because of stomatal constraints associated with less than optimum conditions, and because of the increasing fraction of carbon allocated below ground in suboptimal environments. Across the Oregon transect, Runyon *et al.* (1994) reported that ε for aboveground NPP ranged from 0.09 to 0.46 g C MJ^{-1} APAR. On the other hand, when Light Use Efficiency was calculated by deducting that fraction of PAR absorbed when photosynthesis was limited by other environmental constraints, ε approached a constant for total NPP (above and below ground) of ~0.65 g C MJ^{-1} APAR. The original definition of epsilon (g NPP$_A$ MJ^{-1} APAR) is valuable in judging the combined limitations that the environment exerts on photosynthesis and restricts growth allocation above ground, as will be shown in landscape and regional analyses presented in Chapters 8 and 9.

More specific allocation indices are also available; within a species, subtle shifts in carbon allocation may have diagnostic value. As noted, stem growth has relatively low priority in comparison to root growth. On the other hand, many secondary compounds, such as protective chemicals, are less essential than diameter growth because new foliage requires supporting sapwood. Because carbon allocation to stem wood reflects the carbon uptake per unit of foliage, and allocation below ground, the annual growth of stem wood per unit of foliage, termed "Growth Efficiency," is a general index of tree vigor that can help managers interpret the benefits of various silvicultural options and anticipate the response of forests and individual trees to attack by insects and pathogens (Chapters 5 and 6).

TABLE 3.5

Computation of Light Use Efficiency Conversion of Visible Radiation into Aboveground Net Primary Production and Net Ecosystem Production[a]

Input	Derived input	Equation	Prediction
Incoming solar radiation (I_s)	Photosynthetically active radiation (PAR)	PAR = 0.5 × solar radiation	PAR
Leaf area index (LAI) (with seasonal variation)	Fraction of PAR absorbed (FPAR) (*apply Beer's law*)	PAR × FPAR = APAR (MJ PAR m^{-2} year^{-1})	Absorbed PAR (APAR)
Quantum efficiency (α; *function of chlorophyll content in canopy*)	max (α) ≈ 1.8 g C MJ^{-1} APAR (*decreases with reduction in chlorophyll content*)	GPP$_{pot}$ = APAR × α	Potential gross primary production (GPP$_{pot}$)
Temperatures, humidity, precipitation, and soil water storage capacity	Environmental constraints on photosynthesis	GPP$_{actual}$ = APAR × $\alpha f(H_2O) f(D) f(T)$ [Eq. (3.14)]	Actual gross primary production (GPP$_{actual}$)
All the above inputs for Eq. (3.14)	GPP$_{actual}$	Net primary production (NPP) = $K_{(NPP/GPP)}$ × GPP$_{actual}$	NPP[b]
All the above inputs	Aboveground net primary production (NPP$_A$)	NPP$_A$ = f(GPP$_{actual}$/ GPP$_{pot}$) (see Fig. 3.16)	NPP$_A$[c]
I_s, LAI, NPP$_A$	Light Use Efficiency conversion (ε)	ε = NPP$_A$/APAR	ε[d]
Annual litterfall (LF) when litter accumulation on forest floor (FF) is near steady state	Net ecosystem production (NEP) Heterotrophic respiration (R_h)	Leaf litter, $R_h = f$(LF/ FF) [Eqs. (4.2) and (4.3)] Root litter R_h = 0.5 NPP$_B$ (see Figs. 3.6 and text) NEP = NPP − Σ R_h	NEP[e]

[a]This table summarizes concepts and relationships presented in the text.

[b]The proportion of gross primary production converted into biomass (carbon) annually, $K_{(NPP/GPP)}$, is generally a conservative value within a biome, averaging 0.25 for boreal forests and 0.5 for temperate forests. Comparable data are not yet available for tropical forests.

[c]The annual ratio of aboveground to belowground NPP increases from 0.75 to 0.35 for forests as constraints on photosynthesis increase (GPP$_{actual}$/GPP$_{pot}$). Soil fertility also affects allocation but is incorporated in the amount of foliage present (LAI) and its chlorophyll content, which together determine FPAR and α.

[d]The light use efficiency conversion (ε) defined by Monteith (1972) ranges from a maximum of 0.7 to 0.1 g C MJ^{-1} APAR as conditions becomes less favorable for photosynthesis and aboveground growth (see Section VI,C on allocation indices).

[e]Net ecosystem production = GPP$_{actual}$ − R(leaf litter)$_h$ − R(root litter)$_h$. In Chapter 4, equations are presented that correlate annual leaf (and twig) litterfall with forest floor litter accumulation to estimate mean residence time of aboveground litter and the fraction of carbon (or mass) lost annually through heterotrophic respiration, R(leaf litter)$_h$. Belowground heterotrophic respiration associated with root decay also can be estimated at steady state from measurement of annual litterfall with the equation presented with Fig. 3.6, assuming that the total below-ground allocation of carbon is proportioned as 50% to NPP$_B$, 25% to autotrophic root respiration R_a, and 25% to R(root litter)$_h$ (Ryan, 1991b).

The growth efficiency index is akin to the ratio of nonphotosynthetic tissue produced per mass of foliage (Briggs *et al.,* 1920; Burger, 1929) but expresses foliage in units of area rather than mass. This is an important distinction because the allometric relations are generally defined in units of mass to mass. If the same mass of leaves is distributed into more area, more light may be intercepted, but the total photosynthesis by the canopy may or may not be increased, depending on how nitrogen is distributed to chlorophyll pigments and the Rubisco enzyme. Alternative indices of growth efficiency have been calculated for other organs (Axelsson and Axelsson, 1986). When the referenced product has a special value, such as fruits, nuts, bark, or a specific wood product, the ratio of production to leaf area is termed a *harvest index.*

The pattern of structural growth changes with the availability of certain critical resources and with the imposition of specific kinds of physical stresses. For example, shoot extension on individual branches, as we might expect from analysis of photosynthesis, is closely related to the proportion of light intercepted and the availability of nitrogen (Dougherty *et al.,* 1994). It follows that more shaded branches export proportionally less photosynthate per unit leaf area. When the canopy of a tree is fully shaded, lower branches die and an umbrella-like canopy results (Steingraeber *et al.,* 1979; Kohyama, 1980). Root growth and seed production are also greatly reduced in plants growing in shade. These more subtle kinds of shifts in allocation have special diagnostic value that will be discussed in more detail in Chapter 6.

VII. COMPARISON OF FOREST ECOSYSTEM MODELS

There are now a score of models that integrate carbon, water, and elemental cycles into simulations of forest ecosystem behavior. Some models such as PGEN (Friend, 1995), MAESTRO (Wang and Jarvis, 1990), and MBL/SPA (Williams *et al.,* 1996) provide detailed simulations of instantaneous canopy light absorption and photosynthesis but are not complete ecosystem models. TREGRO (Weinstein and Yanai, 1994) explicitly models tree growth from carbon balance principles and the influences of atmospheric and nutritional stresses, but it does not incorporate a water balance nor allow for multidecadal changes in LAI and stand composition.

Where the objective is to simulate stand growth, many models incorporate only those variables that are important locally in limiting forest production. For example, the models TREEDYN3 (Bossel, 1996) and FORGRO (Mohren and Ilvesniemi, 1995) were developed for nutrient-limited European forests where water stress rarely limits growth. As a result, these models emphasize soil elemental cycles and nutrient limitations and largely ignore energy and water budgets. Similarly, the SPM model (Cropper and Gholz, 1993) incorporates many details of carbon uptake and allocation for slash pine forests that grow only in areas that do not experience drought. In contrast, Bonan (1993) emphasizes energy and water budgets in his simulation of boreal forest photosynthesis and productivity. The latter type of model partitions energy into sensible and latent heat components and thus is easily coupled to land-surface climate models (Bonan, 1995). A whole class of models developed to describe land–atmosphere interactions by the physical sciences and climate modeling community treats canopy energy, water, and carbon exchange with common sets of equations (Sellers *et al.,* 1996a).

Clearly, models that define only a limited but detailed set of variables can quite accurately predict the behavior of a restricted range of forest ecosystems. To extend predictions, more general ecosystem models such as GEM (Rastetter *et al.*, 1991) are required. These general models demand a number of simplifying assumptions to expand their range of application, and, in the process, they lose some accuracy. For example, the PnET model of Aber and Federer (1992) is based on only two principal relationships: (1) maximum photosynthetic rate is a function of foliar nitrogen concentration, and (2) stomatal conductance and transpiration are related to the actual photosynthetic rate. Rather than applying a sophisticated carbon allocation scheme to distribute NPP, the model constrains leaf growth by water availability, which in turn is correlated allometrically with fine-root growth, as the residual is partitioned into woody biomass. In spite of this simplified model structure and the use of monthly climate data, PnET was able to simulate NPP_A surprisingly well across a range of forest stands (see Fig. 5.20). Additionally, as will be illustrated in Chapters 7–9, these simplified models with minimal parameter requirements (such as the LUE model presented in this chapter) can be implemented across landscapes (Coops *et al.*, 1998), whereas models with detailed tree and stand data requirements cannot (Aber *et al.*, 1995).

A powerful approach to address the multiple levels of ecosystem complexity is a set of nested models linked by common logic. McMurtrie *et al.* (1992) applied the refined MAESTRO hourly canopy photosynthesis model to extract critical simplifications for the daily BIOMASS forest production model (McMurtrie and Landsberg, 1992). Properties extracted from BIOMASS were further distilled into a Generic Decomposition and Yield model, G'DAY (Comins and McMurtrie, 1993). The G'DAY model requires only 10 differential equations to represent tree and soil carbon and nitrogen dynamics over time. These three models were run sequentially to evaluate the effect of elevated atmospheric CO_2 on forest growth. Doubled CO_2 was predicted to increase leaf photosynthesis by 30–50% depending on air temperature, which might lead initially to a 27% increase in annual stand productivity. Eventually, however, changes in soil carbon and nitrogen were projected to limit N availability and result in a sustained increase in forest production of only 8%. No single model could apply to all the desired time scales. A common theoretical foundation, however, provided the means to seek and test simplifying assumptions derived from more detailed models. Once tested, the simplifying assumptions provided a basis for extrapolation into conditions for which no direct experimental data yet exist.

Forest biogeochemical models of general ecosystem processes can be applied to a wider range of vegetation than forests. All of the processes identified in Table 1.1 are shared by all terrestrial ecosystems. Consequently, a forest model such as FOREST-BGC has been modified to represent a grassland as BIOME-BGC by simply redefining certain canopy gas exchange and carbon turnover parameters (Running and Hunt, 1993). In a like manner, the CENTURY grassland model (Parton *et al.*, 1993) was redefined to simulate forest ecosystems and other broadly distributed terrestrial vegetation. The TEM model of McGuire *et al.* (1993) originated as a general multibiome biogeochemical model. TEM, CENTURY, and BIOME-BGC are useful in global biogeochemical simulations because they include common ecosystem processes in a logical framework and demand only a minimum amount of detail on stand and site characteristics (see Chapter 9). On the other hand, if provided detailed stand data, these models can also simulate forest productivity and other properties often as well as more specialized forest models.

The large variety of forest ecosystem models illustrates that selection of a model should carefully be matched with objectives. Models provide a versatile means to quantify how ecosystem processes may vary and affect forest growth and other properties. No single model, however, should be expected to apply to all situations, and, when supporting policy decisions, any specific model should be tested to assure its appropriateness to answer selected questions in an acceptably reliable manner.

VIII. SUMMARY

Carbon exchange from forest ecosystems represents a balance between uptake, storage, and losses. The upper limits on photosynthesis are set by the amount of photosynthetically active radiation absorbed by the entire canopy. The amount of radiation absorbed is a function of canopy LAI, chlorophyll pigments, and other related properties. Photosynthesis is the best understood of all ecosystem processes. Fundamental equations are available to describe the process, and models now predict rates close to those measured with eddy-flux analyses.

Autotrophic respiration is also a relatively well-understood process. By separately accounting for carbon content and synthesis cost of all components of NPP and the carbon expended in the maintenance of living cells, a forest carbon balance can be made. Such component analyses, combined with field experiments on plantations, have greatly increased our ability to interpret how changes in resource availability affect carbon allocation. A variety of conflicting theories on allocation exists, but a series of field experiments have greatly increased our understanding of the allocation process. This is fortunate because allocation is a key toward understanding competitive relations among species (Chapter 5).

Heterotrophic respiration is one of the most difficult processes to model, even with a detailed understanding of microbial biology, because microbial respiration takes place in a spatially heterogeneous environment. Carbon to nitrogen ratios or more refined indices that represent substrate quality are helpful, but these must be coupled to hydrologic models to predict seasonal variation in heterotrophic activity.

At present, we rely heavily on simple annual allocation indices related to the efficiency with which carbon is incorporated into biomass per unit of light absorbed (ε) or stem growth per unit of leaf area (growth efficiency), as general indices of stress. We still lack an adequate theoretical base for developing more general phenological models and must rely at present on empirical correlations with direct (or satellite) observations. A number of new indices look promising for improving our ability to predict seasonal allocation patterns under changing environments; in particular, the ratio actual GPP/potential GPP and the total canopy N content are candidates against which to test a variety of theories. The simplified models and indices presented in this chapter are the foundation for further extrapolations in time and space.

CHAPTER 4

Mineral Cycles

I. INTRODUCTION

The cycling of minerals through forest ecosystems is closely linked with those of water and carbon. Precipitation washes minerals from the atmosphere and deposits them on leaves and other surfaces. Water carries dissolved minerals into the soil where they are taken up by roots and transported in the transpiration stream. Water also carries minerals out of the system through erosion and by leaching. Plants respire carbon obtained through photosynthesis to convert minerals from elemental to biochemical forms, and to recycle nutrients internally from older to newer tissues. Heterotrophic and symbiotic organisms rely on carbon supplied from roots and that extracted from detritus to acquire their energy supply and nutrients. Low molecular weight acids produced as metabolic products enhance the release of additional minerals from soil and rock. Other products of microbial decomposition contribute to the accumulation of soil humus.

A general picture of the processes involved in the cycling of minerals through forest ecosystems is presented in Fig. 4.1. In this chapter we start with a discussion of plant processes by identifying the essential elements for growth and then explain how these are acquired, stored, and internally recycled within vegetation before being returned through detritus production and leachate to the forest floor. Next, we consider the relative importance of atmospheric inputs, biological fixation, and geologic weathering for supplying various nutrients to forest ecosystems. We then focus on litter processes to appreciate the slow rate at which detritus is converted to soil humus and the environmental factors that affect this important soil component. A discussion of soil processes follows to explain how soil profiles form and how shifts in the relative availability of nutrients may occur seasonally. Finally, we discuss those processes that hold minerals and nutrients within the soil profile, buffering the system against losses, and those that enhance losses through volatilization and leaching. Much larger losses of nutrients occur through erosion, harvesting, and fire, but these subjects are related to disturbance and will be deferred for discussion until Chapters 6 and 8.

Direct measurement of all the variables important in mineral cycling requires a large investment in equipment and chemical analyses, as indicated in Fig. 4.2. Above the forest canopy, instruments must be installed to collect chemistry samples of atmospheric deposition in fog, precipitation, and dust along with the normal meteorological variables. To assess the exchange of minerals from the canopy, litterfall, throughfall, and stemflow must be measured. In the soil, solution chemistry needs to be monitored through extracts acquired from lysimeters (Johnson and Lindberg, 1992) or ion-exchange resin columns (Giblin *et al.,* 1994). To complete a balance sheet, some measurements of solution losses below the rooting zone and gaseous losses from the soil and canopy are also required. Although a network of installations now exist to monitor atmospheric inputs in some countries, our ability to predict changes in mineral cycling rates through a wide variety of ecosystems requires that we incorporate as much mechanistic understanding as possible into ecosystem simulation models.

In some cases we can derive estimates of chemical fluxes and turnover rates through the analyses of isotopes of nitrogen, carbon, and strontium (Sr, a surrogate for calcium). Stable isotope analyses, and those associated with radioactive decay of carbon-14, have the added advantage of providing a historical record of changes in the rates of chemical deposition, uptake by vegetation, and export into lake sediments, against which model assumptions and predictions can be compared (see Chapter 5).

II. PLANT PROCESSES AFFECTING NUTRIENT CYCLING

A. Essential Elements

In addition to C, H, and O, all plants require certain macronutrients. Nitrogen (N) is a major constituent of proteins, nucleic acids, and chlorophyll; phosphorus (P) is most

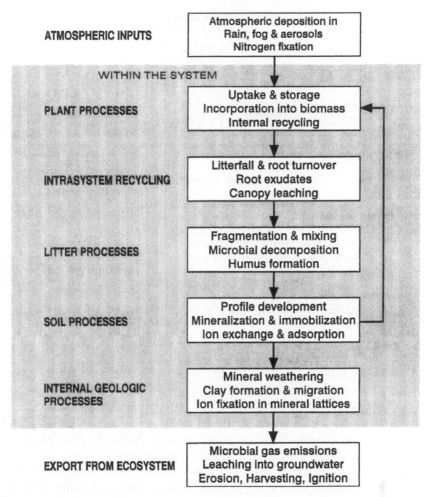

FIGURE 4.1. Minerals that cycle through a forest ecosystem have variable sources. Many are sequestered from the atmosphere; others are derived from geologic weathering of minerals. Plants modify the cycling of many elements through their selective uptake, internal redistribution, and the fraction returned annually to the forest floor. Litter on the forest floor is utilized by many soil organisms, but eventually a small fraction accumulates as soil humus. During the decomposition process, minerals are converted from organic to inorganic forms. Whether the elements are immobilized in microbial biomass, made available on soil exchange sites, adsorbed to clay surfaces, or fixed permanently into mineral lattices depends on a variety of soil and geologic processes that differ within the soil profile. Eventually some minerals are again taken up by plants and recycled through the system, while others may be lost as gases or in leachate. When disturbed, ecosystems may lose large amounts of elements through erosion, harvesting, and ignition, discussed in Chapter 6.

important as a component of the energy currency in biochemical reactions, and sulfur (S) is found in many amino acids. Specific roles are known for potassium (K) in controlling stomatal function and the charge balance across plant membranes, for calcium (Ca) as a constituent of cell walls, and for magnesium (Mg) in chlorophyll. These nutrients also stimulate the rate of various enzymatic reactions. The micronutrients iron (Fe), copper

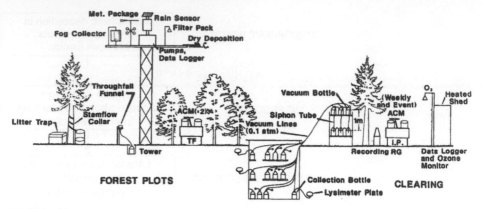

FIGURE 4.2. Schematic representation of equipment installed at 17 Integrated Forest Study (IFS) sites to evaluate the effects of atmospheric deposition and ozone levels on nutrient cycling. The IFS included a thorough analysis of air chemistry, hydrology, meteorology, and canopy, soil, and litter characteristics. Chemical changes in precipitation (with recording rain gauge, RG) and other forms of deposition were separately measured above and below the canopy, and at various depths in the soil. In addition, event and weekly accumulated measures of chemical (ACM) deposition as wet- and dryfall were collected as throughfall (TF) and precipitation intercepted by the canopy (I.P.). (Modified from *Atmospheric Deposition and Forest Nutrient Cycling,* Ecological Studies, Analysis and Synthesis, Volume 91, D. W. Johnson and S. E. Lindberg, eds., p. 3, Fig. 1.1, 1992, © 1992 by Springer-Verlag.)

(Cu), zinc (Zn), and manganese (Mn) are widely involved as coenzymes, whereas the essential roles of boron (B) and chlorine (Cl) are still poorly known. Grasses and some other plants accumulate silicon (Si) in cell walls, which provides strength and reduces tissue palatability to herbivores. Molybdenum (Mo) is essential for N metabolism in plant tissues, as well as for N fixation by symbiotic bacteria. Cobalt (Co) is essential for the microorganisms involved in N fixation. Although higher plants all require the same macronutrients, they differ in their selective accumulations, their microbial associations, and in the exudates and residues they produce. As a consequence, the amount and balance of nutrients sequestered in plant biomass affects soil fertility and forest productivity.

Nutrients must be available in appropriate forms and in sufficient amounts to meet growth requirements. Most plants exhibit rapid growth at the beginning of the growing season. To maintain a constant relative growth rate, even for a short period, requires that the flux of nutrients to growing points match the rate of expansion. Any reduction in the rate at which nutrients are supplied causes nutrient concentrations to drop in expanding tissue (Ingestad, 1982). A decrease in concentration may in turn alter carbon allocation patterns to roots, stem, and leaves (Chapter 3). To assess the nutrient balance of plants, four processes must be considered: uptake, storage, internal recycling, and return in litter. We will consider each in sequence.

B. Plant Uptake

Under field conditions, the concentration of nutrients in the soil solution is reduced during the period of exponential plant growth. Nutrients are supplied to plant root surfaces through three mechanisms: (1) the growth of roots and mycorrhizae into the soil; (2) the mass flow of ions with the movement of soil water as a result of transpiration; and (3) the diffusion of ions toward the root surface when uptake rates exceed supply (Eissenstat and Van Rees, 1994). The relative mobility and concentration of nutrients in soil solution and the rate of plant uptake determine which of these mechanisms predominates. Uptake of Ca is often the result of the interception of ions in newly exploited soil zones. Mass flow is important for Mg, SO_4^{2-}, and Fe. Plant demand for N, P, and K often exceeds delivery by mass flow, such that diffusion is the dominant process that supplies these macronutrients (Eissenstat and Van Rees, 1994).

Mycorrhizal fungi provide plants special advantages in accessing some forms of nutrients. Mycorrhizae can access films of water on soil particles not available to the smallest diameter roots because fungal hyphae have an average diameter of about 5 μm whereas the smaller diameter tree roots average 0.5 mm in diameter (Yanai *et al.*, 1995). Assuming the same construction cost, investment in a gram of mycelia provides 10^4 greater length than a gram invested in small-diameter roots and a hundredfold increase in surface area to extract water and nutrients. In temperate regions, many trees are infected by *ectotrophic mycorrhizae*. These form a hyphal sheath that surrounds the active fine roots of trees and many other plants and extend an additional network of hyphae into the soils. Most tree roots are infected by *endotrophic mycorrhizae* in which fungal hyphae penetrate the cells of the root cortex but do not form a sheath around the roots.

Mycorrhizal fungi, like other fungi, have their own extracellular enzymes, so they are able, with carbohydrates provided by plants, to extract at least some nitrogen and phosphorus directly from organic matter and convert and store these elements in forms available to plants (Martin *et al.*, 1983; Jayachandran *et al.*, 1992; Turnbull *et al.*, 1995). Mycorrhizae also produce organic acids that help aid in the breakdown of organic phosphorus. Mycorrhizae provide the greatest value to higher plants in the acquisition of nonmobile ions, particularly the least available forms of phosphorus and nitrogen (Bolan *et al.*, 1987; Northrup *et al.*, 1995). As the availability of nitrogen and other nutrients increases, mycorrhizal associations provide less competitive advantage and require twice as much carbon to maintain as nonmycorrhizal roots (Rygiewicz and Andersen, 1994).

Although some nutrients may enter the plant passively following the flow of water, many are actively transported with enzyme-mediated reactions across root membranes (Ingestad, 1982). Nonessential or toxic elements may similarly be metabolically excluded. One can easily imagine that plants from infertile habitats might possess adaptations to enhance nutrient uptake by root enzymes. However, little natural selection for enhanced enzymatic uptake among native species has occurred because diffusion limits the supply of most nutrients to root surfaces (Chapin, 1980; Chapin *et al.*, 1986). In conditions where nutrient deficiency slows plant growth, excess carbon is likely to be available to support additional investment in acquiring nutrients (Marx *et al.*, 1977). On the other hand, where nutrients are readily available, less exudates may be produced to stimulate mycorrhizal inoculation and development (Blaise and Garbaye, 1983).

C. Storage and Internal Recycling

Total nutrient demands are highly variable from species to species. Within a species, nutrient concentrations also vary depending on growth rates and the availability of nutrients.

When nutrients are added to deficient soils the growth rate of trees usually increases, often without inducing a change in foliar nutrient concentrations. When one nutrient or other factors limit growth, nutrients may be taken up in excess of immediate metabolic requirements. This results in high concentrations in foliage—a condition that is called *luxury consumption*. Differences in leaf nutrient concentrations form one basis for diagnosing deficiencies, but foliar analyses must be interpreted with care because much variation occurs with season, canopy position, and growth rate (Linder and Rook, 1984; Van den Driessche, 1984). More insight is gained when foliar composition is compared seasonally than judged from a single sampling and when account is taken for changes in carbohydrate reserves that alter the specific leaf weight over time (Linder, 1995; Fig. 4.3).

The optimum balance of nutrients is determined experimentally by varying nutrient concentrations in hydroponic solutions or sand cultures and observing the relative concentrations found in plants with maximum growth rates. At maximum growth rate the balance of nutrients in solution and those in leaf tissue are the same (Ingestad, 1979). The optimum nutrient balance differs only slightly among tree species when referenced to nitrogen content (Ingestad, 1979; Ericsson, 1994; Linder, 1995; Table 4.1). Because most of the nutrient demand in seedlings goes to foliage, these *nutrient balances* are optimum for that organ but not necessarily for other tissue such as wood or bark, which may be far higher in Ca and much lower in N and P (Table 4.2).

From data presented in Table 4.2, the foliar nutrient ratio calculated for Douglas-fir is N 100 : P 28 : K 62 : Ca 73. Calcium concentrations are nearly 10-fold higher than required (Table 4.1). Birch, with the ratio N 100 : P 5 : K 57 : Ca 51, also has acquired a surplus of Ca, but it is deficient in P. Although the concentration of N in the foliage of Douglas-fir is only one-third of that in birch, the total N content is one-third higher (102 compared to 76 $kg\,ha^{-1}$) because of the greater biomass (9180 compared to 2586 $kg\,ha^{-1}$). Assuming that not more than one-third of the conifer's needles are replaced each year, canopy requirements for N are about 35 $kg\,ha^{-1}\,year^{-1}$, less than half that required by the deciduous birch. For diagnostic purposes, departure from the optimum ratio during periods of rapid growth is indicative of nutritional problems. Beaufils (1973) developed a general index based on these principles which provides a statistical estimate of the degree to which each element is below or above the optimum value and assesses the implications on growth (Leech and Kim, 1981).

Physiological studies indicate that when plants are able to maintain good nutritional balance, a positive relation should exist between the total content of nitrogen in foliage and growth rates (Ågren and Ingestad, 1987; Cromer *et al.,* 1993a; Fig. 4.4). When an imbalance exists, growth may still increase linearly with nitrogen content, but at a reduced rate. These principles have been applied in Germany to separate nutrient-deficient spruce stands from healthy ones (Oren *et al.,* 1988a,b), as well as in New Zealand.

Total plant nutrient contents reflect long-term nutrient uptake but tell us little about seasonal nutrient circulation. Mature foliage and other organs may exhibit relatively stable ratios of nitrogen with other elements, but this balance is often accomplished through internal reallocation. Reallocation of nutrients from twigs and older foliage helps sustain rapid shoot elongation when root uptake is inadequate to meet the demand. The actual flux, however, is difficult to estimate accurately unless isotopic tracers are used (Mead and Preston, 1994). Drought and other stresses reduce growth demand and the amount of

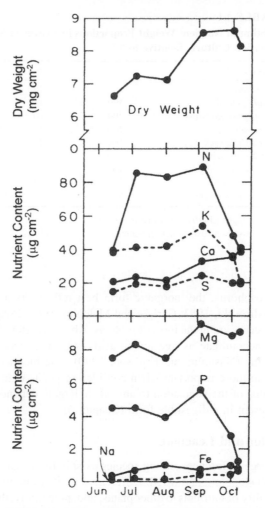

FIGURE 4.3.   Seasonal changes in the dry weight and nutrient content of the leaves of scarlet oak (*Quercus coccinea*) in the Brookhaven Forest, Long Island, New York. (From Woodwell, 1974.)

nutrients reallocated (Nambiar and Fife, 1991). Ray parenchyma cells in sapwood store mobile nutrients such as N, P, and K, which can be transferred via the sap stream to meet growth demands until conversion to heartwood occurs (Van den Driessche, 1984; Van Bell, 1990). Small-diameter roots change little in their nutrient content over time (McClaugherty *et al.,* 1982; Nambiar and Fife, 1991; Aerts *et al.,* 1992). Large-diameter roots, however, store considerable reserves, which in some deciduous species are mobilized to support leaf expansion as well as new root growth (Van den Driessche, 1984; Wendler and Millard, 1996).

The reserves of nutrients available for export from a particular tissue can often best be measured by assessment of metabolically active forms (Attiwell and Adams, 1993). For

TABLE 4.1

Optimum Nutrient Weight Proportions in Leaves of Seedlings Grown in Solution Culture, Relative to N[a]

Species	N	P	K	Ca	Mg
Alnus incana	100	16	41	10	14
Betula pendula	100	14	55	6	10
Eucalyptus globulus	100	10	37	10	9
Picea abies[b]	100	16	50	5	5
Picea sitchensis[b]	100	16	55	4	4
Pinus sylvestris[b]	100	14	45	6	6
Populus simonii	100	11	48	7	7
Salix viminalis	100	14	45	7	7
Tsuga heterophylla[b]	100	16	70	8	5

[a]From Ingestad (1979) and Ericsson (1994).
[b]Expressed on basis of nutrient contents of entire seedlings.

phosphorus, the inorganic form best reflects its availability for allocation (Chapin and Kedrowski, 1983; Carlyle and Malcolm, 1986; Polglase *et al.,* 1992). For nitrogen, most reserves are in the form of proteins. Above a certain level, however, excess nitrogen begins to accumulate as free amino acids, which reflects a significant change in physiological status (Näsholm and Ericsson, 1990). The biochemical composition of plants becomes even more important when considering plant–animal interactions because the nutritional value of the vegetation to animals is largely dependent on the extent to which nitrogen is present in a digestible form (Chapter 6).

D. Return in Litter and Leachate

Large differences among species exist in the extent to which nutrients are concentrated in foliage, bark, and wood. Differences in litter quality affect decomposition rates, the availability of nutrients to other plants, and, potentially, the development of soils under different types of vegetation (Turner and Lambert, 1988; Gower and Son, 1992). Species adapted to disturbance often grow rapidly and have nutrient-rich tissues. Their high nutrient requirements, associated with high photosynthetic rates per unit leaf area and short leaf life spans, result in nutrient accumulations in biomass and litter that might otherwise be lost after forest cutting or fires (Pastor and Bockheim, 1984). Even within the same genus, large differences exist in the quality of litter produced from some components. Thus, although *Eucalyptus grandis* and *E. sieberi* have similar Ca concentrations in foliage (0.5%), their bark contents differ from ~2.0 to <0.05% (Turner and Gessel, 1990).

Small amounts of most nutrients are leached from living plant tissues. Potassium, an element which is highly soluble and concentrated in stomatal guard cells, is particularly easily removed through leaching. In general, K > P > N > Ca in regard to leaching losses from foliage. Differences in the rates at which nutrients are leached from foliage and bark may explain variation in epiphyte loads on forest species (Schlesinger and Marks, 1977). Fine roots also lose nitrogen and potassium through exudation and leaching.

TABLE 4.2

Comparison of Nutrient Distribution in Stands of Douglas-fir and Birch with Similar Total Biomass[a]

Species	Element or biomass	Foliage		Branches		Bole bark		Bole wood		Roots		Total content
		% Pool	% Weight	% Pool	% Weight	% Pool	% Weight	% Pool	% Weight	% Pool	% Weight	(kg ha^{-1})
Douglas-fir	N	32	1.13	19	0.28	15	0.26	24	0.06	10	0.1	320
Birch	N	14	3.04	31	0.59	—	—	27	0.11	28	0.3	543
Douglas-fir	P	44	0.32	19	0.06	14	0.05	14	0.01	9	0.02	66
Birch	P	12	0.16	35	0.04	—	—	32	0.01	21	0.01	34
Douglas-fir	K	28	0.68	17	0.17	20	0.24	24	0.04	11	0.07	220
Birch	K	22	1.76	23	0.16	—	—	32	0.05	23	0.09	200
Douglas-fir	Ca	22	0.81	32	0.48	21	0.37	14	0.04	11	0.11	333
Birch	Ca	6	1.56	28	0.68	—	—	42	0.20	24	0.31	651
Douglas-fir	Biomass	4.5	—	10.7	—	9.1	—	60	—	16	—	204,000
Birch	Biomass	1.2	—	13.3	—	—	—	62	—	23	—	215,500

[a]From Van den Driessche (1984). All values calculated on the basis of dry weight.

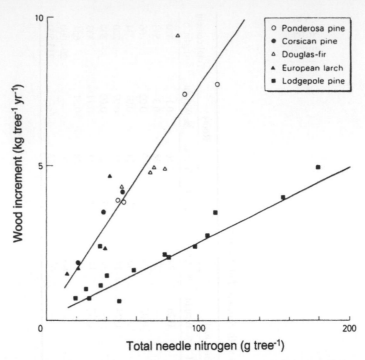

FIGURE 4.4.   Conifer trees produce wood in proportion to the peak seasonal nitrogen content of their canopy. Wood increment is greater when nitrogen is in nutritional balance with other essential elements (upper line) than when it is not (lower line). Lodgepole pine performs similarly to other coniferous species on nutritionally well-balanced soils. (From Waring, 1989; after Benecke and Nordmeyer, 1982, and Nordmeyer *et al.*, 1987.)

Return of nutrients in litterfall is the major route of recycling from vegetation to soil. Aboveground litterfall can be measured through periodic collection, weighing, and chemical analysis of twigs, leaves, fruits, and other products that fall into nets or trays positioned just above the ground surface. Annual additions of coarse woody debris can be estimated by recording the amount that falls across string lines laid out annually in a large grid under a forest canopy. Nutrient return in litterfall can vary seasonally from year to year depending on forest composition and the leaf abscission process. In a temperate deciduous forest, Gosz *et al.* (1972) found that premature abscission of leaves in summer storms resulted in a small amount of litterfall with relatively high nutrient concentrations because nutrient reabsorption had not occurred.

In plant systems, nitrogen, phosphorus, and potassium are particularly mobile whereas calcium, which is bound within cell walls, is the least mobile nutrient. Calcium concentrations often increase as a percentage of dry weight in leaf litter because carbohydrate reserves are depleted before normal leaf abscission. For an absolute basis of comparison, nutrient concentrations should be expressed on a per unit of leaf area basis to take into account seasonal changes in specific leaf mass (Fig. 4.3). Although the fraction of nutrients withdrawn (or leached) from fresh foliage before abscission varies considerably among species (Table 4.3), the concentrations of nutrients in leaf litter are closely correlated with

TABLE 4.3

Nutrient Withdrawal and Leaching from Foliage before Abscission in Conifers and Broad-leaved Species, Expressed as Percent Change of Dry Weight[a]

Species	N	P	K	Ca	Mg
Conifers					
Picea abies	−7.5 to −22	−35 to −50	−68 to −72	−3 to +15	−9 to −26
Picea excelsa	—	−2	−17	+2	+35
Pinus sylvestris	−69	−81	−80	+18	—
Broad-leaved deciduous trees					
Alnus rubra	−30	−50	−6	+22	−5
Betula alleghaniensis	−55	−42	−59	0	—
Fagus sylvatica	−41	−77	−52	−13	−14
Populus tremuloides	−78	−67	−58	−23	+14
Broad-leaved evergreens					
Eucalyptus marginata	−64	−75	−54	+28	—
Eucalyptus regnans	−50	−58	−80	−3	−40
Nothofagus truncata	−45	−63	−81	+38	+26

[a]Selected from reviews by Ericsson (1994) and Van den Driessche (1984).

those in fresh foliage for a given species and site (Miller and Miller, 1976; Hunter *et al.,* 1985). Below ground, nutrients returned annually, as fine roots die, may match or exceed the amount contributed through leaf litter (Vogt *et al.,* 1986). This may seem surprising because nutrient concentrations in fine roots are often half that in fresh foliage; concentrations remain stable, however, and turnover rates and subsequent decomposition of fine roots are high.

E. Nutrient Use Efficiency

The absolute amount of nutrients returned in leaf litter tends to increase with soil fertility and total foliage production, as quantified in a series of fertilizer experiments in pine and eucalypt plantations (Crane and Banks, 1992; Cromer *et al.,* 1993a,b). Because the annual transfer of leaf biomass in litterfall mirrors aboveground production, as described by allometric relations introduced in Chapter 3, a measure of nutrient use efficiency can be derived by calculating the ratio of leaf litter production to its nutrient content. Vitousek (1982, 1984) used aboveground litterfall patterns to compare the relative nutrient use efficiency in world forests. He found that the ratios mass : N and mass : P in litterfall rapidly declined with increasing return of N and P to the forest floor, and that temperate and boreal forests were more likely to be N-limited whereas tropical forests were generally P-limited.

Bridgham *et al.* (1995) provide a thorough review of the nutrient use efficiency concept. They developed a general model that predicts the amount of nutrients returned in litterfall. Nutrient return should increase rapidly and then approach an asymptote with measured litterfall. They tested the model in a peat bog in North Carolina where productivity and species composition varied (Fig. 4.5a). They found general agreement with model predictions; the most nutrient-demanding species produced more litter but required a disproportionate

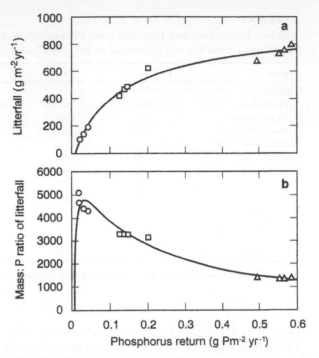

FIGURE 4.5.   (a) Litterfall and nutrient return in litterfall in three peat-land communities in North Carolina illustrate that the least demanding species (○) increased production rapidly in response to an increment of phosphorus, whereas the more demanding (□) and most productive species (△) were progressively less nutrient efficient. Solid line represents model predictions. Similar responses were demonstrated for Ca and N. (b) The mass of litterfall : nutrient return ratio relative to phosphorus indicates from the model predictions that a rapid increase in nutrient use efficiency occurs above some minimum threshold and is followed by an exponential decrease with increasing litter return of phosphorus or other nutrients. (After Bridgham *et al., American Naturalist,* published by the University of Chicago Press. © 1995 by The University of Chicago.)

amount of nutrients compared with less demanding, less productive species. From values derived from the curves in Fig. 4.5a, they modeled and replotted the conventional relationship between nutrient return and the mass : nutrient ratio in litterfall (Fig. 4.5b). The analysis differs from earlier ones in that a minimum return of nutrients in litterfall (or uptake from the soil) is specified below which no production is predicted.

The nutrient use efficiency concept can be expanded to compare species differences in total aboveground production on equivalent sites. In Table 4.4, production efficiencies are compared for similar aged stands of *Pinus* and *Eucalyptus* growing on phosphorus-poor soils in Australia (Turner and Lambert, 1983; Baker and Attiwill, 1985). The two species have fairly similar production efficiencies in regard to gross annual demand for N and P (Efficiency I in Table 4.4), but the native eucalyptus extracts only about one-quarter of the P from soils demanded by pine for a comparable rate of production (Efficiency II). The native eucalypt requires only about half the P content in a kilogram of wood compared with the introduced pine. Lower leaf turnover and higher reabsorption of nutrients before leaf abscission are additional means by which species obtain high nutrient use efficiency.

TABLE 4.4

Comparison of *Pinus radiata* with *Eucalyptus grandis* for Several Measures
of Production and Nutrient Efficiency[a]

Property	Component	*P. radiata*	*E. grandis*
Age, years		22	27
Mass, $kg\,m^{-2}$		27	39
Mean annual accumulation, $kg\,m^{-2}\,year^{-1}$	Total aboveground biomass	1.4	1.5
	Stem wood	1.1	1.2
Current annual production (C), $kg\,m^{-2}\,year^{-1}$	Total aboveground biomass	1.7	2.4
	Stem wood	1.0	1.5
Efficiency I: (C)/gross annual demand for nutrients, $kg\,g^{-1}$	N	0.28	0.24
	P	2.9	3.9
Efficiency II: (C)/annual uptake from soil, $kg\,g^{-1}$	N	1.3	1.8
	P	9.0	35
Efficiency III: Nutrient, cost of stem wood $g\,kg^{-1}$	N	0.57	0.85
	P	0.08	0.04

[a]From *Forest Ecology and Management,* Volume 6, J. Turner and M. J. Lambert, "Nutrient cycling within a 27-year-old *Eucalyptus grandis* plantation in New South Wales," pp. 155–168, 1983, and *Forest Ecology and Management,* Volume 13, T. G. Baker and P. M. Attiwill, "Above-ground nutrient distribution and cycling in *Pinus radiata* D. Don and *Eucalyptus obliqua* L'Herit. forests in southeastern Australia," pp. 41–52, 1985, with kind permission from Elsevier Science–NL, Sara Burgerhartstraat 25, 1055 KV Amsterdam, The Netherlands.

Evergreen species which display these attributes also tend to lose fewer nutrients through leaching. As a result, conifers are generally more efficient in nutrient use than deciduous species (Cole and Rapp, 1981). The efficient use of nutrients by trees, however, is not necessarily a desirable ecosystem property, because heterotrophic organisms are inhibited when carbon to nutrient ratios are high and, as a result, the release of nutrients from litter may be slowed to further limit productivity (Shaver and Melillo, 1984).

In summary, uptake of nutrients from the soil rarely meets demand during periods of rapid growth. In evergreen forests, reabsorption of nutrients from older and senescent foliage becomes increasingly important for maintaining growth as forests age and accumulate leaf area. Although the optimal balance of nutrients may not be maintained, aboveground growth is generally a linear function of the total N content in foliage. This relationship offers a simplification for the development of forest growth models designed to scale across landscapes (Chapter 8). The mass and nutrient content in leaf litterfall are integrating measures of ecosystem function that link not only to aboveground production but, as we shall see, to decomposition of organic matter and release of nutrients into the soil.

III. SOURCES OF NUTRIENTS

A. Atmospheric Deposition

We discussed earlier that carbon from atmospheric CO_2 is sequestered into forest ecosystems through photosynthesis (Chapter 2). In addition, nearly all of the nitrogen and sulfur pools in forest ecosystems is derived from the atmosphere (Placet *et al.,* 1990). The major transfer of nitrogen is through biological fixation, although lightning also produces nitrogen oxides that are transferred to the ground. The production of inorganic fertilizer and atmospheric transfers from agricultural fields, feedlots, and fossil fuel combustion now match or exceed N acquired from the atmosphere through natural processes (Vitousek and Matson, 1993). Volcanic eruptions release SO_2 (Schlesinger, 1991, 1997), but these contributions are relatively infrequent and small compared to those derived from current fossil fuel combustion (Johnson and Lindberg, 1992). Although Ca, Mg, and K are derived from mineral weathering, dust extracted from the atmosphere in precipitation and dryfall is a major source of these elements as well (Hedin and Likens, 1996).

Rainfall constituents are also derived from the ocean. As winds blow across the ocean surface, sea salt rich in Na, Mg, Cl, and S is injected into the atmosphere and carried over great distances as aerosols smaller than 1 μm. The ratio of Na to other ions in precipitation may provide an indication of the direction of storm fronts. For example, the calcium to sodium ratio in seawater is 0.04. For rainfall with Ca:Na ratios close to this value, one would deduce that most of the Ca was of marine origin (Schlesinger *et al.,* 1982). In the eastern continental United States, the typical Ca:Na ratio is 1.3 (Likens *et al.,* 1977) because the airflow that brings precipitation to this region carries calcium derived from soil dust and other inland sources.

The deposition of nutrients in precipitation is called *wetfall*. Clouds and fog may also provide an additional source of wet deposition. *Dryfall* is the result of gravitational deposition of particles and the absorption of gases during periods without rain. With the design of collectors that are electronically sensitive to rain, partitioning of nutrient deposition between wet and dry sources can be estimated (Fig. 4.2). Dryfall is the major atmospheric source in ecosystems with limited precipitation. Even in the humid temperate forests of the eastern United States, however, dryfall makes up an important fraction of the total annual atmospheric deposition of N, P, K, Ca, and Na (Lindberg *et al.,* 1986; Driscoll *et al.,* 1989). In mountain landscapes where forests are often immersed in clouds, sulfur deposited through fog condensation may represent as much as 50% of the total atmospheric deposition (Johnson and Lindberg, 1992).

Accurate estimates of atmospheric inputs are difficult to obtain over forests, in part because their large leaf area effectively captures fog droplets and aerosols from the airstream. Some reactive gases such as SO_2 and NO_2 are absorbed directly from the atmosphere by plant leaves and soil. Nearly one-third of the atmospheric S deposited in northern hardwood forests was estimated to be dependent on the presence of vegetation (Eaton *et al.,* 1978). Coniferous evergreen forests are two to three times as efficient as broadleaf deciduous forests in filtering wet- and dryfall from the atmosphere (Johnson and Lindberg, 1992). These differences are associated with tree leaf area, the seasonal duration of display, and aerodynamic properties of the canopy discussed in Chapter 2. A number of models predict canopy capture of aerosols in wet- and dryfall (Lovett *et al.,* 1982; Baldocchi, 1988; Lovett, 1994). Most of these models assume one-dimensional vertical transport, flat terrain, and homogeneous canopies. These assumptions make them less applicable to complex terrain and patchy vegetation (Lovett, 1994). More realistic models, however, are difficult to parameterize and cannot as yet be validated because of errors in measuring the various sources of deposition (Lovett, 1994).

B. Nitrogen Fixation

Over 78% of the atmosphere is composed of nitrogen as dinitrogen (N_2). Various gases, including oxygen, argon, carbon dioxide, and trace constituents, comprise the remainder. Gaseous N_2 is inert as far as biological processes are concerned. Thus, plants are bathed in a "sea of nitrogen" that they cannot use (Delwiche, 1970). Several types of bacteria and blue-green algae (cyanobacteria) possess the enzyme nitrogenase that converts atmospheric N_2 to NH_3 (ammonia), which is transformed into ammonium (NH_4^+), a form readily available to biota. The N-fixing organisms exist as free-living forms (*asymbiotic*), either on the surface or in soil (Woldendorp and Laanbroek, 1989), and in *symbiotic* association with fungi (*lichens*) and roots of some higher plants.

Symbiotic N fixation is an energy-consuming reaction, with a carbon cost between 4 and 10 g per g N fixed, averaging ~6 g C/g N (Vance and Heichel, 1991). Nitrogen fixation is highly dependent on the production and temporary storage of current photosynthate; thus, activity completely stops within a few days after the cessation of photosynthesis (Ekblak *et al.,* 1994). Because of the sensitivity of the photosynthetic process to light, symbiotic N fixation by shrubs and herbs is progressively limited as the forest canopy closes (Silvester *et al.,* 1979). Nitrogen fixation is also limited if N is freely available or when P, Mo, Co, and other trace nutrients required for the enzymatic reaction are in short supply. Linkage between primary production and soil fertility is thus a prerequisite for the development of any model of symbiotic nitrogen fixation.

Asymbiotic N fixation is generally higher in forest soils rich in organic matter (Granhall, 1981) and also occurs in coarse woody debris. Typical rates are 0.1 to 4 kg N ha^{-1} year^{-1} in boreal forests, 0.1 to 5 kg N ha^{-1} year^{-1} in temperate coniferous forests, 0.1 to 6 kg N ha^{-1} year^{-1} in temperate deciduous forests, and 2 to 20 kg N ha^{-1} year^{-1} in tropical forests (Boring *et al.,* 1988). Where rates of N accumulation are higher in ecosystems lacking symbiotic N fixers (Bormann *et al.,* 1993), asymbiotic N fixers may be obtaining energy from root exudates directly (Attiwill and Adams, 1993). This type of "associate-fixation" is more efficient than when bacteria are free-living, but it is probably less than 10% as efficient as the symbiotic arrangement (Marschner, 1995).

Biological nitrogen fixation is a major source of N in many types of ecosystems. In postfire development of Douglas-fir forests of Oregon, Youngberg and Wollum (1976) found that N fixation by the nodulated colonizing shrub *Ceanothus velutinus* contributed up to 100 kg ha^{-1} year^{-1} of N for a number of decades on some sites. Similar rates of fixation have been reported for pure plantations of *Casuarina* in Puerto Rico (Parrotta *et al.,* 1994) and even higher rates for *Alnus* (Binkley *et al.,* 1994). For comparison, rainfall adds 1–2 kg ha^{-1} year^{-1} in unpolluted forested regions (Hedin *et al.,* 1995) but more than 70 kg ha^{-1} year^{-1} in some parts of heavily industrialized central Europe (Schulze, 1989).

Various methods have been developed to assess N fixation rates. Nitrogenase activity can be measured with the acetylene-reduction technique, which is based on the observation that the enzyme also converts acetylene to ethylene under experimental conditions. Plants or root nodules are placed in small chambers and the conversion of injected acetylene to ethylene is measured over a known time using gas chromatography. The conversion of acetylene in moles, however, is not always equivalent to the potential rate of fixation of

N_2 because the enzyme has different affinities for these substrates under varying conditions. As an absolute measure of N fixation, investigators have injected air enriched in $^{15}N_2$ into chambers and measured the increase in organic ^{15}N in test plants or soil through time. The direct approach to assessing N fixation is largely confined to small-scale studies of single plants or nodules (McNeill *et al.,* 1994).

As an alternative to direct measurements of N fixation, small amounts of ^{15}N-enriched fertilizer can be applied to the soil and the isotopic composition of leaves of a potential symbiotic N-fixing species compared against a nonfixing species. Nonfixing species do not have direct access to the atmospheric N supply, and thus the resulting $^{15}N/^{14}N$ ratio of tissue N will differ between the two species in proportion to the nitrogen fixed, with the absolute amount related to total N incorporated annually in biomass (Parrotta *et al.,* 1994). At times it is difficult to find a suitable nonfixing species with comparable rooting depth and affinity for a particular form of nitrogen; in such cases the $\delta^{15}N$ of extractable soil N may serve as a reference (Handley and Raven, 1992; Handley *et al.,* 1994).

The abundance of ^{15}N relative to ^{14}N in plant tissue or xylem sap may also serve to document the relative importance of N fixation to the available nitrogen pool (Shearer and Kohl, 1986). Dinitrogen fixed from the atmosphere has a similar isotopic composition to its source, which is the standard of reference ($\delta^{15}N = 0.0 \pm 2.0\%o$). As nitrogen is mineralized from organic matter in soils, the heavier ^{15}N accumulates as the lighter ^{14}N is volatilized or leached. As a result, both soil nitrogen content and the fraction of ^{14}N decrease with depth, and they can be predicted on the basis of Rayleigh distillation kinetics (Mariotti *et al.,* 1981; Schulze *et al.,* 1994b). Evans and Ehleringer (1993) applied this approach in a semidesert ecosystem where N-fixing lichens and related forms cover undisturbed soil surfaces. Soils collected beneath trees and shrubs and between canopies showed lower N concentrations and proportionally higher $\delta^{15}N$ values, which implied from the model that essentially all of the N present in the system was derived through fixation rather than from other sources.

In most ecosystems, our understanding of the importance of N fixation is still incomplete. There are few studies of N fixation among the leguminous species of tropical forests. Collectively, these might not only provide an important source of N in these ecosystems, but also account for the abundant N circulation in many tropical forests. In other systems, N fixation is highly variable seasonally and also changes with stand development. Symbiotic N fixation, which depends directly on carbon assimilation by the host plant, is limited by the same set of factors that control photosynthesis but, in addition, is significantly reduced when N is too readily available and when other critical nutrients are in short supply (Liu and Deng, 1991).

C. Weathering

Except for nitrogen, most nutrients have their origin from the mineral composition of rocks. The inherent fertility of forest soils, their texture, and their buffering capacity to acid precipitation are all related to the type and age of the parent material from which the soil is derived. Nearly all rocks contain primary minerals that were formed under conditions of greater temperature and pressure than found at Earth's surface. On uplift and exposure, the rocks undergo *weathering,* a general term that encompasses a variety of geological processes by which parent rocks are broken down. Mechanical weathering is fragmentation of materials without chemical change. Chemical weathering occurs when parent rock materials react with water and mineral constituents are released as dissolved ions. Chemical weathering also includes the formation of new, secondary minerals that are more stable under the physical conditions at Earth's surface. Weathering is closely involved with the formation of soils, because the bulk of the physical structure of soils is composed of fragmented and weathered rock materials.

Chemical weathering is the main process by which nutrients are released from rock. From 80 to 100% of the inputs of Ca, Mg, K, and P is derived from weathering in many forest ecosystems (but see review on dust inputs by Hedin and Likens, 1996). Only in unusual conditions does the parent rock contain measurable amounts of N (Dahlgren, 1994). The release of ions occurs most rapidly under warm temperatures and with large amounts of precipitation. Thus, on a worldwide basis, climate is the major determinant of weathering rates. Chemical weathering is more rapid in tropical forests than in temperate or boreal forests and, likewise, progresses more rapidly in forests than in grasslands or deserts.

It is traditional to think of weathering of the underlying bedrock as the source of nutrients and soil development in forest ecosystems. Over large areas, however, soil profiles are developed from materials that have been transported and not weathered in place. For example, many forests in the northeastern United States occur on glacial deposits. In other areas, volcanic ash, wind-borne material (loess), or stream-water alluvium have resulted in deep and fertile soils. In all these cases, weathering may largely be from minerals in the deposited horizons and not from parent bedrock.

Rates of chemical weathering and nutrient release are dependent on rock type. Metamorphic rocks (e.g., gneiss, schist, quartzite) and many igneous rocks (e.g., granite, gabbro) formed deep in the Earth consist of primary silicate minerals that are crystalline in structure. Quartz is the simplest silicate mineral, consisting of only silicon and oxygen in a tetrahedral crystal. Quartz is extremely resistant to chemical weathering. Other primary minerals are silicates in which various *cation* (positively charged ion) components (e.g., Al^{3+}, Ca^{2+}, Na^+, K^+, and Mg^{2+}) and ions of trace metals (Fe, Mn, etc.) are substituted in the crystal lattice. These minerals include feldspars, mica, olivines, and hornblende. The elemental substitutions make these primary minerals less stable and more prone to chemical attack. In rocks of mixed composition, such as granite, chemical weathering may be concentrated on the relatively labile minerals, while others such as quartz are left unchanged. In the process of chemical weathering, primary minerals are altered to more stable forms, ions are released, and secondary minerals are formed.

Sedimentary rocks underlie 75% of the land area and include shales, sandstones, and limestones. These rocks were formed relatively close to Earth's surface, usually as sediment under water. Shales and sandstones may consist largely of secondary minerals eroded during earlier weathering epochs, and they are often rather weakly cemented rocks. In many instances, sedimentary rocks are prone to erosion, but the component minerals may be stable end products and not subject to further chemical attack. Rapid erosion, thus, does not necessarily imply abundant nutrient availability in the soil profile.

The most important chemical weathering process in forest ecosystems is that of *carbonation*. The reaction is driven by the formation of carbonic acid, H_2CO_3, in the soil solution:

$$H_2O + CO_2 \rightleftharpoons H^+ + HCO_3^- \rightleftharpoons H_2CO_3. \qquad (4.1a)$$

Because plant roots and decomposing soil organic matter release CO_2 to the soil air, the concentration of H_2CO_3 is often much greater than that in equilibrium with atmospheric CO_2 at 0.035% concentration. Carbonation reactions are dominant in the weathering of limestone and yield soluble Ca:

$$CaCO_3 + H^+ + HCO_3^- \Rightarrow Ca^{2+} + 2HCO_3^-. \qquad (4.1b)$$

Silicate minerals also weather by carbonation, for example,

$$K\text{-}Al\text{-}silicate + H^+ + HCO_3^- \Rightarrow K^+ + H_4SiO_4 + Al\text{-}silicate + HCO_3^-. \qquad (4.1c)$$

This simplified reaction represents the weathering of potassium feldspar. During the process, a primary mineral is converted to a secondary mineral by removal of K and soluble silica. Except in unusual circumstances, the dominant *anion* (negatively charged ion) in seepage or runoff water is bicarbonate ion, HCO_3^-.

Other abiotic weathering reactions include simple dissolution of minerals, oxidations, and hydrolysis. Biotic weathering may involve organic acids released by plant roots that can weather biotite mica, a primary mineral that contains K. Lichens are important in rock weathering through the release of phenolic acids (Ascaso *et al.,* 1982). Fungi release oxalic acid and other organic acids which weather soil minerals and affect the concentration of phosphorus and other nutrients in solution (Comerford and Skinner, 1989).

Weathering of silicate rocks yields positively charged nutrient cations (e.g., K^+, Ca^{2+}, Mg^{2+}, and Fe^{3+}) in varying proportions depending on the initial rock composition and the environmental conditions for weathering reactions. Weathering of carbonates is more rapid and yields soils and stream waters that are dominated by Ca and Mg. In these neutral to alkaline soils, the availability of other soil nutrients, particularly P, may be limited. Weathering of ultrabasic igneous rock (peridotite) or its metamorphic derivative (serpentine) yields soils with unusually high concentrations of Mg, Fe, and trace minerals relative to Ca. On soils derived from such rocks, forest growth is stunted or impossible (Proctor and Woodell, 1975).

Many types of secondary minerals can form in soils through weathering processes. Temperate forest soils are dominated by layered silicate or clay minerals. These exist as small particles (<0.002 mm) that provide a great deal of structural and chemical properties of soil. In general, two types of layers characterize the crystalline structure of these minerals: Si layers and layers dominated by Al, Fe, and Mg. These layers are held together by oxygen atoms. Clay minerals and the size of their crystal units are recognized by the number, order, and ratio of these layers. Moderately weathered soils are often dominated by secondary minerals, such as montmorillonite, which have a 2:1 ratio of Si- to Al-dominated layers. More strongly weathered soils, such as in the southeastern United States, are dominated by kaolinite clays with a 1:1 ratio of layers. Some of the secondary minerals formed in soils can incorporate other elements such as K and P into their crystalline structure. When this happens, those nutrients are locked in a form unavailable to plants

(Walker and Syers, 1976; April and Newton, 1992). In tropical rain forests, high temperatures and rainfall cause most of the silica to be leached from soils. The highly weathered soils of tropical forests are dominated by crystalline oxide minerals and hydrous oxides of Fe and Al. Removal of Ca, K, and other basic elements is nearly complete as a result of long periods of intense leaching through the soil profile. At the other extreme, recently glaciated lands have soils enriched with powdered fragments of freshly exposed mineral surfaces, which provides a buffer against acid precipitation often lacking in nonglaciated areas (Johnson and Lindberg, 1992).

Weathering rates cannot be measured directly. Within the soil profile, however, much can be inferred from changes in the solution chemistry collected with porous-plate lysimeters installed at various depths (as shown in Fig. 4.2). Table 4.5 presents lysimeter analyses that demonstrate changes in the weathering processes throughout a soil profile developed under a subalpine forest in the central Cascades of Washington. By the time rain enters the soil it has become acidic as a result of organic acids leached from vegetation and decaying organic matter in the forest floor. This solution mobilizes Fe, Al, and other cations from weathering at 15 cm. These are largely precipitated in a lower soil layer (30 cm), but carbonation weathering commences again at 60 cm, resulting in losses of cations and silica to groundwater.

In undisturbed ecosystems, gross weathering rates from the soil and underlying material are often estimated from the difference between measured atmospheric additions and losses of the elements in stream water (Likens et al., 1977). The approach overestimates weathering rates in the portion of the soil tapped by roots and ignores exchange processes in the riparian zone (Waring and Schlesinger, 1985). Alternatively, the natural abundance of strontium ($^{87}Sr/^{86}Sr$) has been used to quantify rates of weathering associated with different soil horizons and from bedrock (Miller et al., 1993; Bailey et al., 1996). Strontium is an element that reacts chemically similarly to Ca and, by inference, other cations.

TABLE 4.5

Chemical Composition of Precipitation, Soil Solution, and Groundwater in a 175-year-old *Abies amabilis* **Stand in Cascade Range of Northern Washington**[a]

Solution	pH	Total cations, meq liter^{-1}	Soluble ions, mg liter^{-1}			Total, mg liter^{-1}	
			Fe	Si	Al	N	P
Precipitation							
Above canopy	5.8	0.03	<0.01	0.09	0.03	0.60	0.01
Below canopy	5.0	0.10	0.02	0.09	0.06	0.40	0.05
Forest floor	4.7	0.14	0.14	3.50	0.79	0.54	0.04
Soil solution							
15 cm	4.6	0.12	0.12	3.55	0.50	0.41	0.02
30 cm	5.0	0.08	0.08	3.87	0.27	0.20	0.02
60 cm	5.6	0.25	0.25	2.90	0.58	0.37	0.03
Groundwater	6.2	0.26	0.26	4.29	0.02	0.14	0.01

[a]From Schlesinger (1991). Data from Ugolini et al. (1977b), *Soil Science* **124**, 291–302. Copyright (1977) Williams & Wilkins.

Aerosols from the atmosphere have $\delta^{87}Sr$ values in coastal regions that mirror that of oceans, and the $\delta^{87}Sr$ values change predictably as continental sources of dust are added. Rocks develop characteristic $\delta^{87}Sr$ values based on initial rubidium/strontium ratios and rock age. Strontium-87 is produced by the radioactive decay of ^{87}Rb (half-life, 49 Gyear), whereas ^{86}Sr is stable and does not decay radiogenically. By determining the Sr isotopic composition of vegetation, various soil horizons, underlying till and bedrock, as well as stream water flowing from a spruce–fir forest in northeastern United States, Miller *et al.* (1993) estimated the weathering contributions, cation reservoirs, and atmospheric contributions with a simple isotopic mixture model which specified the $\delta^{87}Sr$ of the two extremes: atmospheric inputs and underlying glacial till (Fig. 4.6).

Most models of weathering assume a long-term stable climate and lack responses to changes in atmospheric inputs. With concerns about acid rain deposition on forests, however, more refined models are being developed and applied to interpret the contribution

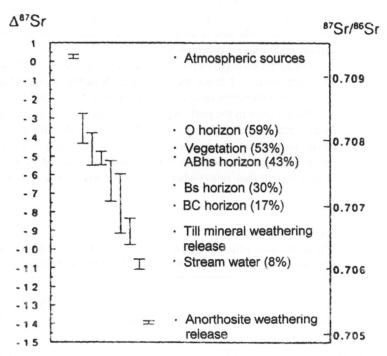

FIGURE 4.6.   Weighted averages (dots) and one standard deviation (range bars) of the $^{87}Sr/^{86}Sr$ ratio of cation reservoirs in spruce–fir forests in New Hampshire. Shown in parentheses are percentages of Sr derived from atmospheric sources, the local comparative reference ($\delta^{87}Sr = 0‰$ versus glacial till at $-9‰$). The $\delta^{87}Sr$ of stream water is lower than that released from till, which indicates weathering from anorthosite bedrock. Mineral weathering, according to this analysis, contributed about 20% of the Sr incorporated into vegetation annually. The cation-exchange reactions contributed an average of 30% and new mineral weathering an average of 70% of the Sr exported in stream water. The pool of exchangeable cations in the organic soil horizon represents <1% of net annual losses, and the soil is accumulating cations from the atmospheric inputs equivalent to about $0.5\,kmol\,ha^{-1}$ $year^{-1}$. (From Miller *et al.*, 1993. Reprinted with permission from *Nature*. Copyright 1993 Macmillan Magazines Limited.)

of mineral weathering toward buffering cation losses on an annual basis (Liu *et al.,* 1991; April and Newton, 1992; Arthur and Fahey, 1993).

IV. SOIL AND LITTER PROCESSES

The soil in forest ecosystems usually consists of a number of layers, or horizons, that collectively comprise the complete soil profile. Recognition of the processes that occur in these horizons is an essential part of understanding nutrient cycling in forest ecosystems. A characteristic property of forest soils is a nearly permanent cover of leaf litter and woody debris. Beneath this surface organic layer, distinct soil horizons usually develop with different chemical, physical, and biological properties. Humans have altered the development of soil horizons by changing the natural sequence of disturbance, the kinds of plants, animals, and microbes present, and the nutrient capital in forest soils. The basic processes, however, remain the same by which nutrients are made available in the soil, taken up by plants, and eventually returned in organic residues.

In this section we first describe patterns of soil development in temperate-zone forests and then briefly contrast these to boreal and tropical regions. We then consider the decomposition process by which detritus is converted to humus and inorganic forms of minerals are released. Once nutrients are in solution they may be transformed, held on soil exchange sites, immobilized in plant and microbial biomass, or adsorbed and made unavailable to plants. All of these processes interact in complicated but, as our knowledge grows, increasingly predictable ways.

A. Soil Profile Development

The forest floor is often easy to separate from the underlying layers of mineral soil, but these two major categories may be further subdivided. The forest floor often consists of L, F, and H layers (Fig. 4.7). The L layer consists of fresh, undecomposed litter. The F layer lies immediately below the L layer and consists of fragmented organic materials in a stage of partial decomposition. This layer is dominated by organic materials in cellular form, and fungi and bacteria are common. Beneath the F layer lies the H or humus layer, primarily consisting of amorphous, resistant products of decomposition and with lower proportions of organic matter in cellular form. The lower portion of the H layer often shows an increasing proportion of inorganic mineral soil constituents, but organic components still dominate.

The upper mineral soil is designated as the A horizon. It may vary in thickness from several centimeters to 1 m. The A horizon is recognized as a zone of removal or *eluvial* processes. Soil water percolating through the forest floor contains organic acids derived from the humic materials. These waters remove iron, aluminum, and other cations by weathering of the mineral components of the A horizon. Iron and aluminum are complexed with the water-soluble fluvic acids in the soil solution and percolate to the lower horizons. Clay minerals are also removed from the A horizon. When the removal is extreme, a whitish A_2 horizon is easily recognized and may consist of nearly pure silica, which is relatively insoluble in acid conditions (Pedro *et al.,* 1978).

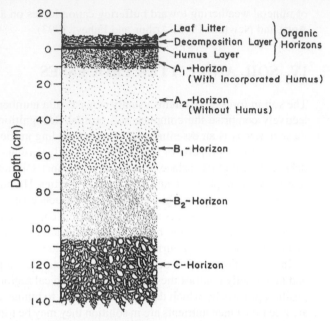

FIGURE 4.7.   Diagram of the profile of a podzolic soil under an old stand of shortleaf pine (*Pinus echinata*) in North Carolina showing the three organic horizons over the A, B, and C horizons of the mineral soil. (From *Plants and the Ecosystem,* Third Edition, by W. D. Billings. © 1978 by Wadsworth Publishing Company, Inc. Reprinted by permission of Wadsworth Publishing Company, Belmont, California 94002.)

Substances leached from the A horizons are deposited in the underlying B horizon. This is defined as the zone of deposition or *illuvial* horizon. Iron and aluminum precipitate in the B horizon, and secondary clay minerals are formed (Ugolini *et al.,* 1977a). Soluble humic materials are complexed with the clay from the A horizon, and their deposition in the B horizon is known as *podzolization.* Below the B horizon, the C horizon consists of coarsely fragmented soil material with little organic content. When the soil has developed from local materials, the C horizon shows mineralogical similarity to the underlying parent rock.

The distribution of forest floor and soil groups forms a continuous gradient over broad geographic regions, in response to parent materials, topography, climate, vegetation, and time (Jenny, 1980). In the more extreme latitudes, decomposition in the forest floor is very slow as a result of cold and waterlogged conditions. Thick forest floor accumulations may comprise the entire rooting zone. In these areas the underlying mineral soil may be effectively isolated from the nutrient cycling processes in the forest ecosystem, particularly where peat bogs and permafrost occur. In much of the boreal region, the mineral soil is derived from glacial deposits. Weathering of primary minerals is slow in cold conditions, and there is little leaching of the soil profile when it is frozen much of the year; thus, the A and B horizons are poorly differentiated.

Within the general description of temperate forest soils, ecologists have long differentiated between mor and mull forest floors. In broad geographic terms, mors develop in cooler climates, often characterized by coniferous vegetation. Decomposition in the forest floor is slow and incomplete, resulting in a thick organic layer. Moreover, the litter of coniferous species contains high concentrations of phenolic substances and lignin that yield acid decomposition residues. As a result, the soil solution often has a pH as low as 4.0. In these conditions, fungi predominate over bacteria. Earthworm populations are low in mor forest floors, which results in little fragmentation and mixing with the underlying soil (Phillipson *et al.,* 1978).

Mull forest floors are typically found under deciduous forests in warm temperate climates. Most of the characteristics of mulls are in contrast to those of mors. Decomposition is more rapid, residues are less acidic, and earthworms are more abundant. Bacteria play a greater role in decomposition processes in mull forest floor, and the pH is higher. Fragmentation and mixing often make differentiation of the forest floor difficult and obscure sharp boundaries between the mineral horizons. Under pH 5.0–7.0, which are typical of these soils, Si is relatively soluble. Thus Si, Fe, and Al are removed in relatively equal proportions from the A horizon minerals, and there is no sharply defined A_2 horizon (Pedro *et al.,* 1978).

As one travels from the temperate zone to the lowland tropics, decomposition is progressively more rapid and complete. There is little forest floor mass, and there are lower concentrations of humic acids in the percolating soil waters. In this environment, Si is more soluble than Fe and Al. Long periods of weathering under high rainfall have removed Si and cations from the entire soil profile. Resistant soil materials are hydroxides of Fe and Al. Aluminum hydroxides may produce acidity (H^+) in soil waters, depending on the state of hydration, and P forms highly stable complexes with Fe and is unavailable. Thus, over large portions of the lowland tropical rain forest region, soils are infertile and show P deficiency (Sanchez *et al.,* 1982a).

Soils in tropical forests may be many meters in depth, because in many areas they have developed over millions of years without disturbances such as glaciation. In the absence of clear zones of eluviation and illuviation, distinction of the A and B horizons is difficult. The profiles are also well mixed by earthworms. The absence of a thick forest floor does not imply that these soils are low in soil organic content. Throughout the lower profile, light-colored fluvic acids are complexed with the mineral soil materials and represent a significant storage of soil C in organic form (Sanchez *et al.,* 1982b).

A useful chemical index of the regime of soil formation and the degree of weathering is seen in the ratio of Si to sesquioxides (Fe and Al) in the soil profile (Table 4.6). In boreal forest soils, Si is relatively immobile and Fe and Al are removed, which results in high values for this ratio in the A horizon. The accumulation of the secondary mineral montmorillonite in the moderately weathered soils of the glaciated portions of the United States yields Si : sesquioxide ratios of 2 to 4, as a result of the ratio of Si to Al in the crystal lattice. Silicon : sesquioxide ratios are lower in more highly weathered soils. In the southeastern United States, lower ratios characterize soils in which kaolinite has accumulated as a secondary mineral. Tropical soils show very low values for this ratio in all horizons: they are dominated by iron and aluminum hydroxides, and in highly weathered conditions relatively little Si remains.

Differences in soil profile development locally and regionally have long served as a convenient way to classify and map local and regional variation. Today, models that

TABLE 4.6

Silicon : Sesquioxide ($Al_2O_3 + Fe_2O_3$) Ratios for A and B Horizons of Some Soils in Different Climatic Regions[a]

| Region | Number of sites | Mean Si : sesquioxide ratio | | Reference |
		A-horizon	B-horizon	
Boreal	1	9.3	6.7	Leahey (1947)
Cool temperate	4	4.1	2.3	MacKney (1961)
Warm temperate	6	3.8	3.2	Tan and Troth (1982)
Tropical	5	1.5	1.6	Tan and Troth (1982)

[a]Note that the removal of Al and Fe results in high values in boreal and cool temperate soils, especially in the A horizon. Lower values characterize tropical soils, and there is little differentiation between horizons as a result of the removal of Si from the entire profile in long periods of weathering.

attempt to expand estimates of NPP, trace gas emissions, and other ecosystem processes depend heavily on landscape delineation of major soil groups (Melillo *et al.,* 1993; Potter *et al.,* 1996).

B. Fragmentation and Mixing

In many forests, leaf litter is fragmented and mixed into the lower layers of the soil within 1 year of abscission. The physical breakage of logs, after fungal attack has weakened the original cellular structure of wood, may largely determine the rate of their disappearance from the forest floor. The physical reduction and mixing of litter is largely carried out by an abundance of soil animals, ranging from microscopic nematodes to large earthworms (Hole, 1981). The result is an increased surface area for microbial attack and the movement of litter to more constant conditions of temperature and moisture in the lower profile.

There is an enormous diversity of soil invertebrates involved in the fragmentation or comminution of litter (Fig. 4.8). The functional roles of each are poorly known, partly because of the interactions between these organisms. For example, destruction of wood cellulose by termites is due to the action of symbiotic protozoa and bacteria contained in their lower digestive tracts. In the forest floor, some litter microarthropods are primarily predators on species that feed directly on fresh litter. Others feed on fungal hyphae that invade the forest floor layers. Dead soil animals and their fecal materials may be eaten by other animals, which, in turn, may be consumed by predators. Such interactions are an important aspect of the animal community in soil detritus in contrast to the more linear trophic relationships among the animals that feed above ground. The result is a rapid reduction in the size of litter material and a corresponding increase in area for microbial attack (Waring and Schlesinger, 1985).

Nematodes are among the smallest soil animals. Their populations range from 1,700,000 to 6,300,000 m^{+2} among various forest types, and nematodes tend to be particularly

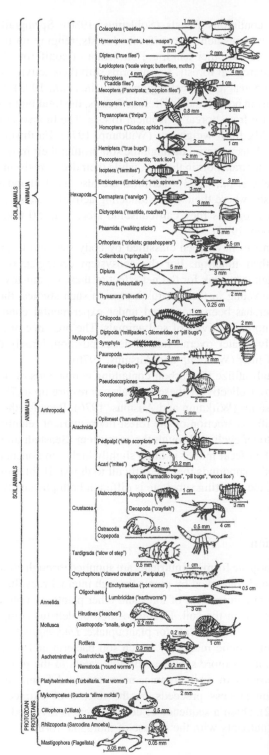

FIGURE 4.8. Classification of some of the groups of small soil animals important in processing forest litter. (From Hole, 1981.)

123

abundant in coniferous forests (Sohlenius, 1980). Springtails (Collembola) and mites (Acari) are also prevalent in coniferous forests, ranging up to $400,000\,m^{-2}$ (Hole, 1981). Earthworms, on the other hand, are intolerant of coniferous litter and more abundant in temperate and tropical forests. Earthworms shift their activity downward in the soil profile moving from boreal to tropical regions (Lavelle *et al.,* 1993). Despite the large number of soil organisms in boreal and coniferous forests, the total biomass of soil animals increases by a factor of 6 from boreal to tropical forests. The biomass of earthworms is sometimes as high as $250\,g$ fresh weight m^2 (Witkamp, 1971). Earthworms are especially effective in mixing the litter material through the soil profile; their movements leave open channels for air and water transport in the soil, while their decay-resistant casts preserve soil organic matter (Lavelle *et al.,* 1993). Termites are also most abundant in warm temperate and tropical forests, where their wood-tunneling activities cause a relatively rapid disappearance of fallen logs, as they can consume more than 50 times their weight per year in biomass (Wood, 1976).

Soil organisms have been classified by body size and trophic structure in attempts to understand their function. For the ecosystem ecologist, however, the most practical approach often is to look for the net effect of these organisms on the rates of energy flow and nutrient cycling in the detritus layers. In such studies, the decomposition of freshly abscised litter has been monitored under experimental conditions that either allow or exclude the normal activity of soil animals. The presence of soil organisms often allows a significantly greater overall rate of decomposition, largely through a more rapid fragmentation of litter (Witkamp and Ausmus, 1976).

Soil animals utilize the carbon-containing compounds of detritus as an energy source, respiring CO_2. Collectively these animals can respire up to 7% of the mass of litterfall in forest ecosystems (Witkamp and Ausmus, 1976). Because the biomass of soil animals is never a significant fraction of the forest floor, nutrient retention in these organisms does not slow nutrient cycling through the ecosystem (Seastedt and Tate, 1981). Compared to fresh litter, their fecal matter is often slightly higher in concentrations of N and P, as well as in lignified compounds that are difficult to digest. If litter material is first processed by soil animals, the resulting substrate is different both physically and chemically from fresh litter.

C. Microbial Decomposition

Although inorganic forms of nutrients are supplied through atmospheric inputs and mineral weathering, and conserved by plants through internal recycling, much, if not most, of the annual nutrient requirement of trees and other plants is supplied from the decomposition of dead materials in the soil. Decomposition of detritus completes the intrasystem cycle by releasing nutrient elements for plant uptake. *Decomposition* is the general term for the breakdown of organic matter by which soil bacteria and fungi obtain metabolic energy. *Mineralization* is a more specific term that refers to the processes that release carbon as CO_2 and nutrients in an inorganic, more oxidized form, such as sulfur in SO_4^{2-}. The decomposition process proceeds by the release of extracellular degradative enzymes (Burns, 1982). Often a sequence of decomposers is involved in the breakdown of organically bound nutrients with the release of inorganic forms.

During decomposition, significant loss in mass may occur through leaching of low molecular weight sugars, polyphenols, and amino acids, a process we all observe in making tea. Some of these compounds are also readily decomposed. Cell-wall materials, cellulose, hemicellulose, and lignin, which make up the majority of nonsoluble material in litter, are differentially attacked by microbes. Lignin is most resistant to attack, but it is so intimately entwined with cellulose strands in most materials that some must be degraded to allow microbial access to the more easily processed cellulose and hemicellulose. In the decomposition process, some degradation byproducts are formed that are "ligninlike," which may cause an increase in the total amount of lignin in the decaying material (Aber and Melillo, 1991; Johnson, 1992).

The rate at which different components in litter decompose is therefore not identical. Nevertheless, a general index of decomposition is often obtained by assuming comparable rates for all materials in broad categories. For example, the mean residence time (MRT, defined as the ratio of pool size to the rate at which material is added or removed from it, per unit of time) of all nonwoody forest floor material can be estimated from the ratio of forest floor mass to annual litterfall once forest floor accumulation stabilizes. The MRT decreases exponentially with increasing annual litterfall across biomes, varying from more than 40 years in some boreal needle-leaved evergreen forests to less than 1 year in some wet tropical forests (Landsberg and Gower, 1997; Fig. 4.9). A prediction of MRT, under steady-state conditions, can be derived Eq. (4.2), purely as a function of the dry mass of annual litterfall (x), in $Mg\,ha^{-1}\,year^{-1}$, where

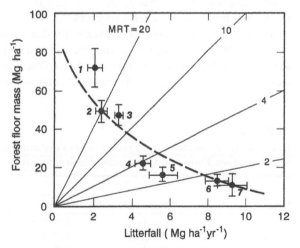

FIGURE 4.9. Average forest floor mass plotted against litterfall for (*1*) boreal needle-leaved evergreen, ($n = 16$), (*2*) boreal broad-leaved deciduous ($n = 7$), (*3*) temperate needle-leaved evergreen ($n = 73$), (*4*) temperate broad-leaved evergreen ($n = 11$), (*5*) temperate broad-leaved deciduous ($n = 40$), (*6*) tropical broad-leaved evergreen ($n = 31$), and (*7*) tropical broad-leaved deciduous forests ($n = 2$). Assuming the forest floor is in steady state, the average mean residence time (MRT, years) can be calculated as forest floor mass/litterfall mass. The straight lines indicate these values. The dashed curve is fit to the mean values and is the basis for deriving a related equation [Eq. (4.2)] that predicts MRT from equilibrium annual litterfall. (From Landsberg and Gower, 1997.)

$$\text{MRT (year)} = 55.4\, e^{-0.443x} \qquad r^2 = 0.93. \tag{4.2}$$

The mean residence time of forest floor litter is an important ecosystem property that is likely to change substantially with climatic variation and with the type of vegetation present. Through remote sensing, the maximum seasonal change in leaf area index can be monitored from space (Spanner *et al.*, 1994), which, with supplemental data on specific leaf mass (Pierce *et al.*, 1994), provides an opportunity to assess MRT across a wide range of spatial and temporal scales (Chapter 7). This estimate of forest floor MRT, however, excludes the turnover of roots and coarse woody debris (Harmon *et al.*, 1986; Vogt *et al.*, 1986).

Alternatively, decomposition rates can be determined using the litterbag technique, in which fresh litter is confined in a large number of mesh bags and placed on the forest floor or buried in the soil. At periodic intervals, some of the bags are retrieved and the loss in mass of litter measured. Simple models are based on an exponential pattern of loss where the fractional weight loss (m_1) over time (t, in days, months, or years) is given by

$$m_1(t) = c_1 e^{-kt} \tag{4.3}$$

Here, k is the decomposition (or decay) constant, which is determined from the plot of ln m_1 against t, from which $k = -(\ln m_1 + \ln c_1)/t$, where ln is the natural logarithm and c_1 is the intercept; when c_1 is set to unity to correspond with the initial weight, then $k = -(\ln m_1)/t$. The MRT and the time required to achieve 63, 95, and 99% of loss in mass (under steady-state conditions) can be calculated as $1/k$, $3/k$, and $4.6/k$, respectively (Olson, 1963). Thus, a litter sample that has a mass of 0.50 of its original after 200 days (0.55 of a year) would have a k value of 0.0035 day^{-1} or 1.26 year^{-1}, with a MRT of 0.8 year and 1% of its original mass present after 3.65 years.

The range in decomposition constants for deciduous and evergreen foliage, as well as fine and coarse woody debris, is variable within a biome (Table 4.7). In general, k values are higher in tropical forests than in other biomes, but there is no simple function with temperature or moisture, for the chemical composition of litter influences decomposition

TABLE 4.7

Decomposition Coefficients (k, years) for Foliage and Fine and Coarse Wood for Deciduous and Evergreen Forests in Contrasting Environments[a]

Biome	Leaf habit	Tissue component		
		Foliage	Fine wood	Coarse wood
	Deciduous	0.39–0.70	0.06–0.12	0.02–0.30
Boreal	Evergreen	0.22–0.45	—	—
	Deciduous	0.28–0.85	0.10–0.38	—
Cold temperate	Evergreen	0.14–0.69	—	0.01–0.06
	Deciduous	0.44–2.46	—	0.03–0.27
Warm temperate	Evergreen	0.16–0.75	—	0.04
	Deciduous	0.62–4.16	—	—
Tropical	Evergreen	0.16–2.81	—	0.12–0.46

[a]After Landsberg and Gower (1997).

rates as much as do climatic variables (Berg *et al.*, 1993). If the chemical composition of the litter is indexed by its initial lignin:nitrogen ratio, weight loss is more predictable within a particular climatic zone (Melillo *et al.*, 1982; Nadelhoffer *et al.*, 1983; Fig. 4.10). This chemical index of litter quality, which is widely applied in models of decomposition, reflects the fact that protein and amino acids are high in nitrogen and easily assimilated by microorganisms, whereas lignin, a complex polymer of aromatic rings, is exceedingly difficult for most microorganisms to attack and decompose. For a given quality of substrate, the decomposition rate increases exponentially with temperature with a respiration quotient (Q_{10}) between 1.5 and 2.5, assuming similar moisture conditions (see Chapter 3, Fig. 3.7).

We have obtained most of our understanding of the effects of temperature and other factors on decomposition from short-term studies of forest floor materials. The larger pool of organic matter stored as humus in soils has a much longer turnover time than the forest floor. This pool has grown in importance with the concern about climatic warming (Chapter 9). The net ecosystem exchange to the atmosphere is likely to increase, because decomposition increases exponentially with temperature, whereas photosynthesis is likely to be less responsive, ignoring change in species composition and the rise in atmospheric CO_2 (Chapters 3 and 5). To address these concerns, ecosystem models designed for longer term analyses separate soil organic matter (SOM) pools into at least three categories: (a) a small fraction (such as microbial biomass) that is extremely labile (5%), (b) an

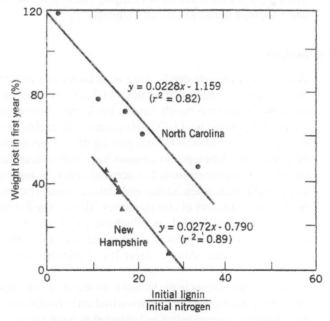

FIGURE 4.10.   Decomposition of forest leaf litter as a function of lignin:nitrogen ratio in fresh litterfall of various species in New Hampshire and North Carolina forests. The warmer climate typical of North Carolina allows litter there with a lignin:nitrogen ratio below 10 to be decomposed in less than 1 year. (From Melillo *et al.*, 1982.)

intermediate pool (60–85% of the total), and (c) a passive SOM pool with turnover times of thousands of years (Parton *et al.*, 1987; Comins and McMurtrie, 1993).

The fraction of SOM in each of the various pools can be determined as a result of nuclear bomb testing in the early 1960s which nearly doubled the radiocarbon concentration of atmospheric CO_2. Eventually, the increased radioactive ^{14}C was incorporated into plant and soil organic matter and now serves as a basis on which the residence times of various categories of soil organic matter may be determined. Fractions are separated by differences in density and resistance to treatment with acids and bases. The low density fraction contains relatively undecomposed vascular plant material and sometimes a fairly large component of charcoal. The higher density fraction is associated with the mineral soil and increases with clay content (Trumbore *et al.*, 1996).

How rapidly the intermediate pool turns over is of particular concern because it represents a major fraction of total SOM and is most likely to affect net ecosystem exchange in the twenty-first century. Trumbore *et al.* (1996) showed from a ^{14}C analysis of archived (pre-1963) and contemporary soils taken from undisturbed forest ecosystems in California that 50 to 90% of the carbon in the upper 20 cm of soil (A horizon soil carbon) was in the intermediate category. The estimated turnover time of the intermediate SOM component ranged from 7 to 65 years, increasing exponentially with current mean annual temperature. Townsend *et al.* (1995) conducted a similar study across an elevation gradient in Hawaii using ^{14}C analyses and made additional comparisons of changes in $\delta^{13}C$ composition of soil organic matter following the conversion of forests (C_3 plants) to C_4 pasture grasses. The long-term implications of changing rates on the decomposition of soil organic matter are discussed in more detail in Chapter 5.

D. Humus Formation

As decomposition proceeds, there is an increasing content of amorphous organic matter, humus, that is produced from microbial activity associated with lignin degradation. The structure of humus is poorly known, but it appears to contain aromatic rings with phenolic (-OH) and organic acid (-COOH) groups. Molecules have no consistent molecular mass (range 5000 to >1,000,000) or repeating units. Humus accumulates in the lower horizons of forest soils. Although most humus has a high concentration of nitrogen in amino ($-NH_2$) groups, it is highly resistant to microbial attack. A long turnover time may apply even in tropical climates, when humus molecules acquire a random coil shape that facilitates protection in micropores of clay structures. Minerals with high specific surface areas and high ion-charge densities such as allophane and some iron oxides are particularly good at complexing large amounts of organic material in a form inaccessible to microbes; in contrast, kaolinite and other 1:1 layer-lattice minerals are much less efficient (Lavelle *et al.*, 1993).

Annual increments in the humus pool are small but continuous during soil profile development. Schlesinger (1991) summarized carbon accumulation rates measured in numerous soil chronosequence studies and reported average rates of 8.7, 5.6, and 2.4 g C m^{-2} year^{-1} for boreal, temperate, and tropical forests, respectively. If we assume that NPP might average 5 Mg C ha^{-1} year^{-1} (Chapter 3), then humus production is generally <1% of annual NPP. Because both the accumulation and turnover of humus are so slow, it is unlikely that this important carbon pool is ever really in steady state. Nevertheless, the dynamics are important to quantify because of the size of the organic and mineral pool involved.

Although the structure and formation of humus substances remain elusive, the importance of humus in forest ecosystems is clear. Humus contains an overwhelming proportion of the total N, P, and S in forest soils. In coarse-textured soils, humus greatly increases the water holding capacity; in fine-textured soils, humus reduces the bulk density, which improves conditions for drainage, root growth, and soil organisms. In most forests, humus in the soil profile exceeds the combined content of organic matter in the forest floor and aboveground biomass (Waring and Schlesinger, 1985).

E. Mineralization and Immobilization

The release of nutrients from decomposing organic matter does not proceed at the same rate as decomposition, because microbes, like plants, are growth-limited if either their energy source or nutrient supply is restricted. The rate at which litter loses mass is only a rough measure of the amount of carbon available to decomposing organisms. Nutrients present in excess of microbial requirements will be released (mineralized) as decomposition progresses, while those limiting growth will be preferentially retained (immobilized) in live microbial biomass. For example, during the decomposition of pine needles in Sweden, microbial growth is more limited by N, P, and S, which are shown to increase in reference to their initial litter content over time, compared with other nutrients that are lost at rates equal to or greater than that at which decomposition proceeds (Berg and Staaf, 1981; Fig. 4.11).

The rate of nutrient immobilization is highest in the first stages of decomposition of fresh litter, when the most easily decomposed compounds are available as an energy source. As the carbon and energy yield from decomposition declines, microbial growth slows, and there is little further nutrient immobilization. It is the death of the microbial populations (<5% of SOM) that accounts for most nutrients released in their inorganic form from litter (Marumoto et al., 1982; Diaz-Ravina et al., 1995). After a year or so, the concentrations of N and P in the partially decayed litter increase substantially. As a general rule, initial net mineralization will occur at C:N ratios below 20:1, C:P ratios less than 200:1, and C:S ratios less than 200:1. In contrast, net immobilization will begin to occur at C:N ratios of 30:1, C:P ratios exceeding 300:1, and C:S ratios over 400:1 (Stevenson, 1986). Immobilization is most significant for N and P, which are generally limiting to microbial growth, and much less significant for Mg, K, Ca, and Mn. The critical carbon to nutrient ratios vary slightly, depending on the substrate and assimilation efficiency of the decomposer (Paul and Clark, 1996).

A total reduction in the absolute amount of nutrients in the litter represents net mineralization and increased availability to plants. On the other hand, a net immobilization of nutrients may also occur through the incorporation by microbes of nutrients taken from older decaying material, captured in throughfall and soil solution, or derived from free-living nitrogen fixers. Asymbiotic N fixation is particularly important in the decay of coarse woody debris, which often has initial C:N ratios between 200:1 and 500:1. Immobilization of nutrients predominates in the layer of fresh litter on the soil surface, whereas mineralization of N, P, and S is usually greatest in the lower forest floor. During soil

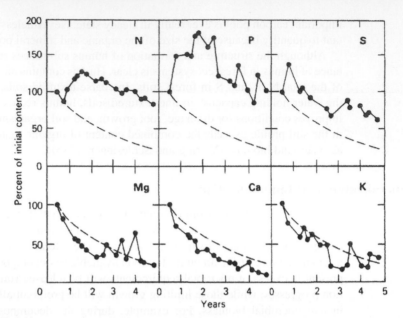

FIGURE 4.11. Loss of nutrient elements from the litter of Scots pine (*Pinus sylvestris*) during the first 5 years of decomposition in Sweden. For each nutrient, the solid line indicates the percentage of the initial content remaining at various intervals. Note that loss of Mg, Ca, and K is more rapid than the disappearance of the organic mass of litter (dashed line), whereas N, P, and S are retained during the period of litter decay. (From Staaf and Berg, 1982.)

development, nutrient-rich fulvic acids with low C:N, C:P, and C:S ratios are transported to the lower soil horizons where the nutrients may be quickly mineralized and made available to plants. Sollins *et al.* (1984) also found that the lighter density fraction of soil organic matter, representing fresh plant residues, generally has a higher C:N ratio and lower mineralization than the heavier density fraction comprising humic substances. Across biomes, N is likely to be most immobilized in temperate and boreal forests, whereas P is likely to be immobilized more commonly in tropical forests (Vogt *et al.*, 1986).

Seasonal fluctuations of soil temperature and moisture can stimulate microbial activity more than in constant favorable conditions (Biederbeck and Campbell, 1973). These variations cause greater turnover of the microbial populations and mobilization of nutrients (Diaz-Ravina *et al.*, 1995). Similarly, the periodic addition of fresh litter materials high in degradable carbon can sometimes stimulate the degradation of resistant substrates in the soil (Sørenson, 1974; Janzen *et al.*, 1995). The balance between mineralization and immobilization thus shifts continually from season to season through a soil profile. This variation can be modeled and measured, but it requires detailed knowledge of spatial variation in environmental conditions and substrate quality (White *et al.*, 1988).

The mineralization of nitrogen is a particularly good example of a dynamic process (Fig. 4.12). It begins with the release of ammonium (NH_4^+) from decomposing material. Some NH_4^+ may undergo nitrification, in which oxidation of NH_4^+ to nitrate (NO_3^-) is coupled to the fixation of carbon by chemoautotrophic bacteria in the genera *Nitrosomonas*

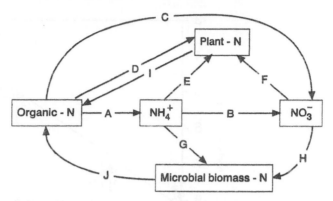

FIGURE 4.12. Schematic diagram of important transformations of N within the soil and forest floor. A, Gross mineralization; B, gross nitrification with NH_4^+ as the substrate; C, gross nitrification with organic N as the substrate; D, mineralization of organic substrate through mycorrhizal fungi directly into plants; E and F, plant uptake of NH_4^+ and NO_3^-, respectively; G and H, gross immobilization (microbial assimilation) of NH_4^+ and NO_3^-, respectively; I and J, organic N inputs from plants and microorganisms, respectively, via death, sloughing, and exudation. Note that net mineralization measured by laboratory incubations and buried litterbags is $(A + C) - (G + H)$. Similarly, net nitrification is $B + C - H$. (From Davidson *et al.*, 1992.)

and *Nitrobacter*. These autotrophic nitrifying bacteria obtain their carbon directly from CO_2 rather than from plant material. Some heterotrophic organisms can also convert organic N directly to NO_3^-. In addition, it is recognized that some mycorrhizal fungi can shortcut the mineralization process by acquiring N directly from dissolved organic nitrogen (DON) associated with humic compounds in extremely N-poor soils (Northup *et al.*, 1995; Ruess *et al.*, 1996). Gross mineralization and gross immobilization, as shown with ^{15}N studies, can occur in a matter of days, so that net mineralization may represent less than 10% of the total N released in a 30-day period (Davidson *et al.*, 1992). Gross mineralization is therefore a preferred measure of N available to plants and microbes (Myrold, 1987).

Stable isotopes of nitrogen offer a means of assessing the relative availability of nitrogen to plants. During each step in the process of converting organic N to NH_4^+, and to NO_3^-, the heavier isotope (^{15}N) is discriminated against, which results in the product of each transformation being less enriched in ^{15}N than its substrate (and exhibiting a lower $\delta^{15}N$). The difference between the natural abundance of ^{15}N found in foliage and that in soil may thus serve as an integrative measure of N availability to a particular tree species. Garten and Van Miegroet (1994) applied this principle in the Great Smoky Mountains of eastern Tennessee, where atmospheric deposition of N has been documented to increase with elevation. They demonstrated that $\Delta^{15}N$ values (between foliage and surface soils) progressively narrowed as coniferous species encountered soils with higher amounts of available N (Fig. 4.13). The correlation was also good for deciduous hardwood species, but varied more because of differential preferences for NH_4^+ and NO_3^- (Garten and Van Miegroet, 1994). Differences in rooting depth and in the form that N is acquired increase the variation in $\delta^{15}N$ observed among species growing on infertile sites. Once N becomes readily available, however, species tend to show similar $\delta^{15}N$ values (Schulze *et al.*, 1994b).

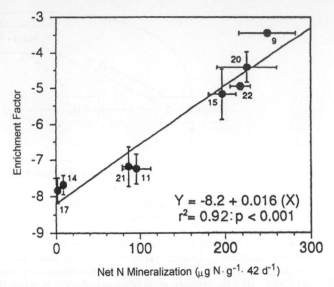

FIGURE 4.13. Atmospheric deposition of nitrogen increases across an elevation gradient in the Great Smoky Mountains National Park in Tennessee. As a result, assays of net mineralization of N in the surface mineral soil show a parallel increase (plots 17 to 9). Stable isotope composition was determined on foliage samples from coniferous species and from the soil on which they grew. The difference in isotopic composition (enrichment factor) decreased proportionally as net mineralization of N in the soil increased. Data are means ($n = 6$), and error bars represent ± SE. (From Garten and Van Miegroet, 1994.)

Forest composition and nutrient availability are often closely interlinked, as we inferred from the earlier discussion on nutrient use efficiency. In the case of nitrogen, tree species with extremely low requirements are adapted to soils where DON is the major form present (Northrup *et al.*, 1995); most conifers are adapted to intermediate conditions where NH_4^+ is the dominant form, whereas riparian zone species are particularly suited to high levels of NO_3^- through their ability to induce the formation of nitrate-reducing enzymes in their foliage (Smirnoff *et al.*, 1984; Smirnoff and Stewart, 1985). Ecosystem retention of N, as we shall see, is also progressively reduced with increasing availability, so that high losses of NO_3^- usually indicate an excess availability while a dominance of DON in leachate reflects extreme scarcity of N.

The following summarizes some key points and implications from the last three sections:

- Fragmentation and mixing are essential steps that a host of small soil animals perform on litter before it is subjected to microbial decomposition. The introduction of pesticides or other toxic chemicals may reduce the efficiency of various groups of animals and could substantially alter the normal rates of decomposition and mineralization.
- Decomposition of woody debris, leaf litter, and belowground components of detritus proceeds at significantly different rates. For this reason, it is generally recommended that these organic pools be separately recognized.

- Moisture and temperature conditions strongly affect decomposition and mineralization rates. The hydrologic and energy balance models introduced in earlier chapters provide a means of defining these environmental variables without requiring direct measurements in the litter and soil.
- The chemical quality of the organic substrate, which can he quantified by C:N ratios and other related indices, strongly affects the mineralization and immobilization processes.
- As a scaling principle, a decreasing amount of detail is required to estimate process rates at progressively longer time intervals. Thus, on an annual basis, decomposition rates can be relatively easily assessed, on the basis of decomposition constants acquired from litterbag studies, forest floor/litterfall mass comparisons, and even from satellites by monitoring the annual transfer of leaf litter from the canopy to the ground. The general reliability of these annual estimates, however, rests on an understanding of key variables and interactions acquired at much shorter time steps.

F. Cation and Anion Exchange

Ion exchange in soils is a process based on the surface charge of clays and organic particles. By virtue of this charge, ions released from weathering or decaying organic matter, or those added through atmospheric deposition, are held on the particle surfaces and resist leaching. The ions, however, are not held with such force that they cannot exchange rapidly with other ions from the soil solution nor be taken up by plant roots.

The layered silicate clay minerals that dominate temperate zone soils possess net negative charge that attracts and holds cations dissolved in the soil solutions. Silicate clay minerals possess several types of negative charges which contribute to soil cation-exchange capacity (CEC). Exchange capacity is defined in *relative equivalent mass:* the amount of the substance, in grams, that will supply the same total charge as is supplied by 1 g of H^+. Equivalent mass is then computed by dividing the mass of the ion by the valence. Thus a solution that contains 1 g of H^+, 23 g of Na^+, or 20 g of Ca^{2+} contains 1 equivalent of each ion. Exchange capacity is usually expressed in milliequivalents (meq) per 100 g or, with standard international (SI) units, cmol (+) per kilogram of soil.

At the edges of clay particles, hydroxide radicals (-OH) are often exposed to the soil solution. As the pH of the solution increases (i.e., becomes less acidic), H^+ ions become less strongly bound to oxygen in the -OH radicals, increasing the negative charges on clays. In addition, there is another source of negative charge that arises from ionic substitutions within silicate clays. For example, when Mg^{2+} substitutes for Al^{3+}, there is an unsatisfied negative charge in the internal crystal lattice. Unlike the first source of negative charge, this second source is permanent because it originates inside the crystal structure and cannot be neutralized by covalent bonding of H^+ from the soil solution. Organic compounds in soils also create cation-exchange sites on phenolic (-OH) and organic acid radicals. Organic matter is the only source of cation-exchange sites in coarse-textured sandy soils and in peat bogs.

Cation-exchange capacity and base saturation, the percentage of the exchange sites occupied by nutrient cations (Ca^{2+}, Mg^{2+}, K^+, NH_4^+, Mn^{2+}), increase during initial soil development on newly exposed parent materials. As weathering of soil minerals continues, cation-exchange capacity and base saturation decline (Bockheim, 1980). Temperate forest soils dominated by 2:1 clay minerals have much higher cation exchange capacity (100–150 meq/100 g) than soils dominated by 1:1 clay minerals (5–40 meq/100 g). Most fertile temperate forest soils have high cation-exchange capacity, which allows more nutrient cations to be held on clay particles and reduces losses by leaching. A high base-exchange capacity in the surface soil and subsoil also provides a buffer against pH changes associated with acid precipitation, and thus is a critical ecosystem property throughout the temperate forest region in the northern hemisphere (Johnson and Lindberg, 1992). Tropical forest soils are general poorly buffered, first because they are dominated by iron and aluminum oxides which provide extremely low cation-exchange capacity (2–4 meq/100 g) and, second, because most of the base cations have been leached as a result of long periods of intense weathering and leaching.

In contrast to the permanent negative charge in soils of the temperate zone, tropical soils dominated by oxides and hydrous oxides of iron and aluminum show variable charge on soil exchange sites, depending on soil pH (Johnson and Cole, 1980; Schlesinger, 1991, 1997). At low pH, soil surfaces absorb H^+ from solution and become positively charged (attracting anions). At some intermediate pH, the previously absorbed H^+ dissociates into solution, which leaves no net charge (zero point charge) on the exchange complex. At extremely high pH (i.e., above pH 9), an additional H^+ dissociates, leaving the surface negatively charged (attracting cations).

The pH at which zero point charge (ZPC) occurs varies. Anion-exchange capacity is typically greater in soils where the crystalline surfaces of Fe and Al minerals are highly fractured, exposing a large surface area. Anion-exchange capacity is also present in the soils of warm temperate forests where iron and aluminum hydroxides are abundant in the soil profile (Johnson et al., 1979). This anion-exchange capacity helps explain how much of the sulfate in acid rain does not readily leach through Fe- and Al-rich weathered soils in the southeastern United States (Johnson et al., 1986).

Similarly, tropical forest soils with high anion-exchange capacities are able to adsorb a fair proportion of NO_3^- in their lower profile. Following a forest clearing, the mineralization of fresh litter may add an additional 300–600 kg N ha^{-1} to that normally produced. Once perennial vegetation with deep root systems is reestablished the adsorbed nitrate is again accessible and may be recycled through the ecosystem (Matson et al., 1987).

Exchange of cations and negatively charged anions from the surface of secondary soil minerals and organic matter occurs as a function of chemical mass balance with the soil solution. Elaborate models of ion exchange have been developed by soil chemists (Tan, 1982). In general, cations are held and displace one another in the following sequence on cation-exchange sites:

$$Al^{3+} > H^+ > Ca^{2+} > Mg^{2+} > K^+ > NH_4^+ > Na^+.$$

Anion exchange follows the sequence

$$PO_4^{3-} > SO_4^{2-} > Cl^- > NO_3^-$$

Either sequence can be altered by the presence of large quantities of the more weakly held ions in the soil solution. Liming to reduce the effects of acid rain, for example, displaces and neutralizes H^+ from the exchange sites by providing an excess of Ca^{2+}. In addition, acidic conditions mobilize Al^{3+}, the cation most likely to displace others from exchange sites. Hydrogen ions affect the solubility of most other ions in soil solution; thus, pH is often considered the master variable that controls soil nutrient availability.

In time, most forest soils become more acid because plant roots selectively absorb more nutrient cations than anions, releasing H^+ in organic acids to maintain an internal balance of charge. In addition, root respiration releases CO_2, forming carbonic acid in the soil solution (Fig. 4.14). The increase in H^+ ions in solution reduces base saturation through time and increases nutrient losses by leaching to lower parts of the soil profile and into the groundwater. Conventional measurements of nutrient availability on exchange sites are performed on small samples of soils collected from various parts of the soil profile. Alternatively, sampling the chemistry of solutions collected from above the canopy to below the rooting zone can provide much additional insight (Table 4.5). The information provided from such detailed analysis of soil solution chemistry is often required to assess the longer term implications of various management policies regarding slash disposal, fire control, fertilization, and the manipulation of species composition in an effective manner.

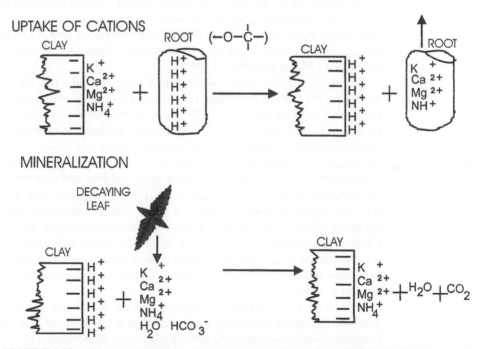

FIGURE 4.14. The cation cycle begins with positively charged nutrient ions held on negatively charged clay particles. The cations are exchanged for hydrogen ions released as organic acids by plants. Roots take up the nutrients and leaf litter returns them annually to the soil. Through decomposition, nutrient cations are released to return again to the soil exchange sites. (From Glatzel, 1991.)

G. Adsorption and Fixation

There is a gradation in nutrient availability between ions held weakly on exchange sites, those absorbed onto surfaces of clay and humus, and those fixed more permanently through substitution directly into the layer lattice of a clay. As previously noted, iron and aluminum hydroxides have a surface positive charge at normal soil pH. Nutrient anions, especially PO_4^{3-} and SO_4^{2-}, are often held under these conditions much more tightly than in normal exchange reactions. This stronger chemical bonding is termed *adsorption*. Because the reaction is sensitive to the concentration of ions in solution, the process may be reversed and *desorption* can occur. Starting at low ion concentrations in solution, the adsorption process increases rapidly, then plateaus as the soil approaches its maximum capacity to adsorb ions (Schlesinger, 1991). Under acidic conditions, a large total amount of anions of phosphate and sulfate may be permanently held on the surfaces of iron sesquioxides through the release of -OH or H_2O to the soil solution (Binkley, 1986).

The cations NH_4^+ and K^+ may also be fixed so strongly on clays that they cannot be recovered by exchange reactions. Both of these cations can be entrapped in the intermicellar regions of expanding lattice clays. On closure of the space, NH_4^+ and K^+ are fixed. The total amount of these ions that can be fixed can be quite high, but the fixation capacity is limiting; thus, above a certain concentration, ions will be kept in solution (Tan, 1982). Ammonia (NH_3) and amino acids can also be fixed through physical condensation reactions of phenolic by-products from partly degraded lignin (reviewed by Johnson, 1992). Nonbiological incorporation of ammonia into humus is enhanced by high pH and high concentrations of NH_3 and NH_4^+. Thus adding nitrogen as urea (CON_2H_4), a condensation product of NH_3 and CO_2, often results in >40% being bound in a fixed form (Preston *et al.*, 1990; Nason and Myrold, 1992).

Biological activity affects adsorption–desorption and even the fixation process by modifying pH through the release of organic acids and by the selective extraction of ions from solution. Graustein *et al.* (1977) suggested that the production of oxalate compounds by fungal hyphae increases PO_4^{3-} availability by complexing iron in the soil solution. Soluble organic compounds, including polyphenols and humic acids, also mobilize or complex ions in solution. In general the adsorption and fixation potential of a soil horizon is increased by the presence of hydrous oxides of iron and aluminum and decreased by the presence of organic matter (Ae *et al.*, 1990). Adsorption and fixation of anions are generally highest in lower soil horizons where iron and aluminum oxides accumulate and organic content is low. Fixation of ammonia and ammonium is more likely in the upper profile where organic content is high. The speed at which adsorption–desorption reactions occur is relatively rapid (hours), in contrast to mineral weathering which requires centuries to convert primary to secondary minerals and to transfer clays, Si, Al, and Fe from one horizon to another.

Models of the adsorption–desorption process are useful in explaining differences observed in nutrient availability and in losses in solution associated with both upland and bottomland soils (Kafkafi *et al.*, 1988; Yanai, 1991; Masscheleyn *et al.*, 1992; Prenzel and Meiwes, 1994). Principles derived from these detailed models provide a basis for judging what soils are most sensitive to disturbance from logging or to chronically high additions of sulfur and nitrogen from anthropogenic sources (Chapter 6).

H. Volatilization and Leaching

In undisturbed forests, losses of nutrients and organic carbon occur mainly through microbially induced volatilization and through leaching. On an annual basis, these losses are small and often in near balance with inputs from weathering, atmospheric deposition, and fixation of N and C. Methane (CH_4), nitrous oxide (N_2O), dinitrogen (N_2), and hydrogen sulfide (H_2S) are examples of gases volatilized from anaerobic soils. Leaching is important in carrying dissolved organic carbon, nitrogen, and other elements downward through the soil and subsoil to the groundwater, where it may seep into adjacent ecosystems, and again become available to plants, or flow directly into streams and lakes.

Nitrogen is one of the nutrients most subject to loss by volatilization, particularly when added as urea fertilizer which can be quickly converted to NH_3 gas when soils are warm and moist (Nason and Myrold, 1992). Some fraction of the volatilized NH_3 may be captured on foliage and incorporated into organic compounds, so the net loss from the ecosystem may be less than that estimated from the soil (Pang, 1984; Nason and Myrold, 1992). During the process of nitrification, some NO and N_2O gas may be released, but the loss from upland temperate and boreal forest soils is usually small, <2 kg N ha^{-1} $year^{-1}$ (Robertson and Tiedje, 1984). In some recently disturbed tropical soils, rates may be much higher (Matson *et al.*, 1987). Often, the net N exchange from undisturbed upland forest is negative, as microbes take up more N from the atmosphere than they release (Aber and Mellilo, 1991).

Nitrogen is also volatilized through *denitrification*, the reverse process that converts nitrate to N_2O and N_2:

$$NO_3^- \rightleftharpoons NO_2^- \rightleftharpoons NO \rightleftharpoons N_2O \rightleftharpoons N_2. \qquad (4.4)$$

Denitrification is performed by soil bacteria that are aerobic heterotrophs in the presence of O_2, but it also operates under anaerobic conditions (Schlesinger, 1991). The ratio of N_2O to N_2 produced during denitrification is not constant; increased concentrations of nitrate, nitrite, molecular O_2, and soil acidity in the presence of NO_3^- stimulate the production of N_2O relative to molecular N (Schlesinger, 1991). For a long time, denitrification was thought to occur only in flooded, anaerobic soils; however, soil scientists have shown that oxygen diffusion to the center of soil aggregates is sufficiently slow that anaerobic microsites are present even on upland sites (Tiedje *et al.*, 1984).

In a closely related process, SO_4^{2-} can also be converted to sulfur dioxide (SO_2) or hydrogen sulfide (H_2S), which occurs commonly in marshes and other types of wetlands but rarely in forests. Again, heavy deposition of sulfate on forests may increase sulfur volatilization losses, but leaching losses are more important because they contribute to the removal of other cation nutrients in solution and the acidification of soils (Mitchell *et al.*, 1996). Trace gas emissions are a particular concern in wet tropical climates where, following forest clearing or windthrow, uptake of water and nutrients by vegetation is reduced while soil temperature and the pore space filled with water increase (Keller and Reiners, 1994). Under these conditions large amounts of NO, N_2O, and other trace gases may be released until the available C and N pools are fully utilized (Steudler *et al.*, 1991; Chapter 6).

All heterotrophic microbial activity is highly dependent on the availability of an adequate labile carbon source, the appropriate substrate, and the temperature. Trace gas emission is particularly sensitive to free oxygen levels, which are highly correlated with the fraction of water-filled pores and the soil texture. In the case of CH_4 production, tropical wetlands and boreal peat bogs are the major contributors, while most upland forest soils are net sinks (Schlesinger, 1991). Generalized models for predicting N_2 and N_2O production from nitrification and denitrification, as well as methane production and consumption, are now available with monthly or finer resolution (Crill, 1991; Peterjohn *et al.*, 1994; Parton *et al.*, 1996). Because these generalized models were based on extensive laboratory and field experiments, they may be widely applied across landscapes. Through remote sensing, seasonal estimates of surface moisture and temperature conditions can be acquired that are closely correlated with trace gas production from specified vegetation units (Waring *et al.*, 1995b).

Leaching losses from forests are usually much higher than those associated with trace gas emissions. In previous sections, we have identified those forms of nutrients and organic matter that are highly soluble and not held strongly on clay exchange sites, or otherwise adsorbed or fixed. These include dissolved organic carbon (DOC), DON, NO_3^-, K^+, Cl^-, H^+, and HCO_3^-. Not surprisingly, these forms are among those most likely to be leached from forest ecosystems, even in undisturbed conditions. The total flux of elements carried in solution can be derived by combining models that predict water flow (discussed in Chapter 2) with those that predict ionic concentrations in solution (referenced in this chapter).

In most undisturbed temperate and boreal forest systems, concentrations follow the order DON > NH_4^+ > NO_3^- in the soil solution, and the total export of N in any form is low (Northrup *et al.*, 1995; Currie *et al.*, 1996). In temperate deciduous forests, conditions favor nitrification in the spring before budbreak; however, once leaves begin to expand, demand for nitrogen usually exceeds the available supply, so again leaching losses are minimum. In autumn, on the other hand, fresh litter has a high C : N ratio, so most nitrogen is immobilized during early phases of decomposition before soil temperatures become limiting. Flooding, of course, removes nitrate from the soil, but anaerobic conditions also halt NO_3^- production. When nitrate appears in seepage and stream water, particularly in temperate and boreal forests, it is often indicative of ecosystem disturbance associated with an outbreak of defoliating insects, root disease, blowdown, or chronically high rates of atmospheric deposition (Chapter 6). In the wet tropics, nitrification is a dominant process, but even there disturbance favors an increased rate of nitrification and losses through leaching (Matson *et al.*, 1987).

V. MASS BALANCE AND MODELS OF MINERAL CYCLES

Mineral cycling through an ecosystem includes transfers in and out, as well as the internal cycles through the vegetation and soil. One approach to account for these transfers is to perform a mass balance analysis. We provide examples of mass balance analyses from an isotope-tracer study and from a more conventional approach. Both studies rely on calculations of primary production and estimates of annual litterfall.

A second approach to account for the movement and transformation of element through forest ecosystems is through the application of detailed mineral cycling models. Unlike ecosystem models presented in Chapter 3, mineral cycling models must incorporate more soil processes, and often this requires separation of the surface and subsoil into many layers. By reviewing one mineral cycling model, we provide an example of the kind of process information incorporated in these types of simulation models.

A. Mass Balance Analysis

The annual circulation of nutrients may be assessed with a mass balance approach. Because annual changes in the soil nutrient pool are small, and difficult to measure, most mass balance analyses exclude belowground measurement, beyond quantifying the total pool of nutrients in the rooting zone. The first mass balance analysis we present was obtained by adding small amounts of isotopically enriched fertilizer to a spruce forest in Germany. Buchmann et al. (1996) demonstrated that less than $1 \, kg \, ha^{-1}$ of $^{15}NH_4^+$ and $^{15}NO_3^-$ was required to obtain a quantitative measure of the circulation of these two forms of N from the soil, through the vegetation, and back in the litterfall. Table 4.8 presents the total amounts and percentages of labeled N recovered from each identified ecosystem component 8 months after initiation of the experiment. From these analyses, which account within <1 to 3.7% for all of the labeled ^{15}N applied, we see that >80% of labeled N was retained in the soil, with more than two-thirds in the organic horizon. Although spruce trees represented over 5 times the biomass of the understory vegetation, they acquired only one-third (3.4 versus 9.1%) of the ammonium and less than half (6.5 versus 14.8%) of the nitrate ^{15}N tracer.

The estimated retention of ^{15}N additions into different ecosystem components was determined with the following mass balance equation:

$$m_{label} \approx m_f(\delta^{15}N_f - \delta^{15}N_i)/(\delta^{15}N_{label} - \delta^{15}N_i) \qquad (4.5)$$

TABLE 4.8

Total Nitrogen Budget and Percent ^{15}N Label Recovered from Major Ecosystem Components after Application of $0.6 \, kg \, ha^{-1}$ of Isotopically Enriched NH_4^+ and NO_3^- to a German Spruce Forest[a]

Ecosystem component	Nitrogen in biomass		^{15}N retained, % of total	
	$g \, m^{-2}$	% of total	$^{15}NH_4^+$	$^{15}NO_3^-$
Picea abies	10.5	0.65	3.4	6.5
Understory plants	10.0	0.63	9.1	14.8
Litter	0.2	0.01	0.03	0.04
Roots	3.0	0.18	1.0	3.5
Plant totals	23.7	1.47	13.5	24.8
Soil organic horizon	164	10.2	62.6	46.3
Mineral soil, 0–65 cm	1412	88.3	24.5	32.6
Total ecosystem	1600	100	100.6	103.7

[a]After Buchmann et al. (1996) with kind permission from Kluwer Academic Publishers.

where m_{label} is the mass of ^{15}N-labebed compound incorporated into the component, m_f is the final mass of N in the component, $\delta^{15}N_f$ is the final $\delta^{15}N$ abundance in the component, $\delta^{15}N_i$ is the initial $\delta^{15}N$ abundance in the component, and $\delta^{15}N_{label}$ is the $\delta^{15}N$ abundance in the labeled compound. Similar mass balance analyses have been derived with the natural abundance of Sr isotopes used as a surrogate for Ca to separate the contribution of atmospheric deposition from mineral weathering and quantified that acid precipitation is causing a net depletion in the available Ca pool in a 53-ha forested watershed in New Hampshire (Bailey et $al.$, 1996).

A more conventional and complete nutrient analysis is illustrated for an 80-year-old beech forest in Germany (Table 4.9). As with the isotopic analyses, measurements of total biomass and its mineral content were required to estimate internal storage and the annual transfer of nutrients. Mineral uptake is equivalent to the annual storage in wood plus the replacement of losses in litterfall and leaching, minus that added from atmospheric deposition or through N fixation. Note that >90% of the annual requirement of N is allocated to foliage, whereas <10% is allocated to wood. The annually recycled nutrient capital returned to the soil in litterfall or leachate was a relatively small percentage of the pool in tree biomass (N, 5%; P, 3%; K, 9%; Ca, 10%; and Mg, 9%). The short-lived organs, however, store a smaller percentage of total nutrients accumulated in biomass. For most nutrients, storage in wood increased by about 1% each year.

TABLE 4.9
Storage (kg ha^{-1}) and Annual Circulation of Nutrients in (kg ha^{-1} year^{-1}) in a Beech Forest, Solling, Germany[a]

Ecosystem components	N	P	K	Ca	Mg
Annual storage					
Foliage	96	6	31	18	3
Wood and branches	309	30	171	285	24
Roots	580	11	32	26	6
Forest litter layer	1050	72	123	118	44
Soil–root zone	9452	2310	550	280	45
Total storage kg ha^{-1}	11,487	2429	907	727	122
Annual demand					
Foliage (A)	96	6	31	18	3
Wood and branches	11	1	7	11	1
Roots	2	0.5	1	1	0.5
Total demand (B), kg ha^{-1} year^{-1}	109	7.5	39	30	4.5
Annual return in solution					
Throughfall and stemflow (C)	22	0.5	23	18	4
Precipitation (D)	22	7	4	13	3
Net loss from canopy leaching (C − D)	0	−6.5	19	5	1
Net uptake (B + C − D), kg ha^{-1} year^{-1}	109	1	58	35	5.5
Return in litterfall (E)	54	4	18	18	2
Total return (E + C − D)	54	−2.5	22	23	3
Reabsorption (A − E)	42	2	13	−7	1
Demand from soil (B + C − E − D), kg ha^{-1} year^{-1}	55	−3	40	17	0.5

[a]From Cole and Rapp (1981).

The beech forest received $22 kg N ha^{-1} year^{-1}$ in precipitation, with additional unmeasured contributions in dryfall. This excess N was derived from pollution, which may help explain the excessive amounts of N and Ca recorded in throughfall and stemflow, as well as the apparent lack of P required from the soil. In any case, excess N has accumulated in the foliage to the extent that other nutrients, with the exception of Ca, are suboptimal in relative abundance (Table 4.1). Nutrient budgets for nearly 40 sites were summarized by Cole and Rapp (1981) as part of the International Biological Program during the 1970s. These analyses serve as benchmarks against which to measure the effects of changes in atmospheric deposition since the 1970s (Chapter 6).

B. Mineral Cycling Simulation Models

The need for extrapolation to conditions where atmospheric inputs, vegetation, litter, and soil conditions may all have changed has encouraged the development of ecosystem simulation models that attempt to couple water, carbon, and mineral cycles (see Chapter 3). Most ecosystem simulation models consider nitrogen as it affects decomposition, net primary production, and soil organic matter accumulation but generally disregard other elements and processes that involve exchange, adsorption, fixation, and weathering. Soil scientists have developed detailed models of weathering and ion exchange, but most are not linked to changes in forest composition or to litter quality. Concern about the long-term implications of acid precipitation has fostered a collaborative effort to produce more comprehensive nutrient cycling models that include the necessary interactions to predict changes in Al:Ca ratios in soil solution, acidity, and base saturation that affect nutrient uptake and primary production (Kros and Warfvinge, 1995).

Although a large number of nutrient cycling models exists, only a few are well balanced, in the sense that they describe all aspects of the forest ecosystem with a comparable level of detail. Tiktak and van Grinsven (1995) in their review of 16 models also found little consistency in the manner in which nutrient uptake and reallocation within vegetation were treated. In fact, the biggest challenge to modeling nutrient cycling appears to lie in predicting how roots respond to changes in soil chemistry and the subsequent effects on tree growth and nutrient use efficiency (Mohren and Ilvesniemi, 1995). Less than a third of the models reviewed were well documented and available for independent testing; only three were designed to consider forests with a mixture of species (Tiktak and van Grinsven, 1995). Even in estimating the water balance of a pure spruce forest, models differed substantially in partitioning evaporation, transpiration, and interception (Bouten and Jannson, 1995).

We briefly review a mineral cycling model described by Liu *et al.* (1991) to encapsulate the components and interactions that must be considered in a complete mineral cycling model. This particular model was applied to compare intensively monitored ecosystems in an acid rain study (Johnson and Lindberg, 1992). The model explicitly tracks mineral fluxes between the vegetation and soil and also considers interactions with atmospheric deposition and mineral weathering. These latter components were essential to assess changes in solution chemistry, soil fertility, and transfers into groundwater and streams. The model routes precipitation through the canopy and soil layers and simulates evaporation, transpiration, and deep seepage, as well as lateral flow out of the soil. The model allows soils to be separated into up to 10 strata, each of which may have different physical and chemical characteristics. The movement of water through the system is simulated with appropriate equations that take into account continuity between strata, saturated and unsaturated flow, and surface runoff. Percolation occurs between layers as a function of differences in soil permeability and water content. Lateral flow occurs when a stratum becomes saturated (Chapter 2).

Canopy chemistry is the first component associated with deposition. The model allows for separate interception of wet- and dryfall and stores a fraction of the precipitation on leaf surfaces up to a defined capacity. Chemical reactions that occur on canopy surfaces allow for direct uptake, increase in surface concentrations by evaporation, and foliar leaching in proportion to LAI. The products of dry- and wetfall interact with the surface canopy to produce throughfall that may be markedly different chemically from incident precipitation. Nutrient reabsorption is simulated as partial withdrawal of nutrients from leaves before leaf fall.

Within the soil, nutrient pools associated with soil solution, the ion-exchange complex, mineral, and soil organic matter are separately tracked. Interactions among these pools include mineralization, nitrification, anion adsorption, cation exchange, and mineral weathering. Release of nutrients from leaf litter that accumulates on the forest floor is primarily dependent on the C:N ratio of that substrate. Other decay products include nutrient and organic constituents, both solid (e.g., humus) and solution phase (e.g., organic acids). The nutrients released enter the solution phase where they are available for uptake by vegetation or the exchange complex, and for transport through the litter and soil horizons by percolation and lateral flow. The dissolved organic matter, which has a preponderance of acidic functional groups, generally depresses solution pH but also serves as a buffer by minimizing subsequent changes in pH. Organic acid concentrations are modeled to decrease with depth as the dissolved material is adsorbed, precipitated, and mineralized from upper to lower horizons.

Nitrogen is tracked as it is mineralized as ammonium through its possible conversion to nitrate. The series of reactions results in the loss of acid-neutralizing capacity. The movement of mobile nitrate ion through the profile as a strong acid accelerates cation leaching. Because the model allows for seasonal variations, it accounts for uptake of nitrogen during the growing season and the possible accumulation of nitrate in surface horizons during the winter under a protective snowpack. Snowmelt may flush some of the nitrate through the soil, although plant uptake may shortly thereafter deplete the same horizons.

Anion adsorption of sulfate, phosphate, and organic acids is also modeled. Sulfate adsorption is particularly important in altering acid–base chemistry when present in large amounts. Adsorption of sulfate increases the acid-neutralizing capacity of the soil solution, thereby reducing hydrogen ion exchange with base cations and the dissolution of base cations through weathering. Much of the sulfate is adsorbed into lower soil horizons, depending on the concentrations of aluminum and iron oxides and hydroxides present. Adsorption increases with decreasing pH.

High base cation exchange serves as an effective solution-phase buffer by limiting changes in soil solution pH through exchange with hydrogen and aluminum ions. If excess base cations are introduced, they are absorbed and hydrogen and aluminum ions released.

Released hydrogen ions are neutralized by reactions with bicarbonate and carbonate, and the aluminum is precipitated as aluminum hydroxide. The cation-exchange process proceeds rapidly and provides effective short-term buffering. Without weathering, however, many soils lose their buffering capacity within 50 to 200 years (Schnoor and Stumm, 1984).

Mineral weathering is normally a slow process, but one which releases base cations and silica into solution. The model predicts these reactions as a function of the mass of minerals present and the solution-phase hydrogen ion concentration, largely buffered through base-exchange processes against the direct impact of acid rain. In soils with low cation-exchange capacities, weathering rates are particularly important in assessing long-term nutrient status (Chapter 5).

In the uppermost soil layers, seasonal variation in soil solution chemistry is likely where atmospheric inputs are high (Fig. 4.15), whereas in the lower soil profile, buffering is sufficient so that decades are required before significant changes can be perceived (Liu *et al.*, 1991; Johnson and Lindberg, 1992). Once confidence is acquired in the model assumptions by monitoring seasonal variation in solution chemistry, the effort can be shifted toward predicting longer trends on the basis of annually integrated estimates of nutrient and organic matter loss or accumulations, changes in soil profile features, and other subjects considered in later chapters.

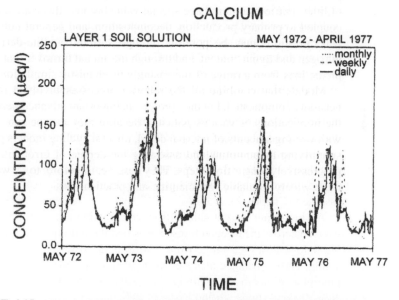

FIGURE 4.15. Mineral cycling models are able to predict changes in soil solution chemistry at frequent intervals in response to changes in the flux of nutrients in precipitation or from snowmelt. In this example, a sensitivity test shows that calcium concentrations in the uppermost horizon of a heavily leached soil can be predicted nearly as well at monthly intervals as at daily or weekly time steps. (From Liu *et al.*, 1991.)

VI. SUMMARY

The extent to which plants are able to acquire and maintain a balance of nutrients strongly controls their growth rate and ultimately the composition of forests. Nutrient balance can conveniently be defined in reference to the mineral nitrogen content in leaves. Imbalances in the availability of nutrients lead to a decrease in nutrient use efficiency. In some cases, the form in which a nutrient appears is also significant because species differ in their selective acquisition and in the way that surplus nutrients are stored. Moreover, in the soil, some forms of phosphate (organic) and nitrogen (NH_4^+) are more available to plants than others that are adsorbed, fixed, or more easily leached. The extent to which a nutrient is derived from soil, recycled within the vegetation, or is dependent on atmospheric inputs is an important consideration in the long-term management of forests. Likewise, the species composition of a forest is important because evergreen and deciduous species differ in their ability to capture nutrients from the atmosphere, extract them from the soil, and maintain a litter substrate with a favorable nutrient balance and rapid turnover.

Throughout this chapter we have illustrated how various stable isotopes (and ^{14}C) provide a quantitative basis to separate the sources of some nutrients, and to follow differential uptake and cycling of selected nutrients through the ecosystem. We advocate broader application of this methodology in testing models and in evaluating historical and future changes in ecosystem responses.

From a scaling standpoint, the distribution of broad soil groups associated with different climates and parent materials serves as a first-order stratification of the mineral cycling potential of forest ecosystems. The ratios forest floor litter mass : annual litterfall and mass of litter : nutrient content have special value because these are easily obtained and closely coupled to primary production, decomposition, and general nutrient availability. Broader generalizations may be possible through various satellite-derived measures of canopy nitrogen and lignin content, and through the annual turnover that may be compared against predictions from a range of increasingly mechanistic simulation models.

Models that combine all the relevant processes involved in mineral cycling are by necessity complicated, but they present an important advancement in our ability to evaluate the implications of various policies. Because they can be run at a variety of time steps with varying amounts of internal detail, mineral cycling models provide a means of testing simplifying assumptions and assessing the degree of error introduced by integrating at progressively longer time steps. They deserve, therefore, to be well documented and made more widely available for scientific and practical use.

SECTION II

Introduction to Temporal Scaling

WE HAVE laid the groundwork of forest ecosystem analysis in Chapters 2–4 by following the seasonal dynamics of water, carbon, and mineral cycles through individual stands. Stand-level models such as FOREST-BGC incorporate many of the principles derived from an understanding of the basic processes that control these cycles (Table II.1 and see Fig. 1.2). We now extend this understanding to longer time scales to address the concerns of managers who search for various harvest options, knowledge about changes in species composition, and the probability and impact of various types of disturbance on the structure and function of specific forest ecosystems. We start with an analysis of how structure and function change through a single forest life cycle by studying historical stand records, fire scars, tissue isotope ratios, and variation in annual ring increments. To gain insights into multiple forest life cycles that extend over many centuries we rely on information obtained from the analysis of pollen and wood samples extracted from bogs. On the basis of such analyses, models have been constructed that project forest development and vegetation dynamics well into the future (Fig. II.1). With these models, forest management may gain considerable insights into the probable consequences on future stand growth and composition of decisions made today. The degree of confidence that one can place in these kinds of models rests to a large extent on how well they incorporate principles validated in the study of individual stands.

TABLE II.1

Input Data Requirements for FOREST-BGC, with Example Values for Coniferous and Broadleaf Forests[a]

Parameter and units	Conifer forest	Broadleaf forest
Leaf C, kg ha^{-1}	2400	2400
Soil C, kg ha^{-1}	300000	300000
Litter C, kg ha^{-1}	100000	100000
Soil N, kg ha^{-1}	30000	30000
Litter N, kg ha^{-1}	2000	2000
Leaf turnover, year^{-1}	0.25	1.00
Stem turnover, year^{-1}	0.01	0.01
Coarse-root turnover, year^{-1}	0.01	0.01
Fine-root turnover, year^{-1}	0.5	0.5
Leaf lignin fraction, kg C lignin kg^{-1} C	**0.25**	**0.125**
Fine-root lignin fraction, kg C lignin kg^{-1} C	0.25	0.25
Specific leaf area, ha kg^{-1} C	**0.0025**	**0.0045**
Canopy light extinction coefficient, dimensionless	−0.5	−0.5
Precipitation interception coefficient, m H$_2$O projected LAI^{-1} day^{-1}	0.0002	0.0002
Snowmelt temperature coefficient, m H$_2$O °C^{-1}	0.005	0.005
Snow albedo decay coefficient, °C^{-1}	0.004	0.004
Snowpack energy deficit, °C	5.0	5.0
Spring minimum leaf water potential, −MPa	0.5	0.5
Leaf water potential at stomatal closure, −MPa	1.65	1.65
Maximum stomatal conductance, m H$_2$O s^{-1}	0.006	0.006
Cuticular conductance, m H$_2$O s^{-1}	0.00005	0.00005
Leaf boundary-layer conductance, m H$_2$O s^{-1}	0.001	0.001
Canopy aerodynamic conductance, m s^{-1}	0.2	0.2
Optimum stomatal conductance temperature, °C	20.0	20.0
Maximum stomatal conductance temperature, °C	40.0	40.0
Vapor pressure deficit, Pa		
Start of conductance reduction	750.0	750.0
Completion of conductance reduction	3000.0	3000.0
Leaf maintenance respiration coefficient, day^{-1}	0.0012	0.0012
Stem maintenance respiration coefficient, day^{-1}	0.0001	0.0001
Coarse-root maintenance respiration coefficient, day^{-1}	0.0001	0.0001
Fine-root maintenance respiration coefficient, day^{-1}	0.0012	0.0012
Q_{10} for plant maintenance respiration	2.0	2.0
Ratio of all-sided to projected LAI	2.3	2.0
Ratio of sapwood, kg C m^{-2} LAI^{-1}	**0.25**	**0.35**
Leaf retranslocation fraction	0.50	0.50
Q_{10} for soil maintenance respiration	2.4	2.4
Maximum leaf/(leaf + root) ratio for allocation	**0.70**	**0.80**
Maximum coarse root/stem ratio for allocation	0.25	0.25
Critical litter C:N ratio	25.0	25.0
Critical soil C:N ratio	10.0	10.0
Fine-root N fraction, %N$_{fine\ root}$ %N$_{leaf}^{-1}$	0.5	0.5
Stem and coarse-root N fraction, %N$_{stem}$ or %N$_{croot}$ %N$_{leaf}^{-1}$	0.01	0.01
Leaf growth respiration fraction, kg^{-1} C	0.35	0.35
Stem/coarse-root/storage growth respiration factor, kg^{-1} C	0.30	0.30
Fine-root growth respiration fraction, kg^{-1} C	0.35	0.35

TABLE II.1 (*Continued*)

Parameter and units	Conifer forest	Broadleaf forest
Maximum storage fraction	0.20	0.20
Living fraction of sapwood	**0.08**	**0.10**
Water stress integral factor	−0.15	−0.15
Fraction N_{leaf} in Rubisco	0.10	0.10
Maximum soil decomposition rate, $year^{-1}$	0.03	0.03
Maximum N uptake by roots (V_{Nmax}), $kg\,N\,kg^{-1}\,C_{fine\ roots}$	0.05	0.05
N uptake Michaelis–Menten constant (K_n), $kg\,N\,ha^{-1}$	20.0	20.0
Maximum average N_{leaf} concentration, $kg\,N\,kg^{-1}\,C$	0.04	0.04
Minimum average N_{leaf} concentration, $kg\,N\,kg^{-1}\,C$	0.02	0.02
Leaf on, year day	**0.0**	**140.0**
Leaf off, year day	**365.0**	**260.0**
Fine root on, year day	60.0	60.0
Fine root off, year day	304.0	304.0
Volumetric saturated soil water content	0.44	0.44
b parameter for soil water potential	−6.09	−6.09
Soil matric potential, −MPa	0.002224	0.002224

[a]For very generalized simulations only the variables in bold type may need to be specifically defined; these default values can be used. When more site- and/or species-specific data are available, the model can incorporate the additional details. From J. D. White, P. E. Thornton, and S. W. Running, *Global Biogeochemical Cycles* **11**, 217–234, 1998, published by the American Geophysical Union.

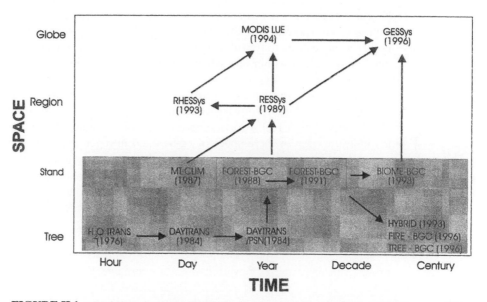

FIGURE II.1.   Evolution of a family of related ecosystem simulation models for application at variable time scales, showing the model name and initial publication date. Each new model shows a progression from its precursor in extending the time scale. Reference papers for the seasonal dynamics of individual stands include the following: H2OTRANS (Waring and Running, 1976), DAYTRANS (Running, 1984a), DAYTRANS/PSN (Running, 1984b), FOREST-BGC (Running and Gower, 1991), HYBRID (Friend *et al.*, 1993), BIOME-BGC (Running and Hunt, 1993), FIRE-BGC (Keane *et al.*, 1996a,b), and TREE-BGC (Korol *et al.*, 1995). To extend the time scale these models progressively add seedling recruitment, stand development, tree mortality, and disturbance processes that are the focus of Chapters 5 and 6.

CHAPTER 5

Temporal Changes in Forest Structure and Function

I. INTRODUCTION

In Chapters 2–4, we presented details about the way forest ecosystems operate at the stand level throughout the year. We now add a longer time dimension to our perspective. Most of our understanding of forest dynamics comes from retrospective analysis from looking at the way past events and stand structure lead to the development of the forests that we can see and measure today. To capture and to quantify that understanding is the role of various forest models, of which a variety exist with different assumptions and limitations. By careful application of these models we can project our knowledge forward in time to visualize how forests may appear and function in the future.

Historically, stand growth models have been developed by statistical analyses of large quantities of forest inventory data (Stage, 1977). These models assume implicitly that conditions in the future will be the same as those under which the measured forests developed. More theoretical but less quantitative models were introduced by ecologists such as Cowles (1899) and Clements (1936) who observed trends in species composition in chronosequences that began on fresh substrates. They noted that soil conditions improved in time as nutrients and organic matter accumulated, and they envisioned that early combinations of plants were replaced by other "communities," one after another in succession until a semistable *climax* type of vegetation appeared. The climax vegetation was thought to replace itself continuously in the absence of disturbance or a major change in climate. The entire sequence from bare substrate to steady-state climax vegetation was termed *primary succession.* If the steady-state community was disturbed, the sequence was not set back to the beginning because of soil development. Instead, a *secondary succession* of plant communities was envisioned that paralleled later phases in primary succession. Although ecologists debated the concept of plant communities (Gleason, 1926) and the idea of a single climatic climax (Tansley, 1935; Whittaker, 1953), the general theory of succession toward a stable end point was accepted in most textbooks until fairly recently (see review by Cook, 1996).

Since the 1960s, however, it has become apparent that the concept of a predictable succession of vegetation types is too simplistic, to the extent that modern ecological texts tend to avoid the term succession and replace it with "vegetation dynamics" (Cook, 1996). There are now many new theories that include more mechanistic and stochastic processes than found in earlier schemes (Pickett, 1976; Connell and Slatyer, 1977; Grime, 1977; Tilman, 1985; Huston and Smith, 1987). Most modern theoretical models recognize that the type of disturbance (fire, flood, windstorm, insect outbreak, etc.), and its intensity and timing, have important consequences that affect vegetation dynamics and ecosystem performance. Modern theories recognize the importance of (a) timing of species establishment, (b) differential abilities of species to compete for space and resources, and (c) tolerance of species to persist under adverse conditions (Cook, 1996).

Models that correctly predict historical patterns of forest growth and vegetation dynamics may still be unable to predict future responses where new conditions affect both growth and the mix of species (Bossel, 1991). During the twentieth century gypsy moth, chestnut blight, Dutch elm disease, and kudzu vine have altered the composition of forests throughout much of the eastern United States, with significant effects on forest stand dynamics. Over the same period, pesticides, herbicides, and atmospheric pollutants have modified conditions from those that occurred previously (Braun, 1950). Also, climatic conditions appear to be changing. Clearly, statistically based forest growth models or classic succession models would be inadequate for predictions; models that can represent ecosystem processes and the changing biosphere are required.

In this chapter, we begin by analyzing how forest ecosystems change in the course of undisturbed stand development and how those changes may be measured retrospectively. We then consider an array of new, mechanistic forest models that provide a means of projecting forest development forward in time. Models such as FOREST-BGC compute the changes in stand biogeochemistry through time. *Gap-phase* vegetation models incorporate patch-sized disturbance events, recruitment, and mortality processes and project their effects on the future composition and structure of the forest. Models which combine biogeochemistry with the dynamics of competition and tolerance among trees hold promise for more realistic projections of forest stand dynamics (Huston, 1991; Cook, 1996). Of course, disturbance is part of any realistic analysis of forest development, as considered in detail in Chapter 6. We do not know specifically what future climatic conditions will be like, nor what natural disturbances or human activity will occur, so we can only test scenarios with these models. These modeled scenarios, however, when done carefully, provide the most realistic method available for assessing the implications of management decisions in a changing future environment.

II. STRUCTURAL STAGES IN STAND DEVELOPMENT

As trees grow, they alter their environment and demands for resources; some trees gain resources at the expense of others. Competition is particularly intense when only one species is present because requirements for resources and the adaptations available to obtain them are nearly identical for all individuals. When more than one species is present, access to resources differs as a result of variable rooting depths and selective patterns of nutrient accumulation (Schulze *et al.,* 1994b). A few colonizing species grow rapidly to become dominant, while the majority of species, particularly in tropical rain forests, remain almost static until a gap in the canopy opens above them. Although mortality occurs, the leaf area index of such stands remains almost constant from year to year. In terms of ecosystems, it is important to recognize changes in canopy level competition that occur over long periods because these affect nutrient cycling, carbon uptake and allocation, as well as water flow through forests.

In only a few places have long-term records of species composition and growth been obtained from repeated measurements on the same plots of ground (Leak, 1987; Harcombe *et al.,* 1990; Fain *et al.,* 1994). More often, changes in forest structure and growth have been inferred from contemporary surveys that represent a wide range of forest age classes (Christensen and Peet, 1981). Scientists have made stem analyses of many trees (including dead standing and fallen stems) to provide a reconstructed history of particular stands (Oliver and Stephens, 1977; Bradshaw and Zackrisson, 1990; Abrams *et al.,* 1995). Changes in aboveground carbon and nutrient stores can be calculated from repeated measurements of stem numbers and diameters using allometric relationships (Chapter 3). Estimating complete forest life cycles, however, requires the construction of dynamic models that can predict tree growth and changes in composition over centuries.

The initial sequence of stand development can be extended to another generation of trees. If the replacement process occurs slowly, LAI and other properties, such as decomposition rates and stand water use, may remain stable from one generation to another (Bormann and Likens, 1979). Underlying the smooth progressive replacement of one species by another are important phases in ecosystem development that reflect subtle differences in the efficiency of operation and the relative importance of various processes. At least four idealized stages in stand development can be defined when a new forest replaces a previous one: (1) initiation, (2) stem exclusion, (3) understory reinitiation, and (4) an old-growth phase (Oliver, 1981; Fig. 5.1). We will describe each of these in some detail and then relate changes in ecosystem processes to each stage, and to repeated sequences to provide a long-term perspective.

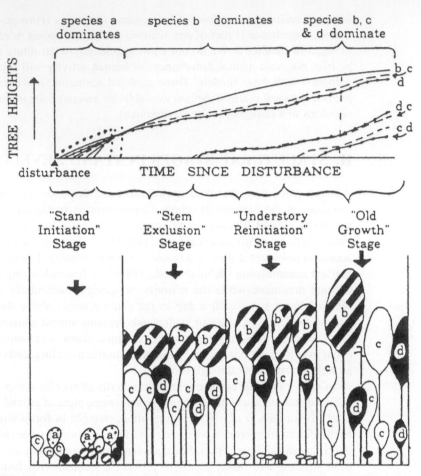

FIGURE 5.1.   Four idealized stages in stand development can be recognized from structural features alone. At initiation, seedlings are rarely affected by competition with other trees. At canopy closure, stem exclusion begins. As the overstory trees approach maximum height, the canopy opens and an understory is reinitiated. In the old-growth stage a multilayered canopy develops. (Modified from *Forest Ecology and Management,* Volume 3, C. D. Oliver, "Forest development in North America following major disturbances," pp. 153–168, 1981, with kind permission of Elsevier Science–NL, Sara Burgerhartstraat 25, 1055 KV Amsterdam, The Netherlands.)

A. Initiation Stage

After a major disturbance, seeds, spores, or other regenerating parts of plants are usually available to germinate or sprout. Seedlings quickly respond to improved light conditions and reduced competition for moisture and nutrients. This is true whether the disturbance results from the death of a single tree or arises from destruction of the entire stand. Where fresh substrate is deposited from floods, landslides, or volcanic eruptions, the biotic links from one generation to another are reduced but may still contribute to the rapid reestablishment of vegetation (Franklin *et al.,* 1985).

Animals are present in all stages of forest development, but their activities in the initiation stage are particularly significant. Wide-ranging birds and large vertebrates play an important role in distributing seeds across areas separated by inhospitable environments. Animals also improve the chances of seed germination by partially digesting seed coats and by preparing the soil in various ways. Some tree species have seeds that only certain kinds of vertebrates can disseminate successfully. The tree *Calvaria major,* a native on the island of Mauritius in the western Indian Ocean, has had no natural regeneration (without the assistance of humans) for 300 years because its seeds needed to be processed through the extinct dodo bird, *Raphus cucullatus* (Temple, 1977). In Africa, elephants consume great quantities of the fruits of *Balanities willsoniana;* when elephants are absent, the fruits, which are toxic to smaller animals, rot. Janzen (1979) observed that *Simaba cedron* trees growing in tropical forests of Central America share fruit characteristics with *Balanities* and concluded that the present restricted distribution of *Simaba* is related to the extinction of mastodons in the last 10,000 years. Some seed coats are so hard that only a few animals can crack them. The reintroduction of the horse into Central America by Spaniards increased dispersal of the seeds of a number of species with extremely hard seed coats (Janzen and Martin, 1982).

Rapid recolonization of forest sites following fire or other major disturbance may well depend on the presence of certain vertebrates. Truffles, fungi that produce potato-shaped fruiting bodies underground, form important symbiotic mycorrhizal associations with tree roots (Chapters 3 and 4). These fungi are almost exclusively hunted for, stored, and dispersed by small mammals (Maser *et al.,* 1978). Loss of small mammals from an ecosystem would reduce the likelihood of barren areas being revegetated with mycorrhizae of the truffle group.

Besides humans, birds have the greatest impact on long-distance seed dispersal. Blue jays (*Cyanocitta cristata*) cached more than 50% of the entire mast crop of an oak forest in Virginia an average distance of more than 1 km from the source (Darley Hill and Johnson, 1981). The ability of birds and small mammals to cache large nuts may help explain how hazelnut (*Corylus*) was able to colonize the landscape after retreat of the continental glaciers in Europe as fast and sometimes faster than species with wind-dispersed seeds such as birch and pine (Walter, 1954). Insects, birds, and bats also play important roles as pollinators (Crawley, 1983). Animals, through their eating habits and movement may concentrate nutrients within one system or distribute them between ecosystems (McClelland, 1973). Large vertebrates also affect forest ecosystems by changing the microhabitat through trampling, burrowing, or, in the case of animals such as beavers, felling trees. The absence of large animals can, therefore, indirectly limit the populations of other animals and plants. For example, when 10-ha segments of Amazon forests were isolated, peccaries no longer provided wallows, and three species of frogs disappeared, along with other types of flora and fauna (Lewin, 1984).

Although the initiation stage in stand development is simple to define on the basis of the limited height and age of trees present, predicting which species of tree seedlings will survive is extremely difficult, because of local variation in resource availability and selective herbivory by animals.

B. Stem Exclusion Stage

When only one tree species is present, competition is particularly intense as stand development enters the "stem exclusion" stage. In extreme cases, stem numbers may be reduced by nearly 90% in 50 years (Tadaki *et al.,* 1977). A "self-thinning" rule defines an upper limit for the number of stems as a function of the average biomass of individual trees (Drew and Flewelling, 1977; White, 1981; Westoby, 1984). This empirical relation is usually presented with a log–log plot that has a nearly constant slope of −3/2 that defines a decrease in maximum stem biomass ($W_{s.max}$, kg) as stem population per hectare (p) increases:

$$W_{s.max} = k_s p^{-3/2} \tag{5.1}$$

where k_s is a coefficient that increases with the maximum stem biomass of individual trees, obtained either from empirical data such as yield tables or from simulation models that calculate maximum accumulation of (live) stem biomass.

To take an empirical example, a stand at maturity with 150 stems and maximum stand biomass of $600 \, \text{Mg ha}^{-1}$ gives $k_s = 7.3 \times 10^6$. To apply the formula, initial stand mass and stocking density must he provided, and the mean stem mass (W_s) generated by a growth model tested against $W_{s.max}$ for the current population [$p(W_{s.max})$]. If $W_s > W_{s.max}$, Eq. (5.1) solves for $p(W_s)$ and predicts that stem numbers will be reduced by $\Delta p = [p(W_s) - p(W_{s.max})]$, where Δp is stem mortality. By the end of the next year, if total stem biomass has increased, the procedure is repeated. The biomass in annual mortality (or harvest) is an important ecosystem variable required to estimate the return (or removal) of organic matter to the soil in large woody debris. The results of applying the −3/2 self-thinning rule to three hypothetical pine plantations with initial stocking at 1000, 2500, and 5000 stems ha^{-1} show that self-thinning occurs first in the stand with the highest initial stocking; eventually, however, all three stands reach similar stem numbers in later stages of stand development (Landsberg and Waring, 1997; Fig. 5.2).

The LAI of the overstory reaches its maximum during the stem exclusion stage. Because there is no net addition of foliage to the canopy, only a transfer of leaf area from less competitive to more competitive individuals can occur. Leaf area, not biomass of stems, is the underlying principle behind the operation of the self-thinning rule (Westoby, 1977; Landsberg, 1986a). The self-thinning rule is widely applied in forestry (Tang *et al.,* 1994), mainly because it does not require information on leaf area, but this makes the rule empirical (Weller, 1987). Ecologists have extended the application of the rule to simulate the optimum stocking and thinning regime required to produce large woody debris as quickly as possible from young, fast growing forests (Sturtevant *et al.,* 1997).

Growth efficiency, defined as stem wood production per unit of leaf area (Chapter 3), is another widely applied measure of the intensity of competition among individual trees that foresters and ecologists share. Between the initiation stage and the stem exclusion stage, growth efficiency often decreases by more than 90% (Fig. 5.3). At canopy closure, trees readily separate into classes (dominant, codominant, intermediate, and suppressed) that reflect the amount of sunlight captured by individual crowns (Oliver and Larson, 1996). Trees of mean diameter (average basal area) usually represent codominant individuals. Trees of below average diameter display below average growth efficiencies because of their less favorable competitive position (Oren *et al.,* 1985; O'Hara, 1988). The tallest,

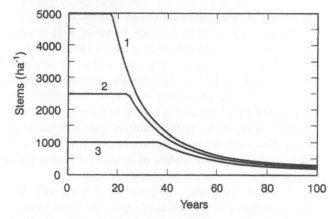

FIGURE 5.2.   Application of the self-thinning rule in hypothetical pine stands with initial stem numbers of (1) 5000, (2) 2500, and (3) 1000 stems ha^{-1} shows that stem mortality is proportionally much higher in (1) than (2) or (3). After 50 years, however, tree numbers begin to approach similar values regardless of initial stocking. (Modified from *Forest Ecology and Management,* Volume 95, J. J. Landsberg and R. H. Waring, "A generalized model of forest productivity using simplified concepts of radiation-use efficiency, carbon balance and partitioning," pp. 205–215, 1997, with kind permission of Elsevier Science–NL, Sara Burgerhartstraat 25, 1055 KV Amsterdam, The Netherlands.)

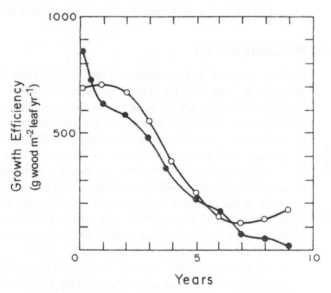

FIGURE 5.3.   Two 20-year-old Scots pines with approximately similar dimensions showed more than a 90% reduction in growth efficiency in less than a decade. During the last 3 years, canopy closure occurred and one tree (●) was overtopped. The other tree (○) achieved a position of dominance and increased its growth efficiency. Allometric equations developed from trees in the same stand were used to estimate growth from changes in stem diameter. Leaf area was calculated from sapwood area present each year. (Data from R. H. Waring, B. Axelsson, and S. Linder, Swedish University of Agricultural Sciences, Uppsala, Sweden. Figure from Waring and Schlesinger, 1985.)

dominant trees in unthinned forests usually produce the most growth per unit of leaf area. Dominant trees in thinned stands, however, may develop extensive crowns, which result in below average growth efficiencies (O'Hara, 1988).

If frequent thinning is practiced within the stem exclusion stage, canopy closure is avoided and the LAI of the overstory remains below the maximum. With a more open canopy, however, understory vegetation, including other tree species, can become established and create conditions similar to the next stage in stand development. The combined LAI of all vegetation remains similar over a broad range in stocking density of overstory trees. This tendency to display a common LAI from initiation through later stages in stand development is a key feature of forest ecosystems that will have important implications for scaling as discussed in Chapters 7–10.

Practices that modify the potential of a site will alter the maximum LAI. This axiom was demonstrated in a series of replicated experiments on young pine forests in Sweden, which received a combination of irrigation and nutrient additions for more than a decade (Axelsson and Axelsson, 1986). In all treatments, including the control, mean tree growth efficiency showed an exponential decrease with time (Fig. 5.4a). Treatment effects could be compared (1) in relation to growth efficiency at a common and necessarily low LAI (1.0) or (2) as a function of total aboveground stem wood production at canopy closure (Fig. 5.4b). The upper limits on wood production tend to approach an asymptote at an LAI of 5 to 6, which corresponds to interception of more than 90% of all visible radiation (Chapters 2 and 3). Considerable flexibility exists in how forests are managed through the control of the spatial distribution of leaf area on trees and understory vegetation.

C. Understory Reinitiation Stage

As trees approach their maximum height, growth slows; however, mortality still occurs, and gaps are created in the canopy that cannot be filled completely by branch extension. The gaps allow sunlight to penetrate to the forest floor and stimulate the reinitiation of understory vegetation that was largely excluded in the previous stage. The reintroduction of an understory provides more opportunity for diverse animal populations to increase as both cover and forage are available. The increase in diversity of habitats and food supply has been suggested as an important feature for supporting a host of controlling agents (arthropods, ants, small rodents, and birds) that reduce the dangers of insect herbivory so common to the previous stage in stand development (Holmes *et al.*, 1979; Price *et al.*, 1980; Doane and McManus, 1981; Schowalter, 1989; Torgersen *et al.*, 1990; Perry, 1994). More complex forests would, according to this reasoning, have greater inherent resistance to insect herbivory, a subject discussed in Chapter 6.

Stem mortality in this stage of stand development provides the first large pieces of organic debris to the forest floor that are likely to persist through the next stage. Foresters are often concerned about the accumulation of woody debris at this time as a potential fuel hazard, breeding site for bark beetles, or habitat for other potentially damaging insects and pathogens. Ecologists, on the other hand, have begun to investigate the significance of woody debris as a potentially important contributor to species diversity and nutrient cycling, and a component of long-term carbon storage (Thomas, 1979; Harmon *et al.*, 1986; McComb *et al.*, 1986; Perry, 1994; Cohen *et al.*, 1996).

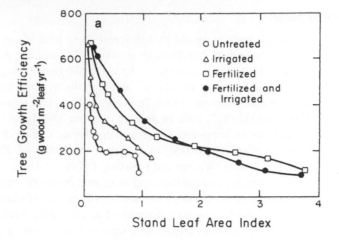

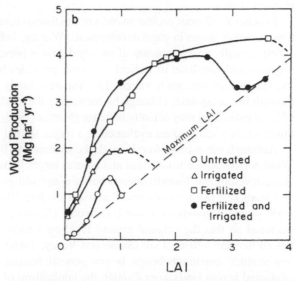

FIGURE 5.4.   (a) Saplings of Scots pine grown over a decade under different nutrient and water regimes showed large differences in growth efficiency at a comparable LAI of 1.0. As the stands approached their maximum leaf area after 10 years, however, growth efficiencies approached similar values. At that time LAI differed by more than threefold. (b) As Scots pine plantations approached their maximum LAI determined by experimental treatments, stem wood production per hectare became closely related to maximum LAI because stem wood production per unit of leaf area was similar, regardless of the environment at canopy closure. (From Waring and Schlesinger, 1985.)

D. Old-Growth Stage

Dominant trees in an old-growth forest show little if any height growth. In many cases, treetops have died, but diameter growth still continues. As a result, the total live biomass usually reaches a maximum in the early old-growth stage. In boreal forests, maximum biomass approaches $400\,Mg\,ha^{-1}$ although the average biomass is only about $40\,Mg\,ha^{-1}$ (Botkin and Simpson, 1990). In dry to wet tropical regions, forest biomass ranges from 160 to $690\,Mg\,ha^{-1}$, with an average of about $330\,Mg\,ha^{-1}$ (Brown *et al.,* 1991). Standing biomass in the temperate old-growth coniferous forests of the Pacific Northwest averages between 800 and $1000\,Mg\,ha^{-1}$, with a maximum of $2500\,Mg\,ha^{-1}$ reported in coastal redwood forests (Waring and Franklin, 1979).

Tree species composition and structural diversity may reach maximum levels in the old-growth stage, although the number of species of plants and animals together may be higher at stand initiation (Oliver and Larson, 1996). Species of trees adapted to growing in shaded conditions require less sapwood to support a given amount of leaf area than those adapted to more exposed situations (Chapter 3). The reduction in sapwood volume may reduce stem and branch maintenance respiration by as much as 50% (Edwards and Hanson, 1996).

In general, all trees present in old-growth forests tend to have lower growth efficiencies than at earlier stages in stand development (Waring, 1983; Waring *et al.,* 1992; Kaufmann, 1996). Usually a large amount of woody debris is present in various decay classes, which together with variation in vertical structures provides habitat for many organisms, so that food chains are extremely varied. This variation tends to favor an even flow of resources through the ecosystem, although by increasing the total number of species, the probability of local extinction may actually increase (Moffat, 1996; Tilman, 1996). Species extinction, however, is a subject best evaluated at a larger spatial scale (Chapter 9).

Although old-growth forests might be assumed to be free of disturbance, they are not. Wind periodically uproots trees and creates larger gaps than present in previous stages of stand development. Fires frequently burn through old-growth eucalyptus forests in Australia and are a key factor in sustaining nutrient cycling (Burgman, 1996; Chapter 6). Even in 1000-year-old stands of coastal redwood (*Sequoia sempervirens*), many floods have occurred so that the original ground level on which trees were first established is now buried beneath >10 m of silt (Stone and Vasey, 1968). The presence of an overstory that has attained maximum height is one general feature of old-growth forests that can be evaluated across landscapes through the application of various remote sensing techniques (Wallin *et al.,* 1996).

Some forest stands, particularly in arid climates with incomplete canopy closure and frequent historic ground fires, develop multiple age classes growing interspersed on the site. O'Hara (1996) studied ponderosa pine stands in Oregon and Montana and found that they naturally contained two to five cohorts, or age classes of trees, ranging from 22 to 220 years. These multiaged stands were found to have the same volume increment productivity as equivalent even-aged stands in the area. O'Hara also concluded that volume productivity was better predicted when stand structure was included in the algorithms than with LAI alone.

III. FUNCTIONAL RESPONSES OF STANDS AT DIFFERENT STAGES IN DEVELOPMENT

Studies that provide measures of functional changes over a complete range in stand development are rare. It is necessary, therefore, to piece together information from a variety of studies. In this section we present examples which emphasize production, nutrient uptake, and accumulation in biomass and organic matter over time.

A. Production of Biomass

Production of biomass, as shown in earlier chapters, can be predicted as a function of stand structure, the availability of critical resources, and climatic restrictions. Mencuccini and Grace (1996) compared structural and functional changes in tree carbon, water, and nutrient status of 10 plantations of Scots pine which ranged in age from 7 to 59 years and grew in a deep sandy soil in southeast England. Although the study included only the first three phases of stand development, it provides important insights.

At age 7 the pine plantation had a leaf area index far below the maximum that the site could support (Fig. 5.5a). Bracken fern (*Pteridium aquilinum*) formed a dense understory. By age 20, stem exclusion began and density decreased from above 3200 trees ha^{-1} to less than half that number. As LAI peaked, the fern understory was completely shaded out. A plateau in LAI was maintained until about age 35, by which time trees had attained nearly 90% of the height recorded at age 60 (Mencuccini and Grace, 1996). Beyond age 40, overstory LAI began to decrease so that, by age 60, LAI was about half of that at age 20. The opening of the canopy at ages beyond 50 years permitted ferns to reappear (understory reinitiation stage).

Aboveground net primary production (NPP$_A$) initially increased in parallel with canopy leaf area index but decreased more rapidly as the stands aged (Fig. 5.5b). This pattern is general for most even-aged forests (Ryan *et al.*, 1997b). An increase in maintenance respiration cannot account for more than a 10% change in growth rates, according to reviews by Ryan *et al.* (1997b) and Gower *et al.* (1996). The study by Mencuccini and Grace offers an alternative explanation for the rapid decrease in NPP$_A$ with stand age. They measured water conducting properties in stems and branches for the full range of age classes and from these data calculated stand hydraulic conductance (G_{st}) (Fig. 5.5c). The relation between NPP$_A$ and G_{st} is linear, with $r^2 = 0.88$. This analysis supports the hypothesis that hydraulic limitations on photosynthesis and the amount of LAI that can be supported are sufficient to account for most of the observed decrease in tree and stand growth with age (Ryan and Yoder. 1997). An alternative hypothesis, namely, that nutrient limitations arise from a reduction in the rates at which minerals are recycled, may apply in some systems (Gower *et al.,* 1996), but no variations in foliar nutrient concentrations were reported across the full range of stand ages by Mencuccini and Grace (1996).

B. Accumulation of Nutrients and Soil Organic Matter

Patterns of nutrient accumulation are roughly similar to those of biomass because nutrient concentrations in wood do not change appreciably as trees age (Fig. 5.6). Some species

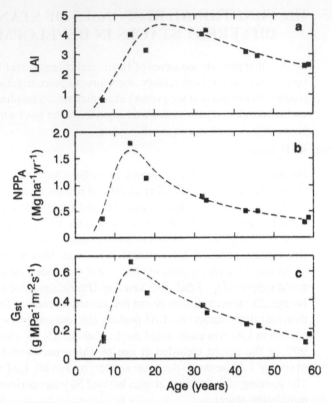

FIGURE 5.5. (a) Leaf area index (LAI) in plantations of Scots pine increases rapidly following establishment, peaking at stand closure, about age 20. In later phases of stand development, the overstory LAI decreases slowly, which permits the understory ferns (and other life forms) to increase their LAI. (b) Aboveground net primary production (NPP$_A$, Mg ha^{-1} year^{-1}) tends to follow overstory LAI with stand development but decreases much more abruptly in later phases. (c) Total stand hydraulic conductance (G_{st}, g H$_2$O MPa^{-1} m^{-2} s^{-1}) was based on average dimensions of height, diameter, and branch length for all trees present in each stand. Changes in stand hydraulic conductance directly parallel changes observed in NPP$_A$ as pine stands develop. (After Mencuccini and Grace, 1996.)

in the initiation stage, however, have particularly high concentrations of nutrients in their biomass after forest burning or cutting (Marks, 1974). As canopy leaf area approaches a maximum, fine-litter accumulations on the forest floor approach equilibrium. The relatively rapid decomposition of these nutrient-rich materials allows recirculation through the ecosystem.

There is a similar pattern of nutrient storage as the forest floor and soil organic matter accumulate when forests develop on fresh substrate (primary succession). Sometimes the organic C and N in the forest floor peak early and then slowly decrease to an equilibrium value after a few centuries. This was the case for forest soils that developed on volcanic mudflows in northern California (Fig. 5.7). A rapid accumulation of N and C in the soil was presumably aided by early colonizing vegetation, which included N-fixing shrubs. On the other hand, C and N may accumulate more rapidly in soils where decomposition rates

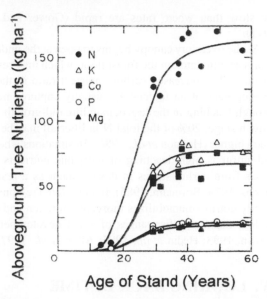

FIGURE 5.6.   Accumulation of nutrients in tree biomass during the postfire development of jack pine (*Pinus banksiana*) in New Brunswick, Canada. (From MacLean and Wein, 1977; drawing from Waring and Schlesinger, 1985.)

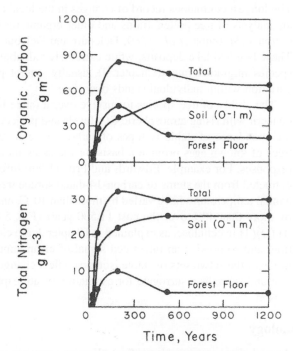

FIGURE 5.7.   Accumulation of organic C and N in soil developed on volcanic mudflows of varying age on Mount Shasta, California. (Modified from Dickson and Crocker, 1953; drawing from Waring and Schlesinger, 1985.)

are slow than where rates are rapid (Gower and Son, 1992; Binkley and Valentine, 1991).

As the overstory canopy begins to open at the understory reinitiation stage, more woody debris accumulates on the forest floor, which may result in slowing the turnover of organic residues. The increased carbon: nutrient ratio in the organic substrate, combined with a development of an understory, usually captures nutrients that might otherwise be lost through leaching in the less demanding old-growth stage. When a forest reaches the old-growth stage, 20% of the total N in litterfall may be woody material, with the Ca fraction much higher (Harmon *et al.,* 1986). In situations where atmospheric deposition of nitrogen and sulfur is high, leaching of inorganic nutrients from old-growth stands may exceed losses from earlier stages in development as a result of reduced uptake (Bormann and Likens, 1979; Bormann, 1985). In unpolluted regions, however, the large store of decaying woody detritus immobilizes inorganic nitrogen and cations, to the extent that losses may be less than in earlier stages of stand development (Sollins *et al.,* 1980; Binkley and Brown, 1993; Hedin *et al.,* 1995; Allen *et al.,* 1997; Clinton *et al.,* 2002).

IV. LOOKING BACK IN TIME

A. Paleoecology

The longest continuous record of changes in the local composition of forests derives from an analysis of tree pollen grains and other organic residues collected from bogs and lake bottoms (Solomon *et al.,* 1980; Delcourt and Delcourt, 1985; Foster and Zebryk, 1993). These bog and lake deposits, while extremely valuable in recording climatic changes and species migration rates (Chapter 9), usually cannot provide information that describes variation within individual stands over time.

The ability of tree species to survive over a range in climatic variation is of particular concern with rising atmospheric CO_2 levels and projected climatic warming. Variations in the abundance of stable isotopes of carbon, oxygen, and hydrogen bound in the rings of trees of known ages provide a basis for the assessment of climatic variation and growth responses. For example, Edwards and Fritz (1986) inferred from analysis of δD and $\delta^{18}O$ extracted from the stems of carbon-14-dated spruce trees buried in a single bog that mean annual temperatures have varied by more than $10°C$, and growing season relative humidity by more than 40% over the past 11,500 years (Fig. 5.8). The isotopic ratios $^{18}O/^{16}O$ and ^2H (D)/^1H in cellulose, as explained in Chapter 2, are closely correlated to that in precipitation, and so provide an integrated signal of mean annual temperature. Differences in the ratios of these two sets of stable isotopes reflect changes in fractionation during transpiration and serve as a surrogate for estimation of atmospheric humidity during the growing season.

B. Dendrochronology

Annual rings in trees also reflect climatic variation in more recent years, as a result of changes in the amount and pattern of carbon allocation to stem increment (Chapter 3). Tree-ring data must be subjected to careful scrutiny in collection and analysis (Fritts and

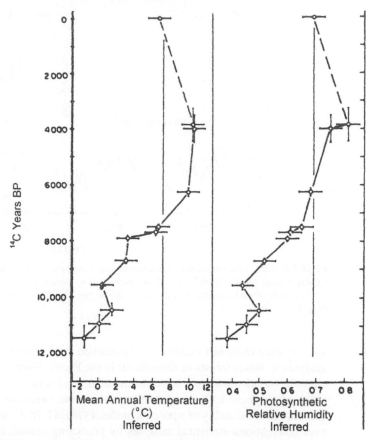

FIGURE 5.8. The abundance of heavier and lighter forms of hydrogen and oxygen isotopes in cellulose from current and fossil wood samples from a bog in Canada indicate that major changes occurred in climate over the last 11,500 years. (Modified from *Applied Geochemistry*, Volume 1, T. W. D. Edwards and P. Fritz, "Assessing meteoric water composition and relative humidity from $\delta^{18}O$ and $\delta^{2}H$ in wood cellulose: Paleoclimatic implications for southern Ontario, Canada," pp. 715–723. Copyright 1986, with kind permission from Elsevier Science Ltd, The Boulevard, Langford Lane, Kidlington OX5 1GB, UK.)

Swetnam, 1989). Ideal trees for analysis grow in exposed situations, often on infertile soils, where they lack competition with other trees and therefore show a more direct growth response to climatic variation. If conditions become too harsh, however, false and missing rings may occur; these must be recognized by analysis of replicate cores or cross sections. Finally, an adjustment must be made to account for the reduction in ring width as trees increase in girth. Once these corrections are made, recent variations in ring widths are correlated with recent climatic measurements, and inferences are made by extrapolation of older ring chronologies to climatic variation in the distant past (Fig. 5.9).

Cook *et al.* (1991) applied dendrochronological techniques to estimate climatic changes from a 1000-year tree-ring chronology on huon pine (*Lagarostrobus franklinii*) in Tasmania. The analyses suggested that during the twentieth century warming has exceeded

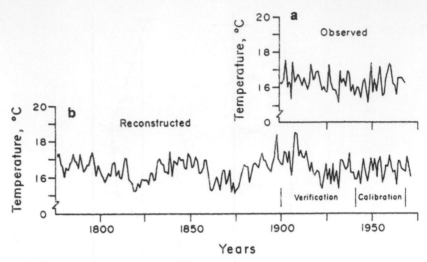

FIGURE 5.9. Correlations between tree-ring width and temperature obtained for the period 1940 to 1970, validated using the data from 1900 to 1940, were used to extrapolate the climate in Tasmania in the interval 1780 to 1900. (From LaMarche and Pittock, in *Climate from Tree Rings,* 1982, courtesy of Cambridge University Press.)

any recorded in the last millennium. Graumlich *et al.* (1989) attempted to extend tree-ring analyses to whole forests of Douglas-fir in the Pacific Northwest and concluded that, once climatic variation was accounted for, no additional response could be found in relation to rising atmospheric CO_2 concentrations during the twentieth century. Hunt *et al.* (1991) demonstrated the ability of a process model, FOREST-BGC, to achieve improved accuracy over conventional statistical analyses by predicting annual variation in growth recorded in a 50-year dendrochronological record extracted from pine trees in Montana (Fig. 5.10). To take account of likely continual changes in atmospheric composition, it is almost a requirement that more sophisticated analyses be developed to interpret dendrochronological records.

Assessments of the effect of rising CO_2 concentrations on tree growth have been inferred from comparisons of carbon-13 and carbon-12 deposited in the cellulose of annual rings. As discussed in Chapter 3, the stable isotope ^{13}C is discriminated against in photosynthesis and, as Farquhar *et al.* (1982) have shown, is quantitatively related to variation in the ratio of CO_2 within the leaf to that in ambient air. Because the $\delta^{13}C$ composition of the atmosphere has been diluted with carbon enriched in ^{12}C from fossil fuel consumption, adjustments must be made to account for these changes, which are derived from analyses of CO_2 trapped in gas bubbles of glacial ice (Polley *et al.,* 1993). Marshall and Monserud (1996) analyzed $\delta^{13}C$ from the annual rings of three species of conifers growing in Idaho and found that the calculated ratio of intercellular CO_2 to ambient CO_2 remained constant at 0.75 for the last 80 years. This implies that stomatal conductance has been reduced and that net photosynthesis has increased by 30%, in proportion to the rise in atmospheric CO_2. Similar responses have been reported for other gymnosperms in the western United States (Leavitt and Long, 1989; Stuiver and Braziunas, 1987) and in Europe (Freyer and Belacy,

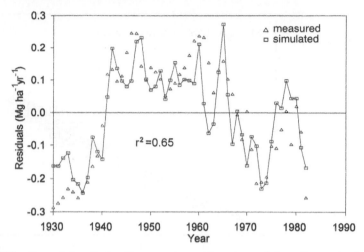

FIGURE 5.10. Detrended residuals of interannual stem biomass growth for a 52-year simulation of a ponderosa pine stand in Montana using FOREST-BGC show that process simulation models link growth with climate more closely than do more empirical dendrochronology models. (After Hunt *et al.,* 1991.)

1983), but not in Tasmania where huon pine appears to have maintained a constant intercellular CO_2 as atmospheric CO_2 levels have risen (Francy, 1981).

Walcroft *et al.* (1997) analyzed variation in the $\delta^{13}C$ within annual rings from *Pinus radiata* trees in New Zealand and showed that it is possible to discern seasonal variations in drought. It is sometimes difficult, however, to isolate CO_2 effects with $\delta^{13}C$ analyses alone because the discrimination of ^{13}C is controlled both by stomata and by the N status of leaves (Chapter 3). Nitrogen could become less available as carbon-rich substrates accumulate in litter and cause mineralization rates to decrease. On the basis of allocation models described in Chapter 3, more carbon resources would be shifted toward roots, as has been documented in at least one CO_2-enrichment experiment (Tissue *et al.,* 1993). On the other hand, atmospheric deposition of nitrogen associated with anthropogenic activities might compensate for deficiencies in N, if not other minerals (Chapter 6). No general rules are yet available to predict which species would benefit most, or how changes in C:N ratios might affect herbivory on trees, although important trends have been identified (Bazzaz, 1990; Woodward, 1991; Field *et al.,* 1992).

Norby (1996) showed the potential value of obtaining a retrospective estimate of leaf area to normalize the reported stem growth observed in eight CO_2-enrichment experiments. Coyea *et al.* (1990) demonstrated that in at least one species, *Abies balsamea,* retrospective estimates of leaf area could be made because sapwood, with a known correlation to leaf area, converted to heartwood after a fixed number of years (Fig. 5.11). The discovery that sapwood converts to heartwood after a fixed number of years allows a retrospective analysis of tree leaf area, which together with stem diameter growth provides a reconstruction of individual tree and stand growth efficiencies backward in time. The technique has important implications for interpreting the periodicity of outbreaks of insects, fire, and other agents of disturbance (Coyea and Margolis, 1994; Chapter 6).

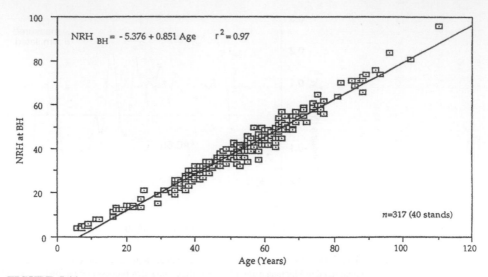

FIGURE 5.11.   In balsam fir (*Abies balsamea*) the number of annual rings in heartwood at breast height (NRH$_{BH}$) is a linear function of tree age. The sapwood rings represent the outer 6.0 years of growth (equivalent to the intercept). By determining the sapwood area and diameter growth at any given age, historical reconstruction of stand LAI and tree growth efficiency is possible. (After Coyea and Margolis, 1994.)

C. Pedology

Most of the carbon in ecosystems is incorporated into soils. Conventional analyses of soil carbon are not sufficiently accurate to assess important changes in turnover rates or whether undisturbed ecosystems are truly in steady state. Isotopic analyses of soil organic fractions have much to contribute in answering these questions and can serve as a baseline against which to measure future changes.

As a result of thermonuclear weapons testing from 1955 to 1963, the amount of ^{14}C in the atmosphere was approximately doubled. As plants incorporated the radioactive carbon into biomass, and that biomass became litter, the bulk soil organic matter increased in ^{14}C in proportion to the rate of turnover of older carbon substrates. Trumbore *et al.* (1996) compared organic matter ^{14}C in present and in archived (pre-1963) soils collected across an elevational transect in the Sierra Mountains of California. On the basis of these analyses, they calculated the turnover time for three increasingly recalcitrant extracts of soil organic matter in six "steady-state" forest ecosystems. Only two fractions showed significant change over the 30 years. Turnover rates of these two less recalcitrant fractions varied from <10 to >200 years across the elevational gradient. When the rates were plotted against mean annual temperature, a general relation was derived that also applied to data from Brazil and Hawaii (Fig. 5.12). From these relationships Trumbore *et al.* (1996) estimated that organic matter turnover rates increase under warming climatic conditions. Calculating the net gain or loss of soil carbon, however, requires that primary production as well as decomposition be accounted for. There are obvious advantages in combining dendrochro-

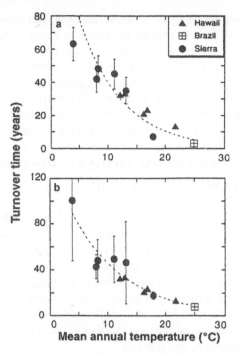

FIGURE 5.12.   (a) Turnover times for fast-cycling fractions of soil organic matter in soils from transects in the Sierra Mountains of California, in Hawaii, and in the Amazon. The exponential curve with mean annual temperature is $Y = 151 \exp(-0.134T)$, where T is in °C. (b) Combined turnover times for the fast- and moderate-cycling fractions of soil organic matter described above; $Y = 138 \exp(-0.110T)$. (Modified with permission from S. E. Trumbore, O. A. Chadwick, and R. Amundson, 1996, "Rapid exchange between soil carbon and atmospheric carbon dioxide driven by temperature change." *Science,* Volume 272, pp. 393–395. Copyright 1996 American Association for the Advancement of Science.)

nological and soil carbon analyses with ecosystem modeling and analysis in future studies.

Conversion of tropical forest to pasture in Hawaii provided Townsend *et al.* (1995) with a chance to extend carbon isotope analyses to disturbed situations and to compare estimates of soil carbon turnover derived from ^{14}C measurements with those obtained from $^{13}C/^{12}C$ analyses. The latter comparison was possible because tropical trees and pasture grasses differ in their photosynthetic pathways; trees discriminate twice as much against the heavier ^{13}C isotope than do grasses. With the dates of forest conversion to pasture known, the turnover rates could be calculated on the basis of the degree of ^{13}C enrichment observed in three organic fractions for which the potential turnover rates were different by orders of magnitude. Predictions made with the two isotopic techniques agreed closely with turnover rates predicted with the CENTURY ecosystem simulation model (Parton *et al.,* 1987), and they confirmed that in order to estimate the residence time of soil carbon it is necessary to separate organic matter fractions into at least three major components (Townsend *et al.,* 1995).

Isotopic analyses of sulfur, nitrogen, and strontium (a surrogate for calcium and other cations) in soils provide the means to assess variation in sources and rates of atmospheric deposition on forests (Winner *et al.,* 1978; Peterson and Fry, 1987; Miller *et al.,* 1993; Durka *et al.,* 1994; Hedin and Likens, 1996). All of the techniques discussed in this section permit retrospective testing of models. In addition, they represent a long-term, widely dispersed source of data that, if properly dated, can serve as a continuous benchmark against which to measure change and compare model predictions of soil chemistry and elemental cycling through ecosystems (Chapter 4).

V. ECOSYSTEM MODELS, PROJECTIONS FORWARD IN TIME

Two classes of forest ecosystem models have been well developed, those that focus on the physiology and biogeochemistry of the ecosystem, such as FOREST-BGC, and those that concentrate on the life cycle dynamics of trees in the ecosystem, best exemplified by the family of canopy gap models (Bossel, 1991; Huston, 1991; Dale and Rauscher, 1994). Both classes are process models; the biogeochemical models compute growth from the seasonal dynamics of canopy carbon balances, while the gap models emphasize disturbance, recruitment, and mortality processes that affect individual trees. Ideally, both types of models should be used together to interpret and predict future changes in stand structure, composition, and function.

A. Biogeochemistry Models

The most direct initial test of a biogeochemical model is accurate simulation of an observable quantity such as stem growth over a short time. Successful simulation of stem growth requires that photosynthesis, respiration, carbon allocation, and tissue turnover all be balanced realistically. A number of models have illustrated the capability to simulate 1- to 5-year stem growth aggregated for a stand (Korol *et al.,* 1991; Aber and Federer, 1992; McMurtrie and Landsberg, 1992; Cropper and Gholz, 1993; Running, 1994).

Korol *et al.* (1991) simulated the 5-year growth for 176 Douglas-fir trees with different levels of canopy dominance growing in stands that contained trees of mixed age classes between 30 and 80 years. The trees were distributed across five sites that encompassed a productivity gradient in the dry interior forests of British Columbia. Individual tree stem growth increments, measured by stem analysis, were in close agreement with those modeled by FOREST-BGC (Fig. 5.13). Aber and Federer (1992) simulated the aboveground NPP for forests at ten sites across a broad climatic range of North America with the PnET model (Fig. 5.14). The high correlation shown between predicted and observed NPP is particularly impressive because PnET does not purport to be a comprehensive ecosystem analysis, but concentrates on a critical synthesis of the relationship between leaf nitrogen content, photosynthetic capacity, stomatal conductance, and leaf longevity, as discussed in Chapter 3. These models illustrate the important philosophical point that the most useful models are not the most complex but are the ones that cleanly and efficiently represent only the most critical processes and interactions operating in forest ecosystems.

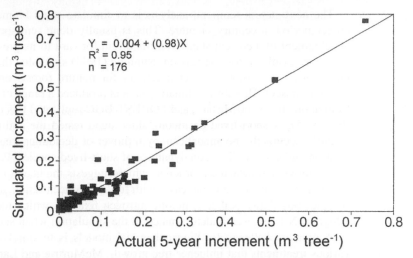

FIGURE 5.13. Comparison of measured 5-year growth increment with FOREST-BGC simulated growth for Douglas-fir trees 35–85 years of age in British Columbia. In this application, FOREST-BGC was defined for individual trees, and intertree competition for radiation and precipitation was added, based on canopy dominance. All other model parameters (Table II.1) represented stand-level conditions normally used by the model. (After Korol *et al.,* 1991.)

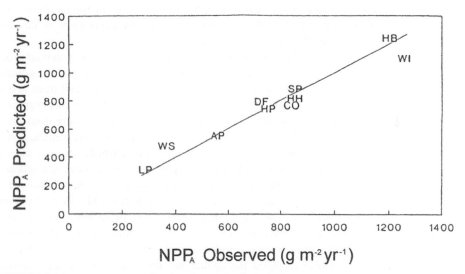

FIGURE 5.14. Comparison of observed aboveground net primary production (NPP$_A$) and simulated NPP$_A$ with the PnET ecosystem model for ten forest ecosystems around the United States. LP, Wyoming; WS, Alaska spruce; AP, Alaska aspen; DF, Oregon; HP, Massachusetts pine; CO, North Carolina; HH, Massachusetts hardwood; SP, Florida; HB, New Hampshire; WI, Wisconsin. (From *Oecologia,* "A generalized, lumped-parameter model of photosynthesis, evapotranspiration and net primary production in temperate and boreal forest ecosystems," J. D. Aber and A. Federer, Volume 92, p. 496, Fig. 2, 1992, © 1992 by Springer-Verlag.)

The next step in temporal analysis is to simulate stand biomass development over a longer period, a century or more. This is usually done retrospectively by defining the development of a current stand from its origin forward to its present age when measurements are available for comparison. Running (1994) simulated the annual NPP and accumulation of stem biomass over a century for mature forests on seven sites across the Oregon transect. This strong climatic gradient produced a range of measured stem biomass from about 10 to 700 Mg ha^{-1}, and FOREST-BGC replicated this range in production well (Fig. 5.15). A short-lived deciduous alder stand tested the ability of FOREST-BGC to quantify accurately the annual canopy turnover of deciduous trees, and the differing respiration and carbon allocation dynamics of short-lived species. Predictions of NPP$_A$ were >50% above that measured for alder, which suggests the need to confirm whether photosynthetic capacity actually increases with foliar N concentration, as assumed in the model. Another possibility is that symbiotic nitrogen fixation, which requires about 6 kg C per kilogram N fixed, was underestimated in the simulation when set at 50 kg N ha^{-1} year^{-1}.

A greater challenge for biogeochemical models is to simulate stand responses after various treatments that influence tree growth. McMurtrie and Landsberg (1992) used the BIOMASS model to analyze the variable growth response of *Pinus radiata* stands in Australia after irrigation, fertilization, and a combination of both treatments over a 5-year period. Foliage mass and stem growth responded to both water and nutrients and increased by 20–40% relative to untreated stands, and the BIOMASS biogeochemical model was able to replicate the responses (Fig. 5.16). Cropper and Gholz (1993) simulated the carbon dynamics of a slash pine stand, with special emphasis on the possible importance of a labile carbon pool. Field measurements illustrated that increased stem growth after fertilization could be attributed entirely to an increase in foliage mass as a result of reduced allocation to roots. No change in photosynthetic capacity or labile carbon was required to account for the measured growth response.

Site quality is a term used by foresters to represent the potential productivity of forested land, usually for wood production (Tesch, 1981a). Site quality is commonly estimated by *site index* (SI), defined as the height achieved by dominant trees at a specified age which have grown in even-aged stands. An SI(50) = 25 signifies that dominant trees will be 25 m tall at 50 years of age. The site index measurement suffers from the same shortcomings as forest growth models based on inventory data, namely, reliance on historical growth of established stands, with no way of adjusting estimates to new site or stand conditions. Additionally, where disturbance or past harvesting have eliminated stands, site index cannot be determined directly from tree measurements. Simulations of annual stand growth or photosynthesis at canopy closure offer a more biophysically sound estimate of site quality that does not require direct knowledge of current stand age or condition. McLeod and Running (1988) showed that annual estimates of net photosynthesis for a range of ponderosa pine forests correlated well with measured site indices ($r^2 = 0.96$). Milner *et al.* (1996) modeled the annual net photosynthesis of the forested area in the state of Montana.

B. Gap Models of Forest Dynamics

Much of the discussion in the previous sections has been concerned with changes in the biogeochemistry of forest stands through their life cycles. To simplify the treatment and

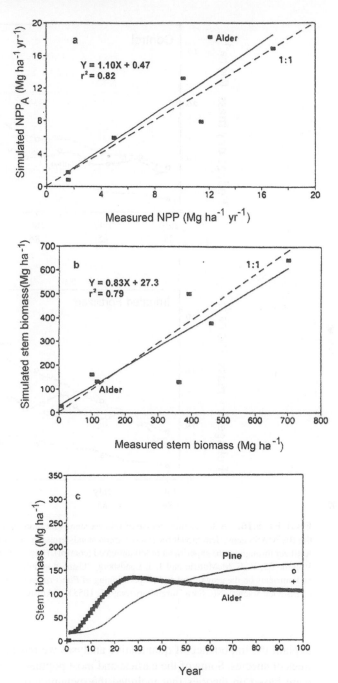

FIGURE 5.15. Observations and FOREST-BGC simulations from the Oregon transect study for (a) annual aboveground NPP$_A$, (b) stem biomass dynamics for a deciduous stand of alder and a fully stocked stand of ponderosa pine, and (c) stem biomass accumulation over 100 years. (From Running, 1994.)

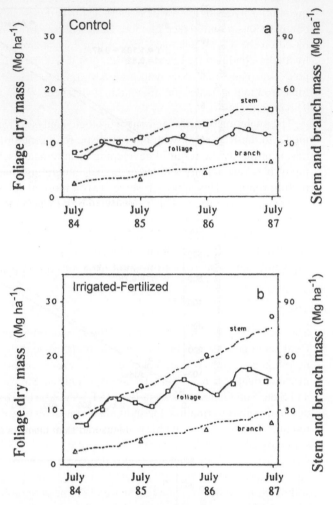

FIGURE 5.16. A 3-year time course of foliage, stem, and branch production measured and simulated with the BIOMASS ecosystem model for *Pinus radiata* stands growing in Australia that were provided irrigation and fertilizer throughout the experiment or left untreated (control). (Reprinted from *Forest Ecology and Management,* Volume 52, R. E. McMurtrie and J. J. Landsberg. "Using a simulation model to evaluate the effects of water and nutrients on the growth and carbon partitioning of *Pinus radiata,*" pp. 243–260, 1992, with kind permission of Elsevier Science–NL, Sara Burgerhartstraat 25, 1055 KV Amsterdam, The Netherlands.)

explain the importance of quantitative analyses we have paid little attention to individual trees or species. Some of the earliest and most popular forest ecosystem models, however, were based on theories that included the dynamics of individual trees. Determining the number of tree seedlings that are present in the stand initiation stage is difficult because the type of disturbance that precedes the establishment of a new population has an enormous effect.

Historically, forests in arid climates, such as those in the western United States, Mediterranean countries, and Australia, have been periodically replaced by large-scale wildfires, often preceded by major insect epidemics which killed trees and made stands more inflammable. These wildfires produced extremely large areas of bare ground from which new forests arose. Major windstorms, such as hurricanes, do not clear the forest but may cause excessive mortality. In wetter climates, where only a small percentage of mature trees die in a year, the death of a single tree may create a gap in the canopy, below which seedlings have an opportunity to become established. This latter type of mortality which creates "gaps" gave inspiration for a wide variety of models designed to predict forest dynamics over long time spans.

D. Botkin, who produced the JABOWA model (Botkin *et al.*, 1972), provided the first example of a gap model which simulated vegetation dynamics. This has since been widely expanded through the efforts of H. Shugart and associates (Shugart, 1984; Shugart *et al.*, 1992). The central components of all gap models are (1) definition of site variables, which includes climate, (2) definition of stand variables, which includes species lists and maximum tree sizes, (3) a growth submodel that computes annual increments of diameter and height growth of each tree in a small (often 0.1 ha) simulation plot, (4) a recruitment submodel that calculates entry of new young trees into the simulation, (5) a mortality model that kills trees, and (6) a resource submodel that calculates the growth-limiting potential of various resources, expressed as growth multipliers that range from 0 to 1 (Fig. 5.17).

1. Recruitment

Whether the initial site is bare, burned, flooded, or covered with detritus after disturbance is important in determining the potential for establishment of various species. Information about the reproductive behavior of species must be stored in the model, for example, which species produce seeds or sprout from roots, how far and by what agent seeds disperse, and how long the propagules remain viable. When conditions are deemed suitable for establishment of certain groups of species, their presence on the imaginary plots is a matter of statistical probability. In early gap models, new seedling recruitment was computed as a simple random probability from an initial tree list with no seed dispersal constraints, even though some species might not be present at that time in the simulation. Newer gap models connect seedling recruitment with tree occupancy, represent seed dispersal surrounding a mother tree, compute herbivory losses, and make success of seedling germination a function of climate, so that, for example, seedling recruitment can be higher in a wet year than during a drought year (Pacala *et al.*, 1993, 1996).

2. Growth Submodel

Once a tree is established, its growth is then predicted, making allowances for limitations on light intercepted by adjacent trees and site-related variables (moisture, temperature, fertility). Plants are grouped with regard to their sensitivity to temperature, light, moisture, browsing, pollution, and other stresses. These groupings are broad; for example, most species are considered to be either shade tolerant or intolerant. Sensitivity to temperature is often inferred from the present distributional patterns in latitude and elevation.

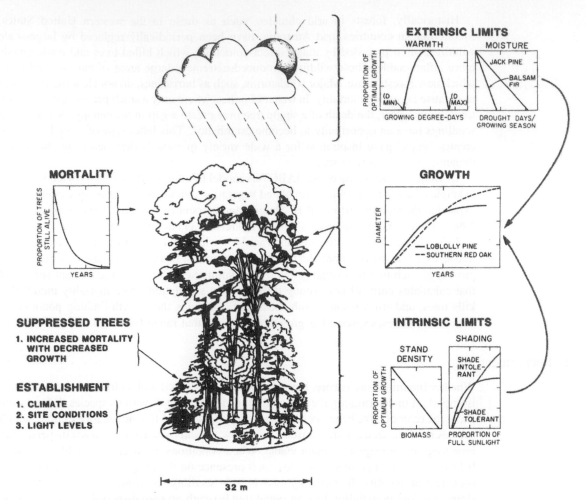

FIGURE 5.17. Summary diagram of the logic and main components of a forest succession or gap model. (From Solomon and Bartlein, 1992.)

The original gap models used a number of species-specific variables that mimic physiological processes, such as temperature and light controls on the growth equation, rather than a specific photosynthesis equation. Standard forest inventory data, which include stem diameters, tree heights, and growth increments by individual and species, are used to initialize the gap model. At the center of all gap models is a diameter-growth equation that quantifies the influence of climate on growth of each tree on the plot. There are many variations, but the equation generally includes a normalized height and diameter term and a series of empirical multipliers defining the effects of light, temperature, soil water, and nutrient availability. Height growth is also computed as a polynomial function of increases in stem diameter for each species (Dale *et al.,* 1985). In the earlier gap models, many of the variables were difficult to quantify; newer models have been modified so that they can be initialized with relatively easily collected field data (Paccala *et al.,* 1993, 1996).

Simple annual climate statistics are used to compute the growth multiplier factors, which relate growth to the maximum recorded within the range of the species. Species differences are reflected in the growth multipliers; thus, the minimum temperature limits of a subalpine tree are lower than for a coastal valley tree. In the moist hardwood forests of the eastern United States where this model was developed, light availability in the gap was the primary climatic determinate of initial seedling germination success. Hence, these early models tended to have rather complicated canopy light penetration and shading subroutines, where the vertical position of each individual tree crown in the stand was carefully followed to determine its competitive success for light. Soil fertility multipliers were added later to define nutritional limitations on tree growth. Growth multipliers to approximate enhancements of photosynthesis associated with higher levels of atmospheric CO_2 have also been added more recently (Solomon and Bartlein, 1992).

3. Mortality

Depending on the environment, some groups of trees are predicted to grow rapidly while others lag behind. Once trees are overtopped, light becomes limiting to their growth and the probability of death increases markedly. Death of a tree may be a simple function of age, sensitivity to fire, wind, snow breakage, insects, diseases, or specified activities of humans. The last point is important; these gap-phase models differ from most previous ecological schemes by incorporating both natural and human effects. When a tree dies, its fate is often critical in determining the future composition and growth rates of the stand. If left on the site, a tree may serve as shade, as a nesting site for birds that disseminate seeds, as a substrate for germinating seedlings, and eventually as a component of soil humus. When trees are harvested, on the other hand, minerals and organic resources are lost from the ecosystem. Whether foliage and branches are removed is often more critical to the assessment than the amount of bole wood because of the difference in nutrient content and resistance to decay (Chapter 4).

Mortality was initially computed as a random event whose likelihood increased as a tree reached its defined maximum size, or slowed in growth rate as a result of competition. Later models have incorporated stochastic disturbances such as fires, hurricanes, windthrow, and floods as mortality factors by adding the probability of these occurrences to the mortality probability of each tree (Shugart *et al.,* 1992).

Gap models are well suited for the evaluation of stand life cycle dynamics, specifically the recruitment, germination, and mortality of trees that define the vegetation dynamics of a stand, although it is necessary to remember that the predictions of gap models cannot be compared with individual stands because they generate averages from hundreds of hypothetical plots. The combination of multifactor growth multipliers and variable disturbance sensitivities allows the models to predict substantial variation in composition by accounting for dynamics and interspecies competition. Bonan (1989), using a gap model, successfully replicated the microclimate and topographically induced pattern of black spruce, white spruce, aspen, and birch in the North American boreal forest. His model incorporated two key controlling variables in that region: soil moisture content and depth of permafrost. White spruce, birch, and aspen occupy well-drained, south-facing slopes, whereas black spruce grows in saturated soils where the depth of thaw is very shallow (Fig. 5.18). Solomon and Bartlein (1992) simulated a diverse forest community in Michigan, which included a total of 19 deciduous and evergreen tree species. One thousand-year simulations run with multiple scenarios of CO_2 and climate change predicted substantial changes in species composition (Fig. 5.19).

Cattelino *et al.* (1979) offered a different set of principles on which to base models of vegetation dynamics. Their approach defined each species in relation to three critical attributes: (1) the persistence or viability of seeds, (2) the conditions required for seedling

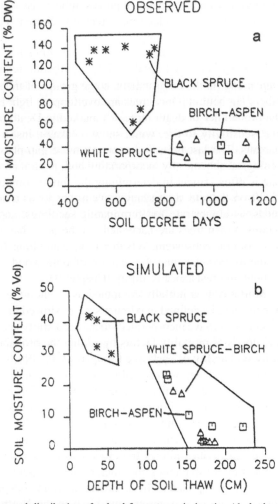

FIGURE 5.18.   Observed distribution of upland forest types in interior Alaska in relation to soil moisture and temperature and distribution of forest types simulated with a gap succession model. (From Bonan, 1989.)

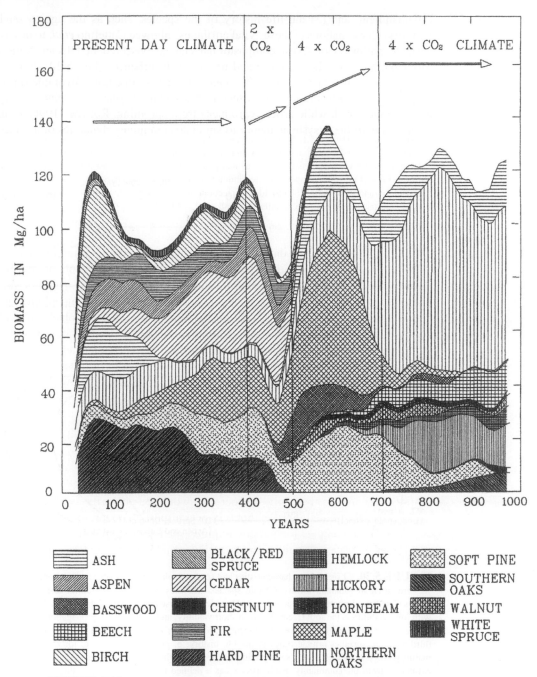

FIGURE 5.19. A one thousand-year simulation of a gap model (FORENA) illustrating dynamics of the species interaction and the response predicted by enhanced CO_2 and climatic change for a mixed Michigan forest. (From Solomon and Bartlein, 1992.)

establishment, and (3) the life history of the species, such as the age to seed-bearing maturity. Combinations of these "vital attributes" allowed Cattelino *et al.* to describe many variable pathways of forest compositional change observed in Glacier National Park, Montana, based on the sequence and timing of disturbances (Fig. 5.20). For example, a single wildfire in a mature spruce/fir stand establishes conditions favorable for aspen and lodgepole pine regeneration, but a second fire within 20 years kills pine saplings before they produce seed, while aspen reproduces from sprouting. Roberts (1996a,b) used vital attributes with fuzzy systems theory to simulate community dynamics of a mixed conifer

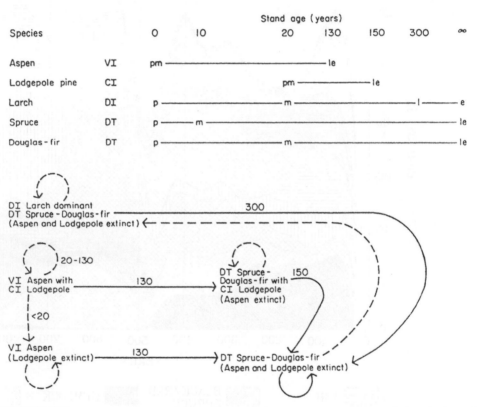

FIGURE 5.20.   Life history characteristics and species replacement sequences for aspen, lodgepole pine, and western larch in Montana forests. Certain species attributes determine differential responses to disturbance events. The first attribute is seed persistence: D, widely dispersed, always available after a disturbance; S, stored, with long viability in the soil; C, canopy, with short viability; and V, vegetative propagation. The second attribute defines necessary conditions for establishment: T, tolerant of competition; or I, intolerant of competition. The third attribute specifies critical times in the life history: p, propagules available for regeneration; m, time to maturity when a species begins seed production; l, loss of the species from senescence and mortality; and e, extinction from the community. Each species has a defined life history incorporating relevant attributes. The collective development of the ecosystem then depends on the timing and magnitude of disturbances producing varying forest successional pathways as influenced by the available species, which change over time. Solid lines indicate pathways with large changes in forest composition over time; dashed lines indicate pathways that lead to repetition of one type of forest community. (From Cattelino *et al.,* 1979.)

forest in Utah under different fire frequencies. He found species diversity to be highest with a fire return interval of 250 years in these forests, and lowest with complete fire exclusion.

An important insight derived from analyses with gap models is the realization that tree species, and often other life forms, can be assigned to similar functional groups (or guilds). Model analyses further suggest that, as a consequence, ecosystem operation continues in a highly predictable manner in terms of the accumulation of biomass and leaf area as long as at least one representative from each major guild is present (Tilman, 1996). The removal of all representatives of a given functional group is therefore a signal that ecosystem and community composition could change significantly in the future.

C. Hybrid Models

Beginning in 1990, a number of groups began developing a new generation of forest ecosystem models that incorporated the best features of both biogeochemistry and gap succession models (Bossel, 1991; Huston, 1991). Models like FOREST-BGC compute carbon and water balances from physiological principles but do not grow individual trees, so they are unable to simulate a real forest stand. Gap models simulate individual tree and life cycle dynamics but do not represent growth allocation mechanistically. Use of annual climate statistics and simple growth multipliers precludes gap models from explicitly representing seasonal carbon balance physiology.

The combined models use a daily biogeochemistry model to simulate canopy processes like photosynthesis and respiration with fairly realistic physiology, then transfer the amount of photosynthate fixed annually to a dynamic vegetation model where carbon allocations are computed and growth increments (or mortality) distributed to individuals on the basis of the sum of carbon resource available and specified differences in the light environment (Friend *et al.*, 1993; Korol *et al.*, 1995; Keane *et al.*, 1996b).

Korol *et al.* (1996) applied the hybrid modeling approach with TREE-BGC and successfully predicted the distribution of basal area and volume growth on nearly 1000 trees in 24 stands across British Columbia. Plot-level measurements of basal area and volume growth were highly correlated (r^2 of 0.94 and 0.96, respectively) with TREE-BGC simulations for a 20-year growth period. In another analysis, Korol *et al.* (1995) demonstrated how thinning a stand from 2100 to 553 trees ha^{-1} reduced LAI from 3.3 to 2.6 and GPP from 19.5 to 16.1 Mg C ha^{-1} year^{-1}. Net primary production, however, was only reduced from 6.3 to 6.2 Mg C ha^{-1} year^{-1} because maintenance respiration of foliage and other aboveground living tissue was reduced by one-third following thinning (Table 5.1). Over the following 5 years, the relative growth efficiency (annual photosynthesis per unit leaf area) increased for trees in the thinned stand and decreased for those in the unthinned stand (Fig. 5.21a). With the hybrid model Korol *et al.* also were able to calculate carbon allocation patterns for trees in relation to their exposure to sunlight (classified as dominant, intermediate, understory regeneration, and overtopped suppressed individuals). More dominant individuals with large crowns acquired more photosynthate than individuals with smaller crowns and less direct exposure to sunlight. The ratio of growth to maintenance respiration, however, was higher for intermediate and understory regeneration than for dominant trees in the stands (Fig. 5.21b).

TABLE 5.1
Allocation of Simulated Stand Carbon in Year 5 for the Stands in Fig. 5.21[a]

Plot	Carbon (Mg C ha^{-1} year^{-1})				
	LAI	GPP	R_m	R_s	Growth
Open (thinned, 553 trees ha^{-1})	2.6	16.1	7.2	2.7	6.2
Dense (unthinned, 2100 trees ha^{-1})	3.3	19.5	10.7	2.5	6.3

[a]From Korol *et al.* (1995).

Bossel (1996) captured the essential dynamics of forest stand growth in a model of 14 ordinary differential equations describing tree growth and soil processes. The model, TREEDYN3, does not require explicit daily climatic data, yet is still able to represent diurnal and seasonal physiological dynamics and multiyear tree growth. Luan *et al.* (1996) developed a forest ecosystem simulation that is hierarchical in both space and time. Their FORDYN model represents forest ecosystem dynamics at four space/time levels. The first level treats cellular CO_2 assimilation on time steps of seconds, the second level computes hourly leaf photosynthesis, respiration, and transpiration, the third computes soil carbon and nitrogen processes and carbon allocation, and the final level computes tree establishment, growth, and mortality. These examples show the continuous innovations in modeling forest stands integrating progressively more ecosystem processes.

VI. SUMMARY

In this chapter we have demonstrated that significant changes in ecosystem function occur as forests develop, even in stands composed of a single species. Four idealized stages in stand structure, which relate to changes in function, can be identified: initiation, stem exclusion, understory reinitiation, and old-growth. A unifying principle derived from analysis of stand development is that the total canopy LAI remains relatively stable in a given environment while the overstory LAI varies. Another scaling principle emerges from the recognition that species may be classified into broad functional groups, and that ecosystem operation may not be adversely affected until most, if not all, of the representatives of a guild are lost from the system. Because animals play such an important role at the initiation stage of forests, studies of their population dynamics might beneficially be concentrated on this stage to identify potentially dangerous trends in local extinction of plants and animals. On the longer term, changes in the regional flora and fauna must also be incorporated in the analyses.

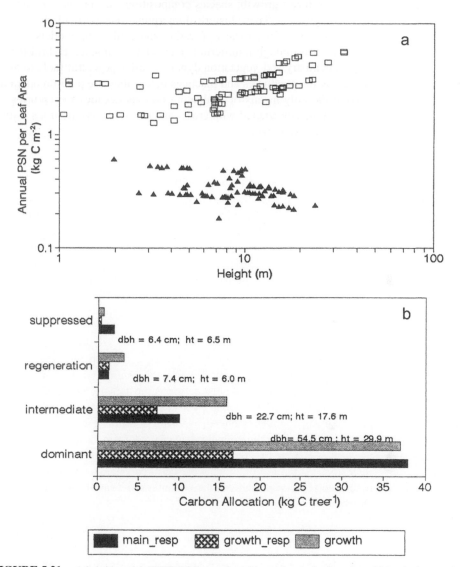

FIGURE 5.21.   A hybrid model, TREE-BGC, contains a biogeochemical subroutine which calculates carbon uptake by thinned and unthinned stands (as described in Table 5.1). (a) A tree growth subroutine allocates the carbon generated by the biogeochemical model and distributes it differentially to dominant, intermediate, suppressed, and regenerating trees for 5 years following treatment. During this time, dominant (□) and suppressed (▲) trees showed opposite responses in terms of photosynthesis (PSN) per unit of leaf area. (b) The total photosynthate available for dominant and suppressed trees differed by 10-fold and was reflected in aboveground growth. Model calculations of growth/maintenance respiration ratios, which can be derived from the diagram, illustrate significant differences for dominant (~1.0), intermediate (~1.5), regeneration (~3.0), and suppressed (~0.5) trees. (From Korol *et al.,* 1995.)

From a historical perspective, a variety of techniques developed in paleobotany, dendrochronology, and pedology provide us with insights into the extent and frequency of changes in forest growth, species composition, soil organic matter dynamics, and atmospheric deposition. These historical interpretations, particularly those gained through isotopic analyses, offer a means of testing and calibrating models more widely than can be accomplished through long-term studies or experiments. Hybrid models able to integrate biogeochemistry and vegetation dynamics offer powerful tools to assess the implications of various forest policies and practices before they are put into operation. Because models provide the only means of evaluating options decades to centuries into the future, they deserve to be constructed with great care and to receive rigorous testing before being presented to decision makers.

CHAPTER 6

Susceptibility and Response of Forests to Disturbance

I. INTRODUCTION

In previous chapters we identified properties of ecosystems that are sensitive to alterations in the availability of resources. Of these, leaf area index was the most general structural variable. During stand development, LAI of the initial complement of species generally increases rapidly and remains relatively stable for some time, particularly when shade-adapted species fill in gaps that develop in the overstory canopy (Chapter 5). We define a *disturbance* as any factor that brings about a significant reduction in the overstory leaf area index for a period of more than 1 year. This definition parallels one by Oliver and Larson (1996), who define a disturbance as an event that makes growing space available for surviving trees.

Often it is difficult to identify the underlying cause of a disturbance. For example, in the selective harvesting of trees, soil compaction and direct injury to residual stems may create conditions favorable for the spread of pathogens that otherwise would not occur. Likewise, attempts to protect forests against fire for long periods may result in insect outbreaks (Johnson and Denton, 1975; Fellin, 1980; McCune, 1983). For managers, the most useful analysis provides an indication when an ecosystem is beginning to perform "abnormally" but has not yet been "disturbed." At such times the forest is "predisposed" to change in structure and composition (Waring, 1987). The actual agents of disturbance, be they windthrow, insects, disease, fire, or, if we choose, selective silvicultural practices, may all lead to recovery of ecosystem function, but with marked differences in potential losses of resources and options for the future.

Although many factors may interact to cause a disturbance, they may be broadly classified as either biotic or abiotic in origin. Making distinctions among an array of biotic and abiotic forces is important because the biota present in an ecosystem have adapted to types of disturbances that have previously occurred at predictable frequencies. By recognizing selective adaptations of the biota we are in a better position to predict changes in the future composition of forests and in the rates of ecosystem functions.

If disturbance is required to perpetuate a certain type of forest, we may wish to mimic the historical "natural" sequence, but the historical sequence may be unrepresentative or hard to duplicate with increased atmospheric deposition of chemicals, introduction of nonnative biota, and changes in management practices. Ecosystem and stand development (succession) models described in preceding chapters provide a means of projecting into the future, but their accuracy depends on the validity of a number of important assumptions and requires projections of climatic variation.

In this chapter we identify structural and chemical indices that reflect changes in the susceptibility and response of individual trees and forests to various kinds of disturbances. Some of the indices verify the historical frequency of various types of disturbances; others indicate shifts in the availability of carbon, water, and nutrients that predispose ecosystems to disturbance. To test the reliability of these indices we report how they change under experimental conditions and across biotic, physical, and chemical stress gradients.

II. BIOTIC FACTORS

The extent to which biodiversity provides a buffer against various types of disturbances is debatable (Perry, 1994). Species diversity within some tropical forests is amazingly high, with as many as 473 tree species having been recorded on a single hectare in Ecuador (Valencia *et al.,* 1994). In addition, thousands of invertebrates fill important niches with uncountable species of microorganisms. Some redundancy in functional groups is always desirable, but large numbers of species in a given guild do not necessarily make the ecosystem more buffered against disturbance. Slight variations in the resistance of species to a common stress, however, should permit more rapid recovery of primary production, as demonstrated in grassland experiments following drought (Tilman and Downing, 1994).

First-order ecosystem processes, such as photosynthesis, transpiration, and decomposition, are often relatively insensitive to forest species composition. In wet tropical forests of Costa Rica, the number of species was regulated following clearing of the original stand, and soil organic matter and nitrogen levels returned to the original status about as fast with 12 species of trees present as they did with 120 (Fig. 6.1). A diverse mix of species is of less consequence if a particular type of disturbance rarely occurs. For example, large areas of highly diverse temperate rain forests were converted to tussock grass on the South Island of New Zealand in a short time after the Maori people arrived and introduced extensive fire (Newnham, 1992; Evison, 1993). However, the abundance of tree species in many tropical forests may provide those ecosystems a buffer against disturbance by the herbivores and pathogens, which abound in warm and moist conditions. Spatial isolation

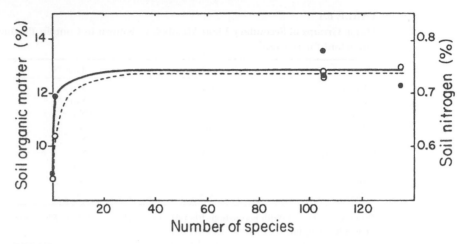

FIGURE 6.1.   In an experiment where the number of plant species was controlled after removal of a tropical forest in Costa Rica, recovery of soil nitrogen (O) and organic matter (●) after 5 years was nearly as rapid with 12 species as with more than 100. (From Vitousek and Hooper, 1993; after Ewel *et al.*, 1991.)

of individual tropical tree species has been shown to improve the chances of survival against some pathogens (Gilbert *et al.,* 1994).

Spatial isolation is, however, a disadvantage when it comes to reproduction. Most tropical tree species make heavy investments in flowers and fruits to attract butterflies, bats, birds, primates, and other mammals which pollinate, disseminate, ingest, and fertilize the propagules (Chapter 5). In addition to investment in attracting pollinators and seed disseminators, perennial plants must expend additional energy to produce defensive chemicals to deter attacks by herbivores and pathogens. Nicotine, caffeine, cocaine, and tannins are all natural products that help plants defend themselves.

A. Biochemical Defenses in Plants

Defensive chemicals present in plants are broadly classified into nitrogen- and non-nitrogen-containing compounds. Compounds that contain N include cyanogenic glucosides, alkaloids, and nonprotein amino acids. Defensive compounds without N include tannins, terpenes, phytoalexins, steroids, and phenolic acids. Each kind of compound may serve in a variety of ways against various organisms (Table 6.1).

Some plants produce fungistatic and bacteriostatic compounds that prevent colonization by pathogens. Other compounds act as physical barriers, such as waxes on the leaf surface or resins or lignin in cell walls. Increasing fiber and lowering water content decrease digestibility of plant tissue and reduce herbivore growth rates and survival (Scriber and Slansky, 1981). Tannins precipitate protein, which inhibits most enzyme reactions and makes protein present in plant tissue nutritionally unavailable to most animals and microbes (Zucker, 1983). Phytoalexins are lipid-soluble compounds, which are activated following an attack by pathogens, and exhibit antibiotic properties (Harborne, 1982). The alkaloids found in many angiosperms are particularly toxic to a variety of mammals (Swain, 1977).

TABLE 6.1

Major Groups of Secondary Plant Metabolites Known to Contain Products Important for Defense[a]

Class	Number known	Contains N	Protection against
Alkaloids	1000	Yes	Mammals
Amino acids	250	Yes	Insects
Ligans	50	No	Insects
Lipids	100	No	Fungi
Phenolic acids	100	No	Plants
Phytoalexins	100	No	Fungi
Quinones	200	No	Plants
Terpenes	1100	No	Insects
Steroids	600	No	Insects

[a]From Swain (1977). With permission, from the *Annual Review of Plant Physiology,* Volume 28, © 1977, by Annual Reviews Inc.

Changes in host biochemistry may also affect the colonization of organisms helpful to the host plant. These include protective ant colonies, macroorganisms that graze on bacteria, and symbiotic associations of N-fixing bacteria and mycorrhizal root fungi. These beneficial organisms may directly infect or prey on attacking organisms, release antibiotics, or provide essential nutrients. Many of the compounds released as exudates, which include a variety of polysaccharides, organic acids, and amino acids, are essential to beneficial associates, but when these organisms are absent they can also be assimilated by herbivores and pathogens.

In general, defensive compounds that lack N reach rather high concentrations in cells, often 10–15% by weight, whereas N-containing compounds are usually at concentrations below 1%. Plants expend less total energy in the synthesis of small amounts of N-containing defensive compounds than when producing large amounts of C-rich compounds, but variations in the turnover rates of defensive compounds and differences in relative growth rates must be considered in assessing relative costs (Bryant *et al.,* 1991). Nitrogen-containing compounds are most frequently found in deciduous, fast-growing vegetation, whereas defensive compounds without N are more characteristic of slow-growing plants, particularly evergreens with long leaf life spans (Bryant *et al.,* 1986; Coley, 1988). Regardless of the defensive compound synthesized, no plant is completely immune to attack. Specialized insects and pathogens have evolved that not only detoxify toxic compounds, but actually require them for optimal growth (Bernays and Woodhead, 1982). These highly evolved specialists are restricted to a few host species, but they may attack vigorous as well as weak individuals (McLaughlin and Shriner, 1980). Other less specialized organisms accommodate a wider range of biochemical challenges and attack a wider variety of plants.

Plants that depend on defensive compounds rich in N are at a competitive disadvantage where N is in short supply. On the other hand, plants producing C-rich defensive compounds are at a disadvantage when growing in shade with an abundant supply of N. Those plants adapted to more fertile soils may be expected to build a variety of defensive compounds from N. Thus, alkaloids predominate in the foliage of trees in many lowland tropical forests where N is relatively abundant (McKey *et al.,* 1978). Plants growing in areas where N is scarce generally produce only tannins and related C-based compounds. Foliage is so unpalatable in some tropical forests growing on sterile sands that primates survive mainly by eating fruits (Gartlan *et al.,* 1980). A similar pattern in distribution of vegetation with N-based or C-based defensive compounds to that observed in tropical forests has been observed in boreal and temperate forests (Rhoades and Cates, 1976; Bryant *et al.,* 1991). At the time of foliage elongation, when N is relatively available, even plants growing in nutrient-poor habitats may produce a few N-based defensive compounds (Dement and Mooney, 1974; Prudhomme, 1983).

Insects differ from other animals in the way they locate host plants. Birds and larger animals depend on sight to recognize flowers and fruits. Insects rely much more on odors of compounds volatilized or exuded by plants. Adult insects seeking to lay eggs on a suitable host may use their antennae to sense volatile compounds at levels as low as $10^{-12} \, g \, cm^{-3}$. By direct tasting, insects may discriminate nonvolatile compounds at concentrations of 1 mg per 1000 cm^3 in tissue, which is far below toxic levels (Swain, 1977). To meet the challenge of insects, many plants are able to produce toxic compounds quickly and to construct barriers that consist of dead or resin- or gum-filled tissue almost immediately following attack (Schultz and Baldwin, 1982; Raffa and Berryman, 1983). In response to localized insect activity, foliage throughout an entire tree may become less palatable (Haukioja and Niemelä, 1979; Karban and Myers, 1989). Morphological responses such as stiffer thorns on *Acacia* trees may also be induced by grazing (Seif el Din and Obeid, 1971). Bark sloughing is a response to attack by the woolly aphid (*Adelges piceae*) that only reaches epidemic populations when balsam fir (*Abies balsamea*) replaces native forest species (Kloft, 1957). The implication is that the woolly aphid is at a low-level equilibrium with its native host as a result of long-term evolutionary adaptations but reaches epidemic populations on the relatively defenseless introduced balsam fir.

Induced responses to attack can only be effective if sufficient resources can be quickly mobilized. The rate at which stored carbohydrates or protein can be converted to mobile forms (sugars and amino acids) and transported to sites of attack may limit the capacity of trees to respond (McLaughlin and Shriner, 1980), although this may not affect canopy responses to partial defoliation if photosynthetic rates are high. Changes in the allocation of current photosynthate to remote organs, such as the lower bole or roots, however, cannot be accomplished rapidly because of the distance involved and limitations imposed by phloem transport (Chapter 3). For this reason, concentrations of stored reserves in roots, stems, and twigs are a good indicator of a tree's potential to survive localized attack by insects or pathogens (Ostrofsky and Shigo, 1984). Wargo *et al.* (1972) demonstrated that defoliation of sugar maple (*Acer saccharum*) greatly reduced starch content in roots at the end of the growing season. Low starch reserves in the roots predisposed trees to attack by root pathogens (Wargo, 1972).

Tropical forests are rarely heavily defoliated because the insects and pathogens are highly specialized and their host trees are few and widely separated. In temperate and boreal forests, on the other hand, only a few species of trees may be present. Even with genetic diversity within a population, major outbreaks of insects are common in higher latitude forests. Population outbreaks of insects can completely defoliate a large fraction of trees in a forest within a single year, but, depending on the physiological status of the trees, mortality may be low, as shown in two photographs taken during and 5 years after an outbreak of tussock moth in northeastern Oregon (Fig. 6.2). It is critical to understand whether a forest is resistant or susceptible to defoliating insects or other biotic agents before making management decisions. In the following sections we draw on ecosystem-level experiments to test the reliability of various stress indices and follow changes in susceptibility and response to changing environmental conditions created by biotic disturbance.

B. Experimental Tests of Theories

Most ecological theories are based on observations following change over time without recourse to experimentation. Because many properties of ecosystems change concurrently following a disturbance, theories are often difficult to reject or to confirm. Theories involved with forest insect outbreaks are particularly difficult to evaluate because the physiological status of host and herbivore both change rapidly following a disturbance. To understand the interactions more fully, the canopy structure and environmental resources available to trees under attack may be experimentally altered at the start of an outbreak and followed through its duration. Four examples bring out different aspects of interactions between the physiological state of the vegetation, the availability of resources, and the population dynamics of (1) defoliating insects, (2) bark beetles, (3) pathogens, and (4) browsing mammals.

1. Defoliating Insects

In the boreal forests of Canada, outbreaks of eastern spruce budworm (*Choristoneura fumiferana*) occur at regular intervals in the extensive nearly pure forests of *Abies balsamea*. Using a correlation established between tree age and the conversion of sapwood to heartwood (Fig. 5.11), Coyea and Margolis (1994) determined the dry matter production of stem wood per unit of leaf area per annum (growth efficiency) of trees which survived or succumbed to an extensive stand defoliation (Fig. 6.3). They found that the insect outbreak occurred as the stand reached its lowest mean growth efficiency, and that mortality was concentrated on trees with growth efficiencies significantly below average. Following the death of many of the less resistant trees, the remaining trees reattained their previous levels of wood production per unit of leaf area. These observations suggest that native defoliating insects play an important role in stand development by reducing competition, which allows surviving trees to maintain moderate growth efficiencies and the ecosystem to produce near maximum NPP, a result similar to a silviculturalist prescribing a precommercial thinning.

A somewhat different series of events has led to extensive outbreaks of western spruce budworm (*Choristoneura occidentalis*) in forests with mixed populations of coniferous species. Throughout much of the western United States selective harvesting of ponderosa pine trees, combined with fire protection, has created conditions that favor the establishment of grand fir (*Abies grandis*) and Douglas-fir (Fig. 6.4). These new stands support the maximum leaf area index possible for the environment, which may be twice as high as that observed when frequent ground fires occurred. The additional leaf area requires more

FIGURE 6.2. (a) Outbreaks of defoliating insects occur more frequently in the western United States where fire protection has enabled large expanses of fir trees (*Abies grandis*) to replace much of the original pine. (b) Mortality from insect outbreaks varies, depending on the length of the outbreak and physiological status of host trees. This scene records nearly full recovery 5 years after the first photograph was taken. Any mortality improved diameter growth of surviving trees (Wickman, 1978). (Photograph from Boyd Wickman, La Grande, Oregon.) See Color Plate.

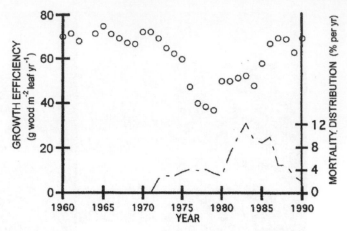

FIGURE 6.3.   Dynamics of a spruce budworm outbreak in a balsam fir forest in Quebec, Canada, was reconstructed from extracted wood cores. The growth efficiency for the majority of trees decreased before an outbreak of spruce budworm occurred in the 1970s. As trees were selectively killed between 1971 and 1990, surviving trees regained efficiency in wood production per unit leaf area. (After Coyea and Margolis, 1994.)

nutrients and intercepts more of the annual precipitation. As a result, trees growing over extensive areas have growth efficiencies that average <30 g wood produced m^{-2} leaf area year^{-1} (Waring *et al.,* 1992).

During an outbreak of western spruce budworm, a thinning and fertilization experiment with four replications was installed in a dense, 80-year-old grand fir forest under a few scattered overstory pine. The resulting defoliation was recorded, along with measures of tree growth, foliar chemistry, and the availability of soil nitrogen and water (Mason *et al.,* 1992; Waring *et al.,* 1992; Wickman *et al.,* 1992). Fertilization with nitrogen increased amino acid levels in foliage by nearly threefold (Waring *et al.,* 1992). As a result, budworm larvae needed to consume less foliage to develop to maturity. Increased availability of nitrogen, combined with better illumination, induced trees in the thinned plots to produce additional leaves, which enhanced both terminal and diameter growth, and resulted in more than a twofold increase in tree growth efficiency for the thinned and fertilized treatment (Fig. 6.5a,b). With increased foliage and decreased herbivory, the proportion of foliage consumed by insects during the peak of the epidemic in 1987 dropped from >98% in unfertilized treatments to about 75% in the fertilized treatments (Fig. 6.5c). The outbreak ended in 1989 when pathogens caused the budworm population to return to near minimum levels on all treatment areas (Mason *et al.,* 1992; Wickman *et al.,* 1992).

As a result of insect defoliation, N fertilization without thinning (in the same experiment) also eventually resulted in a significant increase in growth efficiency over untreated stands. Summer drought did not exhaust all reserves of water in the rooting zone. Predawn Ψ values recorded on twigs fell to only −0.7 MPa (Waring *et al.,* 1992). Drought, however, did restrict most nitrogen uptake during the growing season to the spring and autumn months when the upper soil horizons were moist (Waring *et al.,* 1992). On other sites where water and nutrients are more limiting, epidemic outbreaks of defoliating insects

FIGURE 6.4.   (a) Around 1900, many pine forests in western North America lacked an understory of young trees as a result of frequent ground fires. (b) With fire protection and removal of many overstory pine, dense stands of mixed conifers became established. The younger forests require more nutrients and are more fire-prone than the old-growth pine stand. (From Gruell, 1983.)

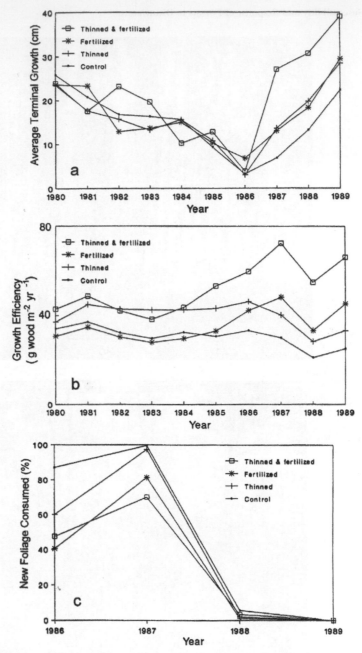

FIGURE 6.5. (a) Average terminal growth of trees started to decrease before a major spruce budworm outbreak in eastern Oregon occurred in 1985. Experimental treatments were applied in autumn of 1984 and affected growth subsequently. (b) Tree growth efficiency improved significantly during the peak of the insect outbreak for trees which received N fertilizer; thinning alone did not improve growth efficiency. Growth response was less, however, when nitrogen was added without thinning. (c) The percentage of new growth consumed by insects was significantly lower with fertilized than with unfertilized treatments in 1987. The insect population fell to endemic levels in 1988. (From B. E. Wickman, R. R. Mason, and H. G. Paul, 1992, "Thinning and nitrogen fertilization in a grand fir stand infested with western spruce budworm. Part II. Tree growth response," *Forest Science,* Volume 38, pp. 252–264.)

cause more extensive mortality, correlated with extremely low values of tree growth efficiency (Wickman *et al.,* 1992).

When mortality is extensive, the LAI in these kinds of forests may not return to maximum values for half a century. During that time, water, nutrients, and light are more readily available to surviving trees. In the arid summer environments characteristic of most of the western United States, if dead trees are not harvested, they increase the probability of fire. If a seed source is available, a young pine forest, which is unpalatable to spruce budworm and most other defoliating insects, will develop.

In many cases, managers consider applying biocides or pesticides during an outbreak to reduce damage from insect defoliators. These practices rarely halt outbreaks and can perpetuate them by slowing the natural buildup of pathogens in the insect population unless great care is taken (Cadogan *et al.,* 1995). Attempts to control outbreaks also prevent the benefits of nutrient recycling which result from insects concentrating elements in their frass at more than twice that found in fresh foliage (Rafes, 1971). Through monitoring of a simple structural index such as tree growth efficiency, a judgment can be made regarding the ability of the forest to withstand an epidemic attack of defoliating insects. Whether judicious application of fertilizer might be beneficial once an outbreak commences would depend on the availability of nitrogen (which has changed with increases in atmospheric deposition rates) and its balance with other nutrients (Chapter 4).

2. Bark Beetles

Different species of bark beetles attack a wide variety of conifers. Female beetles select susceptible trees based on the presence of terpenes that are generated by the conifers in increasing amounts as temperatures rise (Christiansen *et al.,* 1987). Bark beetles deposit their eggs in galleries excavated in the phloem, cambium, and sapwood of trees. Successful brood production is contingent on the death of these tissues. Most species of bark beetles can only breed in trees that exhibit severe decline or are already dead, and so they merely promote decomposition and mineralization. A few species, however, are able to attack and kill living, sometimes quite healthy trees. Epidemic outbreaks by these "aggressive" species may greatly alter the state and function of forest ecosystems over large areas.

Aggressive species have developed three ways of conquering living trees: (1) by having the first attacking beetles produce chemical attractants to bring other beetles, (2) by tolerating resin secretions, and (3) by inoculating trees with a pathogenic fungus that kills by halting water transport through the sapwood (Christiansen *et al.,* 1987). The degree to which trees can defend themselves successfully is based on the extent to which they can produce resins and mobilize carbohydrates to wall off areas in the phloem and sapwood where beetles have introduced fast-growing strains of blue-stain fungi. Stored reserves are generally insufficient by themselves to protect trees against mass beetle attacks (Christiansen and Ericsson, 1986). In cooler climates where only one beetle population develops in a year, attacks are synchronized with the expansion of new growth. Some variation exists, however, because budbreak is controlled by soil temperature, which constrains water uptake, more than by air temperature to which insect development is closely keyed (Beckwith and Burnell, 1982). Genetic variation also exists in both tree and insect populations.

In another well-replicated experiment, synthetic pheromones were released to attract mountain pine beetles (*Dendroctonus ponderosae*) to various thinning and fertilization treatments in 120-year-old forest of lodgepole pine (*Pinus contorta*) (Waring and Pitman, 1985). Treatments included (1) N fertilizer, (2) N fertilizer combined with a reduction in canopy LAI of about 80%, (3) additions of sugar and sawdust to limit mineralization by microorganisms, and (4) untreated plots. At the start of the experiment, tree growth efficiency averaged less than 70 g wood produced per square meter of leaf area per year with a stand average LAI of 4.7. As beetles killed trees and foliage was shed a year later, more light, nutrients, and water became available to surviving trees.

Within 2 years of the application of fertilizer, surviving trees increased their efficiencies by more than 55% to values above 100 g wood m^{-2} leaf area $year^{-1}$ (Table 6.2). Surviving trees in untreated stands also increased their growth efficiency by over 40% after 2 years. Only in the sugar and sawdust treatment did tree mortality not result in a significantly improving residual tree growth efficiency. When 100 trees sustaining different levels of attack were compared, tree mortality, measured by the proportion of sapwood observed with blue-stain fungus, was accurately predicted by noting when the ratio of bark beetle attacks (square meter of bark surface) to growth efficiency (grams of wood produced annually per square meter of foliage) exceeded 1.2 (Fig. 6.6). At values above 100 g wood m^{-2} leaf area $year^{-1}$ no successful bark beetle attacks were recorded. The same relationship was demonstrated in other thinning experiments with ponderosa pine (Larsson *et al.*, 1983) and lodgepole pine grown at different densities (Mitchell *et al.*, 1983). A comparable response has also been reported for European spruce bark beetle (*Ips typographus*) cited by Christiansen *et al.* (1987).

In areas where bark beetle outbreaks occur, thinning may improve the resistance of residual trees if sufficient time is allowed to raise their growth efficiency to a safe level, as demonstrated in a photograph taken after a bark beetle epidemic swept through an even-aged pine forest that had been partially thinned (Fig. 6.7). Thinning alone, however,

TABLE 6.2

Growth Efficiency (Wood Production per Unit Leaf Area) under Various Treatments[a]

Treatment	Growth efficiency (g wood m^{-2} foliage $year^{-1}$)			
	1979	1980	1981	1982
Control	51	59	73	66
Sugar and sawdust	76	81	88	72
Fertilized	67	84	108	87
Fertilized and thinned	77	95	120	115

[a]Means (*n* = 12) connected by brackets are significantly different at *p* = 0.05. From Waring and Pitman (1983).

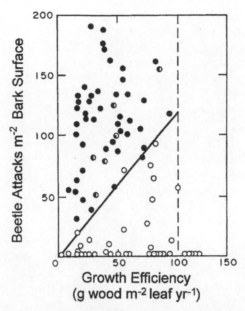

FIGURE 6.6.   Growth efficiency provides an index to the density of bark beetle attack on lodgepole pines. Filled or partly filled circles represent the proportion of sapwood killed on attacked trees. Open circles represent trees able to survive all beetle attacks before any conducting tissue was killed. The dashed vertical line indicates the boundary above which beetle attacks are unlikely to cause tree mortality. (From "Modifying lodgepole pine stands to change susceptibility to mountain pine beetle attack," by R. H. Waring and G. B. Pitman, *Ecology,* 1985, **66,** 889–897. Copyright © 1966 by the Ecological Society of America. Reprinted by permission.)

may not be sufficient to prevent subsequent mortality if other abiotic or biotic factors constrain water, nutrient, and CO_2 uptake. Annual growth efficiency may prove an inadequate index if conditions are highly variable from year to year or if beetle attacks extend throughout the growing season (Lorio, 1986).

3. Pathogens

Many pathogens and parasites are carried by insects, birds, and other vectors from one area to another. Transport of pathogens by humans into forests where native trees lack resistance has caused near extinction of some species such as American chestnut (*Castanea dentata*). Below ground, many native fungi are present with the capacity to break down cellulose and lignin in plant cell walls (Harvey *et al.,* 1987). With large accumulations of woody debris following windstorms or logging, pathogens may spread from dead to living root systems. This fact has encouraged removal of stumps or their treatment with fungicides (Thies *et al.,* 1994). In some cases, root pathogens become so well established that a rotation of resistant species is recommended (Thies, 1984).

In the absence of fire or wind, root pathogens often play an important role by opening canopies and fostering nutrient cycling. When pathogens kill trees in stands with near

FIGURE 6.7. Mountain pine beetles attacked and killed blocks of old-growth lodgepolc pine in eastern Oregon, except where the forest was previously thinned. (Photograph provided by John Gordon, Yale University, School of Forestry and Environmental Studies, courtesy of Boise Cascade.) See Color Plate.

maximum LAI, surviving trees may quickly respond in growth to compensate fully for the reduction in stocking levels (Oren *et al.,* 1985). At times, root pathogens do kill all dominant trees, which allows stand replacement to occur. For example, in some subalpine forests of mountain hemlock (*Tsuga mertensiana*), laminated root rot (*Phellinus weirii*) causes mortality in wavelike patterns followed by replacement with younger trees (McCauley and Cook, 1980). Field studies showed that mineralizable N in the undisturbed soil was extremely low but increased significantly in the recently disturbed zone following death of the overstory trees (Fig. 6.8). Growth efficiency of trees increased from less than 30 in the old-growth forest to nearly 80 g wood m^{-2} leaf area year^{-1} in young regenerating forest. As stand LAI peaked and mineral soil nitrogen returned to the low values common to older stands, growth efficiency was again reduced so that, about 50 m behind the current edge of dying old-growth forests, trees again became susceptible to infection from the extensive inoculum available in decaying roots. Laboratory studies on mountain hemlock seedlings collected from the site indicated increasing susceptibility to the root pathogen when nitrogen was limited or when shade restricted photosynthesis (Matson and Waring, 1984).

In regions where soils are unable to supply trees with a balance of nutrients, an increase in the concentration of amino acids occurs (Ericsson *et al.,* 1993). The concentration of amino acids in the foliage of *Pinus radiata* has been shown to be an excellent index of a tree's susceptibility to the needle-cast fungus *Dothistroma* (Turner and Lambert, 1986).

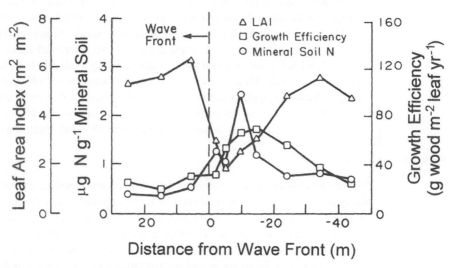

FIGURE 6.8. In a subalpine forest of mountain hemlock, 220-year-old trees are killed by a root disease approaching along a broad front. As older trees are killed, younger ones replace them and eventually regrow into forests at an increasing distance behind the wave front (vertical dashed line). Mineralizable soil N was highest where the leaf area index was lowest in the bare zone where young trees are becoming established. Growth efficiency of trees also peaks in the area at 0–10 m behind the wave front. The pathogen is present in decaying roots but successfully reinfects only trees with low growth efficiency and limited N supply. (From Waring *et al.,* 1987.)

Entry *et al.* (1986) also demonstrated in a laboratory experiment that a balance between N and P was important in determining the resistance of *Pinus monticola* seedlings to the root pathogen *Armillaria mellea* (Entry *et al.*, 1986). Additional laboratory studies with five species of western conifers confirmed that an adequate balance of both light and nutrients was important in reducing infection by *Armillaria* on seedlings. The basis for seedling resistance was further related to a critical ratio between the concentration of phenolic or lignin compounds in the root, which inhibit pathogen growth, and the concentration of sugars, which stimulate growth (Entry *et al.*, 1991a).

Entry *et al.* (1991b) designed a field experiment to test the laboratory-derived biochemical indices. They compared 27 individual Douglas-fir trees with highly variable root biochemistry as a result of having been grown in stands that had been thinned, thinned and fertilized with nitrogen, or left untreated for 10 years. The ratios of phenolic and lignin compounds to sugar in trees roots (expressed in units of energy required to degrade phenolics or lignin to the energy available from sugars) identified those trees that were most susceptible to infection (Fig. 6.9). Subsequently, other field studies have shown the importance of adding a balance of fertilizer, because nitrogen alone often results in decreasing root resistance against pathogen attack (Mandzak and Moore, 1994). Entry *et al.* (1991b) recommended that managers should commence thinning early in stand development when trees are small, so that any infected roots present will decay rapidly. Where forests have evolved with frequent disturbance initiated by fire or insects but have been protected against these agents, pathogens are likely to play an increasing role in disturbance.

Mortality caused by defoliating insects, stem-killing bark beetles, and root pathogens has similar effects in relation to ecosystem responses (Table 6.3). These biotic disturbances all increase the amount and substrate quality of leaf and fine-root detritus. With reduction in LAI, the microclimate for decomposition and mineralization is also improved. Surviving trees, or those that replace the original stand, have improved access to water, nutrients, and light, which result in more photosynthate being available for growth and defense. If recovery in growth efficiency and biochemical balance is not observed following a biotic-induced disturbance, other factors such as atmospheric pollution or climatic variation may be the underlying cause. In such cases, attempts to control the biotic agent of disturbance will have minor long-term benefits and may actually exacerbate the situation.

4. Browsing Animals

Vertebrate animals are generally less selective in their diet than are invertebrates. However, the quality of browse is as important or more important than the amount available, and it changes with plant species, plant age, and environment. This fact is most clearly demonstrated from extensive studies conducted in the boreal forests of Alaska where the species composition is limited and harsh winters force large numbers of mammals to compete for limited resources. From these and related studies elsewhere, a number of important generalizations have emerged and been summarized in reviews by Bryant *et al.* (1986, 1991):

- Food selection is not based on "energetic optimization." A wide variety of browsing animals, which include hare, moose, and deer, avoid eating evergreens such as black spruce although this species contains higher concentrations of lipids

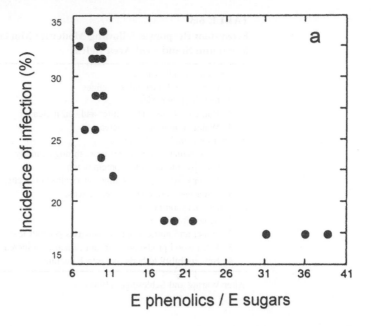

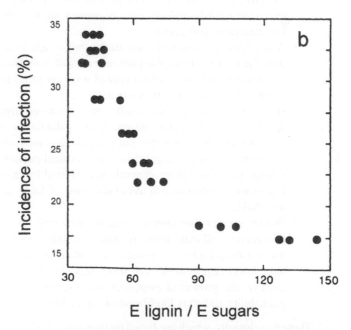

FIGURE 6.9.   (a) Incidence of infection by a root pathogen, *Armillaria,* on Douglas-fir trees increased significantly once the ratio energy required for phenolic degradation : energy available from sugars ($E_{phenolics}/E_{sugars}$) fell below about $15 \times 10^{-4} \, \mathrm{kJ \, mol^{-1} \, g^{-1}}$ root bark. (b) Similarly, when the ratio energy required to degrade lignin : energy available from sugars (E_{lignin}/E_{sugars}) decreased below about $60 \times 10^{-4} \, \mathrm{kJ \, mol^{-1} \, g^{-1}}$ root bark, the incidence of infection increased significantly. (From Entry *et al.,* 1991b.)

TABLE 6.3
Ecosystem Responses following Moderate Mortality at
Maximum Stand Leaf Area Index[a]

A. Structural modifications
   1. Canopy leaf area index declines
   2. Litterfall increases
   3. Content of nutrients in litter and soil increases
   4. Water content of soil increases
B. Environmental modifications
   1. Solar radiation penetrates more through canopy
   2. Canopy intercepts less precipitation
   3. Canopy intercepts less atmospherically borne chemicals
   4. Litter and soil temperatures increase
C. Functional modifications
   1. Mineralization increases
   2. Water and nutrient uptake increases per unit leaf area
   3. Stem wood production per unit of leaf area increases
   4. Susceptibility to herbivore attack declines

[a]After Waring and Schlesinger (1985).

and easily digestible sources of energy than more preferred woody plant species. Moreover, the juvenile form and component plant parts of deciduous species with the highest nutritional and energy content are less preferred than older forms and less nutritious components.

- A few specific chemical constituents, which are usually volatile and unstable, play the major role in plant defense against browsing animals. These defense components are often not correlated with the gross fractions of total phenols, tannins, or resins present in tissues.

- Browsing animals usually require more than one highly palatable species in their diet to meet daily dietary needs. As the palatability of dietary components declines, a greater diversity of plant species, growth stages, and component parts is required to maintain animal weight throughout critical periods when food resources are scarce. Thus the dietary generalism exhibited by many browsing animals is a necessary consequence to avoid ingestion of large quantities of toxic secondary metabolites.

- Woody species that have evolved to grow well on fertile soils or following disturbances allocate fewer resources to defense and have a greater ability to recover from injury by sprouting (or other mechanisms) than those species that have lower requirements and inherently slower growth rates. Fertilization can improve the growth of evergreen species, but it results in increasing their palatability and thus likelihood of injury from browsing animals.

These conclusions, which are based on numerous field and laboratory experiments, suggest that, where browsing animals are native, the vegetation is likely to be well adapted to herbivory. Evolutionary considerations are important because when numerous species of vertebrate herbivores were introduced into New Zealand, a country previously without native mammals (except for bats), large areas of forests were greatly affected by over-browsing of understory trees and shrubs; possums introduced from Australia have also caused extensive damage to New Zealand forests (Howard, 1964; Coleman *et al.,* 1980). Even within a region where the vegetation has evolved with vertebrate herbivores, conversion of large areas to fast-growing plantations of single species has obvious implications for residual animal populations. The native ungulates, according to the theories summarized above, will find less diversity in their diet and are likely to shift their herbivory to the plantations.

An independent test of these principles was made in southeastern Alaska where extensive logging of old-growth hemlock forests has greatly stimulated the growth of a deciduous blueberry, *Vaccinium ovalifolium,* favored by the black-tailed deer (*Odocoileus hemionus* var. *sitkensis*), but deer populations have fallen rather than increased in the region (Hanley and McKendrick, 1985). Previous to logging activities, deer largely survived the winter by browsing *Vaccinium* and *Cornus,* species that grow well in gaps beneath the old-growth canopy. These two shrub species provide deer with high concentrations of readily digestible protein, which is required for lactating females to nurse fawns born in the spring (Hanley, 1993). Because blueberry was present in all stages of stand development, it was hypothesized that the quality of browse may have changed with removal of the overstory canopy. A specific biochemical analysis was developed to define the amount of digestible protein in deer browse (Roberts *et al.,* 1987; Hanley *et al.,* 1992). Little insight into the nutritional value of forage could be gained by analysis of mineral, carbohydrates, or total protein content (Hanley *et al.,* 1992).

Rose (1990) designed a series of laboratory and field experiments to determine how twig growth and leaf biochemistry of blueberry varied with changes in overstory cover and nitrogen availability. The field experiment was conducted on a recently logged area where large numbers of blueberry plants were growing. To obtain a range in incident radiation, zero to three layers of netting were placed over individual blueberry plants; in addition, a range in nutrient availability was provided through application of nitrogen fertilizer with supplements of other nutrients at the start of the growing season. Measurements of twig growth and foliar analyses were made at the end of the growing season before leaves began to senesce.

The results of the field experiment fully supported theoretical predictions. Annual shoot production by blueberry increased with irradiance as expected (Fig. 6.10a). Excess amounts of nitrogen proved damaging to growth, a not uncommon response by ericaceous plants that have limited ability to reduce nitrate nitrogen (Smirnoff *et al.,* 1984). The concentration of tannins in foliage increased sharply with irradiance as theory also predicted (Fig. 6.10b). As a result of the increase in tannins and decrease in amino acid levels (not shown), the specific assay of digestible N for deer (Roberts *et al.,* 1987) indicated a general reduction in palatability as irradiance increased (Fig. 6.10c). Specific leaf weight, an easily measured structural index, varied inversely with the pattern shown for digestible nitrogen (Fig. 6.10d). In the particular region where the field study was conducted, changes in specific leaf weight were closely correlated with the biochemical analyses. Such correlations between structure and biochemistry have proved helpful to managers interested in assessing forage quality that affect winter carrying capacity and the survival of fawns (Van Horne *et al.,* 1988; Hanley, 1993).

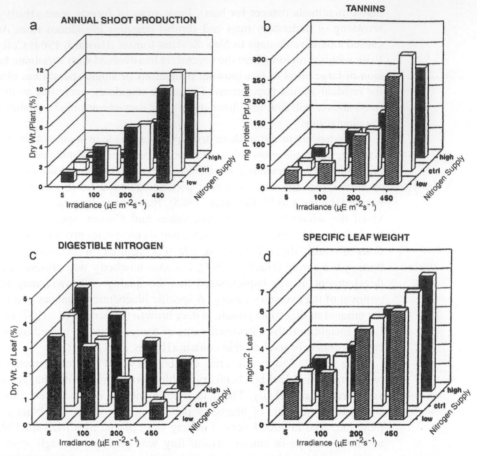

FIGURE 6.10. Field-grown blueberry plants placed under zero to three layers of netting which received increasing amounts of N and other nutrients showed significant changes in (a) annual shoot production, (b) tannin concentrations in leaves, (c) digestible N, and (d) specific leaf weight. These responses explain why open-grown plants were less palatable to deer than those found under more shaded conditions in gaps of an old-growth hemlock forest in southeastern Alaska. (From Rose, 1990.)

Browsing animals exert a number of controls on the rates of important processes in ecosystems. By selectively browsing young plants growing in partial shade, ungulates have effectively removed many deciduous hardwood species from European forests (Wolfe and Berg, 1988). Exclosure experiments in Isle Royale National Park (Michigan) have shown that such selective and intensive herbivory reduce the quantity and quality of litter returned to the soil, and hence N mineralization and NPP (Pastor *et al.,* 1993). Generally, deciduous hardwoods are first removed, with conifers usually less injured by browsing. When animal populations reach levels where they consume the less palatable species, their impact on forest ecosystem function and structure is generally negative. Where populations of ungulates and other vertebrate herbivores are high, the leaves and twigs of aspen and birch that can be reached by browsing animals are less palatable than in areas with significantly lower densities of the animals (Moore, 1994). It is possible that these differences could result in reduced growth and altered rates of decomposition that would have little correlation with environmental variables that drive most ecosystem process models.

In summary, we note that it is essential to consider plant biochemistry when evaluating the susceptibility of plants to biotic agents of disturbance. The search for general indices of plant susceptibility to biotic disturbance is challenging because only a few biochemical constituents account for defense against particular organisms, and these constituents may differ significantly, depending on the herbivore and species (or race) of plant. Nevertheless, we see value in comparing the ratio of the concentrations of broad groups of defensive compounds (tannins, phenolics, lignin, alkaloids) to readily assimilated compounds (sugars and amino acids) against the amount of plant material consumed by insects, digested by browsing animals, or infected by pathogens. Under some conditions, key biochemical properties may be related to structural indices such as tree growth efficiency and specific leaf weight. When this is the case, changing patterns of host susceptibility or palatability may be readily monitored. The absence of the normal complement of palatable and unpalatable vegetation typically found in different stages of stand development may be indicative of unsustainable management practices that will result in fostering excessive variation in animal populations. Extremely high animal populations are particularly damaging as they often lead to a reduction in biodiversity and long-term site productivity.

III. ABIOTIC FACTORS

In earlier chapters we noted how seasonal variation in climate affected ecosystem function and how species differed in their allocation of carbon to leaves, sapwood, and roots depending on their genetics, the availability of resources, and type of stress. The biotic components of ecosystems, and particularly long-lived trees, must adapt to infrequent events such as fires, floods, and windstorms. Throughout human history, attempts have been made to alter the natural frequency of fires and damage caused by floods and windstorms. We have to accept, however, that relatively catastrophic, natural disturbances will continue to occur. In fact, some practices associated with human activity, such as logging, grazing, and road building, have the potential to destabilize the natural resistance and repair mechanisms that reside in ecosystems. In other cases, human activities provide scarce resources to forest ecosystems by increasing the atmospheric concentrations of CO_2 and the deposition of nitrogen and sulfur.

In the following sections we introduce additional functional indices that have diagnostic value and quantify responses to specific kinds of disturbances. In addition, we evaluate how various abiotic disturbances alter the flow of water, carbon, and nutrients through ecosystems. A historical perspective on the frequency of different kinds of abiotic disturbances and their effects on ecosystem processes puts us in a better position to assess the implications of changes imposed by management practices.

A. Fire

Although fires start naturally, human activities have long played a role in their spread and frequency. As Attiwill (1994) points out in a review, the record goes back at least 10,000 years in the Americas, over tens of thousands of years in Australia, and over 1.5 million years in Africa. Fire plays important roles in many forest ecosystems, and there are two main types: (1) ground fires, which consume litter and kill understory trees and shrubs, and (2) crown fires, which usually kill the overstory trees and lead to stand replacement.

The spread and intensity of fire depends on the prevailing climate and on the amount of fuel available. The amount of fuel available varies with stand development as a function of fire frequency and logging practices. In North America the greatest fuel loads accumulate on the forest floor in natural stands of Douglas-fir about 20 years after a crown fire, and 500 years later as western hemlock (*Tsuga heterophylla*) begins to replace dying old-growth Douglas-fir. Agee and Huff (1987) analyzed the circumstances in which fires occur and concluded from their analyses that to lessen the fire danger to a patch of old-growth forest, surrounding stands should be between 100 and 200 years old rather than made up of younger or older age classes. Their recommendation would also minimize wind damage to old-growth stands by establishing tall, younger forests on all sides.

Fire differs in its effect on soil fertility depending on the region. In the boreal zone, fire occurs frequently in upland spruce forests and muskeg bogs, two areas where organic matter with excessive C : N ratios accumulates and nutrient cycling through decomposition is severely limited. As litter accumulates, it shields the soil from solar radiation, permafrost rises closer to the surface, and the productive capacity of the system decreases. Fire removes much of the surface organic matter, concentrates nutrients in the ash, and allows solar radiation to warm the soil and increase the rooting depth to permafrost. As a result, more productive hardwoods return to occupy upland sites (Van Cleve *et al.,* 1983). In muskeg bogs, willows replace poor quality browse provided by spruce and ericaceous shrubs and support an increase in moose and deer herds, as well as their predators (Viereck *et al.,* 1983).

In temperate forests of the Great Lakes region, fire historically regenerated many of the short-lived tree species at intervals of less than 50 years (Heinselman, 1973). Much of the wildlife is dependent on frequent disturbance to the forest to provide high-quality browse to sustain them through harsh winters (Hansen *et al.,* 1973; Jakubas *et al.,* 1989; Gullion, 1990). With protection from fire and logging, longer lived tree species eventually replace short-lived ones. Shade-tolerant trees are often not well adapted to fire because they lack thick bark, the ability to sprout from roots, or the ability to produce epicormic branches. As organic matter continues to accumulate on the soil surface, a multiaged forest develops, which provides a fuel ladder from the forest floor to the canopy. Under these conditions, the likelihood of stand replacement fires increases (Romme, 1982). An analysis of fire histories from dated fire scars on tree trunks and stumps suggests that frequent, moderate burns were typical in fire-adapted forests and that infrequent fires were generally destructive, particularly to thin-barked trees (Keane *et al.,* 1996b).

Periodic fires at intervals of less than a century have been important in maintaining diversity in other regions. Large areas of the pine forest native to the southeastern United States were maintained by fires; in the absence of fire, hardwood species predominate. Nitrogen-fixing species in many of the drier regions of the Pacific Northwest are dependent on fire for their regeneration and the release of a limiting supply of an essential trace element, molybdenum (Mo), which is sequestered over time in accumulations of organic matter (Silvester, 1989). Douglas-fir forests in many parts of the Pacific Northwest and the Rocky Mountains regenerated following fires at intervals between 125 and 400 years (Romme, 1982). Even the giant sequoia (*Sequoiadendron*) that may live for more than 2000 years is fire-dependent for seedling establishment in its native habitat (Kilgore and Taylor, 1979).

If we attempt to limit the area burned annually, large fires which are difficult to control will occur at less frequent intervals (Christensen *et al.,* 1989). Individual stands may still be protected if fuel loads are first reduced and periodic ground fires are set and allowed to burn under prescribed conditions. These practices are widely applied to protect and perpetuate many pine plantations throughout the world (Sackett, 1975; Goldammer, 1983; Covington and Moore, 1994). Ground fires have some additional benefits in consuming volatile organic compounds that inhibit decomposition (White, 1986). Volatile compounds are particularly high in plantations composed of pine and eucalyptus trees, which raise the danger of fires to surrounding native vegetation and to human settlements.

The extensive and species-rich eucalyptus forests of Australia have been among the best studied in terms of the effects of fire on ecosystem processes (Raison *et al.,* 1993). Although most species of eucalyptus are well adapted to fire, fire intensity and frequency greatly affect nutrient availability, which limits productivity on the highly weathered soils typical throughout much of Australia (Raison, 1980). Fire changes the availability of nutrients by volatilizing C, N, and S, while concentrations of K, P, and divalent cations increase in ash. Soil heating leads to an immediate accumulation of ammonium nitrogen (NH_4^+) as a result of chemical oxidation of organic matter. The amount released increases with the degree of soil heating for temperatures up to 400°C and with the content of N in the oxidized organic matter (Walker *et al.,* 1983). Additions of ash to acid soils may lead to a decrease in exchangeable aluminum and an increase in soluble silica; as a consequence, soils are likely to fix less P in an unavailable form.

Fire frequency plays a critical role in the eucalyptus forests of Australia. Too long of an interval between fires results in excessive damage to soils and loss of nutrients, whereas too short of an interval prevents nitrogen-fixing shrubs from restoring N lost in combustion and reduces soil N mineralization rates by 35 to 50% (Raison *et al.,* 1993). In most eucalypt forests the pattern of forest floor fuel accumulation can be described, as shown by Raison *et al.* (1993), by the sum of two negative exponential relationships:

$$X_t = X_0 e^{-k't} + X_s(1 - e^{-kt}) \tag{6.1}$$

where X_t is the mass ($Mg\,ha^{-1}$) of litter accumulated at time t (years), X_s is the mass of litter accumulated under steady-state conditions, k is a decomposition rate constant (year^{-1}), X_0 is the residual litter remaining after the previous fire, and k' is its decay constant. For Australian eucalypt forests, k varies from about 0.1 to 0.3 year^{-1} and X_s varies from about 10 to 30 $Mg\,ha^{-1}$. The accumulation of elements such as N and P in fuels can also be described by similar exponential equations. Because the overstory trees are not generally killed, litterfall rates are maintained, so there is a rapid accumulation of fuels during the initial 5 years after burning with a plateau approached by 15 years (Fig. 6.11a). Nitrogen in the litter and shrubs follows a similar trend as fuel accumulation because symbiotic N-fixing plants, such as the woody leguminous shrub *Daviesia mimosoides* or the cycad *Macrozamia riedlei,* are able to establish themselves after fires and grow rapidly (Raison *et al.,* 1993; Fig. 6.11b). In Australian eucalypt forests, an interval of about 10 years or

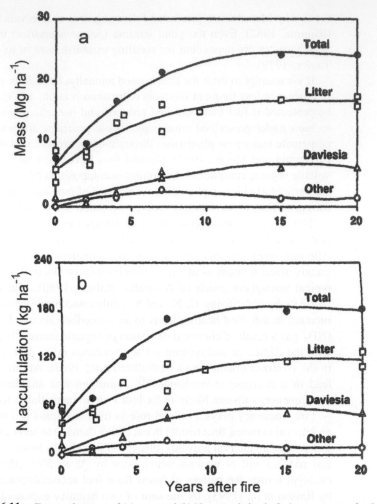

FIGURE 6.11.   Temporal pattern of (a) mass and (b) N accumulation in fuel components after low-intensity prescribed fire in a *Eucalyptus pauciflora* forest. Totals include ground litter derived mainly from eucalypt trees, standing biomass of *Daviesia mimosoides* shrubs, and other components, which represent mostly grasses and forbs. (After Raison *et al.,* 1993.)

more allows natural processes time to replace N loss during burning if volatilization of N is limited to approximately 50% of that in fuel (Raison *et al.,* 1993).

In the tropical regions of Africa, Asia, and South America, fire frequency has increased as forest land is converted to farms and pastures (Matson *et al.,* 1987). Depending on the intensity of fire and the kind of fuel present, large amounts of nutrients may be lost through volatilization, wind dispersal of ash, and surface erosion. Kaufmann *et al.* (1993) reported in studies of Brazilian second-growth tropical dry forest that maximum losses could exceed $500\,kg\,N\,ha^{-1}$ and more than $20\,kg\,P\,ha^{-1}$ following intense fires. In a study across a broader vegetation gradient in the Amazonian Basin of Brazil, Kaufmann *et al.* (1994) concluded that N and S were the nutrients lost in highest quantities during fire. Losses of P were intermediate, and losses of K and Ca were negligible. The total N in the rapidly decaying fuel was <4% of that in the surface 10 cm of soil and should, under natural conditions, be replaced in 1 to 3 years. Because of the extensiveness of fire in the tropics, the full implications must be assessed at a landscape and regional level with other methods (Chapters 7–9).

In many areas, fire plays a dominant role in recycling nutrients and in creating conditions favorable for establishing certain species. The historical frequency of fire varies with climatic conditions and with the amount and type of fuel accumulated. Through fire scar, charcoal carbon-dating, and stand-age analyses, it is possible to reconstruct fire frequencies and compare them with present-day conditions (Bradshaw and Zackrisson, 1990; Duncan and Stewart, 1991; Abrams *et al.,* 1995). Although many stand development models take into account the differential responses of vegetation to fire, fuel loads, and climatic conditions (Shugart and Noble, 1981; Kercher and Axelrod, 1984), ecosystem models that predict the effect of fire on the pool of available soil nitrogen (Fig. 6.12a), LAI (Fig. 6.12b), and evapotranspiration (Fig. 6.12c) are just beginning to appear in the literature (Sirois *et al.,* 1994; Keane *et al.,* 1996b). These combination models, when expanded across landscapes, better assess the implications of forest practices under stable or changing climatic conditions (Chapters 7 and 8). In addition, they have the potential to include new species introduced into the ecosystem that may alter nutrient capital, fuel loads, and fire hazards.

B. Atmospheric Factors

1. Gases

As noted in Chapter 4, solid, gaseous, and dissolved compounds in the atmosphere are captured by forest canopies to a greater extent than by other types of vegetation because of the larger volume of exposed leaf surfaces (Lovett, 1994). Over the past 50–100 years, the concentration of gases such as ozone (O_3), sulfur dioxide (SO_2), nitrogen oxides (NO_x), ammonia (NH_3), and methane (CH_4) have increased substantially above previous natural levels. The increase in atmospheric CO_2, on the other hand, is still within the range of concentrations that encompass the recent evolutionary development of most species (Beerling and Chaloner, 1993; Van de Water *et al.,* 1994). We have discussed the implications of continued increases in CO_2 on photosynthesis, growth allocation, and community structure in previous chapters and will defer further discussion on the global scale significance until Chapter 9. In this section, the implications of increased trace gas concentration are considered as it relates to gradual reduction in stand LAI over time.

Like CO_2, ozone and other gaseous pollutants are absorbed at rates proportional to the concentration gradient between the atmosphere and the leaf, constrained by the leaf stomatal conductance. The same models developed to predict stomatal control on transpiration provide a basis for estimating uptake of gaseous pollutants (Matyssek *et al.,* 1995). Humid atmospheric conditions favor maximum opening of stomata and maximum sensitivity to pollutants. Crop plants and deciduous trees are more sensitive to given exposure to ozone or other pollutants than are most evergreen species due to differences in maximum

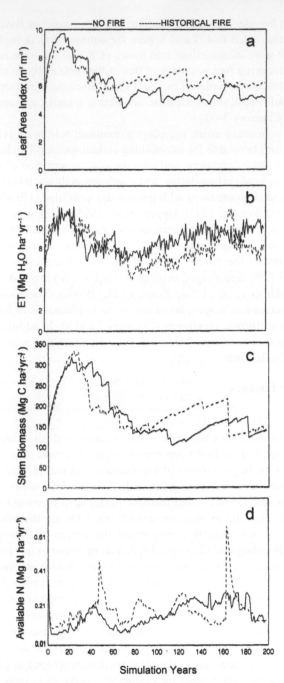

FIGURE 6.12.   Model simulations predict the implications of fire control on (a) leaf area index, (b) evapotranspiration (ET), (c) stem biomass, and (d) nitrogen availability for the northern Rocky Mountains over two centuries. (From Keane *et al.,* 1996b.)

stomatal conductances (Reich, 1987). If ozone damage reduces the photosynthetic capacity of leaves more than stomatal conductance (Matyssek *et al.,* 1992), then the carbon isotopic composition of foliage and wood (adjusted for change in atmospheric composition) should become more depicted in ^{13}C and enriched in ^{12}C. On the other hand, SO_2 has been reported to cause stomata to close more quickly, which results in an increase in the $^{13}C/^{12}C$ ratio in foliage (Martin and Sutherland, 1990). Although evergreen species restrict the diffusion of ozone and other pollutants into their leaves more than deciduous species, the longer life span of evergreen leaves increases their total exposure and often results in premature shedding of foliage (McLaughlin, 1985). Rising concentrations of atmospheric CO_2 are predicted to induce some stomatal closure that could ameliorate the detrimental effects of gaseous pollutants (Field *et al.,* 1992).

Trees may also produce a number of volatile organic compounds that create haze above forest and result indirectly in reducing the solar energy load on the canopy. Isoprene, one volatile compound produced by plants, has been shown to protect chloroplasts as temperatures rise to dangerous levels (Sharkey and Singsaas, 1995). Concentrations of isoprene inside leaves increase at temperatures above 35°C; the efflux from leaves continues to increase although stomata may close completely (Monson *et al.,* 1994; Sharkey *et al.,* 1995). Emissions of monoterpenes also increase greatly with temperature and vapor pressure deficits, which has implications for attracting bark beetles (Waring and Pitman, 1985).

2. Wet- and Dryfall

More cations (Ca^{2+}, Mg^{2+}, K^+) are often derived from atmospheric deposition than from mineral weathering in many forests (Chapters 4 and 5). Since the 1950s, atmospheric inputs of nitrogen and sulfur have become dominant sources of these nutrients in many northern temperate forests (Hedin and Likens, 1996). Airborne deposition of heavy metals (Pb, Cu, Cd, Hg, Ni, Zn) has also increased locally where traffic, mining processing plants, and heavy industrial development are concentrated. Heavy metals are often lethal to mosses, lichen, and other species which collect nutrients exclusively from airborne sources or from stemflow (Tyler, 1972). When heavy metals enter the detritus pool, they may inhibit microbial activity and the release of nutrients to higher plants (O'Neill *et al.,* 1977). Changes in heavy metal deposition rates have been quantified by analyzing the contents of dated wood cores extracted from long-lived trees (Ragsdale and Berish, 1988; Jordon *et al.,* 1990; Eklund, 1995).

Because nitrogen and sulfur are transported further from their source than heavy metals, they have potentially greater impact. Fortunately, deposition rates of N and S have been widely monitored at a number of sites, with one of the best long-term records available at Hubbard Brook Experimental Forest, New Hampshire (Likens and Bormann, 1995). During the International Biological Program in the 1960s, a host of additional sites began to record nutrient inputs in wetfall. In the 1980s, many more sites were established in the United States and Europe, and included elemental deposition in dryfall and cloud/fog precipitation. The more inclusive measurements showed that wetfall often represents less than 50% of total inputs, and less than 25% where cloud/fog precipitation is prevalent (Johnson and Lindberg, 1992). The network of monitoring sites have shown considerable variation in atmospheric deposition over decades which reflect changes in national policies on emissions of pollutants, land use, and weather patterns (Hedin and Likens, 1996).

Where ecosystems have developed on infertile soils, the additions of nitrogen and sulfur, as well as base cations, can be extremely beneficial. Atmospheric deposition or nitrogen may counterbalance any increase in C:N ratios that might be associated with rising CO_2 levels, and by doing so increase litter decomposition rates and site fertility. Such responses have been already observed in some ponderosa pine forests distributed across a deposition gradient in California (Fenn, 1991; Trofymow *et al.,* 1991). Where sulfur and nitrogen are deposited in excess of what vegetation requires or the soil can store, the excess moves into the groundwater, carrying with it acids and toxic concentrations of aluminum that affect stream and lake chemistry in detrimental ways.

Enough is known about cation exchange and leaching processes to construct fairly general models to predict the transfer or immobilization of sulfur, hydrogen, and other cations through soil and into streams (Cosby *et al.,* 1985; Gherini *et al.,* 1985; Norton *et al.,* 1992; Wright *et al.,* 1995). Models of nitrogen cycling are available, but they are difficult to test because some of the variables cannot be easily measured and long-term predictions are not based on fully established principles (Aber *et al.,* 1991; Johnson, 1992).

Beyond certain limits, N deposition will exceed the uptake potential of vegetation and soil storage. Such ecosystems become "nitrogen saturated" as excess N is released into the groundwater and vented to the atmosphere through denitrification (Chapter 4). We might conclude that annual additions of nitrogen in the range from 25 to $50\,kg\,ha^{-1}$ would be entirely beneficial to some types to forest ecosystems. Certainly such rates are well below those applied commercially to enhance forest growth. However, pulse additions of fertilizer are quickly incorporated into the soil or stored in foliage and other tissue to be later utilized by plants. In contrast, chronic, long-term additions of N, even at rates slightly above $10\,kg\,ha^{-1}\,year^{-1}$, have resulted in site degradation in some parts of Europe (Dise and Wright, 1995; Wright *et al.,* 1995).

Johnson (1992) reviewed the results of 60 fertilization trials and ecosystem studies conducted in temperate and boreal forests where the amount of nitrogen deposited, leached, and incorporated in aboveground increment was measured, and that retained in the soil calculated by difference (input − increment − leaching). From the analysis, Johnson suggested that leaching of nitrogen from soils could be predicted quantitatively as a function of atmospheric N input minus N increment in wood (Fig. 6.13) At N deposition levels below that required for increment, no leaching was observed; trees were apparently able to obtain nitrogen from the soil in competition with heterotrophic fungi. At higher deposition rates, a fixed proportion of the excess N was stored in the soil, equal to that lost through leaching. There were no obvious relationships between N deposition rates and calculated soil N retention, nor with tree N increment (Johnson, 1992).

To appreciate how the nitrogen cycle may be altered with increasing deposition of N, we adopt a convenient classification scheme developed by Aber *et al.* (1989). We will identify shifts in the relative importance of various processes at different stages and extract some general indices from the reviews by Abet (1992) and Johnson (1992).

Figure 6.14 depicts four stages that represent a transition from conditions improving forest productivity to ones causing a major decline, associated with a massive reduction in LAI. Stage 0 represents conditions with background levels of N deposition of $<2\,kg\,N\,ha^{-1}$

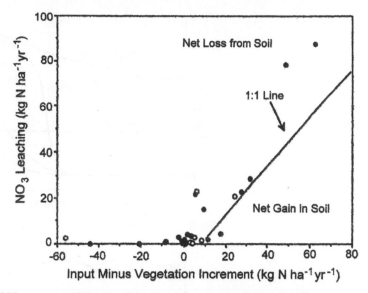

FIGURE 6.13. Nitrogen (NO_3^-) leaching as a function of atmospheric N deposition minus tree N increment indicates that up to 60 kg N ha^{-1} year^{-1} may be sequestered in wood growth without leaching loss. Where growth in increment is low, however, the excess N appears to be partitioned nearly equally between storage in the soil and loss to streams. Rates of deposition, which included wet- and dryfall (○), were often 50% higher than N measured in bulk precipitation only (●). (After Johnson, 1992.)

year^{-1} (Hedin *et al.*, 1995). At these levels, most temperate and boreal forests are limited by the slow nitrogen release rates through decomposition and by the ability of mycorrhizal fungi to take up organic forms of N directly (Chapter 4). The concentration of N in foliage and litter potentially limits both photosynthesis and decomposition. Some of the NH_4^+ released through microbial activity may be accessed by nitrifiers, but no net nitrification occurs in evergreen-dominated systems (Robertson, 1982). In deciduous forests, some net nitrification could be expected, but NO_3^- uptake by plants and microbes would prevent loss of N, except in dissolved organic matter (Hedin *et al.*, 1995).

In the beginning of stage 1 (Fig. 6.14), nitrogen deposition increases above background levels to a constant 20 kg ha^{-1} year^{-1}. Such deposition rates may seem low, compared to single applications of fertilizer, but they appear sufficient to cause N saturation in central Europe (Dise and Wright, 1995). The capacity of soils and vegetation to incorporate large amounts of nitrogen vary, so that stage 1 may continue for only decades or for several centuries (Aber *et al.*, 1991). Foliar N concentrations and foliar biomass peak during the middle of stage 1 as nitrification becomes the dominant process. With a relative excess of nitrogen, total NPP increases, but proportionally less carbon is allocated to roots and mycorrhizae, which reduces the ability of the system to immobilize N in the soil (Stroo *et al.*, 1988) and favors heterotrophic fungi over symbiotic species (Osonubi *et al.*, 1988; Johnson *et al.*, 1991; A. H. Johnson *et al.*, 1994). Less root growth, together with leaching of base cations, is likely to create deficiencies in magnesium and other nutrients (Zinke,

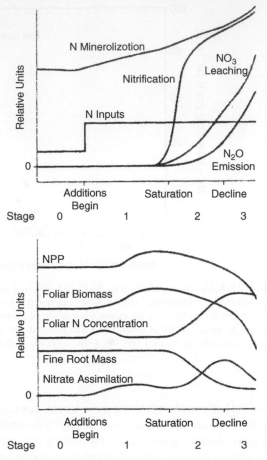

FIGURE 6.14.   Hypothesized time course of forest ecosystem response to chronic nitrogen additions. (Top) Changes in nitrogen cycling and nitrogen loss rates. (Bottom) Plant responses to changing levels of nitrogen availability. Four stages are recognized: stage 0, background levels of pollution; stage 1, additions begin but no changes in forest growth occur, although N availability and uptake increase; stage 2, aboveground NPP increases and N is in excess of requirements; and stage 3, N saturation causes decline in ecosystem structure and function. (From Aber *et al.,* 1989. © 1989 American Institute of Biological Sciences.)

1980; Oren *et al.,* 1988b; Schulze 1989; Cote and Camire, 1995). As a result, growth per unit of leaf area or per unit of N will decrease (Oren *et al.,* 1988a,b).

At the start of stage 2, foliar biomass begins to decrease as continued additions of N have detrimental consequences (Fig. 6.14). Evergreen coniferous species are particularly sensitive because they have limited ability to increase their photosynthetic capacities per unit leaf area (Reich *et al.,* 1995b) or to utilize large amounts of nitrate (Smirnoff *et al.,* 1984). The added NO_3^-, if not reduced when taken up by plants, can be toxic and may cause chlorosis of foliage, premature foliage drop and mortality (Schulze, 1989).

The start of stage 3 is defined by extensive mortality of overstory trees, which results in reduced foliar biomass (Fig. 6.14). Nitrogen-saturated ecosystems typically lose large amounts of nitrogen through leaching of NO_3^- and volatilization of nitrous oxide (N_2O). The ability of soils to consume methane (CH_4) is also reduced because the enzyme system involved in the reaction no longer distinguishes between NH_4^+ and CH_4 (Steudler *et al.*, 1989). Tree species with higher demands for nitrogen and the ability to induce nitrate reductase activity in their foliage can slowly replace less well adapted species. In the interim, both the overstory and total LAI will be substantially reduced (Ulrich, 1983).

Aber (1992) encapsulates the ideas presented above to explain a lag of a decade to a half century or more between the time when N deposition increases (stage 1) and when excessive nitrate losses are observed (stage 2), as the result of competition between plants, heterotrophic microbes, and nitrifiers for the available pool of ammonium. Net nitrification should remain at zero as long as the demands by plants and decomposers for NH_4^+ are not fully met. In addition, ammonium is held relatively strongly on cation-exchange sites and can also be immobilized directly into soil organic matter (Johnson, 1992; Paul and Clark, 1996). The relative strengths of these competing forces for NH_4^+, along with the gross N mineralization rate, determine the residual concentration of inorganic N in the soil solution and leachate (Aber, 1992). Table 6.4 summarizes the principal changes expected in the transition from an N-limited to an N-saturated system.

What management options might delay N saturation? Fast-growing trees, harvested in short rotations, can extract N, but the wood will contain considerable amounts of base cations and phosphorus, which are likely in short supply. Site preparations that included slash burning would transfer N back into the atmosphere, while the more limiting nutrients would remain on the site. Increase leaching could result with reduced cover, however, and the total amount of N in biomass is usually less than 20% of that stored in the soil (Johnson,

TABLE 6.4

Contrasting Characteristics of N-Limited and N-Saturated Forest Ecosystems[a]

Characteristic	N-limited	N-saturated
Soil properties		
Soil C:N ratio	High	Low
Dissolved organic (labile) C in soil	High	Low
Ratio of gross NO_3^- immobilization to gross nitrification	1.0	<0.01
Ratio of gross NH_4^+ immobilization to gross mineralization	~0.95	<0.5
Population of nitrifiers	Low	High
Plant properties		
Form of N taken up by plants	NH_4^+	$NO_3^- > NH_4^+$
Foliar N concentration	Low	High
Foliar free amino acid concentration	0	High
Transfer properties		
Nitrate losses in leaching	0	High
N_2O production	0	High
CH_4 absorption by ecosystem	High	Low

[a]Principally from *Trends in Ecology and Evolution*, Volume 7, J. D. Abert, "Nitrogen cycling and nitrogen saturation in temperate forest ecosystems," pp. 220–223, 1992, with kind permission of Elsevier Science–NL, Sara Burgerhartstraat 25, 1055 KV Amsterdam, The Netherlands.

1992). Grinding up large woody debris and mixing the residue into the soil should immobilize considerable N and slowly release base cations back into the system (Turner, 1977; Waring and Pitman, 1985). We advocate that further experimentation be coupled with selected measurements of N cycling controls and ecosystem response to clarify the relative merits of various practices designed for specific soil and climatic conditions.

C. Forest Harvesting

Forest harvesting involves mechanical extraction of biomass from an ecosystem. In most instances, the removal of woody biomass results in only a small percentage loss of the total content of nutrient elements in forest ecosystems because the largest pool is in the soil (Chapter 4). Sawlog harvest of a mixed oak forest in Tennessee removed from 0.1 to 7.0% of the nutrient pool of N, P, K, and Ca (Johnson et al., 1982). Firewood harvest of a young rain forest in Costa Rica resulted in removal of 3% Ca to 31% S of the pool of various nutrients in the vegetation and surface soil (Ewel et al., 1981). The foliage and branches that are left behind contain a large portion of the nutrient pool in vegetation, because of the higher nutrient concentrations in these tissues than in wood. Leaf litter and small branches are likely to decompose rapidly, releasing nutrients for regrowth. With conventional harvest, the number of years needed to replace nutrients removed through the accumulation of annual inputs from atmospheric deposition, N fixation, and weathering is usually less than the harvest cycle (Johnson et al., 1982; Van Hook et al., 1982; Likens and Bormann, 1995).

Harvest of whole trees for pulpwood or biomass for energy results in substantially greater nutrient removals, although these may still be replaced within a regrowth cycle in many instances (Silkworth and Grigal, 1982; Van Hoob et al., 1982). Often the available forms of N, P, and other nutrients are derived mainly from the rapid decay of fresh organic matter. In such cases, the analysis of loss during harvest might more reasonably be expressed in reference to the available pool in vegetation and litter rather than that in the total soil. Whole-tree harvesting may remove 30% of the available Ca in a northern hardwoods forest in New Hampshire, but only 2% of the total in the soil (Hornbeck and Kropelin, 1982). The consequence of removing nutrients in harvested biomass on different sites can, in principle, be assessed by measuring the rate at which the available pool is replenished from various sources (including atmospheric deposition and fertilization) and by monitoring nutrient balances and return in litter over critical periods. In reality, atmospheric deposition is highly variable as a result of rapid changes in pollution control policies (Hedin and Likens, 1996). Moreover, the availability of two of the most critical nutrients, nitrogen and phosphorus, is often unrelated to the total content of these elements in the soil.

With the removal of the forest, the soil surface warms, and, with less transpiration, more water is available; these are environmental combinations that generally favor increases in decomposition, mineralization, and nitrification. Normally, herbaceous and shrubby vegetation quickly reoccupies a site following removal of the forest overstory so that little loss of inorganic nutrients occurs (Marks and Bormann, 1972). Attempts to control the growth of herbs and shrubs through application of herbicides, introduction of grazing animals, or by fire may contribute to accelerated losses of nitrate and other nutrients if the capacity of microbial organisms and other components of the soil to immobilize nutrients is exceeded (Vitousek *et al.,* 1982).

The problem of nutrient loss is particularly severe in the wet tropics because, following clearing, a rich supply of carbon and organic nitrogen is initially available (Matson *et al.,* 1987). As a result, large pulses of nitrous oxide, nitric oxide, and methane may be produced while NO_3^- is leached downward in the profile (Matson *et al.,* 1987; Luizao *et al.,* 1989; Keller and Reiners, 1994). Trace gas emissions are reduced rapidly once either the available carbon or nitrate substrate is exhausted in the surface soils (Matson *et al.,* 1987; Parsons *et al.,* 1993). In most situations, microbial biomass is rapidly reduced following disturbance, so its contribution to immobilizing N is much less than in temperate and boreal forests. On the other hand, tropical soils have a much larger capacity than temperate soils to capture nitrate on anion-exchange sites. Once deep roots are reestablished his nitrate is available for regrowth of the forest (Matson *et al.,* 1987).

Many tropical forests are converted for growing crops or more permanently into pasture. Conventional soil analyses do not provide reliable measures of soil fertility because so much of the nutrients are bound in various fractions of soil organic matter. A better assessment of available nutrients can be obtained with knowledge of the turnover rates or mean residence times (MRT) of various organic fractions. The pulse of radioactive carbon-14 put into the atmosphere during the peak years of atomic bomb testing provides a radioactive label to all detritus produced in the late 1950s and early 1960s which differs from that produced before or since. Tiessen *et al.* (1994) used carbon-14 analyses to determine the mean residence time of organic matter that holds most of the available nutrients in an extremely poor sandy soil from an Amazon rain forest at San Carlos de Rio Negro in Venezuela and from a sandy soil which supported a semiarid thorn forest in Brazil. With measurements of MRT and nutrient content of the organic matter, Tiessen *et al.* calculated that agriculture could be sustained following slash burning of the two forests only 3 and 6 years, respectively. In contrast, sandy prairie soils from the Canadian Great Plains were cited as being able to support economic agriculture without fertilization for 65 years. Trumbore *et al.* (1996) applied the same type of carbon-14 analyses to confirm the importance of soil carbon turnover at depths below 1 m in an evergreen Amazonian forest. They showed that managed pastures fertilized with phosphorus and planted with highly productive grasses could add large amounts of carbon quickly to the upper meter of soil, which partially offset predicted carbon losses due to death and decomposition of fine tree roots below 1 m in the soil. Such isotopic analyses, when applied to specific fractions of soil organic matter, provide a sound basis for calculating organic matter production and turnover, two rates that are critical in assessment of regional and global carbon balances (Chapter 9).

Erosion is one additional route for nutrient loss when trees are harvested from steep slopes. With tree removal, large-diameter roots begin to decay. As decay progresses slope stability is reduced until a new network of interlocking roots is reestablished (Ziemer, 1981). To remove wood products, roads must be constructed, and soil compaction often occurs in the process of harvesting. Both activities may contribute to slope instability and accelerated erosion (Harr *et al.,* 1979; Reid and Dunne, 1984; Jones and Grant, 1996). Some physically based models are available to predict the risks of erosion and runoff associated with logging and road construction (Rice and Lewis, 1991; Luce and Cundy,

1994). Identification of unstable slopes and unreliable road construction practices can greatly reduce erosion hazards because most erosion related to forest management occurs on a small fraction of the total forest area (Rice and Lewis, 1991).

In general, extraction of wood products from forests removes only a modest fraction of the total nutrients present in soil and litter. If vegetation is not rapidly reestablished, however, or if the permeability of surface soil is significantly reduced following logging, the stabilizing network of roots can be destroyed, increasing the potential for soil erosion and nutrient loss from the site. Under exposed conditions, the net release of radiative gases that affect Earth's energy balance also will increase along with the export of nutrients (and pollutants) into the groundwater and adjacent aquatic systems. By careful location and construction of roads and by adhering to good logging practices, disturbance can be minimized, but the largest benefits may accrue by reducing the fraction of LAI removed in a single harvest, as discussed in Chapter 5.

D. Mechanical Forces

1. Wind

Mechanical forces include the effects of wind, snow, and mass movement of material. Trees that possess flexible branches and leaves suffer little damage at normal wind speeds. Increased stem taper also provides greater resistance to breakage from wind. Where wind is a potential problem, thinning operations must be limited to provide time for residual trees to add more wood to their lower boles and large-diameter roots (Petty and Worrell, 1981). Genetic variation in tree resistance to wind is well documented. Spruce (*Picea*), with typically shallow roots, is easily blown down, whereas pine (*Pinus*) with deep tap roots is a species more resistant to wind. However, variation exists within a genus too; plantations of lodgepole pine in southern Finland are much more susceptible to wind damage than the native Scots pine (Lahde *et al.,* 1982).

The edge of a forest exposed to the prevailing wind is of particular concern. Normally, the edge is shaped so that wind sweeps up and over the canopy and does not penetrate directly through the stand. When roads or logging disturb the edge, wind enters, modifies the microclimate, and may result in extensive blowdown (Chen and Franklin, 1992; Ruel, 1995). By quickly reestablishing the canopy configuration typical of boundaries exposed to wind, the structural integrity and microclimate within a stand can be maintained.

To protect forests, Ruel (1995) concludes that trees should be selected which can adjust to wind by developing more wind-resistant canopies and by growing deeper and larger diameter roots that interlock with one another. Soil and root characteristics determine the sturdiness of the anchoring systems; age, stocking density, and thinning practices modify a stand's resistance to wind, while the topographic exposure and general climate impose ultimate limits on management options. In western Scotland, the choice of species and rotation age are made from an analysis of topographic exposure. Potential plantation sites are evaluated by installation of standard sized cloth flags, on which ablation is observed after a set time (Lines and Howell, 1963). On the basis of these wind hazard analyses, pine or spruce plantations are established and their "safe" rotation age specified.

Many forests are dependent on wind for maintaining diversity and productivity. High-intensity windstorms associated with hurricanes, typhoons, and tornadoes rarely destroy whole forests. For example, in 1938 a hurricane struck 4.5 million ha of forest in the northeastern United States (Smith, 1946), but only about 5% of the forest experienced heavy damage. Damage was mainly restricted to trees infected by pathogens or to those occupying particularly wind-prone sites (Henry and Swan, 1974; Bormann and Likens, 1979). Likewise, tropical windstorms cause frequent treefall (Lugo *et al.,* 1983), averaging about one tree per hectare per year (Hartshorn, 1978). The regularity of these disturbances limits the maximum age of trees in forests of Costa Rica to a range of 80–140 years and fosters, as a result, relatively stable net primary production over time. Comparable rates of windthrow are found elsewhere in the tropics (Putz and Milton, 1982). On the other hand, hurricanes that topple whole forests create conditions where the original composition is maintained through resprouting and seedling establishment. This sequence is not possible through normal gap formation and may explain the composition and structure of tropical rain forests subjected to frequent hurricanes (Yih *et al.,* 1991; Attiwill, 1994).

Wind can play an important role in maintaining soil fertility. For example, in coastal Alaska where a temperate rain forest of Sitka spruce (*Picea sitchensis*) and western hemlock (*Tsuga heterophylla*) dominates, organic matter accumulates rapidly on the forest floor and leads, even on freshly deposited glacial material, to podzolic horizons in the surface soil after only 100–150 years. Wildfires are uncommon so that continued accumulation of organic matter, combined with leaching of Fe and Al, create conditions where the water table rises and roots become largely confined to the upper organic-rich, nutrient-depleted horizons. Historically, periodic windstorms uprooted trees, mixing the soil horizons and thus slowing the degradation process (Bormann *et al.,* 1995). Harvesting practices should encourage the soil horizons to be mixed, and the amount of organic woody debris left on site should be minimized. Without following these practices, many forest ecosystems in northern Europe have been transformed to unproductive bogs (Miles, 1985).

2. Snow and Ice

Snow and ice are mechanical forces that may create extremely high stresses on tree branches and stems. Under some conditions, the load is sufficient to cause breakage of stems as well as branches. On steep slopes, snow moves slowly downhill in the spring, causing young trees to be uprooted, broken off, or forced to develop a "pistol-butt" shaped lower bole. In temperate regions where snow accumulates at predictable depths in response to elevational gradients, the distribution of forest types is often correlated with the average depth of snowpack. Models described in Chapter 2 provide estimates of snow water content and canopy interception which can provide a basis for judging the bearing load that snow can impose on vegetation. In the western United States, the upper elevational limits of Douglas-fir, western hemlock, and ponderosa pine, as well as most hardwood species, are clearly limited by their susceptibility to snow breakage. Subalpine fir (*Abies lasiocarpa*), mountain hemlock (*Tsuga mertensiana*), and Jeffrey pine (*Pinus jeffreyi*) replace less adapted species as the snowpack increases (Waring, 1969).

Lemon (1961) suggested that ice storms could speed up the normal successional sequence by selectively damaging faster growing species that are normally established earlier (Chapter 5). DeSteven *et al.* (1991) confirmed this prediction in a study of a forest dominated by sugar maple (*Acer saccharum*), beech (*Fagus grandifolia*), and basswood (*Tilia americana*) in which an ice storm caused accretions up to 12 cm in diameter. Mortality of beech and other dominant species allowed saplings of shade-tolerant sugar maple to increase their growth. On the other hand, Nicholas and Zedaker (1989) suggest that ice storms may play an important role in maintaining red spruce (*Picea rubens*) in the higher elevation forest in the southeastern portion of the United States in spite of the presence of more shade-tolerant species.

3. Mass Movement

Mass movement of soil and snow avalanches create conditions that favor regeneration of some species. In the Cascade Range of the Pacific Northwest, two nitrogen-fixing species are well adapted to colonizing cleared areas following soil erosion (*Alnus rubra*) or snow avalanches (*Alnus incana*). Similarly, in the Andes mountain range of Chile and Argentina where earthquakes and volcanic eruptions are rather frequent, single-aged stands of *Nor-thofagus* species are maintained up to 300 years and then replaced following another stand-destroying event which exposes bare soil (Kitzbergen *et al.,* 1995; Veblen and Alaback, 1996). Trees growing on unstable soils can often be recognized because they tend to shift their centers of gravity over time, which leads to an irregular buttress and nonvertical alignment.

Some of the residue from massive slope failures creates benches of deep soil deposits at midslope. When soil displaced by mass movement reaches streams it is deposited as alluvium on floodplains. Silt from a single flood may accumulate in deposits more than 50 cm deep around the base of coast redwood trees (*Sequoia sempervirens*) without preventing a new root system from quickly reestablishing (Stone and Vasey, 1968). In some areas, the sequence has been repeated for over 1000 years so that the surface on which trees were originally established now lies nearly 10 m below the surface. In addition, the fresh silt provides a seedbed free of fungi which allows redwood seedlings to germinate successfully and to colonize new areas on the floodplain (Stone and Vasey, 1968).

Native forests are well adapted to mechanical forces typical of the environments in which they are found. Silvicultural handbooks provide a general ranking of species in regard to their tolerances to various environmental stresses, including responses to mechanical forces (Fowells, 1965; Burns and Honkala, 1990). Often, widely distributed species are still planted in zones where they are ill-adapted because they initially show rapid growth. If rainfall intensity or the accumulation of snowpack were to change significantly, the boundaries of present forest types could shift abruptly.

IV. SUMMARY

In this and previous chapters, a number of structural and chemical properties have been identified that have general or specific diagnostic value (Table 6.5). In general, if high levels of amino acids are observed in foliage or other plant tissues, this is a sign of an

TABLE 6.5

Summary of Structural and Chemical Indices for Diagnostic Analysis of Ecosystems Exposed to Different Kinds of Environmental Stresses

Environmental factor	Diagnostic indices
General stress symptoms	Increase in free amino acids in leaves, reduced wood growth per unit leaf area, nitrate leached from system
Herbivory	Loss of most palatable species, evergreens replace deciduous species, low ratio of defense/energy compounds in plant tissue
Soil flooding	Presence of buried stems and soil litter layers in soil profile
Slope instability	Asymmetrical and leaning trees, mixed soil horizons with buried large roots
Soil drought	Low sapwood relative water contents, trees with low leaf area/sapwood area ratios, leaf $\delta^{13}C$ not correlated with branch length
Soil nutrient imbalance	Adverse nutrient ratios in foliage, low wood production per unit of N in foliage
Excess N deposition	Leaf N in excess to other nutrients, leaching of nitrate, volatilization of nitrous oxide
Fire-prone environments	Charcoal in soils, fire scars on trees that have thick bark, serotinous cones, or other adaptations to fire
Wind-prone environments	Stem taper in trees with asymmetrical crowns, historical evidence of uprooted trees, predominant direction of blowdowns on exposed sites
Ozone and SO_2	Loss of lichens, accelerated loss of older evergreen foliage, evidence of sulfur concentrations in annual rings, historical shifts in $\delta^{13}C$ in cellulose unrelated to climatic variation or atmospheric concentrations of CO_2

unstable forest with high susceptibility to biotic-induced disturbance. Chemical and physical analyses of tree rings are particularly useful to assess historical variation in atmospheric deposition of heavy metals and tree responses to variation in atmospheric pollutants and CO_2. The frequency of fire and the intensity of insect outbreaks may also be inferred from scars and historical patterns in growth efficiency. Plant community attributes provide some insight into the populations of browsing animals. Together, biochemical, isotopic, and mineral analyses of foliage, wood, and roots yield clues to the susceptibility or resistance of forests to specific agents of disturbance.

Historical analysis of stands suggests that moderate mortality caused by wind, fire, insects, disease, and other types of disturbances creates gaps in the canopy that provide benefits to residual trees. These benefits include reduced competition for light, water, and nutrients and improved resistance against many agents of disturbance. When assessed over decades, frequent, moderate disturbances seldom increase the normal rates of mortality beyond that expected with self-thinning, nor is the general forest structure greatly altered. Frequent disturbances, however, may favor some species over others and alter expected trends in forest composition.

Large-scale natural disturbances associated with hurricanes, fire, landslides, and insect and disease outbreaks greatly increase mortality but, in the long run, rejuvenate the system and help maintain or increase its productive capacity. When the frequency or kind of disturbance is significantly altered, new conditions are created that favor new mixtures of species from those previously dominant. With insight into basic ecosystem processes and the use of various indices of disturbance, the long-term consequences of disturbances may be better recognized so that remedial, preventive, or adaptive actions have a sound basis.

FIGURE 6.2. (a) Outbreaks of defoliating insects occur more frequently in the western United States where fire protection has enabled large expanses of fir trees (*Abies grandis*) to replace much of the original pine. (b) Mortality from insect outbreaks varies, depending on the length of the outbreak and physiological status of host trees. This scene records nearly full recovery 5 years after the first photograph was taken. Any mortality improved diameter growth of surviving trees (Wickman, 1978). (Photograph from Boyd Wickman, La Grande, Oregon.)

FIGURE 6.7. Mountain pine beetles attacked and killed blocks of old-growth lodgepole pine in eastern Oregon, except where the forest was previously thinned. (Photograph provided by John Gordon, Yale University, School of Forestry and Environmental Studies, courtesy of Boise Cascade.)

SECTION III

Introduction to Spatial Scaling and Spatial/Temporal Modeling

IN SECTION II we developed an appreciation of how ecosystems operate in the temporal dimension and presented an array of models developed to predict their behavior into the future. In Section III, we extend this analysis to the spatial dimension. Most forested landscapes are extremely heterogeneous in both physical and biological characteristics. In this section we introduce ways to quantify the physical heterogeneity of topography, climate, and soils. We further extend the analysis of heterogeneity to include biological features of forests, primarily through the application of remote sensing techniques. Finally we demonstrate how heterogeneous spatial data may be incorporated into simulation models that quantify ecosystem responses across landscapes and project changes in forest structure and composition forward in time (Fig. III.1). These ecosystem models can provide land managers a visual image of how the landscape may change in the future under the range of alternatives being considered today.

Critical to the success of these spatially explicit models is a judicious choice of variables that can be measured and compared over landscapes and regions. In most cases the appropriate variables are derived by simplifying the analysis of

Multi-Scale Ecological Model Evolution

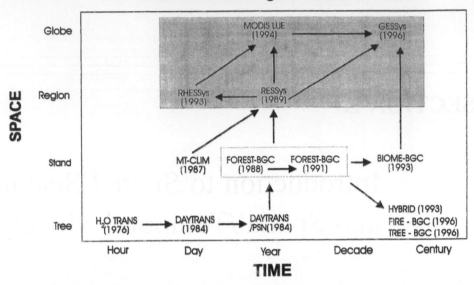

FIGURE III.1.   Evolution of a family of related ecosystem simulation models for a range of space/time applications, referenced by the model name and initial publication date. Model development shows a progression from an earlier precursor with a smaller temporal or spatial scale. Reference papers for stand-level seasonal models include H2OTRANS (Waring and Running, 1976), DAYTRANS (Running, 1984a), DAYTRANS/PSN (Running, 1984b), and FOREST-BGC (Running and Coughlan, 1988). Extensions of forest dynamics through a stand life cycle are represented by FOREST-BGC (Running and Gower, 1991), HYBRID (Friend *et al.*, 1993), BIOME-BGC (Running and Hunt, 1993), FIRE-BGC (Keane *et al.*, 1996a,b), and TREE-BGC (Korol *et al.*, 1995).

In this section, a number of georeferenced models that extrapolate across increasingly large areas are introduced. They include MT-CLIM (Running *et al.*, 1987), RESSys (Running *et al.*, 1989), RHESSys (Band *et al.*, 1993), MODIS-LUE (Running *et al.*, 1994), and GESSys (Hunt *et al.*, 1996). Together these models provide a means of extrapolating climate, hydrology, net primary production, and ecosystem gas exchange across the entire surface of Earth.

individual stands as discussed in Section III (Fig. III.2). Because direct validation at regional to global scales is extremely difficult, our relative confidence in any prediction is heavily weighted toward models that incorporate principles that have been thoroughly tested at smaller spatial scales where direct measurements were possible.

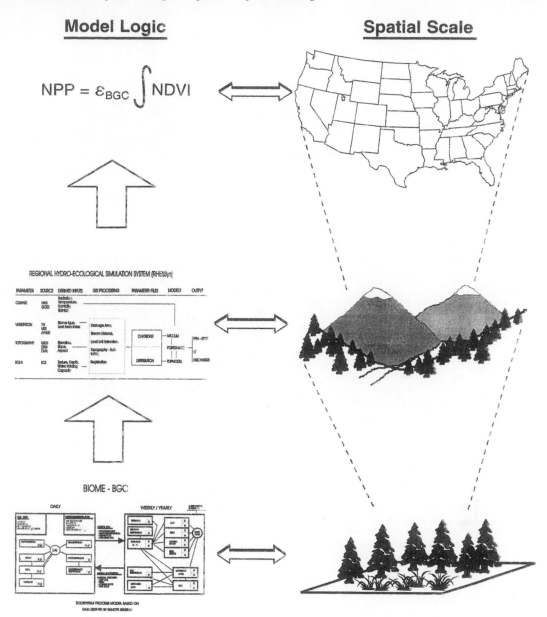

FIGURE III.2. Example of the simplifications required in shifting from local to regional to global scale ecosystem analyses. When computing NPP at the stand level, FOREST-BGC includes over 70 variables to define site and stand conditions. At the regional level, the RHESSys model requires only 20 variables. At the global scale, simple models compute NPP with only three variables: time-integrated satelite estimates of vegetation greenness (NDVI), incoming solar radiation (I_s), and biome-specific light conversion efficiencies (ε). These conversion efficiencies are derived from the more refined models such as FOREST-BGC.

FIGURE 10.1.

CHAPTER 7

Spatial Scaling Methods for Landscape and Regional Ecosystem Analysis

I. INTRODUCTION

Representing the variability of a typical landscape quantitatively is a daunting task. The most common source of spatial information for land management has historically been recorded on maps. Explorers denoted only rivers, mountains, and related landscape features along their routes. Later, the locations of roads, rail lines, and towns were added. In time, specialists made surveys of the types of vegetation and soils present, and mapped major drainage patterns. All of these properties represented what could be directly observed and quantified from a small number of samples. The means of these samples were then extrapolated, with the aid of aerial photographs and topographic maps, to fill entire maps.

Today's landscape analysis requires maps with additional features that are not directly observable. These include estimates of the stores of water, carbon, and nutrients in vegetation and the soil, as well as the cycling rates of various ecosystem processes. In addition, landscape and regional scale ecosystem analysis must document on maps those processes that cause changes in the spatial distribution of flora and fauna, such as the rate and direction of seed dispersal, animal population dynamics, and the spread of wildfires and other disturbances. A fundamental problem is that we cannot rely on the same field techniques that were satisfactory for stand-level analyses; rather, we must relate features that can be obtained from stand-level analyses to more observable landscape features based on completely different methods.

This chapter introduces a number of methods borrowed from other disciplines that have proved valuable, in fact, essential, to conduct broader scale ecosystem analysis. These new methods include application of geographic information system techniques, database management, climatological extrapolations, and remote sensing. We present a brief explanation of each of the major fields, along with some underlying principles that permit them to contribute so essentially to landscape and regional scale analyses. References are provided for those interested in developing special skills in these areas. Applications at the landscape and regional scales will be presented in Chapters 8 and 9.

A. Landscape Patterns and Processes

There is no single approach appropriate for the ecological analysis of landscapes. Different questions require that different variables be estimated at varying space and time scales (Levin, 1992). At a more fundamental level, landscape analysis questions differ dramatically in the level of spatial *interconnectedness* required. Figure 7.1 shows a raster representation of a hypothetical landscape in a grid system with a "target cell" in the center. As an organizing principle, we can think of three dimensions of connectivity around the target cell. The cell is connected vertically upward to the atmosphere, vertically downward with the soil, and horizontally with adjacent ecosystem units. As we simplify our view of the landscape, we may choose to define some of these dimensions weakly, while ignoring others entirely. Such simplifications are necessary to accomplish landscape analysis, but they must be made to accommodate the problem being addressed. The additional dimension of "time" must also be considered along with the three dimensions of connectivity.

Some questions in ecosystem analysis can be dealt with by evaluating the vertical connections of the target cell only, such as those concerned with vegetation and atmospheric interactions through the application of Soil–Vegetation–Atmospheric Transfer (SVAT) models. For example, estimating NPP with a Light Use Efficiency (LUE) model (Chapter 3) requires for a given target cell only knowledge of the incident radiation, the light absorbing capacity of the canopy, and the conversion efficiency of the absorbed radiant energy into biomass. Except for rare cases of topographic shading, properties of the adjacent cells have little impact on the target cell, which permits estimates of NPP to be considerably simplified. When applying this "unconnected" logic, each cell can be evaluated individually, in any order. The earliest regional photosynthesis maps were assembled following this approach (Running *et al.*, 1989).

The spatial connectiveness must be expanded, however, if we ask the question, "How favorable is the wildlife habitat in a specified cell?" To answer this question we must obtain information on forage quality and cover present within a target cell as well as in adjacent cells with other critical attributes that affect wildlife populations: locations of water, breeding grounds, predator populations, roads, and other barriers or distractions associated with human activities. Continuing this line of reasoning, one can pose ecologi-

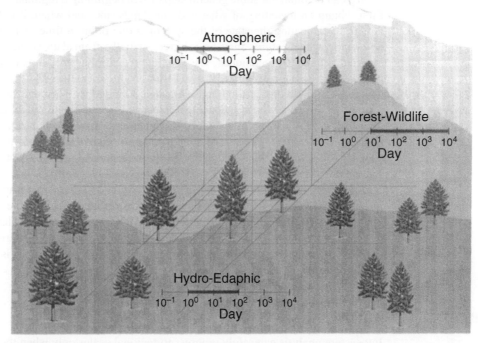

FIGURE 7.1.   Principles of landscape interconnections. Each landscape cell is connected vertically above to the atmosphere, below to hydroedaphic processes, and horizontally to adjacent terrestrial cells. Key ecosystem processes and the inherent time constants (darkened line) vary for these different spatial dimensions. Simplifying regional scale analyses requires that only the essential connective processes be defined.

cal questions associated with animal migration or fire spread where connectiveness beyond adjacent cells is required. The amount of information required to relate properties in the target cell with adjacent (and even distant) landscape cells adds complications to spatial analysis in a geometric way. There is great merit, therefore, in simplifying the degree of interconnectedness by identifying properties that might be ignored in a particular analysis. For example, in a wildlife habitat survey, one might minimize evaluation of the vertical atmospheric and hydrologic connections, and define only a handful of more stable attributes of the vegetation.

Hydrologic analyses of whole watersheds were among the first problems in forestry, beyond road location, that required explicit information from horizontally connected cells. Hydrologists originally dealt with landscape complexity by concentrating on the stream network and properties of the soil that affected seepage and runoff. Vegetation was usually treated as a static evaporating surface without any explicit connection to the hydrologic properties of adjacent cells. Stream discharge was of primary interest; most other ecosystem processes were thus ignored. In Chapter 8 we will cover recent projects that incorporate vegetation processes interactively to simulate streamflows with changing vegetation.

We can recommend some general steps when beginning a regional ecosystem analysis. First, obtain an inventory of what is on the landscape, and where it is located. This step is akin to evaluating the state of the system at one point in time. A second step involves describing what features may be inferred spatially from data acquired through remote sensing and related techniques. We soon appreciate that attributes commonly measured by foresters and botanists, such as variation in tree diameters and species compositions, cannot be measured over thousands of square kilometers. Regional forest growth models must instead depend on relationships associated with climatological extrapolations and remotely sensed indices of vegetation greenness and LAI. A final step required in spatial analyses is to develop a means of inferring other variables of critical interest from the few variables that can be measured. This last step usually involves the application of a computer simulation model.

We begin in the following section to provide some rules for translating ecosystem properties into observable variables. Next we cover principles of remote sensing that define what ecosystem and climatic attributes are directly measurable across landscapes. Finally, we summarize the landscape metrics used to quantify landscape patterns, as a prelude to problems that require spatially explicit and dynamic landscape analysis with strong horizontal coupling.

B. Model Simplifications

Ecosystem analysis can safely progress to regional scales only when the principles underlying ecological processes are reasonably well understood and have been synthesized into some type of analytical model. In Chapters 2–6, we provided background on our current understanding of forest ecosystem operation and synthesized much of that understanding into models such as FOREST-BGC. With this background, we can, through sensitivity tests, discover the most critical variables and establish relationships for constructing more simplified models required to extrapolate selected information across landscapes (Burke *et al.*, 1991).

The first requirement of spatial scaling is to reduce the number of variables requiring measurement to a minimum. Early ecosystem models of the International Biological Program (IBP) era in the 1970s required so much information that they could not be extrapolated beyond the highly sampled sites on which they were developed. For example, the CONIFER model of the IBP Coniferous Biome project required information on 109 variables before it could be operated (Swartzman, 1979).

As FOREST-BGC evolved from its origins in the early 1970s, a number of system representations were tested and discarded in the search for a scalable model. Ultimately, five important simplifications were incorporated into the model:

- No demography of individual species was defined, only general physiological attributes.
- Individual trees were not represented, only carbon pools.
- No details on internal physiology concerning water, carbon, or nutrient transport were included. This decision resulted in compression of 16 specified water storage compartments from an earlier version of FOREST-BGC to 1 in the current model.

- No individual canopy strata were represented in FOREST-BGC, only total LAI. In contrast, some photosynthesis models define 8–10 vertical canopy strata, each age class of foliage, and additional physiological attributes (Wang and Jarvis, 1990).
- No belowground details on root distribution or variation in soil profile properties were represented. Root growth was computed as a simple "carbon tax" to meet the demands associated with the development of canopy LAI.

These types of simplifications, although unnecessary for stand-level models (see Korol *et al.*, 1995, 1996; Keane *et al.*, 1996a), are essential for landscape and regional scale analyses. Belowground ecosystem properties, difficult to measure under any circumstances, must be inferred at larger scales from correlations with modeled ratios of actual to potential GPP (see LUE models in Chapter 3) or similar derivations. Even the widely applied CENTURY model, which focuses explicitly on soil processes, computes soil carbon dynamics exclusively from aboveground climate, NPP, and the application of carbon cycling principles (Parton *et al.*, 1988, 1993). Kicklighter *et al.* (1994) simulated regional soil CO_2 fluxes by first using air temperature to estimate soil temperature, and then extrapolated an empirical relation between soil temperature and soil CO_2 efflux obtained from a few plots. Not surprisingly, no example exists where regional analyses of CO_2 fluxes and changes in carbon storage have been able to include detailed information on belowground processes.

The final objective of system simplification is to identify a small set of critical but *observable* system attributes, and key environmental variables that affect them. The model, when complete, still must be able to compute important physiological processes across landscapes. It is essential that the steps in model simplification proceed hand in hand with evaluation of the technology available to make measurements and analyze the results. For example, we chose canopy LAI as a key property of the vegetation for inclusion in FOREST-BGC because LAI is an excellent reference on which to base calculations of CO_2 and water vapor exchange (including surface evaporation), and because it can be assessed at an adequate level of spatial and temporal resolution with present satellite sensors. Other more classic forestry measures such as stem basal area and live-crown ratio neither provide the direct relationship to canopy processes nor can be remotely sensed with present technology.

In summary, we recommend a specific order in compiling data for landscape analyses: (1) Begin with the most stable and available types of data, namely, topography and soil maps. (2) Next, acquire climatic data needed to drive most types of ecosystem models. (3) Finally, obtain information on the more dynamic properties, such as the vegetation. The flow diagram represented in Fig. 7.2 identifies major types of data sets and their sources. From these basic data sets techniques are available to convert the primary information into forms that can be modeled and displayed in a geographic information system (GIS). The spatially organized data are further transformed through a series of specialized simulation models into ecosystem projections of carbon, water, and nutrient cycling across landscapes and regions to answer different types of questions. The most stable variables are those listed under the broad headings of topography and soils. Climatic data are usually archived in a form that can be used to drive most ecosystem models. Attributes of the vegetation are most difficult to obtain because satellite imagery must be acquired and the

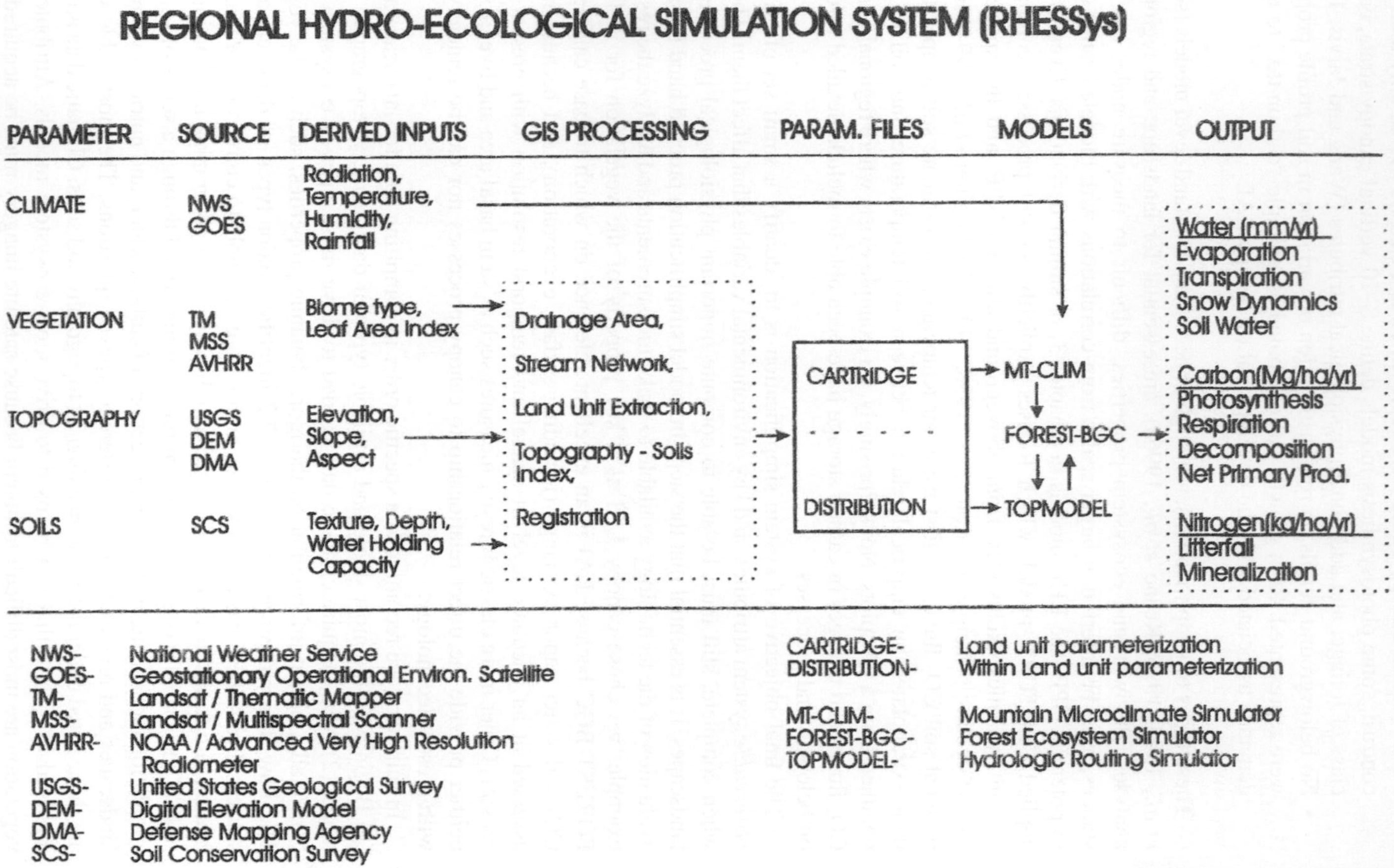

REGIONAL HYDRO-ECOLOGICAL SIMULATION SYSTEM (RHESSys)

PARAMETER	SOURCE	DERIVED INPUTS	GIS PROCESSING	PARAM. FILES	MODELS	OUTPUT
CLIMATE	NWS GOES	Radiation, Temperature, Humidity, Rainfall			MT-CLIM	Water (mm/yr) Evaporation Transpiration Snow Dynamics Soil Water
VEGETATION	TM MSS AVHRR	Biome type, Leaf Area Index	Drainage Area, Stream Network	CARTRIDGE	FOREST-BGC	Carbon (Mg/ha/yr) Photosynthesis Respiration Decomposition Net Primary Prod.
TOPOGRAPHY	USGS DEM DMA	Elevation, Slope, Aspect	Land Unit Extraction, Topography - Soils Index	DISTRIBUTION	TOPMODEL	Nitrogen (kg/ha/yr) Litterfall Mineralization
SOILS	SCS	Texture, Depth, Water Holding Capacity	Registration			

NWS- National Weather Service
GOES- Geostationary Operational Environ. Satellite
TM- Landsat / Thematic Mapper
MSS- Multispectral Scanner
AVHRR- NOAA / Advanced Very High Resolution Radiometer
USGS- United States Geological Survey
DEM- Digital Elevation Model
DMA- Defense Mapping Agency
SCS- Soil Conservation Survey

CARTRIDGE- Land unit parameterization
DISTRIBUTION- Within Land unit parameterization

MT-CLIM- Mountain Microclimate Simulator
FOREST-BGC- Forest Ecosystem Simulator
TOPMODEL- Hydrologic Routing Simulator

FIGURE 7.2. Integration of data layers of physical and biological characteristics of a landscape with process models to yield key ecosystem responses for the Regional Hydroecological Simulation System (RHESSys) (Band et al., 1991, 1993). An earlier, simpler version of the Regional Ecosystem Simulation System (RESSys) does not incorporate TOPMODEL horizontal hydrologic connections (Running et al., 1989).

reflectance properties from various electromagnetic bands must be translated into defined biotic properties, such as LAI.

II. ABIOTIC SITE VARIABLES

A. Principles in Geographic Representation

The simplest mathematical representation of a landscape is accomplished by overlaying a grid of square cells with equal dimensions, known as a *raster* format. Any cell within the grid can be addressed in a computer database by simply defining the x and y coordinates. With any raster GIS package, the size of the cell is arbitrary, so, in principle, any landscape could be defined within a raster database to any degree of accuracy. This approach would likely result in an exceedingly large database. The Landsat Thematic Mapper (TM) with $30 \times 30\,m$ pixel dimensions is often used as a cell size, but describing the 1200-km^2 Seeley–Swan basin in Montana at 30 m resolution would require 1.3 million raster cells. Alternatively, polygons of any shape can serve to define the landscape in what is called a *vector* format. Cell identification within a vector database is more difficult than in a raster format, but a vector format more realistically represents complicated landscape patterns.

Regardless of the spatial representation chosen, the goal is to partition the landscape into cells that maximize both the homogeneity of ecosystem properties within each cell and the heterogeneity between cells. In practice, properties most useful for landscape partitioning are either site variables, such as topographic discontinuities represented by ridge tops, hydrologic drainage paths, and local climate, or distinct variation in plant life forms, such as forests and grasslands. Ideally, a landscape mosaic of forest and grasslands can be defined by a set of homogeneous forest cells intermingled with homogeneous grassland cells. In reality, this clear representation is impossible because transitional situations exist.

1. Topography

Of the many sources of landscape heterogeneity, topography is the probably most important, because it influences both the local climate and hydrologic drainage. It is also most directly quantified. Stratification of the terrain into units with small internal topographic variation in aspect, slope, and elevation is accomplished by building the watershed structure from gridded digital elevation data. Band (1989a,b) developed computer programs that read the raw file of elevation data points into a digital elevation model (DEM). Then, beginning at the lowest point of the landscape, the model progressively computes the area above that will drain into the cell at the bottom of the specified area. When the program completes conputation of a landscape, each watershed is partitioned into uniquely defined hill slopes of specific slope and aspect, from which the hydrologic routing pattern is then produced. The threshold area specification allows a user to vary the level of topographic detail described in the final maps of the watershed. This important capability allows one to zoom in to obtain high detail from subdivisions of a watershed, or to pan out to a coarser description of an entire drainage basin. In this way, Nemani *et al.* (1993a) were able to represent the 1200-km^2 Seeley–Swan basin of Montana at varying degrees of topographic complexity from 170 land units aggregated down to only 6 land units within the same RHESSys modeling system (see Plate 1).

Although the precise computations may differ, nearly all hydrologic analyses now make use of digital elevation data to construct a topographic template (Fairfield and Leymarie, 1991; Paniconi and Wood, 1993; Wigmosta *et al.*, 1994). Software packages for geographic information systems usually contain programs to compute landscape topography and define hill slope units with similar slopes and aspects. The accuracy of these topographic analyses is dependent on the scale and accuracy of the original DEM database. In many countries DEM data are available at 30 m horizontal and 1 m vertical resolution. In some parts of the world, however, only data with 1000 m horizontal and 100 m vertical resolution are available. Spatially coarse DEM data may still represent overall topography well enough to map local climatic variation, but they may be unable to depict the hydrologic networks correctly because small hills and streams were not shown in the original DEM data.

2. Hydrologic Routing

The first essential set of calculations in a hydrologic model involves the accurate computation of the terrestrial water balance, as presented in Chapter 2. A second set of critical calculations are those that deal with lateral flow of excess water from the land surface and subsurface soils into and through the stream channels, an example of horizontal connectiveness. In the earliest raster-based RESSys model, a simple "bucket" analogy from FOREST-BGC was applied to calculate changes in soil water content. When precipitation filled the bucket, excess water was spilled into surface outflow or into groundwater. No attempt was made to model the lateral flow of water into adjacent landscape cells (Running *et al.*, 1989). Although this simple bucket model proved adequate in areas of low relief with little lateral drainage, in mountainous areas explicit soil water routing proved necessary to produce realistic soil moisture profiles across a watershed and to predict streamflow.

The most direct approach to define hydrologic routing is to connect all landscape units to one another. Wigmosta *et al.* (1994) presented an example with a raster-based landscape representation where each square cell was connected to eight adjacent cells. Those cells located at higher elevation than the reference cell defined a potential source of water, whereas those located below it became receptors for any flow out of the cell. Models using vector or polygon representations on a landscape make for efficient computations and result in a reasonable depiction of the flow of water downhill to the boundary of a watershed (Michaud and Sorooshian, 1994). Robinson *et al.* (1995) illustrated that in small watersheds the hydraulic conductivity of hill slope cells plays an important role in determining variation in streamflow, but as the area increases, good definition of the stream network becomes more important. If the scale of analyses is restricted to upland forested watersheds, soil depth and hydraulic conductivity of hill slope cells must be well represented. At the scale of drainage basins, however, the priority shifts toward assuring accurate definition of the stream network.

The TOPMODEL logic of Beven and Kirkby (1979; Beven, 1989) offered an improvement in computational efficiency by developing equations that aggregate areas into zones with similar hydrologic properties. TOPMODEL calculates the upslope drainage area and slope gradient for any hill slope unit and provides an estimate of the saturated flow rate through the soil from that unit. These soil drainage characteristics can either be defined for each hill slope unit or expressed as a frequency distribution for the entire watershed. For example, an area near a ridge top receives little drainage from upslope and would characteristically be situated on steep topography with shallow, coarse soils. Precipitation will rapidly drain downslope from such a location, leaving soils at moisture conditions below saturation most of the time. At the other extreme, an area at the base of the slope area, situated on gentle topography, continues to receive water from uphill sources most of the time and remains near saturation for long periods. Band *et al.* (1993) developed a Topographic Saturation Index (TSI) that spatially distinguishes all hill slope units possessing similar hydrologic flow characteristics in order to substantially reduce the number of computations required (see Plate 2). Daily outflow from any hill slope unit was calculated based on the TSI of each cell, the current soil saturation deficit, and the soil's saturated hydraulic conductivity (Nemani *et al.*, 1993b). This simplified TOPMODEL logic computes seasonal streamflow with accuracy equal to that achieved by more explicit hydrologic routing schemes on mountain watersheds (Franchini *et al.*, 1996; Moore and Thompson, 1996).

The evolution of early regional hydrologic analyses from a simple, raster-based, bucket model to the RHESSys, with more realistic representation of watershed drainage properties, is primarily the result of developing an automated scheme for partitioning the topography into functionally similar units and integrating that information into a hydrologic routing routine. This approach serves two important roles: it provides good estimates of watershed capacitance, which greatly improves accuracy in estimating the timing and volume of stream discharge, and it represents realistically how soil moisture is distributed across landscapes. This latter attribute is important because geologically unstable slopes are identified, and ecosystem model predictions of gas exchange from soils and vegetation supplemented by seepage from uphill positions are improved.

3. Soil Characteristics

Soil variables are among the most difficult ecosystem properties to quantify because of local heterogeneity and the deficiency of broad scale data. Georeferenced databases of soil morphological and structural characteristics are becoming available to resource managers in some countries. In the United States, the National Soil Geographic Date Base (NATSGO) and State Soil Geographic Data Base (STATSGO) (USDA, 1991) provide national coverage and soils information for individual states. Ecological modeling, however, usually requires more specific soil physical and chemical information than are available from these summarized databases. Kern (1995a) tested different published formulas that use particle size distribution, bulk density, and soil organic matter data to derive soil water retention curves. Kern (1995b) then took particle size distribution, organic matter, rock content, and soil depth from the NATSGO database to define geographic patterns of soil water holding capacity for the contiguous United States. Kern's estimates are hard to validate at continental scales, but the research yields a consistent set of important variables for regional ecosystem analysis.

An alternative to statistically interpolated databases is to apply more fundamental geo-morphic principles to construct detailed soil maps of watersheds. Zheng *et al.* (1996) used the STATSGO database and employed TOPMODEL to compute an index of topographic position, which was then used to infer soil depth and water holding capacity of the Seeley–Swan basin; the analysis later was expanded to apply to the entire state of Montana. Standard errors of estimates of water holding capacity derived with the topographic index were within 5 cm of values predicted in the STATSGO statistical database at both the watershed and statewide levels. Zheng *et al.* (1996) found, by applying geomorphic prin-ciples to define drainage patterns from DEM data, that a more logical spatial interpolation of soil depth resulted, with shallow soils concentrated near ridge tops and progressively deeper soils distributed toward the base of a slope. Zhu *et al.* (1996) made further improve-ments by applying fuzzy set theory and expert systems logic to refine the spatial distribu-tion of soil survey data for a 400-km^2 forested area of Montana.

B. Efficient Landscape Representation

Statistical landscape representations provided by TOPMODEL analyses do not include the explicit interconnections required to represent spatial patterns related to fire spread, seed dispersal, or the distribution of wildlife. In part this is because the hydrologic system is driven by gravity, a variable less important to many other patterns observed on land-scapes. In fact, fire spread is accelerated by upslope winds while convective winds inhibit downslope spread. Plate 1 illustrates again that a simple raster representation of a land-scape at 30 m resolution produces more cells than can be efficiently analyzed. However, there is a danger also in choosing too large of a cell size because *aggregation errors* accumulate in the computations (Bartell *et al.*, 1988; Rastetter *et al.*, 1992). These errors have two primary causes: averages are derived from spatially diverse input variables (e.g., local climatic variation), and more nonlinear processes are averaged as the size of cells increases (Fig. 7.3).

The magnitude of aggregation errors is dependent on the spatial heterogeneity on the landscape of both abiotic (topography, soils, climate) and biotic variables (biome type, stand age, and cover distribution). Evaluating the significance of these aggregation errors is not easy because both the specific landscape and the output variable of interest must be designated. Turner *et al.* (1996) evaluated the aggregation errors propagated by describing a mountainous 55,000-km^2 region of Oregon with raster cells 1, 10, and 50-km on a side. Elevations ranging from 0 to 3000 m represented a large range in temperatures and pre-cipitation that were continually averaged together as the width of cell size increased from 1 to 50 km (Fig. 1.4). Simulated NPP was 11% higher with the 50-km cells than at 1 km, but simulated hydrologic runoff was 45% lower. White and Running (1994) found that predictions of annual photosynthesis and evapotranspiration for a 450-km^2 mountain watershed differed by more than 15% with identical input data, depending on the choice of raster, polygon, or TOPMODEL landscape representation.

A common finding in these studies is that landscape aggregation may preserve mean values but progressively reduces the apparent variance (Band and Moore, 1995). A land-scape aggregation that averages a north and south slope into a single cell may compute an accurate average NPP, but the actual variation between the slopes is lost. Consequently,

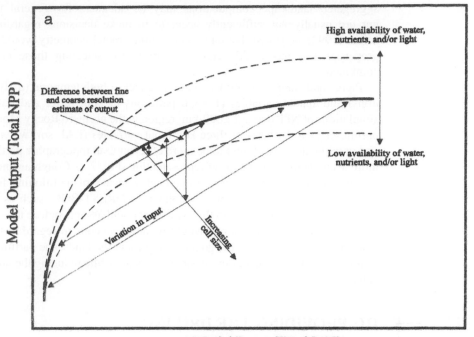

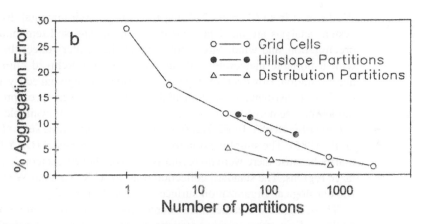

FIGURE 7.3.   (a) Illustration of aggregation errors propagated by representing a landscape with progressively larger cells (Rastetter *et al.*, 1992). In this test, the relationship between LAI and NPP was nonlinear. If a landscape is represented with many cells over the full range of observed LAI, the estimation of NPP should be accurate. When the landscape is represented by progressively fewer aggregate cells assigned averaged LAI values, departure from the correct values increases. (b) The aggregation error in simulating NPP with RHESSys for a 12,100-km² region was reduced with smaller grid cell size. Topographically defined hill slope partitions improved the efficiency of computations (attaining a certain computational accuracy with the minimum number of cells) equal to grid cell representation. Landscape partitions defined by the distribution of LAI, soil water holding capacity, and precipitation provided the greatest efficiency. These variables were chosen for landscape classification based on previous understanding of their importance in affecting NPP in this region. (Redrawn from Pierce and Running, 1995.)

aggregated landscape computations may be adequate for general planning purposes but are usually not sufficiently accurate to make decisions regarding a specific site. Milne (1992) suggested that units with similar fractal geometry would share topographic properties and this could serve as a basis for selecting those cells that could be combined.

Pierce and Running (1995) studied the effect of both temporal (daily to annual simulations) and spatial variables (1–8456 partitions in both raster and polygon formats) on simulations of NPP over a 12,000-km^2 region in the northern Rocky Mountains. Landscape partitioning on the basis of three key input variables (LAI, soil water holding capacity, and precipitation) was found to be more efficient than topographic partitioning, and much more accurate than simple raster representation (Fig. 7.3). Coughlan and Running (1997b) developed from an expert system approach a means of partitioning forested landscapes based on key biophysical input variables for FOREST-BGC, such as LAI, that strongly influenced predicted output. In this way, they could assume before the model computation that the landscape units defined would be homogeneous in regard to the important variables. In all of these cases, the goal is to represent the landscape partitioning as accurately as possible, which produces more accurate results while reducing the number of redundant cells.

III. PROVIDING THE DRIVING VARIABLES, CLIMATOLOGY

We learned from Chapters 2–4 that the critical environmental driving variables for ecosystem analysis are incident solar radiation, air and soil temperature, humidity deficits, precipitation, and wind. Most ecological studies either collect the necessary data on site with a portable meteorological station or rely on archived climatological data from a nearby station. However, for regular execution of a regional ecosystem model a permanent source of consistent meteorological data is required. Again, compromises in the ecosystem model may be necessary to accommodate the realities of available regional data. An early version of FOREST-BGC (H2OTRANS) ran at hourly time steps (Running, 1984a; Knight *et al.*, 1985). The simplification to a daily time step model (DAYTRANS) was made primarily because there were no regularly archived hourly meteorological data available over large regions. The standard collection time for U.S. National Weather Service and World Meteorological Organization surface weather data is daily.

These standard weather data sources have two shortcomings for regional ecosystem analysis. First, most weather stations do not measure solar radiation, humidity, or wind speed. Second, the weather stations are located predominantly in valley bottoms, often at airports, and thus poorly represent the local climatic variability associated with complex terrain. To remedy these problems an array of climatological principles and formulas have been assembled into a computer model that allows nonlinear extrapolation of climatic data into mountainous terrain (MT-CLIM, Running *et al.*, 1987). Rather than reproduce the many equations in MT-CLIM here, we will discuss the principles on which the model is based. For those wishing to see the equations and other details, the complete documentation of MT-CLIM with computer code is available free at http://www.ntsg. umt.edu/textbooks/.

A. Climatology

1. Air Temperature

Canopy processes controlled by stomatal opening such as photosynthesis and transpiration are predominantly affected by air temperature during daylight hours. Other processes such as stem respiration occur continuously, so are best computed from integrated 24-hr average temperatures. The simple sine wave in MT-CLIM allows both these temperatures to be generated from daily minimum and maximum temperature data alone (Fig. 7.4). Although a sinusoidal variation in temperature is not typical every day, on average the assumption is valid.

2. Soil Temperature

Soil temperature is required for calculating most belowground ecosystem processes, including root growth and respiration, decomposition, and nitrogen mineralization. Soil temperature, however, is not a standard variable collected at most weather stations. Because soil temperature responds to the net effect of the daily surface energy balance, it can be estimated by computing a running average of air temperature, with progressively longer integration times as soil depth increases (Parton, 1984; Stathers *et al.*, 1985). Zheng *et al.* (1993) employed an 11-day running average air temperature, modified by daily precipitation and overstory LAI, to predict soil temperatures at 10 cm depth. When tested on sites across the United States, estimated soil temperatures were close to those observed ($r^2 = 0.86–0.95$).

3. Humidity

Climatological studies have long shown that nighttime air temperature typically cools until the dew point, T_{dew}, is reached. The resulting release of latent energy of vaporization (2454 kJ kg^{-1} at 20°C) usually inhibits further air temperature drop. MT-CLIM applies this simple principle to define daily dew point as equal to the minimum daily temperature, T_{min} (Fig. 7.4). This approach to calculating dew point is reasonably accurate, even in climates where dewfall does not regularly occur (Fig. 7.5; Glassy and Running, 1994; Running *et al.*, 1987). Kimball *et al.* (1997b) demonstrated a simple correction based on the ratio of annual precipitation to annual evapotranspiration (ET) that reduced the standard error of estimated vapor pressure to less than 0.4 kPa of measured values, even in arid climates (Fig. 7.6).

The dew point is an absolute measure of water vapor in an air mass. As such, it is not influenced by air temperature, so is fairly constant throughout the daily cycle. Because of the huge capacity of the atmosphere, daily surface evapotranspiration usually raises the water vapor concentration by only a small amount. Consequently, the dew point provides a conservative spatial and temporal measure of atmospheric water vapor. Other measures of humidity, such as the vapor pressure deficit (D), can then be calculated from T_{dew} and air temperature (Chapter 2).

4. Solar Radiation

Given the importance of solar radiation to meteorology, hydrology, and ecology, it is unfortunate that the network of weather stations generally disregards this variable. In

MT - CLIM

(Mountain Climate Simulator)

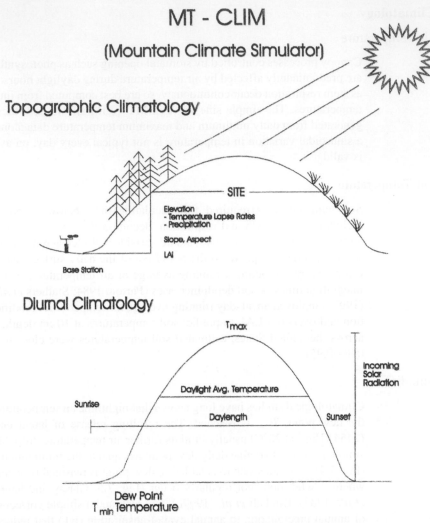

FIGURE 7.4. Depiction of principles behind the mountain microclimate simulator MT-CLIM to extrapolate daily meteorological data across topographically complex terrain and to compute ecologically relevant variables from daily maximum–minimum temperatures (Running *et al.*, 1987; Hungerford *et al.*, 1989). From diurnal extremes in temperature, the incident solar radiation, humidity, and soil temperature are derived. Solar radiation, precipitation, humidity, and air temperatures are further modified to take account of slope, aspect, elevation, and latitude.

MT-CLIM, a two-step approach is required to estimate solar radiation: first, potential solar radiation (often called top of the atmosphere solar radiation) is computed using solar geometry algorithms from Garnier and Ohmura (1968) and Swift (1976). These algorithms allow computation of direct beam potential radiation to a surface at any latitude for every day of the year. Then, by applying principles described in Bristow and Campbell (1984),

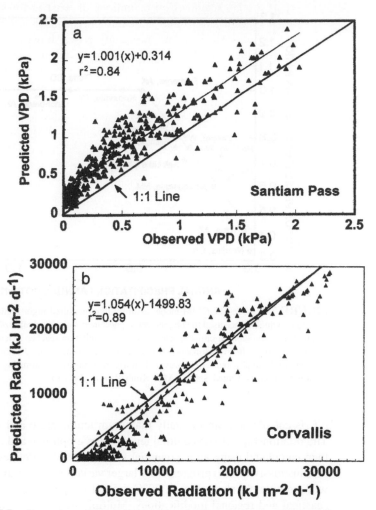

FIGURE 7.5. Comparison of MT-CLIM estimates of daily incident solar radiation and vapor pressure deficit for two sites on the Oregon transect. Most weather stations report only daily temperature and precipitation, so MT-CLIM provides estimates of solar radiation and humidity conditions that may otherwise be unobtainable. (From Glassy and Running, 1994.)

the diurnal temperature range serves as the basis for calculating the clarity (*transmittance*) of the atmosphere. The general principle expressed here is that the range of diurnal temperature is highest on clear days, and that cloudy and/or hazy days result in much smaller temperature ranges. Atmospheric transmittance is a general measure of the attenuation of solar radiation caused by clouds, haze, and aerosols. Elevation determines the atmospheric path length, so is a strong determinant of clear sky transmissivity.

MT-CLIM begins with an optical transmissivity set at sea level of 0.65, then increases clear sky transmissivity by $0.08\,\text{km}^{-1}$ of elevation. Next, the diurnal temperature range,

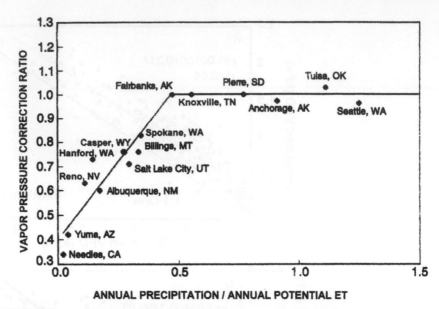

FIGURE 7.6.   The MT-CLIM model assumes that measured night minimum temperature equals the daily dew point, which is then used to compute vapor pressure deficit. Kimball *et al.* (1997b) found this T_{min}-dew point assumption to be valid in climates where annual precipitation reaches at least 50% of annual potential evapotranspiration (ET). In drier climates, where minimum temperatures may not reach the dew point, the ratio annual precipitation/annual potential ET provided an accurate means of estimating vapor pressure at weather stations scattered across the United States. (After Kimball *et al.*, 1997b.)

ΔT, is calculated from the daily maximum temperature and minimum temperature (°C). The relationship between diurnal temperature amplitude and atmospheric transmittance is then calculated using an exponential formulation that defines an asymptotic relationship of transmissivity with progressively larger variation in diurnal air temperature. On average, with completely clear skies, a 20°C diurnal temperature range is observed, with certain seasonal and regional modifications. Multiplying potential (top of atmosphere) radiation by the atmospheric transmissivity gives an estimate of daily solar radiation at the surface. These MT-CLIM estimates of daily solar radiation have been tested repeatedly against measured incident solar radiation, and correlations consistently show $r^2 > 0.85$ (Fig. 7.5; Glassy and Running, 1994; Running *et al.*, 1987).

B. Topographic Extrapolations

The previous sections have illustrated how to compute the necessary suite of daily meteorological variables from limited weather data at a single station. Now we must extend the approach to predict meteorological variables across the landscape. MT-CLIM was designed with the assumption that daily weather data would be primarily available from standard stations usually located in valley bottoms near airports. To represent the microclimates of adjacent, more complex topography requires spatially extrapolating these valley bottom conditions to higher elevations and to various combinations of slope and aspect. A set of procedures now exist to present the spatial variability of solar radiation, air temperature, and precipitation in complex terrain.

1. Solar Radiation

The most obvious microclimatic effect of topography, especially at high latitudes, is on incident solar radiation. Many studies have shown how extreme microclimate and forest responses differ on north and south facing slopes (Lee and Sypolt, 1974; Tesch, 1981b; Running, 1984b). Fortunately, the solar geometry calculations used in MT-CLIM from Garnier and Ohmura (1968) and Swift (1976) provide a means for computing the incident radiation to any combination of slope/aspects each day of the year (Fig. 7.7). One obvious application of solar geometry is to predict the easily observed differences in the disappearance of snow from different aspects. In the mountains of Montana (44°–47°N latitude), it is common to see south facing slopes melt free of snow 1–2 months earlier than adjacent north slopes at the same elevation (Wigmosta et al., 1994).

2. Air Temperature

Climatologists have long measured the direct change in average air temperature associated with decreasing atmospheric pressure at higher elevations (Baker, 1944; Barry, 1992). These elevational *lapse rates* usually average −7° to −9°C km^{-1}, although considerable day-to-day variation occurs in response to air mass movement and precipitation (Fig. 7.8). The first step in MT-CLIM for extrapolating a valley bottom temperature to a mountain hill slope is to subtract an average lapse rate based on the elevational difference between the two points.

The next step of computing aspect-related temperature differences relies on the computed solar radiation differences as described above. Hill slope air temperatures, however, are also influenced by wind patterns, surface vegetation, and energy budgets. The MT-CLIM program uses empirical weighting coefficients to add or subtract air temperature from an observed base station temperature as a ratio of the solar radiation received on a slope to that received on a flat surface. Thus, a south slope in the northern hemisphere which receives more solar radiation will be 1°–2°C higher, whereas a north facing slope will be 1°–2°C lower than a flat surface. More sophisticated computations of these microclimate differences are available, but they require much more computer time and demand more data than available to most resource managers (Pielke et al., 1992).

3. Precipitation

Precipitation estimates in mountainous terrain must primarily be able to represent the orographic influences of the topography on the prevailing storm track. In the western United States, annual precipitation in the valley bottoms can be one-tenth that recorded in the surrounding mountains. Daly et al. (1994) used statistical analysis of station data embedded in a digital terrain model to compute regional precipitation. Thornton and Running (1996) and Thornton et al. (1997) developed DAYMET, which embeds the original MT-CLIM in a digital topography to allow weighted interpolation of data between many stations (see Plate 3).

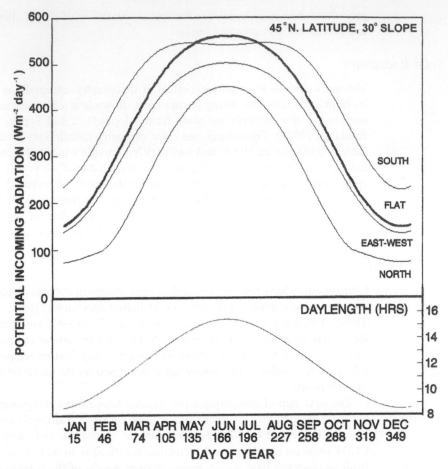

FIGURE 7.7. Comparison of incident potential (or cloud-free) solar radiation with aspect and season computed by MT-CLIM. This example is for a 30° slope at 45° N latitude, but the model can compute any slope/aspect/latitude combination. Note that for this latitude and slope the relative differences in radiation loads on different aspects are highest outside the growing season. In fact, north facing aspects may receive no direct radiation for almost 3 months in midwinter. Differences in annual radiation loads are the primary driving force associated with aspect-related differences in vegetation found throughout arid lands at mid- to high latitudes. Differences in radiation loads on slopes of varying aspect are much less in cloudy climates and at lower, tropical latitudes.

4. Wind

Wind is an important factor in the energy and water budgets of canopies, often expressed through an aerodynamic resistance term (Fig. 2.5). It has been difficult, however, to apply climatological principles that can predict wind speed from our limited selection of micro-climatic variables. Unless direct measurements of wind are available, a moderate wind speed equivalent to a light breeze, around $2\,\mathrm{m\,sec^{-1}}$, is often assumed. The error involved in this assumption is small for most forests, which are well coupled to the atmosphere, but becomes larger with vegetation of shorter stature (Chapter 2).

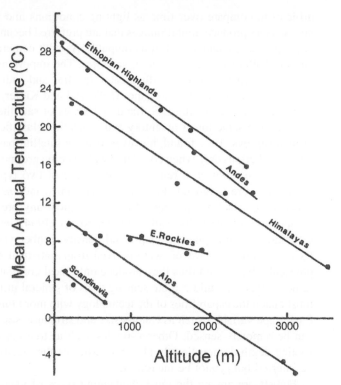

FIGURE 7.8. Observed elevational lapse rates (decrease in air temperature produced by decreased barometric pressure with altitude) for mountain ranges all over the world. (From Friend and Woodward, 1990.)

IV. DESCRIBING THE ECOSYSTEM

Forested landscapes have commonly been described by defining vegetation units or by calculating the commercial potential to grow trees. Neither approach provides a quantitative measure of system attributes important for ecosystem analysis. Vegetation classifications incorporate knowledge about species assemblages but lack information on biophysical attributes. Commercial assessments of standing volume are insufficient by themselves to derive estimates of leaf area or autotrophic respiration. Moreover, both types of information are generally dependent on infrequent ground sampling on relatively small areas. For consistent and continuous monitoring of landscape properties, some broader, more frequent sampling is required, and this leads us into the field of remote sensing.

A. Principles of Remote Sensing

Any serious quantitative study of a landscape must begin with some type of remote sensing; there is no other way to obtain consistent measurements across large areas, Remote sensing originated with aerial photographs taken by cameras from airplanes and balloons, technology still used today for certain applications. Photographs, however, are difficult to compare over time as lighting conditions and film quality vary. An array of new sensors produce digital images that are preferred because the raw data can be manipulated, processed, and stored in a computer. As with cameras, sensors can be mounted on aircraft, balloons, truck booms, or satellites. The important distinction is that digital data provide a superior format with greater spectral and radiometric accuracy required for comparative analyses of landscapes (Lillesand and Kiefer, 1994; Jensen, 1996).

With the launch of the U.S. Landsat 1 series of satellites in 1972, scientists first tried to repeat the same forest inventory and vegetation classifications that were possible from ground surveys. They found, however, that the satellite-based surveys could provide adequate predictions only when correlated with specific ground-based data. Computer-assisted classification of spectral reflectance data were of two types: *supervised*, where spectral signatures were assigned to known landscape units, and *unsupervised,* where the landscape was distinguished by spectral differences without any previous attempt to recognize the composition of the units (Lillesand and Kiefer, 1994; Jensen, 1996). The classification techniques were applied to map stand density, timber volume, and forest types. All of these techniques, however, were derived from statistical associations that were not biophysically based, and thus they produce only local correlations that cannot be transferred to new areas. To build remote sensing tools of general utility for landscape analysis, one must match the capabilities of the technology with more fundamental ecological attributes. Landscape variables such as canopy size and structure, soil color, and surface temperature can be remotely sensed. Other variables such as tree gas emissions, soil chemistry, and rooting depth do not directly alter properties of the electromagnetic spectrum; they may be inferred but cannot be measured.

Forests are among the most challenging types of vegetation to decipher with remote sensing (Peterson and Running, 1989). Trees are present in a wide range of sizes, shapes, and colors. Tree canopies can range in height from 1 to 100 m. Broadleaf trees exhibit marked seasonal variation in canopy display, while evergreen trees show little detectable seasonal variation. Mixtures of evergreen and deciduous trees are common. The ground surface below trees often supports other vegetation such as grasses with different reflectance characteristics, which may alter the average reflectance properties of the ecosystem. Forests are located in regions with complex topography, and occur at high latitudes, factors that induce shadowing. Consequently, only a small number of remotely sensed properties are available to assess the state of forest ecosystems in a dependable manner (Verbyla, 1995).

An important distinction is made in remote sensing between the *radiances* that are measured by the sensor and the *reflectances*, which represent the fraction of incoming radiation in a particular spectral region that is emitted or reflected from the surface (Fig. 7.9). This seemingly subtle distinction is most easily understood by considering an evergreen forest imaged every day by a satellite sensor for several weeks. While no property of the forest canopy may change over this short period, and hence the absolute reflectance would not change, the radiances measured from a satellite may change considerably because of variability in the clarity of the atmosphere, view angle of the sensor, and intensity of solar illumination. To discern real changes in landscape properties, procedures to correct for or normalize these sources of variability in radiance must be applied (Verbyla, 1995).

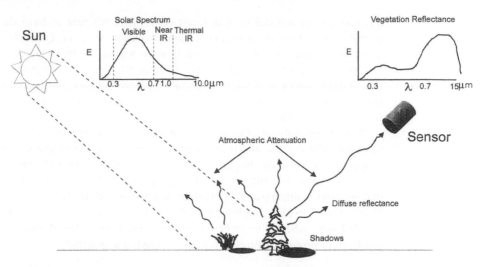

FIGURE 7.9. Important principles of the geometric relationships among the source, target, and sensor that apply to all terrestrial remote sensing. The sun's energy is attenuated by the atmosphere in all directions, but to varying degrees dependent on the wavelength. The angle of solar illumination and view angle of the sensor determine how much sunlit or shaded surface the sensor observes. Remote sensing sensors typically look straight down, from a nadir view, but advanced sensors analyze reflectance differences from various view angles to improve estimates of canopy structure.

In evaluating the potential application of remote sensing it is useful to consider four dimensions: spatial, spectral, temporal, and directional. Each of these dimensions offers special advantages in interpreting landscapes, which we will clarify in the following discussion.

1. Spatial Resolution

To someone unfamiliar with remote sensing, it would seem that a landscape viewed at highest possible resolution would always be the preferred option. Early Landsat satellites provided a square picture element (*pixel*) about 80 m on a side; the Landsat TM sensor now provides 30 m resolution, and the French SPOT satellite carries a sensor that can resolve detail to 10 m resolution. The volume of data, we must recognize, increases geometrically with decreasing pixel size. A first 1-km resolution view of Earth (1.4×10^8 pixels) from the National Oceanic and Atmospheric Administration (NOAA) Advanced Very High Resolution Radiometer (AVHRR) sensor has only relatively recently been attempted (Eidenshink and Foundeen, 1994). A global Landsat data set does not exist, simply because the data volume and number of scenes required to produce a cloud-free image (150 billion pixels) is too great. Thus, very high spatial resolution is desirable primarily for local studies.

A second problem associated with high spatial resolution data is that variability increases as additional small details become visible. For example, a seemingly smooth desktop viewed under a microscope reveals variation in wood structure not visible at the more coarse resolution of the naked eye. For broad studies with a rather simple goal, such as quantifying the national area in forest cover, it is efficient to have the sensor provide a spatially averaged representation at 1 km resolution or larger. Quantifying the same forest cover for a landscape that has been heavily logged in 10-ha patches, however, would require a sensor with 30 m resolution. Even at 10 m spatial resolution, the canopy of individual trees cannot be resolved. Aircraft-mounted sensors, on the other hand, are of particular value for local studies because virtually any level of spatial detail is obtainable from <1 to >1000 m^2.

With these cautions in mind, any regional ecology project that applies remote sensing must consider the size and heterogeneity of the area being studied, as well as the specific objectives of the study. In general, one should choose a spatial scale with a pixel size as large as possible to minimize the cost of acquiring and processing data. We caution readers that geography literature typically uses the term "scale" in the opposite sense of ecologists. "Large scale" to a geographer means that ground features are represented at a larger, more detailed size, whereas to an ecologist the term means a larger area is being represented, resulting in ground features being represented in less detail.

2. Spectral Resolution

Vegetation is typically analyzed with spectral bands in the visible red (Red) and near-infrared (NIR) range (Asrar, 1989). Current satellites carry sensors that measure reflectances between about 0.4 and 10 μm with three to four channels that are selected to avoid water absorption wavelengths (Fig. 7.10). The most common reflectance indices associated with vegetation are the Simple Ratio (SR) of NIR to Red reflectance and the normalized form of those two bands, called the Normalized Difference Vegetation Index or NDVI:

$$SR = NIR/Red \qquad\qquad (7.1a)$$
$$NDVI = (NIR - Red)/(NIR + Red). \qquad\qquad (7.1b)$$

The particular wavelengths selected differ among sensors, and many other spectral ratios have been proposed for various purposes. All make use of the general remote sensing principle that plant canopies absorb much of the radiation in the visible wavelengths and reflect more in the infrared. The predominant wavelengths of interest in the infrared are 0.7–0.9 μm, often called the near-infrared. Representing these radiances as a ratio minimizes atmospheric influences and also differences in sensor geometry (sun–earth–sensor angles) without greatly complicating data processing (Lillesand and Kiefer, 1994).

New sensors have been built and tested from aircraft that provide continuous spectral coverage from 0.4 to 2.5 μm in 200–300 channels, almost equivalent to laboratory spectrometers (Curran, 1994). Spectral studies conducted in laboratories and greenhouses demonstrate that leaf lignin, chlorophyll, and nitrogen concentrations, as well as water content, may be assessed nondestructively with high spectral resolution sensors (Gamon et al., 1992; Field et al., 1994). Increases in each of these leaf constituents tend to enhance near-infrared reflectances and result in raising the SR, NDVI, and other related spectral combinations (Matson et al., 1994; Yoder and Waring, 1994). Carter (1994) expanded the approach by interpreting leaf reflectance changes as indicators of plant responses to environmental and chemical stresses. Because of the complex geometry of canopies, however, and the necessity for sophisticated atmospheric corrections, it has been difficult to extend

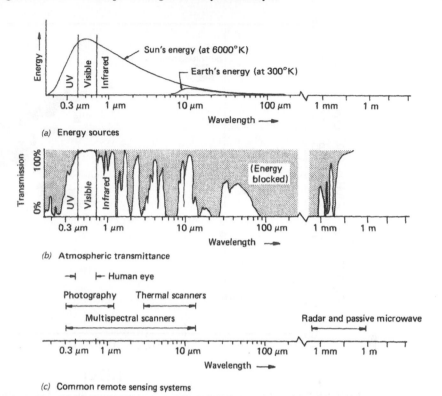

FIGURE 7.10.   Spectral characteristics of solar and terrestrial electromagnetic energy. Wavelengths at which water vapor, aerosols, and greenhouse gases strongly absorb are identified. The spectral channels common to a range of remote sensing systems have been chosen to avoid atmospheric absorption regions and to emphasize regions of reflectance valuable in assessing land surface properties. (After T. M. Lillesand and R. W. Kiefer, *Remote Sensing and Image Interpretation*, 3rd Edition, Copyright © 1994 by John Wiley & Sons, Inc. Reprinted by permission of John Wiley & Sons, Inc.)

fine-resolution spectral analyses to general field conditions (Zagolski *et al.,* 1996; Martin and Aber, 1997).

Some of the solar energy incident on a forest is absorbed, heats the surface, and is then emitted as thermal energy in wavelengths from 8 to 12 μm, as discussed in Chapter 2. Current satellite sensors usually contain one or two channels in the thermal region, and this emitted radiation provides a basis for calculating a radiometric surface temperature (Price, 1989). There are many ecological applications for landscape temperature maps, but one must recognize that these radiometric surface temperatures can be much different than the near-surface air temperatures recorded at weather stations. The largest differences, which can reach temperatures of 30°C above that of ambient air, occur on bare, dry soils under high solar radiation (Fig. 7.11; Norman *et al.,* 1995).

Vegetated surfaces also emit radiation in wavelengths from 1 mm to 150 cm, the microwave and radar region. Passive microwave sensors measure the weak naturally emitted radiation from surfaces in these wavelengths (Engman and Chauhan, 1995). Active radar

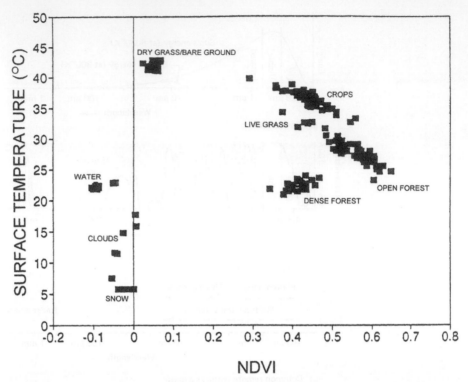

FIGURE 7.11.   Variation in radiometric surface temperature observed by AVHRR for 14 July 1987, 1430 hr local time, for a complex 106,000-km^2 region with farmland, rangeland, and forestland in Montana (Nemani and Running, 1989b; Nemani *et al.*, 1993a). The 40°C range in surface temperatures observed across the region result from varying albedo and energy partitioning to latent or sensible heat, and aerodynamic coupling of the vegetation (see Chapter 2). Surfaces with higher vegetation cover and LAI (as defined by the NDVI) exhibit surface temperatures below 25°C which approaches ambient surface air temperatures. Nearby, recently cleared areas exhibit temperatures above 45°C.

sensors emit a pulse of energy and analyze the return signal (Dobson *et al.*, 1995). A key feature of passive microwave and radar is their ability to receive signals through clouds, haze, smoke, or at night when optical sensors are useless (Waring *at al.*, 1995b). Radar data, however, are difficult to interpret because microwaves are scattered and bounce off surfaces multiple times before returning to the sensor. Processing is particularly complicated when look angles and polarization fields are varied to obtain additional information (Lillesand and Kiefer, 1994). Despite these difficulties, imaging radar has special merit in discerning flooded, frozen, and drought-stressed forests, and for research in perennially cloudy or solar-limited regions at high latitudes.

3. Temporal Resolution

The highest temporal coverage currently available with a satellite is every 30 minutes, provided by the Geostationary Operational Environmental Satellites (GOES) weather satellites. To remain geostationary, these satellites circle 36,000 km above the equator in tne same direction and speed as Earth rotates. Because of the high altitude orbit these geostationary satellites offer limited spatial resolution and at present also are restricted in spectral coverage and radiometric accuracy. The AVHRR sensor, which is primarily used for weather forecasting, provides daily coverage, but on some orbits must look at extreme angles as much as 55° from nadir. AVHRR experts prefer to use data from <30° look angles to minimize atmospheric path length effects and pixel elongation, which means that acceptable data may be available at 3- to 4-day intervals depending on latitude (Goward *et al.*, 1991). Sensors that provide high spatial detail such as Landsat and SPOT have 16- and 26-day repeat cycles because their lower orbits produce a narrower swath across Earth's surface.

Clearly, there is a choice that must be made between acquiring greater spatial detail and temporal coverage. This choice must be made in reference to the scientific objective and the type of satellite-derived data available. Also, the daily overpass time must be considered because the likelihood of encountering cloudy or hazy conditions limits the acquisition of optical data. The day-time AVHRR sensor is synchronized to pass over all sites at mid-afternoon (1430 solar time), while Landsat acquires data at 0945. Controls in some satellites are inadequate to prevent a drift in orbits which results in a progressive shift in overpass times. The Earth Observing System (EOS), launched since 2000, will carry the Moderate Resolution Imaging Spectroradiometer (MODIS) sensor in a similar orbit to AVHRR, with 36 spectral channels and spatial resolution of 250 m, 500 m, and 1 km. By employing two satellite platforms, MODIS collects data at both 1030 and 1330 (Running *et al.*, 1994).

Because of the temporal constraints reviewed above, Landsat and SPOT sensors have shown the most general utility for local to regional analyses. AVHRR has served predominantly for assessing seasonal variation in ecosystem activity at regional to global scales. In combination, one can attain high spatial detail from Landsat or SPOT, then follow the temporal activity of landscapes with AVHRR or MODIS.

4. Directional Options

A new dimension of remote sensing has been gained by allowing sensors to look at the surface from multiple angles (Irons *et al.*, 1991; Ranson *et al.*, 1994). From these off-nadir perspectives dramatic differences in reflectances are observed. For example, visualize the difference of viewing a tree from above on its sunlit side, where the foliage is entirely illuminated, and viewing the same tree from the shaded side where much of the canopy is in shadow (Hall *et al.*, 1995).

Regardless of what combination of the four remote sensing properties are employed to distinguish surface features, much training and specialized equipment are required to process the raw data (Lillesand and Kiefer, 1994; Jensen, 1996; Ryerson, 1996). Fortunately, most image processing programs now run on personal computers. In addition, new programs like NASA's Earth Observing System should further simplify analyses by providing selected sets of fully processed data in a georeferenced form (Running *et al.*, 1994).

B. Remotely Sensed Landscape Attributes

1. Albedo, Land Cover, and Snow

As discussed in Chapter 2, albedo quantifies the fraction of short-wave radiation reflected from the land surface. Albedo affects the energy balance and is thus an important variable for all climate models, as will be discussed in Chapter 9. Albedo can be defined for specific spectral regions (a spectral albedo), but it most often includes the total short-wave reflectance. The total short-wave reflectance provides an excellent means of discriminating between water, bare land, snow, ice, and vegetation on a seasonal basis using data acquired from various satellites (Chapter 9). Because water absorbs short-wave radiation almost completely, surface albedo is particularly valuable for distinguishing the distribution of floodwaters. At the other extreme, snow has a high albedo, so the presence of snow cover and the timing of its disappearance can be monitored from satellites, assuming cloud cover is not prevalent.

2. Satellite-Derived Surface Meteorology

Meteorological satellites can provide an alternative means of obtaining weather data to drive ecosystem models. Meteorological satellites have either continuous or daily global coverage, but they are often limited in their spatial resolution as a result of the high altitude of their orbits. Regular archiving of these data is often required at even coarser resolution. As yet, procedures for processing and distributing these derived data sets are not standardized. Estimates of surface meteorology are also somewhat restricted by cloud cover, although the amount of cloud cover can he monitored and conversions made to predict incoming solar radiation and diurnal variations in temperature. Goward *et al.* (1994) estimated the monthly integrated incident solar radiation across the Oregon transect with high accuracy from ultraviolet reflectance data collected with the Nimbus-7 TOMS (Total Ozone Mapping Spectrometer) (Fig. 7.12a). Annual global incident radiation was mapped by Dye and Shibasaki (1995) using the same methodology. The GOES geostationary satellites are typically used to estimate daily PAR, because cloud cover and aerosol calculations can be done every half hour throughout the day. Daily estimates of incident solar radiation typically agree well with ground observations, with $r^2 > 0.95$ (Frouin and Pinker, 1995; Pinker *et al.*, 1995).

Meteorological satellites also provide good estimates of surface temperatures as illustrated from a summertime mid-afternoon sampling acquired over $106,000 \, \text{km}^2$ (Fig. 7.11). Variation across the scene illustrates the effects of variable aerodynamic mixing and evapotranspiration from surfaces with differing cover of vegetation. The coolest surfaces (22°–25°C) represent dense forests or lakes, and closely correspond with measured air temperatures. Dry grass and bare ground were >40°C. Goward *et al.* (1994) demonstrated

---

FIGURE 7.12.   Satellite sensors provide an alternative to extrapolation of surface weather observations as Goward *et al.* (1994) demonstrated across the Oregon transect (see Fig. 1.4). (a) Comparison of monthly integrated PAR derived from daily observations made with the TOMS sensor, (b) air temperatures estimated with AVHRR, and (c) estimated vapor pressure deficits from a combination of optical and thermal sensors, all compared against ground observations. (From Goward *et al.*, 1994.)

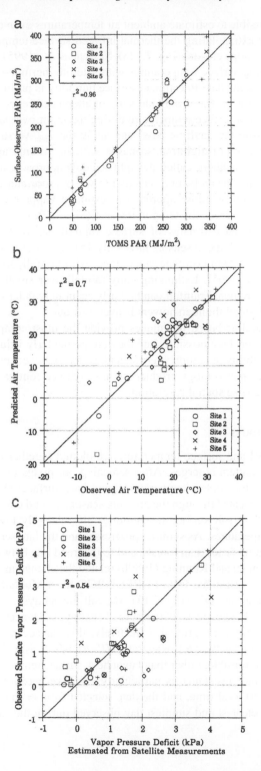

that it is possible to estimate ambient air temperatures even on a range of lightly vegetated surfaces by extrapolating the thermal infrared surface temperatures to conditions equivalent to full canopy closure (Fig. 7.12b). Norman *et al.* (1995) reviewed a variety of methods for computing surface temperatures with satellite thermal–infrared data.

Using satellite data to estimate surface humidity is problematic because water vapor in the atmospheric profile is a major cause of attenuated reflectances. One possible approach is to compute a surface temperature during the night overpass of a polar orbiting satellite (12 hr opposite the daytime pass) and apply the T_{min} versus T_{dew} relationship. However, image navigation of nighttime scenes is difficult, so only approximate locations are possible. Another method, called the split-window approach, analyzes the variable response of two thermal channels, one sensitive to water vapor and one not. Goward *et al.* (1994) applied the split-window approach to estimate vapor pressure deficits during satellite overpass times across the Oregon transect (Fig. 7.12c).

Satellite-based estimates of surface rainfall are best accomplished by microwave sensors, although these sensors have only 15 km spatial resolution. Efforts to use visible/NIR sensors at 1 km correlate cloud cover with rainfall, so have been only partially successful (Petty, 1995). Estimating daily rainfall over small areas is particularly difficult, but monthly regional rainfall can be fairly accurately determined because the random distribution of small thundershowers is averaged out.

These satellite-based techniques provide the most complete spatial representation of surface meteorology possible, and they are particularly valuable for quantifying meteorological variability in complex terrain and over variable cover of vegetation. Unfortunately, these surface meteorology products are not archived routinely and often are available only at a rather coarse spatial resolution.

3. Surface Resistance/Wetness

When NDVI and surface temperature are plotted together in a two-dimensional graph, a predictable pattern emerges. After screening out pixels of clouds and snow, we see that forests have the coolest temperatures, and bare surfaces the warmest (Fig. 7.11). If this analysis is repeated through the growing season, a trend emerges. In the spring, the ground is wet, so bare soil surface temperatures are relatively low and differ little from forest canopy temperatures. As summer progresses and the landscape dries out, bare soil and dry grasses heat up faster than forests, and so the trend line between $T_{surface}$ and NDVI becomes steeper. Nemani and Running (1989b) related the change in $T_{surface}$ versus NDVI to a simulated canopy resistance from FOREST-BGC for a Montana forest landscape that shows severe summer drought (Fig. 7.13). We call this analysis a *surface resistance,* because it operates as a landscape-level measure of canopy resistances as well as that associated with evaporation from the ground. Effectively, this surface resistance is a simple aggregate measure of the landscape Bowen ratio ($H/\lambda E$), where as incident radiation is progressively dissipated to sensible, rather than latent heat, surface temperatures increase. This AVHRR-defined surface resistance is particularly effective because the satellite overpass time is roughly 1430 local time, and the date selected from each biweekly set of daily data represents the clearest and sunniest day observed during the sampling period.

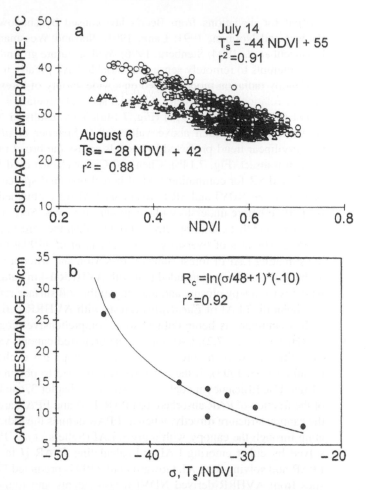

FIGURE 7.13. (a) As the land surface dries, exchange of latent energy declines, and the slope of the scatter of points in Fig. 7.11 becomes steeper as sensible heat exchange and surface temperatures (T_s) increase. (b) This change in the T_s/NDVI scattergram was expressed as a regional surface resistance to evapotranspiration by Nemani and Running (1989b) and Nemani *et al.* (1993a). The change in slope of T_s/NDVI computed for 8 clear days during the summer of 1987 was closely related to canopy resistance simulated by FOREST-BGC.

4. Leaf Area Index and Fraction of Light Absorbed by Canopy

Estimation of LAI across a landscape is essential for regional ecosystem analyses. Because the seasonal variation in LAI is important, considerable effort has been expended to develop nondestructive optical methods to estimate this variable. Initially a simple Beer's law inversion with a fixed extinction coefficient was applied to estimate LAI by measuring the attenuation of PAR through the canopy (Chapter 2; Pierce and Running, 1988). This simple analysis works well for many broadleaf canopies (Burton *et al.*, 1991; Wang *et al.*, 1992a), but for many evergreen species more complicated procedures are required to account for deviations from Beer's law caused by shadows and clumping of needles (Gower and Norman, 1991; Lang, 1991; Nel and Wessman, 1993; Smith *et al.*, 1993; Fassnacht *et al.,* 1994; Stenberg, 1996). With accurate ground estimates of LAI now available, attempts to remotely sense this variable have greatly increased.

Canopy radiation models tested on a wide variety of forests consistently predict a curvilinear relationship between LAI and a spectral vegetation simple ratio or its derivative NDVI (Myneni *et al.*, 1997a) (Fig. 7.14a). This results in NDVI not being an accurate predictor of LAI much above values of 4. Spanner *et al.* (1990, 1994) documented the curvilinear trend predicted by models across the broad range of LAI available on the Oregon transect (Fig. 7.14b). Chen and Cihlar (1996) used Landsat TM to test both the NDVI and SR for computing LAI of boreal pine and spruce stands. The highest correlations between NDVI and SR to overstory LAI were obtained with Landsat data from the springtime before understory plants produced leaves. Soil and understory plants produce a background that can be spectrally brighter than the forest canopies, and cause erroneous satellite estimates of overstory LAI. Nemani *et al.* (1993c) obtained increased precision in estimating LAI in forest stands across Montana when a mid-infrared band was added to the analysis. They concluded that with AVHRR 1-km data, part of the sensitivity of the NDVI is to canopy closure and masking of the soil surface reflectance. Pierce *et al.* (1993) also quantified LAI of *Eucalyptus* forests with AVHRR-derived NDVI.

If a landscape is being defined for a comprehensive biogeochemical simulation such as RHESSys (Fig. 7.2), LAI may be the preferred canopy variable because it is required for estimating so many process rates, from canopy gas exchange to nutrient return in litterfall (Bonan, 1993). If the objective is narrowed to obtain estimates of NPP only, then a Light Use Efficiency model, as described in Chapter 3, can be initialized from a measure of the fraction of PAR absorbed, or FPAR. LAI and FPAR are closely related: LAI defines the canopy structure directly, whereas FPAR defines the radiometric result of light attenuating through the canopy with a given LAI (Sellers *et al.*, 1992a). Equation (2.7) can be solved by either entering LAI and calculating FPAR [I in Equation (2.7)], or entering FPAR and solving for LAI. Goward *et al.* (1994) predicted FPAR for the Oregon transect sites from AVHRR-derived NDVI measurements and found generally good agreement with ground-based estimates (Fig. 7.15). Prince and Goward (1995) extended the FPAR estimates from NDVI to compute NPP for all terrestrial ecosystems with their Global Production Efficiency Model (GLO-PEM).

5. Biomass, Structure, and Density

The ability of remote sensing techniques to quantify forest stand structure has steadily improved from aerial photography through Landsat, SPOT, and now advanced radar and multidirectional sensors (Peterson and Running, 1989; Lillesand and Kiefer, 1994; Sample,

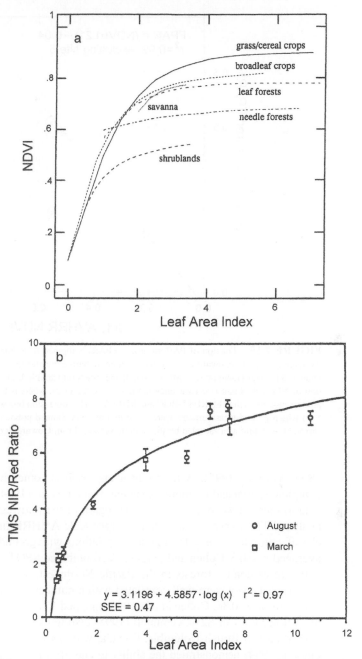

FIGURE 7.14. Calculation of LAI from satellite NDVI data (a) as predicted by a theoretical canopy radiative transfer model for various biome types by Myneni *et al.* (1997a) and (b) as measured for coniferous forests across the Oregon transect by Spanner *et al.* (1994). (From Mynemi *et al.*, "Algorithm for the estimation of global landcover, LAI, and FPAR based on radiative transfer models," *IEEE Transactions in Geoscience and Remote Sensing*, © 1997 IEEE; and from Spanner *et al.*, 1994.)

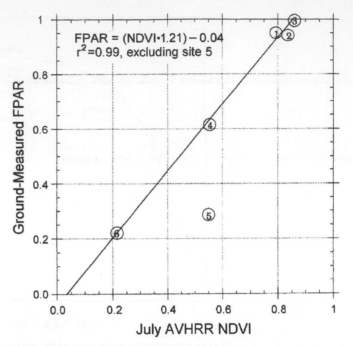

FIGURE 7.15.   Fraction of PAR absorbed (FPAR) computed from NDVI for coniferous forest canopies of the Oregon transect (Goward *et al.*, 1994). Canopy radiative transfer theory predicts NDVI to be more linearly related to FPAR (a radiometric variable) than to the structural variable LAI. FPAR is used primarily for computation of NPP with Light Use Efficiency models (see Chapter 3), whereas LAI is required in more sophisticated process models such as FOREST-BGC and RHESSys. The regression line was defined by excluding plot number 5 on which ground measurements were made in a recently thinned ponderosa pine stand that differed from the surrounding 1-km^2 area sampled by the satellite sensor. (From Goward *et al.*, 1994.)

1994; Verbyla, 1995). A multisensor approach is sometimes valuable to provide more complete spatial and temporal coverage. Iverson *et al.* (1994) used both AVHRR and TM data to map forest cover of a 13-state region of the midwestern United States. They first analyzed TM scenes to calibrate the larger scale AVHRR imagery and then extended the classification to the broader region. Correlations with ground-based forest cover estimates averaged $r^2 = 0.8$. Cohen and Spies (1992) combined SPOT and TM data to estimate forest structure of conifer forests in the Pacific Northwest. They concluded that tree size and overstory density were well correlated with a number of spectral indices and obtained r^2 values of 0.78–0.86. Cohen *et al.* (1995) mapped forest cover over 1.3 million ha with an overall accuracy of 82%. Gemmell (1995) distinguished between stand volumes of 150–300 m^3 ha^{-1} in British Columbia, Canada, with 78% accuracy using TM data. Likewise, Olsson (1994) demonstrated the ability to classify thinned and unthinned stands of Scots pine and Norway spruce in Sweden with multispectral indices, also from TM data. The above-cited classifications provide only statistical associations between forest attributes and measured remotely sensed reflectances; thus, they offer little basis for extrapolation to other sites or conditions. Often stand structure and density are best evaluated with a multivariable approach based on information extracted from a georeferenced database, rather than depending exclusively on what can be inferred from various spectral indices.

Although standing and dead biomass represent the major form in which carbon is stored above ground in forest ecosystems, they are exceedingly hard to quantify from space. Many attempts have been made to estimate this variable with optical sensors and with radar. Unfortunately, all of these sensors, regardless of the wavelengths selected, and the analyses attempted, fail to provide good estimates of standing biomass much beyond $100\,\mathrm{Mg}$ dry mass ha^{-1} (Waring *et al.*, 1995a). Prototypes for satellite-borne laser altimeters have shown promise in greatly increasing the range over which biomass, as well as other stand characteristics, can be obtained from satellites (Weishampel *et al.*, 1996).

V. SPATIALLY EXPLICIT LANDSCAPE PATTERN ANALYSIS

Some regional ecological analyses can be accomplished by evaluating each landscape cell individually. Many ecosystem analyses, however, require information about the size, distance, and characteristics of adjacent cells, necessitating a much more sophisticated landscape description. In general, the more adjacent units affect one another, the more complicated are the analyses. As a result, we must recognize that the scale at which we choose to define individual cells will dictate the degree of detail and complexity of the analysis. Many landscape properties could be computed similarly for a global forest cover map or for a local watershed. At the global level, however, only broad forest classes would be defined, whereas at the local scale, patterns in species distribution and age class structure might be evident if the database provided a sufficient level of detail.

A number of quantitative descriptors have been devised to measure landscape spatial characteristics (Fig. 7.16, Table 7.1). All of these landscape computations are based on the assumption that at some level of spatial resolution, a homogeneous unit of land exists, which is termed a *patch*. Each patch exists within a matrix of other patches, together forming the tapestry of a landscape (Forman, 1995). In forests, a patch is generally equivalent to a stand with a homogeneous mixture of species, ages, sizes, and/or stockings of trees. The delineation of a patch, however, is dependent on the spatial scale and degree of detail specified. A landscape defined in units of square kilometers cannot be discriminated into smaller sized patches.

Once the dimensions and other qualifications of a patch are defined, the geometric analyses required to yield estimates of area, shape, and frequency of occurrence over a specified landscape are easily applied. The *edge* or perimeter of a patch defines the boundary with adjacent patches. A *core area* within each patch may also be defined at some specified distance from the boundary. Various additional geometric properties can be computed, such as the fractal dimension, which relates the perimeter of a patch to its area and represents a measure of the complexity of surrounding boundary.

The population variables defined can relate to any organism present, so each defined patch can be very different for each population on the same landscape. Diversity and dominance indices measure the abundance and proportion of types of patches within the population on the landscape. Diversity measures how many different types of patches

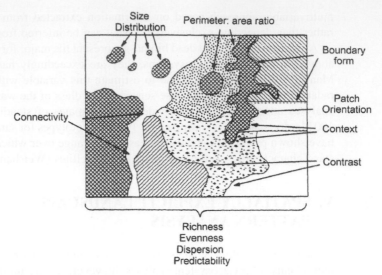

FIGURE 7.16. Hypothetical landscape illustrating various metrics used for quantifying spatial landscape characteristics. Landscape metrics are typically of two types: geometric variables that quantify sizes, shapes, and configuration of land units, and variables that quantify composition, abundance, and locations of populations (McGarigal and Marks, 1995). (Adapted from Wiens *et al.*, 1993. © 1993 *Oikos*.)

TABLE 7.1

Types of Quantitative Metrics Computed to Describe Landscape Spatial Characteristics[a]

Metric	Example variables
Geometric variables	
Area	Area in different landscape classes
Patchiness	Patch density, patch size, variability
Edge	Perimeter lengths, densities, contrast
Shape	Fractal dimension
Core area	
Population variables	
Nearest-neighbor	Proximity indices
Diversity	Species richness, diversity indices
Contagion and interspersion	

[a]From FRAGSTATS, McGarigal and Marks (1995).

occur, and dominance measures the proportion of the landscape occupied by each type of patch. FRAGSTATS computes nine different statistical measures of diversity, because there are variable definitions (McGarigal and Marks, 1995). The spatial arrangement of a patch with other patches and degree of regularity of a pattern is measured by proximity (distance to the nearest neighbor), dispersion (the randomness or uniformity of spatial distribution of patches), and contagion indices (clumpiness of patches).

Excellent summaries of these landscape metrics are provided by Turner (1989), Turner and Gardner (1990), Baker (1989), and Naveh and Lieberman (1994). Public domain software packages such as FRAGSTATS developed by McGarigal and Marks (1995) include an assemblage of algorithms required for most of the desired computations. The challenge is to decide which metric best quantifies the ecosystem property of greatest importance to each resource problem. Also, it should be recognized that different metrics when applied to the same problem can lead to divergent interpretations. Regardless, resource managers can, by following landscape changes in time and space, obtain a measure of the success or failure of particular policies, an approach we will explore in some detail in Chapter 8.

VI. DATA LAYER INCONSISTENCIES

An essential part of building databases for regional ecology is the requirement to maintain quality control because data acquired from a myriad of sources will invariably include gaps and unrealistic values. A single error in a huge data layer, such as relative humidity >100%, can prevent the completion of calculations in some ecosystem models. Also, when data are acquired from a variety of sources, inconsistencies will arise that demand cross-comparisons to discard or question values. For example, a soil data set might indicate a water holding capacity of 3 cm for a particular landscape unit, while a vegetation survey will indicate an LAI for the same unit as 12. Independently, both of these values are reasonable, but in combination they are highly unlikely and must he recognized through an automatic screening protocol. One inherent disadvantage of multilayered data sets is the likelihood that many small errors may be embedded in acquired data sets. Special routines are therefore required to find inconsistencies that exist in the geographically registered information on topography, climate, soils, and vegetation before a GIS analysis is initiated.

VII. SUMMARY

To extend our understanding to the landscape and regional level requires different techniques than those appropriate for stand/seasonal studies. The level of detail required depends on the choice of variables and the extent to which they vary spatially. Remote sensing offers an excellent source of georeferenced data but can provide direct information only on a limited suite of variables that alter the reflectance properties of landscapes (spatially, spectrally, temporally, and directionally). Climatological analyses are required to convert information gathered at a dispersed network of weather stations into more relevant forms (radiation, dew point, vapor pressure deficit) and to provide the basis for extrapolation of climatic variation across landscapes. The more time-invariant properties of topography, hydrology, and soils also must be represented spatially. Finally, the extent to which land cover and the type of vegetation vary in both time and space must be presented in a standardized form consistent with the capabilities of sensors, the frequency of satellite coverage, and the cost and difficulty of obtaining and processing digital data. As a consequence of these constraints, landscape ecological models must be developed in concert with what can be directly measured or inferred from remote sensing platforms.

AVHRR, 1.2 × 10³ pixels
6 polygons

Landsat TM, 1.3 × 10⁶ pixels
33-170 polygons

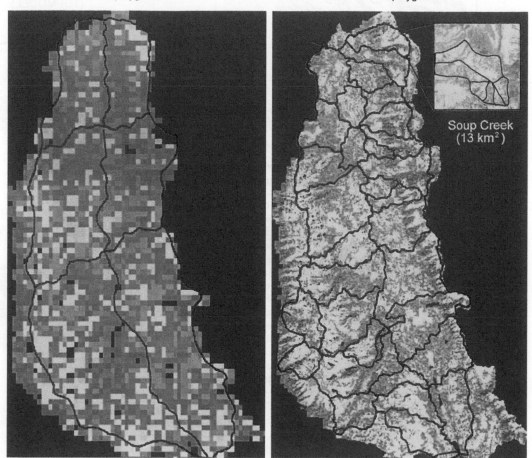

Soup Creek
(13 km²)

PLATE 1. Principles of landscape partitioning illustrating how digital elevation model (DEM) data can be processed to define the landscape at varying levels of topographic complexity. The Seeley–Swan watershed in Montana is depicted with 1-km² raster cells on the left-hand image, 30 × 30 m resolution cells equivalent to the Landsat pixel size on the right-hand image, and by vector cells defined from topographic analysis with RHESSys (inset). For general regional analyses and averages, the landscape defined with 1-km² cells, or six polygons, often provides the most appropriate level of detail to assess the long-term timber harvest potential, watershed yields, and wildlife carrying capacities. For more site-specific decisions, such as the consequences of a specific harvest cycle on other watershed resources, 30 × 30 m cells or homogeneous polygons that delineate areas <10 km² are probably necessary. In general, the more heterogeneous a landscape, the greater is the number of cells required to provide an accurate assessment of spatial variation.

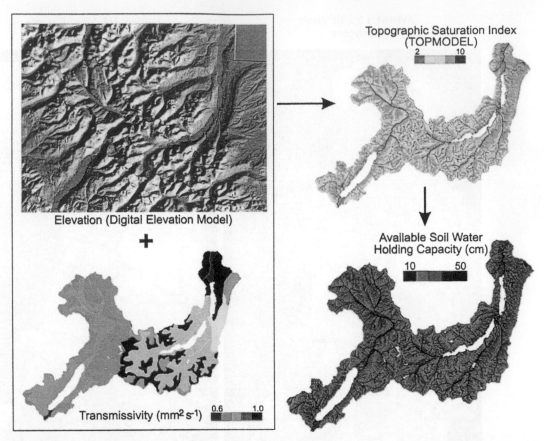

PLATE 2.   Steps involved in landscape partitioning based on physical characteristics are demonstrated for the McDonald watershed in Glacier National Park, Montana. First, digital elevation data are combined with a soil water transmissivity data layer derived from soil texture maps and basin delineations and hydrologic drainage networks are computed from the DEM data (Band, 1989a). Then the Topographic Saturation Index (TSI) is computed from TOPMODEL (Beven and Kirkby, 1979). Finally, raw soil textural data are converted to topographically consistent soil water holding capacities in RHESSys following techniques described by Zheng *et al.* (1996). These methods result in data layers that permanently define hydrologic characteristics of a watershed. RHESSys then computes the daily variation in properties for each cell in the watershed, such as precipitation, snowmelt, soil water storage, and evapotranspiration (see Chapter 2).

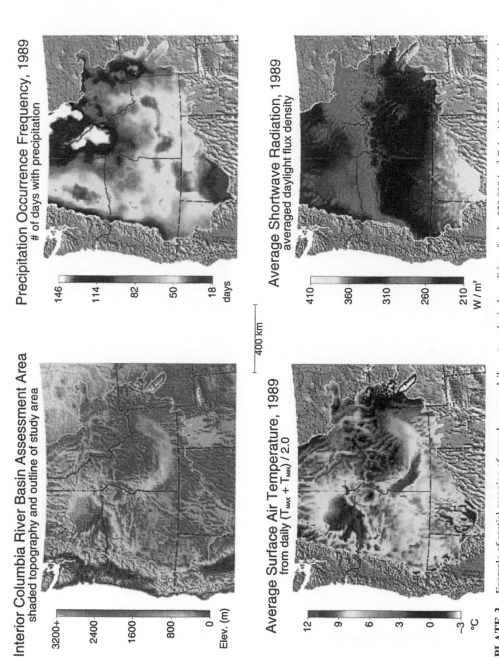

PLATE 3. Example of spatial mapping of annual average daily meteorological conditions for the 820,000-km² Columbia River basin in the U.S. Pacific Northwest for 1989 using a three-dimensional version of MT-CLIM called DAYMET (Thornton and Running, 1996; Thornton *et al.*, 1997). Daily air and soil temperature, incident solar radiation, humidity, and precipitation are computed for each cell as part of the RHESSys model, for input into ecological process models such as FOREST-BGC.

CHAPTER 8

Regional and Landscape Ecological Analysis

I. INTRODUCTION

In Chapter 7 we introduced principles and special techniques that allow the extension of ecosystem analyses to the scale of landscapes and regions. We concentrated on techniques to quantify the abiotic variables of the landscape and meteorological variables that drive the ecosystem. Techniques for quantifying directly observable or inferable state variables of the ecosystem were also identified. In this chapter we assemble these spatial scaling techniques to address landscape-level processes. We also integrate various ecological simulation models into our spatial analysis to provide capability for temporal extrapolation. This chapter presents a variety of landscape ecological studies that vary in the questions being asked and the degree of interconnections required for solution. Returning to the principles illustrated in Fig. 7.1, we evaluate the relative importance of spatial coupling in the vertical and horizontal directions, and the temporal dynamics appropriate to an array of regional scale analyses. Regardless of what spatial connections or time scales are chosen, we must incorporate process-level understanding to extend quantitative predictions to landscape scales.

Of equal importance, the studies demonstrate a capacity for real-time monitoring or future projection of ecosystem processes. Some resource management decisions require real-time information, such as those involving fire management, grazing intensity, or flood control. Other resource management decisions depend on the implications of projecting ecosystem responses decades into the future. How will streamflow and wildlife habitat change if a proposed harvest plan or prescribed fire is implemented in a watershed? Land managers currently make many decisions based on past experiences or local field trials, because those are the only options available when a decision is required. The capacity to demonstrate the potential implications of various policy decisions is perhaps the most valuable contribution ecosystem analysis can provide to decision makers. Making such contributions, however, requires geographically explicit analyses over a range of time frames. Ecological understanding that is not presented in these terms can provide little direct help in land management. We feel that geographically referenced predictive responses, resting on a foundation of process models, provide a consistent and useful summary of ecological information that can be made available to a wide range of managers. Although models will continue to be improved as knowledge increases, the most immediate challenge is to present predictions in a more graphic, understandable form. To meet this challenge, it is essential that model developers work closely with the land managers who must ultimately interpret and be responsible for implementing policies.

The first section of this chapter covers ecological processes where the horizontal connections are strong and the vertical links are weak. Because of the longer time constant associated with many horizontal connections, temporal dynamics are often not emphasized. Detection of land cover changes and wildlife habitat analysis are examples of problems with these connective properties. In the next section, the emphasis shifts to subjects where the vertical and temporal processes are strongly coupled, such as interactions between vegetation and the atmosphere. The final section covers problems where both horizontal and vertical connections are required, and where systems are also temporally dynamic. These analyses quantify both the change in *structure* through time as stands develop, are disturbed, or are eliminated, and changes in rates of *functions,* represented by the movement of water, carbon, and nutrients across the landscape (regional biogeochemistry). These models present land managers a tool to depict future changes in the landscape based on various policy options before implementing them.

II. HORIZONTAL CONNECTIONS: BIOTIC ANALYSIS OF FOREST PATTERNS

In Chapter 6 we discussed how certain types and frequencies of disturbance were a natural and continuous part of the development of forest ecosystems. Forest disturbances at the local level can result from the activities of resident animals, insects, and pathogens. At the landscape scale, forest disturbances are more commonly a human activity, through harvesting, establishment of large exotic plantations, converting forests to farms, air pollution, and urbanization. Some natural disturbances such as insect defoliation, wildfires, floods, and windstorms also can impact large areas. The spatial extent and frequency of various types of disturbances are important to quantify. It is instructive to compare the natural rates and magnitudes of disturbance with more recent human activities because both the timing and spatial extent of human disturbances can frequently be controlled. It may prove desirable to match management-related disturbances to those for which natural systems are well adapted.

It is becoming increasingly difficult to decipher original forest patterns because so little land remains in a relatively pristine state. Parks and wilderness areas can provide some means to compare contrasting management policies on adjacent landscapes. However, if these park areas have had wildfire protection or controlled populations of native animals, then the resulting forest cover is rather atypical because of the interruption of natural disturbances. A satellite image provides vivid documentation of disturbance associated with logging activities on National Forest land that is in contrast with adjacent parkland of Yellowstone National Park (Fig. 8.1). Extensive clear-cut patches are visible in the national forest, but the *natural* parkland was also extensively altered in 1988 when a wildfire burned 321,000 ha of predominantly old pine forest. The large extent of the 1988 fires was caused by a century of fire suppression causing a buildup of spatially continuous fuels that would never have occurred with the annual mosaic of small fires typical of that region. Thus, differing management strategies have produced dramatically different landscapes, but neither has retained the age classes and structural distributions of trees typical of an earlier, more natural landscape.

The significance of a disturbance can only be measured in the context of the landscape of interest and the ecological resilience of the local forests. M. G. Turner *et al.* (1993) designed a general analysis to evaluate the significance of disturbances at any spatial scale, called a state–space diagram (Fig. 8.2). Within this general framework, any time/space combination of landscape disturbance can be evaluated, from a year to a century and from a few hectares to thousands of square kilometers. The critical spatial measure of significance is the fraction of the defined landscape that is influenced by the disturbance (x axis). Different disturbances have varying levels of severity or temporal consequences to the original forest (e.g., the influence of a wildfire lasts much longer than a single defoliation). Thus, Turner *et al.* quantify temporal disturbance as a ratio of the disturbance interval to the time it will take that forest to return to its prior state (y axis). Defoliation of a few hectares of trees in a thousand square kilometer landscape would be quantified with both x and y values approaching zero, which represents a disturbance of limited significance in both time and space. The wildfire in Yellowstone Park in 1988 that burned 36% of the area has a statistical return interval of 300 years, so will leave its imprint on that landscape for centuries.

A. Temporal Changes in Forest Cover

The magnitude of human-induced landscape disturbance is often difficult to determine, and few studies have been able to quantify long-term changes in forest cover. White and Mladenoff (1994) compared the forest cover of a 9600-ha area of northern Wisconsin from pre-European settlement (1860s) to postsettlement (1930s) and the present (1989). Presettlement forest cover was estimated from U.S. Government survey records,

FIGURE 8.1. A Landsat TM image acquired in July 1991, clearly reveals the boundary between Yellowstone National Park and adjacent National Forest lands in the northern Rocky Mountains. Timber harvesting has been extensive on the National Forest lands while this activity was prohibited in the national park land. Wildfires were actively suppressed in both areas for nearly a century, resulting in almost no regular landscape-scale forest disturbance in Yellowstone National Park until the summer of 1988 when a 321,000-ha wildfire burned across much of the fuel-rich landscape within the park boundary.

postsettlement forest cover was derived from a county inventory map in 1931, and 1989 forest cover was interpreted from aerial photography. The area occupied by old-growth forest was reduced progressively from 100 to 23 to 6% from 1860 to 1989 (Table 8.1). Early pioneer species covered 50% of the area in 1931, and by 1989 over 90% of the landscape supported mixed hardwood and conifer second-growth forests.

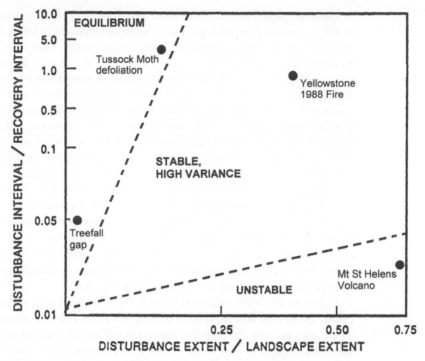

FIGURE 8.2. State–space diagram that evaluates landscape disturbance and dynamics across any spatial and temporal scales. The spatial parameter S, on the x axis, defines a disturbance as a fraction of the landscape of interest, regardless of size. The temporal parameter T is the ratio between the disturbance interval and the time required for the forest to recover to a mature state. When a disturbance interval is long relative to the recovery time and a small fraction of the landscape is involved, the system may be considered stable or at equilibrium. A disturbance interval comparable to the recovery interval, and affecting a large portion of the landscape, can be considered stable but with high variance. Disturbance intervals much shorter than the recovery interval and that influence a large proportion of the landscape may lead to an unstable system that shifts to another state. (From M. G. Turner *et al.*, 1993.)

TABLE 8.1

Compositional Change in Forest Type on a 9600-ha Landscape in Wisconsin over 130 years[a]

Forest type	1860	1931	1989
Pin cherry	—	37.2%	2.3%
Mixed hardwood	—	13.7%	21.8%
Mixed conifer	—	0.5%	10.3%
Hardwood/conifer	—	5.9%	29.4%
Northern hardwood	—	19.4%	30.5%
Old growth hemlock/hardwood	22.5%	—	—
Old growth hemlock	46.2%	7.2%	3.7%
Old growth hardwood	31.4%	16.1%	1.9%

[a]From White and Mladenoff (1994).

Landsat satellite data are now available from 1972 to the present for most areas of the world, so images from multiple dates can detect changes in forest cover during this recent era. Hall *et al.* (1991) compared forest cover change rates over a 10-year period in a managed forest and wilderness areas of a 900-km^2 region of Minnesota. They found that regenerating clear-cuts in the managed forest increased from 15 to 22% of the land area during the 10-year period. Eleven to thirteen percent of the wilderness area was in early stages of forest development, a result of fire and natural tree mortality. Hall *et al.* estimated that the managed forest had an interval between establishment and harvesting one-fifth the length of the interval associated with natural fires in a wilderness area. The frequency of natural fires, however, has also been significantly diminished through fire suppression activities. Frelich and Reich (1995) found that the fire frequency of a 234-km^2 region of cold temperate forest of Minnesota has changed from 1 in 50–100 years to 1 in >1000 years because of fire suppression. In some forests, the natural fire frequency can be mimicked by harvest rates and aid in managing for reasonably natural forest dynamics. Gruell (1983) provided photographic evidence of the invasion of forests over the twentieth century into grasslands in many arid areas of the northern Rocky Mountains of the United States. Fire suppression and control of grazing are the primary management actions that have resulted in this land cover change (see Fig. 6.4).

Cohen *et al.* (1995) estimated that 15.3% of the 10.4 million ha of forested land of the Pacific Northwest was harvested between 1972 and 1991, an average harvest rate of 0.8% of the area per year. In contrast, the much-publicized clearing rates in the Amazon have been documented at only 0.4% of the area per year (Skole and Tucker, 1993). Both of these studies illustrate that satellite data can provide a more repeatable and unbiased estimate than ground surveys of deforestation. Of course, the more critical analysis for forest management is the fraction of landscape *permanently* deforested and converted to urban or agriculture use. Rates of forest regrowth are not so clearly quantified by satellite imagery because the change in land cover is gradual as regenerating trees reoccupy an area.

A challenge for retaining a semblance of natural landscape dynamics may be to manage the spatial size, distribution, and timing of harvesting activities to more resemble natural disturbance processes. By tracking the trajectory of landscape change throughout recent history we attempt to project the rate of future change. Extrapolating these recorded rates of landscape change into the future rests on an implicit assumption that external conditions (population pressure, lumber prices, climate, etc.) will remain similar to those in the past. When this assumption is unwarranted, more complicated analyses are required to make future projections. Thus, past trends and rates of change in landscape patterns are a necessary precursor to evaluating future trends, but insufficient given the likelihood of additional changes in external conditions.

B. Quantifying Spatial Heterogeneity

The landscape displays structural heterogeneity beyond that recognized within groups of stands. Forest cutting is the means by which land managers most commonly affect the structural heterogeneity of landscapes, either purposefully or inadvertently. Franklin and Forman (1987) illustrated the relationship between different idealized stand cutting patterns and the resulting patch sizes and edge lengths, both critical factors in defining wildlife habitat. Their analysis showed that as cutting unit size is decreased, the patch size of undisturbed area also decreases because more land must be entered to sustain a fixed yield of forest products. Additionally, cutting unit border length increases. One cannot state that any specific cutting pattern is "best" because preserving large patches for spotted owl habitat would reduce edge habitat along cutting boundaries favorable for producing forage and cover for deer.

Spies *et al.* (1994) found that the percentage of old-growth closed canopy forest decreased from 71 to 58% between 1972 and 1988 on a 2600-km^2 forested area studied in the center of the Oregon transect area (Fig. 8.3). Spies *et al.* also found that the size of contiguous uncut patches optimal for certain wildlife habitat declined from 160 to 62 ha, and edge density increased from 1.9 to 2.5 km^{-2} during the 16-year period. The disturbance rate for this region was 1.2%/year, or a recurrence time of 84 years. This suggests that although the current harvest rate is reasonably near the natural disturbance frequency, current forest management is increasing fragmentation and reducing the size of uncut patches.

Li *et al.* (1993) developed a simulation model that also allows constraints from streams and roads to be included in the spatial analysis of landscape heterogeneity. In simulating the habitat fragmentation produced by different forest cutting patterns, Li *et al.* (1993) found that the widely used method of placing many small clear-cuts in staggered setting produced the maximum landscape fragmentation. Wallin *et al.* (1994) used a simulation of different stand cutting patterns and timings to illustrate the interaction between clear-cut size and cutting frequency in determining the long-term landscape pattern of age-class heterogeneity in western Oregon transect forests. Gustafson and Crow (1994) evaluated the influence of different sizes and distributions of clear-cuts on bird habitat in Indiana. Their conclusions illustrate that in order to sustain a given harvest level and simultaneously retain bird habitat, clear-cuts must be aggregated to larger units. However, their recommendations may not apply to ungulate species, such as deer and elk, which tend to prefer edge habitat, foraging in recently cleared areas on successional forbs, but hiding and resting in adjacent closed canopy forests.

C. Disturbance Propagation

The propagation of disturbances is rarely random across a forested landscape. Harvested areas are usually adjacent to a road network. Disturbance by fire or wind is driven by topography and microclimate conditions, and insect epidemics spread rapidly to adjacent stands with low growth efficiencies. Models representing the explicit spatial propagation and timing of these disturbances are now being built, although few complete studies are available in the literature (Roberts, 1996a,b). The spread of balsam wooly aphid in the Appalachian Mountains was evaluated by considering population dynamics of both the aphid and the host tree, Fraser fir, and the biophysical conditions that encourage spread (Dale *et al.*, 1991). Johnston and Naiman (1990) evaluated the history of beaver pond alteration on the hydrology and vegetation at Voyageurs National Park, Minnesota. They found that from 1940 to 1986 analysis of aerial photography of a 250-km^2 area showed beaver ponds increased from 1 to 13% of the landscape, and the beaver population in 1986 was 1 colony km^{-2}. This analysis can be used directly for making management

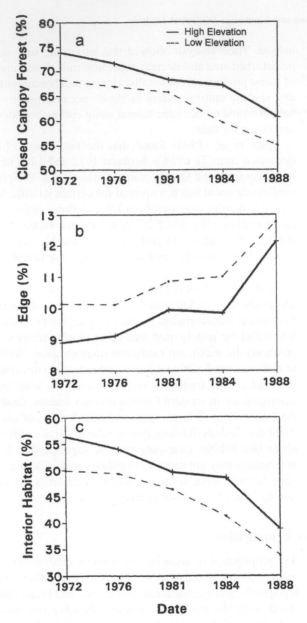

FIGURE 8.3. Reduction in closed canopy conifer forest and interior forest habitat, and resulting increase in percentage of edge, for a 2600-km² landscape in Oregon for the period 1972–1988 (Spies *et al.*, 1994). Landsat imagery served as the basis to classify the landscape into conifer forest versus nonforest status at 4- to 5-year intervals, with an accuracy of 91%. High elevation areas (>914 m) are primarily public land, which include 7% in wilderness areas where forest cutting is prohibited. Low elevation land (<914 m) has a mixture of public and private ownership. Over the 16-year study, conversion of areas supporting mature forests (predominantly by clear-cutting or wildfires) averaged 0.2, 1.2, and 3.9% per year for wilderness, public, and private forestland, respectively. These observed harvest rates equal or exceed reported rates in the Amazon basin (Skole and Tucker, 1993). The rates of *regrowth* are of equal importance in determining the net change in forest cover and NPP, but they are often not reported. The increase in edge resulting from harvesting activity represents a habitat improvement for deer but a detriment for spotted owls, illustrating the differential habitat requirements of wildlife and the conflicts inherent in land management. (From Spies *et al.*, 1994.)

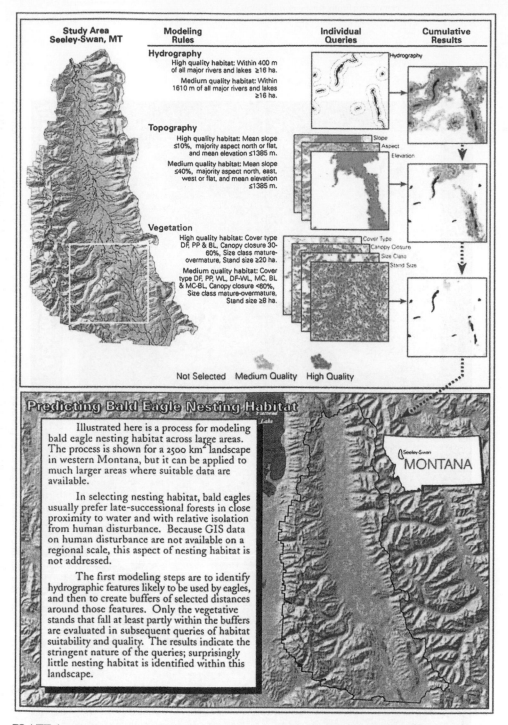

PLATE 4. Geographically explicit framework for modeling bald eagle habitat in a 2500-km^2 basin in Montana. Hydrography, topography, and vegetation data layers which define known species preferences (DF, Douglas-fir; PP, ponderosa pine; BL, broadleaf trees; WL, western larch; MC, mixed conifers) are merged with a logical hierarchy to reflect decreasing habitat priorities. The resulting analysis produces an intersection of required and optimum conditions. The habitat identified represents less than 1% of the study area and would warrant special management consideration. These procedures could be used for other species, but each will have different habitat requirements, necessitating somewhat different analysis. (From Steve Holloway and Melissa Hart, University of Montana Spatial Analysis Laboratory.)

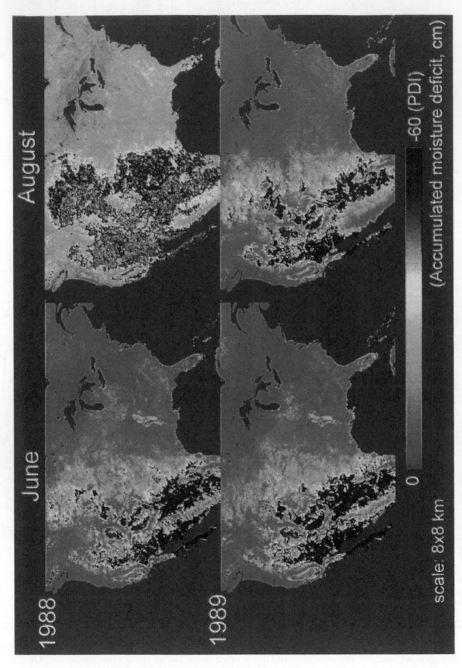

PLATE 5. Seasonal changes in a satellite-derived drought index produced from biweekly composited AVHRR NDVI values and radiometric surface temperatures for the United States, expressed as a moisture deficit similar to the Palmer Drought Index. The hot, dry conditions of the summer of 1988 which resulted in the huge Yellowstone National Park fires are clearly contrasted with the more normal conditions of 1989. This drought index can also be expressed as a fire danger index. Both indices are based on the T_s/NDVI relationship and surface resistance computation explained in Chapter 7 (Nemani and Running, 1989b; Nemani et al., 1993a). The Earth Observing System has produced a Drought Index weekly for the world since 2000.

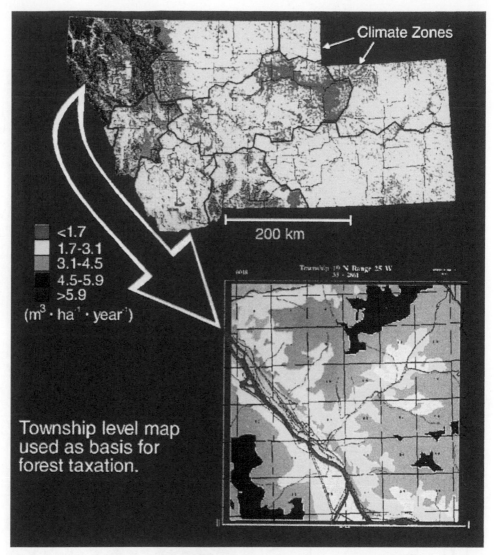

PLATE 6. Map of the potential forest productivity of Montana produced with RHESSys logic for the state forested land taxation system (Milner *et al.*, 1996). MT-CLIM was used to build a topographically sensitive daily climate data file. Potential LAI was computed from the hydrologic equilibrium concepts of Nemani and Running (1989a). Soil water holding capacities were computed from the STATSGO database. FOREST-BGC was then run with fixed physiological parameters representative of lodgepole pine. Rather than executing each of ~2 million cells individually, each climate/topography/LAI/soils combination was simulated and the appropriate result mapped back to each 0.36-ha cell. These computed productivities are now the basis of forestland valuation for property taxation in Montana. It was estimated that the model-based mapping was done for one-quarter of the cost and in half the time required for an equivalent field inventory. (From Milner *et al.*, 1996.)

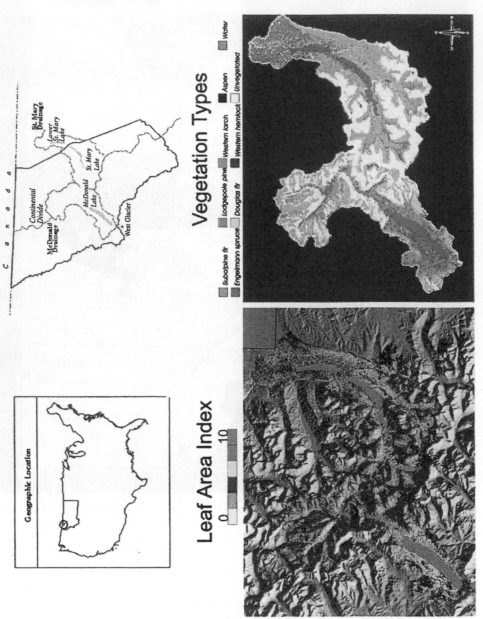

Geographic Location

Vegetation Types

■ Subalpine fir	■ Lodgepole pine	■ Aspen	■ Water
■ Engelmann spruce	■ Western larch	□ Unvegetated	
	■ Douglas fir	■ Western hemlock	

Leaf Area Index

0   10

PLATE 7. Definitions of LAI and forest types for two mountainous watersheds straddling the Continental Divide in Glacier National Park, Montana, for the RHESSys simulations in Plates 8 and 9. This 450-km² region encompasses elevational ranges from 1000 to 3800 m with orographically induced precipitation ranging from 200 to 2000 mm. Mapping of LAI and classification of the diverse montane forest communities was done with a 15 July 1990 Landsat TM data set. Plate 2 shows the topographic and hydrologic partitioning of the landscape produced as input data layers for a RHESSys simulation. (From Joseph White, University of Montana, and Robert Keane, USDA Intermountain Fire Science Laboratory, Rocky Mountain Research Station, Missoula, MT.)

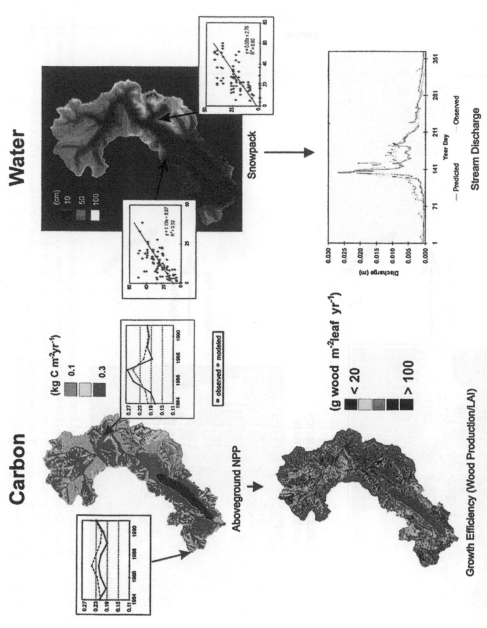

PLATE 8. Seasonal carbon and water budgets simulated by RHESSys for the Glacier National Park study area, with accompanying validation data collected from sites within the watersheds. The 10-year carbon balance simulations were compared to stem growth increment data at two sites in the McDonald Creek watershed (inset graphs). The mid-elevation site showed higher productivity and interannual variability than the southwest facing site. Growth efficiency is simulated to be highest on the south to southeast facing slopes at low elevations where temperatures and incident solar radiation combine to produce the most favorable microclimate in this energy-limited mountain landscape. Stream discharge peaks during the May snowmelt period, and predicted versus observed snowmelt results are inset for two plots in the watershed. (From White *et al.*, 1998)

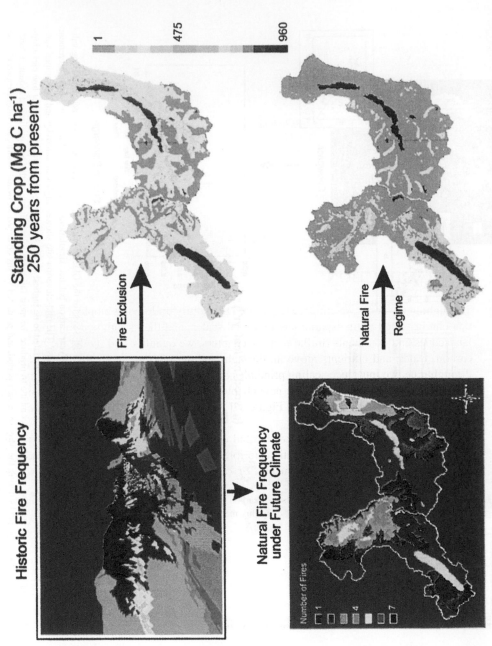

PLATE 9. The stand dynamic model FIRE-BGC (which combines a gap succession model, see Fig. 5.24, and FOREST-BGC) was embedded in RHESSys, and landscape dynamics were studied over a 250-year simulation (Keane *et al.*, 1996a,b). When the model incorporated the historic fire frequency, as high as 7 ignitions as 250 years, much lower standing forest biomass resulted, with higher net primary productivity, compared to a landscape with total fire exclusion. (From Joseph White, University of Montana, and Robert Keane, USDA Intermountain Fire Science Laboratory, Rocky Mountain Research Station, Missoula, MT.)

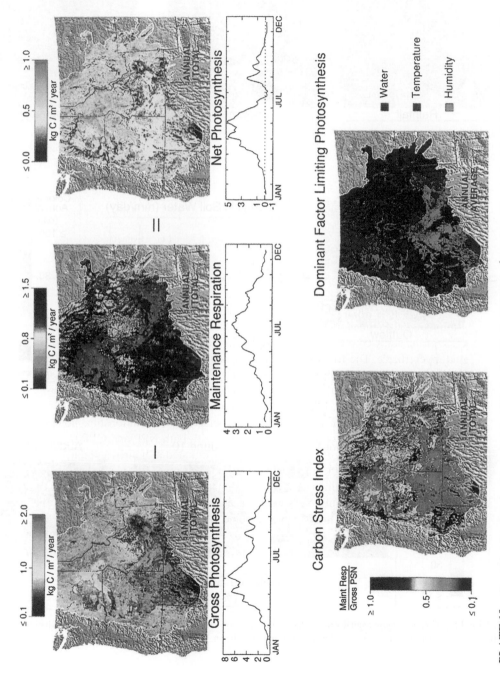

PLATE 10. Comprehensive annual carbon budget for 1989 simulated for the 820,000-km² Columbia River basin of the U.S. Pacific Northwest with RHESSys. Seasonal dynamics of each variable averaged for the entire basin are also shown. The extreme climatic diversity of this region is shown in Plate 3. (From Peter Thornton, University of Montana.)

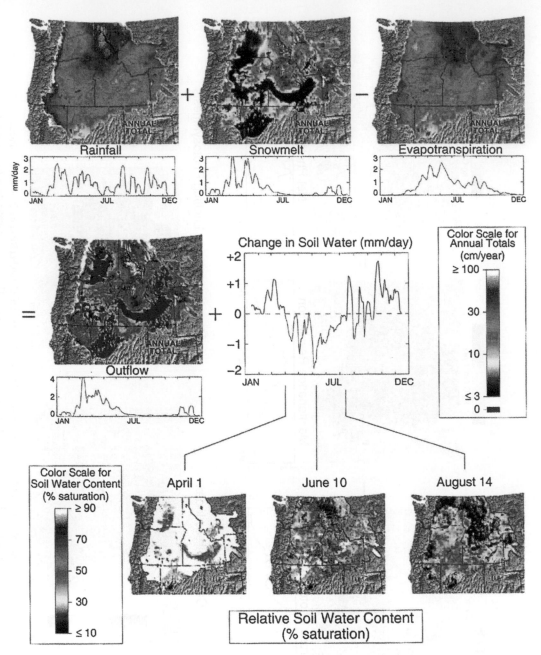

PLATE 11. Comprehensive annual water balance simulated for 1989 over the 820,000-km^2 Columbia River basin with RHESSys. Seasonal dynamics averaged for the basin are also shown. (From Peter Thornton, University of Montana.)

decisions of whether the measured beaver density can be sustained or should be reduced.

D. Wildlife Habitat Analysis and Spatially Explicit Population Modeling

Evaluating wildlife habitat requires consideration of physical features of the landscape including microclimate, structural and community relationships of the vegetation, and the physiology and behavior patterns of the target organism. Habitat criteria have different levels of priority; some factors are essential while others are preferred but not required, and each animal needs different habitat, which further complicates landscape analysis. Possibly the most difficult challenge in landscape analysis involves following the movement of organisms within a complex landscape. Such analyses require "spatially explicit population models" (Dunning *et al.*, 1995). Population models have typically simplified the analysis by concentrating on the horizontal biotic interactions across ecosystem boundaries shown in Fig. 7.1, and by evaluating the vertical atmospheric and hydroedaphic processes in very simple ways or not at all. Emphasis is on the habitat quality for the organism, for example, evaluating winter thermal vegetation cover rather than absolute temperatures. Population-based models define a population for each landscape cell, so are more appropriate for abundant organisms like insects or birds, and simulate their collective movement in the landscape through time (Dunning *et al.*, 1995).

An analysis of bald eagle nesting habitat in a Montana basin began with a requirement of proximity to rivers or lakes where the primary food supplies are found (see Plate 4). A secondary priority was relatively flat terrain at low elevation in the basin, as these are not high mountain birds. The third criterion was mature forests with nearly complete canopy closure in blocks of at least 8 ha. With sequential queries of georeferenced databases of hydrography, topography, and vegetation, the intersection of criteria produced a map of probable bald eagle nesting habitat, identifying key areas of the landscape that may require special management.

A classic resource management conflict has developed around the northern spotted owl in the forests of the Pacific Northwest, including the Oregon transect area (Fig. 1.4). Prime habitat for the spotted owl is old-growth forests, yet harvest rates from 0.9 to 5.4% of the area per year since World War II had measurably reduced the old-growth forests to the point that the owl was legally defined as a "threatened species" in 1990 (Noon and McKelvey, 1996). Lamberson *et al.* (1992) proposed an array of habitat conservation areas to enhance owl populations. Extensive landscape simulations were completed to evaluate the size, shape, and proximity of these habitat blocks that would be necessary to sustain the owl population at various levels (Noon and McKelvey, 1996). Following these analyses, decisions that reduced harvest rates on millions of hectares of forested land were made that had an international impact on wood commodity values.

Hansen *et al.* (1993) combined the landscape fragmentation model of Li *et al.* (1993) with a forest gap model (Fig. 5.17) to explore the consequences of different land management objectives on future bird habitat trends. Four management alternatives were tested on a 3300-ha watershed in western Oregon, including a natural fire regime (no fire suppression), an intensive wood production alternative with 70-year harvest frequency, a multiple-use alternative with 140-year harvest frequency and larger cutting blocks, and finally a no action alternative of letting the current landscape develop without harvesting. Simulations of forest stand development from the present for 140 years into the future were evaluated for stand conditions that provide suitable bird habitat for each alternative. All alternatives except the intensive wood production scenario retained suitable bird habitat after 140 years (Fig. 8.4). It would he difficult to perform this evaluation with so many landscape, forest, and wildlife factors and multiple management objectives without computer models.

Individual-based models track the location on the landscape of each individual at each time step. M. G. Turner *et al.* (1993, 1994) and Pearson *et al.* (1995) report on extended studies of large ungulate populations (elk, moose, deer, and bison) in Yellowstone National Park. Their analyses first quantified the landscape topography, habitat types, fire history, and forage distribution. Then initial population data for the ungulates including age, size, sex, and spatial distribution were defined. A snowpack simulation was then run to allow computation of animal foraging behavior based on snow depth and forage availability through the winter. Model emphasis is on animal energetics for survival, not hydrologic dynamics of the snowpack. Interactions among weather conditions as quantified by snow depth, fire size and spatial distribution, and foraging success of animals were evaluated with simulation experiments that would require decades of field studies to develop similar insights. Simulations showed that fires burning single large blocks, followed by a severe winter, could induce high mortality rates for an elk herd, because of the lack of forage and the difficulty animals would experience in moving through deep snowpack to search for food. The simulations indicated that future prescribed burning in this ecosystem should be distributed as small patches if the primary goal is to maintain a large elk herd. That objective would increase landscape fragmentation but would achieve park management goals of minimizing fire danger and maximizing ungulate populations (Turner *et al.*, 1995).

It is worth again noting that fragmentation of a forested landscape recommended for sustaining large populations of ungulates would be in conflict with management objectives that seek to favor large populations of some species of birds which require unbroken blocks of forests. Most land managers must balance concerns about wildlife with timber production, grazing, and watershed considerations. For these reasons the ability to evaluate the likely impact of various policies through landscape simulations before implementation has special merit. Although the landscape models available are still far from ideal, each assumption and calculation is explicit and so can be challenged and modified. When managers are considering multiple options that will shape the landscape a century into the future, decisions made must be based on the best evaluation possible.

Stable isotopes can offer a different and unique perspective for the analysis of animal habitats. The preferred home ranges and their seasonal or interannual variation can be confirmed through analyses of the isotopic composition of animal tissues such as bone, teeth, horn, and other organs. For example, Ambrose and DeNiro (1986) deduced the long-term habitat distributions of 43 species of East African mammals from the stable carbon and nitrogen isotope ratios extracted from bone collagen. Differences in the Sr, Pb, and N isotopic composition of soils were also reflected in diets of elephants, whose tusks provided a unique multiple-isotopic signature related to the distribution of herds

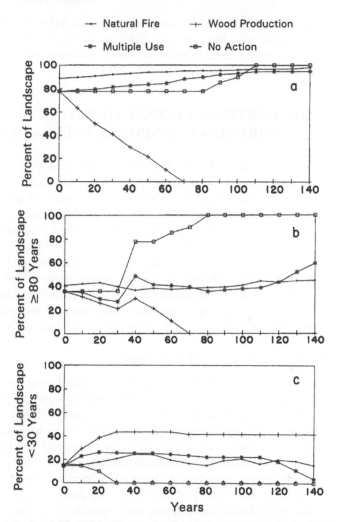

FIGURE 8.4. Simulation of change in forest structure and resulting bird habitat suitability for a 3300-ha watershed in western Oregon under four management scenarios, beginning with the current landscape and extending for 140 years into the future. Variations in fire frequency and the size and frequency of forest harvests were simulated as alternative management policies. The resulting percent distribution of the landscape under different management scenarios is shown in (a). The percentage of the landscape in key age classes representing mature individuals (≥80 years) is shown in (b) and seral stands <30 years old in (c) through the 140-year simulation. The importance of these stand ages and spatial distributions varies with bird species. (From Hansen *et al.*, 1993.)

across much of Africa (Van der Merwe *et al.*, 1990; Vogel *et al.*, 1990). Smaller animals, of course, have less extensive distributions but still often differ in their isotopic composition based on dietary selections related to marine and terrestrial food sources (Grupe and Krueger, 1990; Angerbjörn *et al.*, 1994).

III. VERTICAL CONNECTIONS: FOREST–ATMOSPHERE INTERACTIONS

The first scale of forest–atmosphere interactions are the one-dimensional or stand-level Soil–Vegetation–Atmosphere–Transfer (SVAT) models discussed in Chapters 2 and 3. In Chapter 7 we explained the representation of meteorological conditions across a landscape, using models such as MT-CLIM to extrapolate measured microclimate conditions across complex terrain. However, MT-CLIM and similar climatological models do not analyze the dynamic interaction between the forested landscape and the atmosphere. How forests respond to atmospheric dynamics is well understood at local scales from micrometeorological studies (Chapter 2). We also appreciate that advected energy and momentum are distributed across the landscape in direct response to local topography and surface conditions. Forest responses at regional scales, however, are less well known, and the influences that forests exert back on the atmosphere are particularly poorly understood.

A. Regional Forest–Atmosphere Interactions

New modeling analyses that integrate a regional scale biospheric gas and energy exchange model with a mesoscale dynamic meteorology model are providing evidence that extensive forest cutting produces a highly fractured canopy energy exchange surface; this intensifies atmospheric turbulence and thunderstorm activity over scales of tens to hundreds of kilometers (Pielke *et al.*, 1997). Pielke *et al.* suggest that thunderstorm activity in Georgia may have increased from presettlement times as a result of land conversion from continuous forests to interspersed 10- to 40-ha blocks of forest, fields, and urban areas. Forests consume the majority of incoming solar energy in evapotranspiration while agriculture and urban landscapes generate more sensible heat, which initiates the vertical convection columns that spawn thunderstorms (Fig. 8.5). This appreciation of the role of forest cutting patterns on local severe weather and the reduction of sensible heat transfer into the atmosphere may help justify not only more wide scale reforestation efforts but also tree planting and the establishment of more parks in large cites.

B. Regional Drought Analysis

Satellites allow a landscape view of the forest, but the reflected and emitted signals produced from the landscape and transformed into digital imagery often fall short of quantifying vegetation responses of interest. Vegetation development in the spring and leaf senescence in the autumn can be well represented by simple vegetation indices such as NDVI. However, the onset of seasonal drought may not be directly observable, particularly in evergreen forests. Drought influences canopy gas exchange, productivity, fire danger, grazing capacity, and wildlife survival in many forests of the world. If frequent, geographi-

Current Vegetation Patterns   Homogeneously Forested

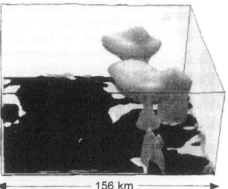

← 156 km →

FIGURE 8.5. Simulation of afternoon convective thunderstorm development for 26 July 1987 at 1400 hr local time in central Georgia simulated with RAMS (Regional Atmospheric Modeling System). The simulation contrasts diurnal mesoscale meteorology of the current fragmented forest/farm/urban landscape mosaic with a homogeneous forest similar to a pre-colonial settlement condition (Pielke *et al.*, 1997). This analysis suggests that mesoscale turbulence and thunderstorm development have been accelerated by increasing the heterogeneity of forest land cover over the last few centuries. (From Pielke *et al.*, 1997.)

cally explicit data were available on the severity of drought, some advanced planning to minimize the impact of climatic variation would be possible.

The principles introduced in Chapter 7 to define a surface resistance from the ratio $T_s/$ NDVI from the AVHRR satellite can be applied to display regional scale drought conditions throughout the growing season. Nemani *et al.* (1993a) developed algorithms for screening out clouds, snow, open water, and other confounding effects from the raw T_s/NDVI relationship shown in Fig. 7.11, and then computed the slope relationship to quantify surface resistance in Fig. 7.13. This automated procedure was then extended to calculate seasonal surface resistance for the continental United States for the extremely dry summer of 1988 and to contrast these values with the more normal summer of 1989 (see Plate 5). The seasonal surface resistance was converted to water deficit units and compared with the Crop Moisture Index, computed from surface meteorological data by the National Weather Service, for the summer of 1990. A high correlation was found between the standard Crop Moisture Index and the satellite-derived drought index ($r^2 = 0.83$), but the satellite-derived index can be computed more quickly and with greater spatial accuracy (Nemani *et al.*, 1993a). This satellite-derived drought index has been produced routinely by the NASA Earth Observing System program since 2000.

C. Fire Dynamics

Wildfire suppression costs at least $700 million per year in the United States alone, so any improvement in fire forecasting and management would have immediate utility and financial benefits. Fire danger, the risk of an uncontrolled ignition, has historically been computed in the United States from surface weather and fuel moisture conditions using the

National Fire Danger Rating System (Burgan, 1988). This system provides computations of fire danger from the stations of measurement and does simple extrapolation to provide continuous maps across the landscape. The satellite-derived surface resistance calculated by Nemani and Running (1989b) and Nemani *et al.* (1993a) can be translated into a fire danger index that provides an automated and spatially continuous mapping of the danger of fire ignition, integrating both the surface meteorology and the vegetation energy partitioning condition related to stress and fuel flammability. Vidal *et al.* (1994) applied a derivation of the algorithm used for Plate 5 to evaluate fire ignitions during the summers of 1990 and 1991 (when 537,000 ha burned in the Mediterranean region). The satellite-derived fire risk corresponded with the dates of observed fire ignitions in two study areas of southern France.

Fire ignition and behavior are controlled by available fuel, current meteorological conditions, and topography and have been well quantified (Rothermel, 1972). Subsequent fire spread is inherently a spatially defined problem that must integrate the georeferenced location of fuels and spatial pattern of the forest on the landscape, with the dynamic meteorology that defines the combustibility of the fuels and moves the fire across the landscape (Keane *et al.*, 1996a,b). Daily AVHRR satellite data have been used to follow fire spread, but the 1 km spatial resolution, severe changes in viewing angle from day-to-day images, and interference by smoke makes this technique useful for large fires only (Chuvieco and Martin, 1994).

Accurate mapping of the extent of large fires is also valuable for determining timber volume losses, reforestation requirements, wildlife habitat changes, hydrologic risks, and atmospheric emissions. The black, charred surface of a completely burned forest can be easily differentiated from green vegetation in visible wavelengths. However, forests with interspersed burned and unburned patches cannot be evaluated so easily. Kasischke and French (1995) were able to detect 78% of the 1.9 million ha burned by fires in Alaska during 1990 and 1991, many in remote and uninhabited areas where ground surveillance was impossible. Turner and Romme (1994) evaluated the role of preexisting stand structure patterns in crown fire spread, and the resulting mosaic produced by fires in Yellowstone National Park. The spatial distribution of burned areas influences winter elk survival, as discussed in the previous section. While acknowledging the tremendous destructive potential of wildfires, we now recognize many beneficial effects of fire on ecosystems that justify prescribing controlled fires or sometimes allowing wildfires to burn to perpetuate the more natural pattern in historically fire-dominated landscapes.

IV. VERTICAL AND HORIZONTAL CONNECTIONS: REGIONAL BIOGEOCHEMISTRY

Analysis of forest biogeochemistry at the regional scale involves vertical connections to the atmosphere as driving variables and drainage dynamics into the soil, horizontal connections between adjacent stands that influence hydrologic partitioning, and finally time dimensions from subdaily intense storms to decadal shifts in hydrology and geochemistry as stands develop. This section explores a range of regional biogeochemistry studies focused on hydrologic processes.

A. Watershed Hydrologic Balance

The most important activity of watershed scale hydrology is accurate calculation of stream and river discharges. Knowledge of river flow dynamics is required for flood forecasting, irrigation scheduling, hydroelectric power operation, and water quality issues such as temperature, oxygenation, and dilution of pollutants. The headwaters of most major river systems are in forested landscapes, and thus the management of those areas influences the quantity, quality, and timing of river flows. Near-term forecasting of streamflow relies on accurate and geographically explicit monitoring of current snowpack and precipitation, information now often transmitted from field sites to office computers instantaneously by satellite. Projections of streamflow require representation of surface heterogeneity of topography, soils, and vegetation in a realistic way. In Chapter 7 we introduced the components of a Regional Hydroecological Simulation System (RHESSys) that represents these ecosystem characteristics in a hydrologic modeling system.

Historically, hydrologists treated the vegetation on watersheds as passive wicklike evaporating surfaces, and calculated evapotranspiration as the residual of precipitation input minus measured streamflow, often computed at only monthly to annual intervals. Once an ET fraction had been established, a simple mass balance computation was applied to predict discharge for a watershed. For example, streamflow records were measured for a 185-km^2 watershed in Montana with annual precipitation of 152 cm and annual discharge of 109 cm giving a ratio of annual ET to precipitation of 0.28, or a runoff ratio of 72% (Running *et al.*, 1989). The RHESSys-modeled annual ET/precipitation ratio ranged from 0.26 to 0.32 for the different hill slopes of the watershed, producing both daily estimates and georeferenced spatial heterogeneity that simple mass balance approaches cannot provide.

Stream discharge is a particularly important variable for regional modeling as it is one of the few spatially integrated variables with direct validation data and documentation, namely, stream hydrograph records. Band *et al.* (1993) simulated peak seasonal stream discharge from the 15-km^2 Soup Creek watershed in Montana with an accuracy within 10 days of observations using RHESSys. White and Running (1994) expanded the approach to simulate daily discharge of the 2922-km^2 Middle Fork of the Flathead River of Montana, a mountainous region with evergreen forests, ranging from 1000 to 3000 m elevation and experiencing a range in precipitation from 750 to 1400 mm year^{-1}. River discharge varies from 0.6 mm day^{-1} in February to 8.6 mm day^{-1} in June. Spring snowmelt provides 78% of annual stream discharge during the April–July melt season. White and Running (1994), using RHESSys, simulated river discharge with a correlation of $r^2 = 0.80$ between predicted and observed daily values. Wigmosta *et al.* (1994) built a similar modeling system with strongly coupled hydrologic and ecological processes, and tested it on the Middle Fork of the Flathead River basin. The predictions of daily river discharge produced an r^2 of 0.91 with observed values, and the estimates of total annual runoff were within 7% of that measured. In these studies, evapotranspiration, photo-synthesis, dynamics of canopy water potentials, and other ecosystem processes were simulated by the mechanistic hydro-ecological models, but only stream discharge can be directly validated against daily measurements.

With RHESSys we can explore the hydrologic consequences of various landscape-level ecological disturbances that may have occurred in the past, be operating now, or be projected for the future. A longtime retrospective analysis was completed by Pierce *et al.* (1993) to evaluate the influence of historic land-use change on regional hydrology of the 1 million km^2 Murray–Darling basin in Australia from pre-European settlement times to the present. Increased localized salinization and waterlogging of soils were hypothesized to be the result of widespread forest clearing, with an estimated 12–15 billion trees removed since the 1700s (Walker *et al.*, 1993). RHESSys simulations for a selected sample area of 7750 km^2 suggested that ET has decreased by 10–45 mm month^{-1} in the Murray–Darling basin as a consequence of deforestation. The areas of largest hydrologic surplus, predominantly riparian valley bottoms, are now being considered for reforestation (Fig. 8.6).

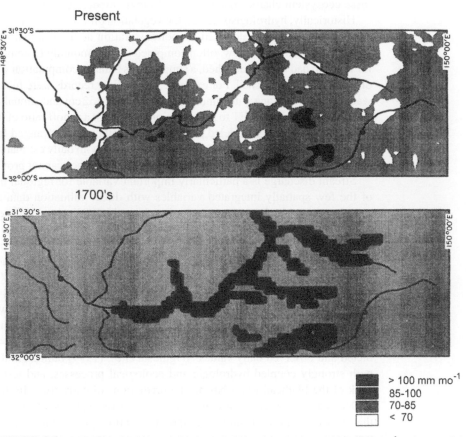

FIGURE 8.6. RHESSys simulation of the change in evapotranspiration within a 7750-km^2 study area of the 1 × 10^6 km^2 Murray–Darling basin of Australia. European settlement of the region 200 years ago replaced the native eucalyptus and acacia woodlands with farmland and pastureland, which greatly reduced evapotranspiration and increased the areas with waterlogged and saline soils. This simulation identified areas planned for tree planting to increase ET, which are predominantly riparian areas situated along the river networks. (Redrawn from Pierce *et al.*, 1993.)

As we think of a landscape in a more integrated and dynamic sense, we want to be able to predict what will happen to streamflow if portions of a watershed are harvested, burned, or planted with nonnative (exotic) tree species. Simulations of the Onion Creek experimental watershed in California predicted how seasonal stream discharge would react to partial cutting of various fractions of the watershed, as well as to complete harvest. The input data layers required by RHESSys (Fig. 8.7) show the topographic partitioning,

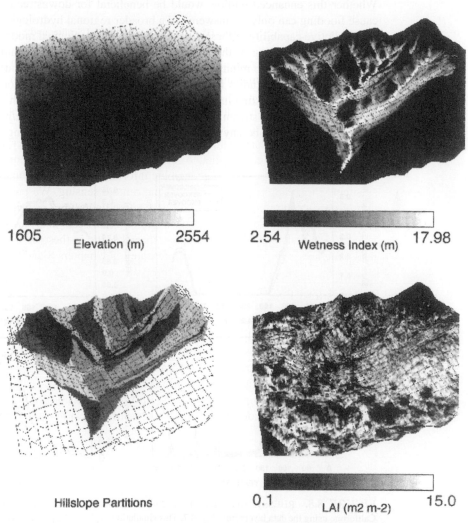

FIGURE 8.7. Topographic, soils, and vegetation data layers developed for the RHESSys simulation in Fig. 8.8 of the 25-km² Onion Creek watershed in the Sierra Mountains of California. The hill slope partitioning was based on algorithms developed by Band (1989a, 1993) to define landscape units with similar slopes, aspects, and elevations. The wetness index, based on the TOPMODEL logic of Beven and Kirkby (1979), represents the drainage capacity of different locations on the landscape. Leaf area index is calculated from Landsat TM data. (By R. Nemani, University of Montana.)

hydrologic drainage network, and current LAI of the watershed. Simulations with 100, 50, and 0% forest cover (Fig. 8.8) illustrated little sensitivity of wintertime ET and snowmelt processes to changes in forest cover. Growing season ET was proportionally reduced, and soil moisture (depicted as percent field capacity in Fig. 8.8) increased with reduction in forest cover. Forest removal was predicted to produce outflow events around year days 20, 50, 250, 300, and 350 that would not occur if full forest cover was maintained. Peak outflow increased from 12 to 18 mm day^{-1} and could contribute to downstream flooding. Whether this enhanced outflow would be beneficial for downstream irrigation or might cause flooding can only be answered in a broader regional hydrologic analysis. However, the predictive capability of these process-based hydroecological models can provide managers an understanding of the present state of landscapes on streamflow and encourage study of alternatives to minimize the possibility of implementing undesirable policies.

Many hydrology studies have illustrated that streamflows temporarily increase after extensive forest harvesting in a watershed, due to reduced ET (Bosch and Hewlett, 1982). The time required for streamflows to return to normal is termed *hydrologic recovery* and has been assumed to be anywhere from 30 to 100 years, depending on the climate and

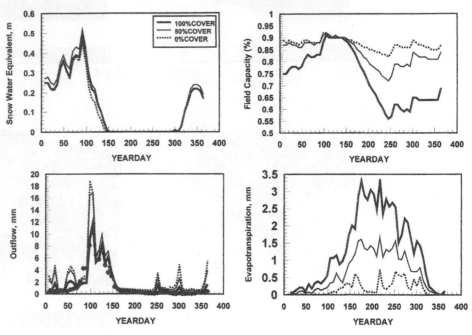

FIGURE 8.8. RHESSys simulation of the 25-km^2 Onion Creek watershed in the Sierra Mountains of California, using the data layers from Fig. 8.7. This simulation explores the potential consequences to the hydrologic budget of clear-cutting various fractions of the watershed, including estimates of seasonal peak outflow that could cause flooding. Points on the outflow graph represent measured daily stream discharge with the current 100% forest cover. Simulation tools such as RHESSys provide managers the ability to evaluate consequences of land management before any forest harvesting might be done. RHESSys can simulate different spatial patterns or temporal sequences of cutting, and check resulting potential streamflow responses for flooding during wet years or minimum flows during dry years. (By R. Nemani, University of Montana.)

postharvest recovery dynamics of the local forests (also see Chapter 2). Analyses with RHESSys have suggested that the true period of hydrologic recovery, generally corresponding to when the transpiring leaf area of the landscape has returned to preharvest conditions, may be much shorter (5–20 years), which could alleviate some management constraints on watersheds. For example, McKay and Band (1997) simulated a clear-cut in the Turkey Lakes watershed of Ontario by reducing the LAI from 7 to 1.5. Annual hydrologic discharge increased from 825 to 980 mm the year after clear-cutting. However, RHESSys simulated recovery of LAI to 7 within 10 years, at which point annual discharge returned to preharvest values (Fig. 8.9). Stem biomass recovery is not mechanistically related to hydrologic dynamics (and required >50 years for regrowth in this study), but it is often inappropriately used as the field measure of stand development for hydrologic recovery.

On longer time horizons, how would climatic change impact streamflows? Running and Nemani (1991) predicted the hydrologic consequences of a future climate change scenario of doubled CO_2, increasing air temperature by 4°C and precipitation by 10%, on Montana watersheds. Simulations suggested that snowpack, the source of most streamflow in the Rocky Mountains, would melt 19–69 days earlier in the spring, depending on topographic location. Average summer streamflows, however, would *decrease* by as much as 30% from current discharge levels, a change in hydrologic activity equivalent to the extreme drought year of 1988 in the western United States when some smaller rivers ceased flow by late summer for the first time in memory. Ironically, these RHESSys simulations showed forest productivity to be enhanced by the climate change scenario, partially because of the direct CO_2 response of photosynthesis and partly because of the longer growing season and higher soil water availability brought on by earlier snowmelt. Clearly, projecting regional ecosystem responses to interannual climatic variability or to more permanent changes requires a consideration of hydrologic and ecological interactions that may not be immediately obvious.

B. Watershed Biogeochemistry

Analysis of the biogeochemistry of a watershed requires not only a good hydrologic balance model as transport medium, but also good models of precipitation inputs, soil elemental cycling, and mobility (Aber *et al.*, 1993). Ollinger *et al.* (1993) built a model of atmospheric elemental deposition for the northeastern United States. Observations of precipitation were extrapolated to mountainous areas with elevational relationships similar to those discussed in Chapter 7. Elemental concentrations of nitrogen, sulfur, calcium, and potassium species were measured in the U.S. National Atmospheric Deposition Program. Spatial maps of elemental input were then produced by relating measured elemental concentrations with the precipitation of each landscape element, and summarizing these predictions into maps extending across the 11-state region.

Integrating watershed biogeochemistry with hydrology requires adding a mechanistic terrestrial nutrient budget to the model analysis. Creed *et al.* (1996) coupled the nitrogen budget of FOREST-BGC with RHESSys to calculate the seasonal dynamics of nitrate export from the Turkey Lakes watershed in Ontario, Canada. Creed *et al.* (1996) identified two mechanisms for seasonal nitrogen release (Fig. 8.10). The first is an N-flushing

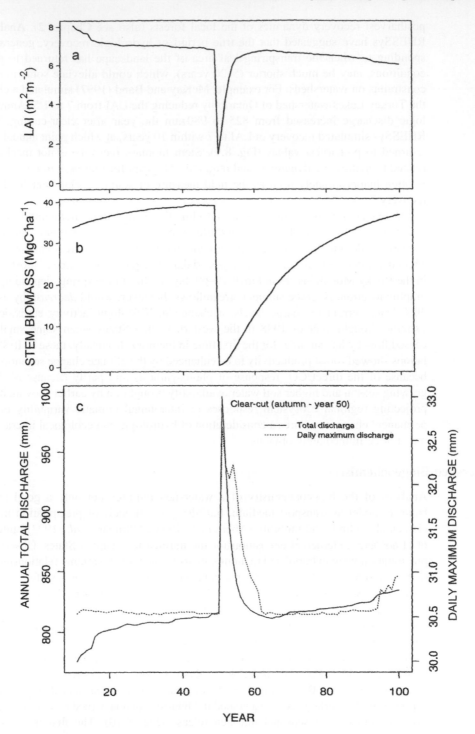

mechanism associated with snowmelt when soils become saturated. After a period of low N demand by forests during the winter, soil N enrichment occurs, so that saturated flows export large amounts of N in the spring. Second, during nonsaturated flow conditions, an N-draining mechanism transfers mobile N to lower soil horizons below the tree rooting zone, where continuous hydrologic base flow slowly releases N throughout the year. Removal of forest cover eliminates canopy interception and reduces uptake from the rooting zone, which accelerate N release from a watershed until the forest canopy is reestablished.

Elements can be transported in solution or in suspended sediments. The concentrations of these elements are a prime determinant of water quality. Hunsaker and Levine (1995) emphasized the necessity of accurately characterizing the entire land cover and land-use pattern of a watershed to represent elemental mass balances and their seasonal dynamics. The analysis confirmed the necessity of including the elemental contributions of land units well beyond the immediate proximity of the riparian zone.

C. Landscape Carbon Balances

Sustainable yield, whether of wood products or domestic and wildlife forage, is ultimately related to the net primary productivity of the landscape. Although our understanding of the components of an ecosystem carbon budget were well developed in Chapter 3, the implementation of that knowledge across a landscape is complicated by varying microclimates, soils, and stand conditions. Running *et al.* (1989) first used an early version of RESSys without hydrologic routing to simulate annual photosynthesis of the 1200-km^2 Seeley–Swan basin of Montana. The net photosynthesis computed across this complex mountainous landscape ranged from 9 to 20 Mg C ha^{-1} year^{-1}; however, CO_2 fluxes cannot be directly validated by observations. Our confidence in these regional calculations is based on comparisons of FOREST-BGC predictions against stand-level measurements (McLeod and Running, 1988; Korol *et al.*, 1991; Running, 1994). Because the early RESSys incorporated no hydrologic routing, these simulations are an example of an unconnected landscape analysis, that is, each cell was computed with complete independence from adjacent cells. Band *et al.* (1993) and White and Running (1994) incorporated the hydrologic routing of RHESSys into landscape simulations of photosynthesis and NPP. Carbon balances were improved at the bottom of hill slopes where water was received from lateral flow draining from above. This source of seepage water alleviated summer water stress and enhanced annual photosynthesis and NPP of the riparian vegetation.

The degree of complexity represented in a landscape simulation can influence the results. White and Running (1994) simulated annual NPP on the 414-km^2 McDonald

---

FIGURE 8.9. A 100-year RHESSys simulation of the 10-km^2 Turkey Lakes watershed in Ontario, Canada, explored the rate of hydrologic recovery after simulating a clear-cut of the native sugar maple forest at year 50 (McKay and Band, 1997). Although stem biomass did not return to precut levels for 50 years, LAI recovered within 10 years, causing annual stream discharge to also recover at a similar rate. Return of peak daily discharge to original levels, the critical variable for flooding, was only slightly slower. See Band *et al.* (1993) for additional details on methods. (From McKay and Band, 1997, *Hydrological Processes* **11,** 1197–1217. Copyright John Wiley & Sons Limited. Reproduced with permission.)

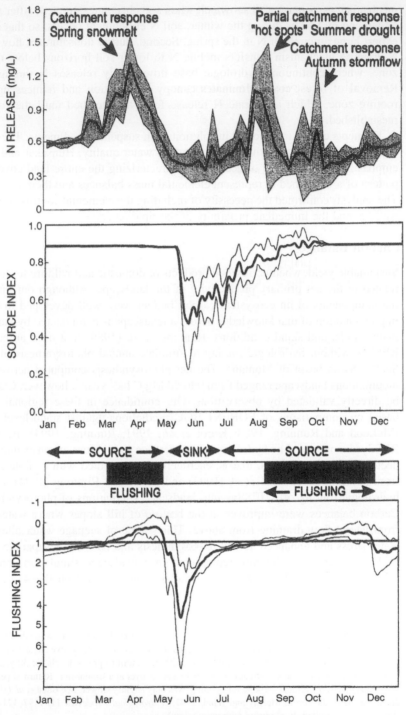

Catchment release of N is a function of the
CO-VARIANCE of the Source and Flushing Indices

watershed of Glacier National Park, Montana, using four techniques that progressively increased the detail defined on the landscape. The simplest landscape representation was unconnected 1-km raster (square) cells, and the most complex defined 855 hydrologically connected polygons in a RHESSys simulation. Final basin average NPP ranged from 7.04 to 7.55 Mg C ha^{-1} year^{-1} with the four methods of calculation, but the within basin NPP ranged from 2.5 to 15.5 Mg C ha^{-1} year^{-1}, with highest productivity on the warm, lower south slopes and lower productivity on high, north slopes with shallow soils. White and Running concluded that for broad landscape averages the simplified satellite-driven NPP calculations based on the Light Use Efficiency model from Chapter 3 would be adequate, but for site-specific management decisions the full RHESSys modeling analysis including hill slope unit definition was recommended. Pierce and Running (1995) also evaluated potential errors propagated by landscape aggregation, and they found that overestimates as high as 30% were possible if NPP was computed from broad $1° \times 1°$ cell attributes, as done by global scale models, compared to 1-km cells based on AVHRR data. Rastetter *et al.* (1992) suggested a number of statistical tools for dealing with such landscape aggregation errors.

These geographically explicit simulations of forest NPP quantify both the heterogeneity of the landscape and the environmental factors that limit tree growth, so could serve as the basis for improved computation of sustainable harvest levels in managed forests. Although direct field measurements are still the best means of determining local forest growth rates, ecosystem models can more accurately extrapolate stand data to the rest of the landscape and offer the additional advantage of exploring landscape responses to alternative management options.

Milner *et al.* (1996) applied a simplified RESSys logic to simulate the potential NPP of private forested lands in Montana for purposes of defining tax rates. Productive *potential,* not current productivity, was sought as a reference by the state government to encourage more intensive management of the more productive forest lands. The biophysical principles incorporated in RESSys were appropriate for defining ecological potential and did not require extensive data of current stand conditions. The final map produced from the regional analysis illustrates five productivity classes ranging from <1.7 to >5.9 m^3 ha^{-1} year^{-1} of stem wood (see Plate 6). Because the evaluation is based on established ecological principles from Chapters 2–4, the introduction of different species, genetically improved trees, fertilization, climatic change, or other changes in the ecosystem could all be incorporated in a reevaluation if desired. Also, it was estimated that the statewide analysis of 14,200 km^2 of forested land was done by biophysical model analysis for a quarter of the cost and half the time that an intensive field inventory based method would have required.

FIGURE 8.10.   RHESSys simulation of seasonal nitrate release in streamflow from the Turkey Lakes watershed in Ontario, Canada. Creed *et al.* (1996) hypothesize two mechanisms of nitrate release, an N-flushing dynamic during saturated soil conditions during spring snowmelt and autumn storm flow, and a second release during midsummer when mobile N has drained below the rooting zone and slowly enters into the groundwater. The source index defines the balance between the supply of available N from mineralization relative to the demand for N by plant uptake. The flushing index quantifies the availability of soil water as a mechanism for N transport. (From Creed *et al., Water Resources Research* **23,** 3337–3354, 1996, published by the American Geophysical Union.)

D. Regional Biogeochemical Applications

1. Glacier National Park

A multiresource analysis in Glacier National Park, Montana, that incorporated RHESSys modeling to link wildlife, hydrologic, and vegetation studies together was designed to obtain a wider range of analytical capability for land management. Site physical data on topography and soils, stand structural data, climatic measurements, and snow and stream dynamics were organized to provide input data layers for RHESSys, and output variables were selected for which validation data were available. The McDonald and St. Mary's drainages represent topographically, climatically, and ecologically diverse landscapes of two connected watersheds which cross the Continental Divide. Plate 2 illustrates how digital elevation data and soil physical data were incorporated in the TOPMODEL analysis to provide soil transmissivity and saturation indices for hydrologic routing in RHESSys, and soil water holding capacity, for initializing FOREST-BGC. Plate 7 shows the geographic location of the study area. An SR (Simple Ratio, Chapter 7) was computed from Landsat TM data, and a nonlinear transformation to LAI was made on the basis of field plot measurements throughout the study area (White *et al.*, 1997a). The forest community types were defined with a fuzzy logic classifier combining observed topographic relationships with Landsat spectral data in red, near-infrared, and mid-infrared wavelengths.

Eight-year RHESSys simulations of carbon and water balances were executed for the McDonald watershed applying the input data layers indicated in Plates 2 and 7, and selected variables were validated with field plot data (see Plate 8). Aboveground NPP was highest at low to mid-elevations on forests that had not been disturbed recently by wildfire or avalanches. High elevations and north slopes were more energy limited and typically had shallow soils and slow snowmelt, conditions that restrict the growing season. A growth efficiency map, defined from the principles in Chapter 5, clearly separated the oldest stands, with lowest wood production/LAI, from the younger stands with higher efficiency. This map could serve to delineate areas with high hazard rating that are particularly susceptible to mountain pine beetle outbreaks. The inserted validation graphs present observed and modeled 8-year stem growth increments, translated into NPP units. All plots measured in the drainage had highest NPP in 1987 as a result of warmer than average temperatures and average precipitation for that year.

The annual snowpack map (Plate 8) shows the areas of highest snow accumulation, and the inset validation graphs show predicted versus observed monthly snow water equivalents from snow course survey data. The northern Rocky Mountains have a strong snowmelt-dominated seasonal hydrology, illustrated by the peak stream discharge around year day 140 and near minimal discharge from year days 1–80 and 280–365 when temperatures are nearly continuously below freezing. The persistent underestimate of simulated stream discharge may be a result of contributions from melting glaciers in the upper reaches of the watershed that were not accounted for by RHESSys.

In Chapters 5 and 6 we learned that projecting a forest forward in time over centuries requires analysis of forest processes in stand development, community dynamics, and disturbance activity. In Chapter 6 we introduced a model that combined the ecophysiology of FOREST-BGC with the stand dynamics and disturbance timing of a gap model,

FIRE-BGC (Keane *et al.*, 1996a,b). The stand-level FIRE-BGC model was embedded in a spatial modeling framework that simulates probability of ignition on the basis of historic fire frequency, fire behavior and spread rates associated with fuel loading, topography, and microclimate, and tree mortality, with mortality being computed from simulations of fire intensity and duration. Seed dispersal was then computed as a probability to estimate the species composition and spatial distribution across the regenerating landscape.

With this multifaceted modeling system we can now ask real-world land management questions, such as what might the Glacier National Park landscape look like 250 years from now as climate changes, and how will fire management policy influence that result? Furthermore, how will the landscape differ if a complete fire suppression and exclusion policy is practiced, as has predominated for the last century, or if managers allow a more natural fire frequency to return for the next 250 years? These important questions are made more difficult to answer because the natural fire frequencies measured in the field by analyzing fire scars are based on historic climate. What would a new natural fire frequency be in a different climate, and how would fire intensity, a highly climate-sensitive variable, change? Keane *et al.* (1997) attempted to answer these kinds of questions by simulating fire and ecosystem responses for the same Glacier National Park landscape depicted in Plate 7 for 250 years first with historical climate and full fire suppression, then with a potential future climate and the resulting natural fire frequency. The current climate for these simulations was defined from 44 years of measured weather data replicated for a 250-year simulation with topographic microclimate variability, including humidity and solar radiation computed from MT-CLIM (see Chapter 7). A climate change scenario was evaluated, based on global circulation model predictions for the twenty-first century that would increase air temperatures by 2°–5°C and precipitation by 25–30%, with a doubling of atmospheric CO_2 levels.

The number of fire ignitions were predicted to not change with the future climate, but fire intensities were predicted to triple and the extent of burning to double under the future climate, with some areas being burnt up to 7 times in 250 years (see Plate 9). This dramatic change in fire dynamics resulted because of higher NPP and forest encroachment into areas that were meadows. The increased accumulation and more contiguous distribution of fuels allowed fires, once started, to burn more intensely over wider areas. The rather complicated interactions and their implications could not be foreseen without comprehensive simulation modeling. Although we cannot confirm these simulations 250 years into the future, the results, once presented, seem reasonable, based on our understanding of this system and the detailed information provided in the multidimensional analysis.

2. Columbia River Basin

The 820,000-km^2 interior Columbia River basin of the Pacific Northwest, which encompasses parts of eight states, was chosen in 1993 to be the test bed for developing ecosystem management principles for federal land managers in the western United States. The Columbia basin is hydrologically the most important river drainage in the northwestern United States in terms of hydroelectric and irrigation activity, and it encompasses ecological diversity from some of the most productive forests to one of the most desolate deserts in the United States. At this broad regional scale, the management questions were as follows: What is the primary productivity of this region; has it changed in the last century with increased exploitation and management, and is current productivity *sustainable*? What risks of accelerated ecosystem disturbance are likely from insect, disease, or fire activity that may influence sustainable productivity? How do land managers balance the multiple interests of hydrologic needs for irrigation, hydroelectric power, and anadromous fish habitat?

A RHESSys simulation of daily carbon and water balances was executed at 2 km spatial resolution for the entire basin. To evaluate interannual variability, 3 years of weather station records were selected for the simulations: 1982, 1988, and 1989, representing a cool-wet, hot-dry, and average year climatically. The DAYMET model, a spatial version of MT-CLIM, provided extrapolation of climate data for the entire basin (Plate 3; Thornton and Running, 1996). Both current and historical land cover of all biome types including forests, shrub lands, grasslands, and crops were defined, to analyze changing productivity resulting from land-use conversions. Forest stand ages, species mixes, density, and structure were estimated for each cell from permanent plot records. We must recognize, however, that 205,213 cells were defined for this regional simulation, so parameter values are based on relationships defined among variables and automatically assigned to a cell; each cell could not be individually considered.

The annual carbon budget for this region (see Plate 10) shows great seasonal variation in photosynthesis and respiration resulting from the cold temperatures and short day length of the winter season contrasting with strong activity in the early spring. The mid-growing season period of July and August represents the time of highest temperatures and maintenance respiration, but not the time of highest photosynthesis, because water stress and low humidity limit stomatal responses and canopy gas exchange (Chapters 2 and 3). Many of the areas of highest gross photosynthesis (dense forests of high LAI) also have high maintenance respiration, so the final annual net photosynthesis (and NPP) is rather similar across biome types in similar climatic zones. The difference between NPP of the historic and current landscape was not great, as NPP is controlled more by climatic regime than the vegetation type unless a site has been severely and recently disturbed. Annual NPP was 8% higher in the cool-wet year of 1982 and 37% lower in the hot-dry year of 1988, compared to the normal year 1989 shown in Plate 10.

This analysis documents that the dominant factor controlling photosynthesis and productivity in this region is water more commonly than energy, except for cooler, wetter sites in the high mountains where temperature limits the growing season. The carbon balance simulations were summarized into a carbon stress index, the ratio of maintenance respiration to gross photosynthesis, as a simple measure of carbon availability for plant growth and defense. This carbon stress index, which is the regional equivalent of the stand-level growth efficiency index, identifies forests most susceptible to insect/disease attack (Chapter 6). Older stands on drier sites, with high respiration costs but water-limited photosynthesis, had highest carbon stress indices, and hot-dry years such as 1988 accentuated stress by both raising respiration and lowering photosynthesis. Grasslands and shrub lands tend to have low carbon stress indices as a result of less investment in respiring tissue.

The seasonal hydrologic dynamics of the Columbia basin are clearly dominated by spring snowmelt (see Plate 11). Although precipitation is fairly evenly distributed seasonally, it is accumulated as snowpack from November to February, snowmelt and hydrologic outflow are concentrated in the period March to May, and virtually no outflow occurs from July through October, except from the high mountain watersheds. Daily evapotranspiration far exceeded precipitation during the early growing season until stomatal restrictions limited ET in July and August. Soil moisture was depleted from April to August, with some autumn recharge before snow began to accumulate (Chapter 2). Most of this region could not support cropland without reservoirs for water storage and irrigation during the summer period, but anadromous fish migration requires late summer water flow in the lower Columbia River which causes a conflict with demands for irrigation water. Average basin river outflow was 38% higher during the cool-wet year 1982, and 20% lower during the hot-dry 1988 year than the normal 1989 climate year shown in Plate 11.

It is evident that analysis of an 820,000-km^2 region can only provide information useful for defining broad policy objectives and should not be used for management of specific sites. The key criteria for whether a regional model analysis can support detailed versus only general interpretations lies in the accuracy with which landscape variables were assigned to individual cells. The assumptions, generalities, and inaccuracies involved in defining site and stand parameters for hundreds of thousands of cells for this simulation reduces accuracy at the local level. Although it may be simple to define the stem biomass for a stand-level simulation of FOREST-BGC from plot inventory data, we lack procedures or data to assign accurate stem biomass for 820,000 km^2. Nevertheless, some valuable analyses are possible. Where are the most and least productive landscapes in the region? How much irrigation water or hydroelectric production may be possible in a record drought year? Where are the most logical areas for irrigated agriculture, intensive forest management, and other land-use alternatives? For example, NPP of forested land in the Columbia basin simulated in Plate 10 could be translated into the sustainable harvest, or "annual allowable cut," of wood products for the region. Details of where the harvest should take place would require more precise watershed scale simulations using field inventory data. The basic principle operating here is that the final accuracy obtained is less a function of the model's formulation than it is of the accuracy with which parameters are defined for the simulations. Planning and management of specific areas are usually more appropriately accomplished at the scales represented in Plates 7–9 because field data for points throughout the watershed can be used to initiate the models and obtain more accurate simulations.

E. Validating Regional Biogeochemical Models

We now have the capability to calculate regional biogeochemistry over millions of square kilometers; however, there are very few large-scale data sources for validating these simulations. Although a wealth of plot-level data is available for small-scale validation tests, as discussed in Chapters 2–6, validation activity specifically designed for large areas is rare. The most commonly available regional scale validation data are stream discharge records from which ET can be computed by mass balance and predictions of stream discharge can be compared, as presented in Figs. 8.7 and 8.8 and Plates 7 and 10. Over scales of a few kilometers, tower-based measurements of micrometeorological fluxes of energy, water vapor, and CO_2 can record diurnal to seasonal changes in forest fluxes (Baldocchi *et al.*, 1988; Desjardins *et al.*, 1992a,b). At larger scales, aircraft mounted with fast response gas analyzers can detect varying flux patterns across a landscape (Matson and Harriss, 1988; Sellers *et al.*, 1992a,b). Ray Desjardins (personal communication, 1996) showed high correlation between a radiometric vegetation index and aircraft-measured CO_2 flux rates across a 500-km transect of the Canadian boreal forest (Fig. 8.11). At still larger scales, the global atmospheric CO_2 station data can be used together with a transport model for terrestrial process validation, and will be covered in Chapter 9 (Hunt *et al.*, 1996). This difficulty of large-scale validation reemphasizes the importance of doing regional analysis with ecosystem models that have been tested at the stand level where the principles of ecosystem operation are well understood and model results can be compared directly against field measurements.

V. SUMMARY

Analysis of forest ecosystem processes is now possible at regional scales by integrating stand-level models with georeferenced databases. Remote sensing is an essential element of regional ecosystem analysis, as it provides complete and consistent measures of the landscape, but only for certain variables that can be measured with this technology. Satellites provide landscape maps of biome type and LAI, as well as regular monitoring of key

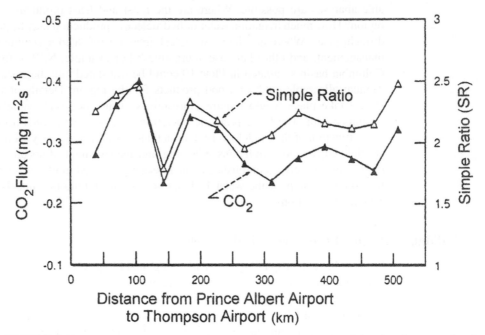

FIGURE 8.11.   Regional scale measure of forest CO_2 flux measured by low-flying aircraft, compared to the Simple Ratio of NIR/Red remote sensing reflectance. The SR here is proportional to the land surface vegetation density (see Chapter 7). The BOREAS study area is in central Canada. Satellite, aircraft, and tower-based measurements will be required to provide validations for regional scale ecosystem simulations. (From Ray Desjardins, Agriculture Canada, personal communication, 1996.)

conditions such as drought and fire danger. Regional models simulate forest biogeochemical responses and stand dynamics at space and time scales relevant to land management. Predictions of wildlife population changes, recovery of hydrologic balance, and shifts in landscape productivity can now be simulated for a range of situations that include past, present, and an array of future conditions. As a result, alternative management policies can be evaluated before actions are implemented. Accurate georeferenced databases, however, are not yet generally available for many forested regions, which makes it difficult to initialize these integrated landscape models. Additionally, we lack ways of validating the current predictions of regional models because of the paucity of ecological data acquired at this scale, beyond that which can be obtained through remote sensing.

CHAPTER 9

The Role of Forests in Global Ecology

I. INTRODUCTION

Forests cover approximately 30–40% of the vegetated area of Earth. If it were not for human and natural disturbances, forests might be expected to occupy most landscapes where annual precipitation exceeds about 25 cm, excepting polar regions where energy limitations constrain vegetation growth (Neilson, 1995). Global ecological studies frequently identify forests as major contributors to a host of important biospheric processes (Vitousek, 1994). Boreal forests are of particular interest because they are expected to undergo the greatest climatically induced change in the twenty-first century (Bonan *et al.,* 1992). Temperate forests are currently hypothesized to provide a major CO_2 sink that helps stabilize the global carbon balance (Tans *et al.,* 1990; Ciais *et al.,* 1995; Keeling *et al.,*

1995; Turner and Koerper, 1995). At present, tropical forests are believed to be undergoing the most rapid conversion to other land uses (Skole and Tucker, 1993; Houghton, 1994). Given this prominence of forests in many aspects of global science, one might expect that much detailed information would be available on their distribution and composition. The inherent physical magnitude of global scale questions, however, makes addressing even such obvious questions difficult.

To illustrate the challenge facing global scale assessments, scientists have yet to measure with accuracy the vegetated land surface of Earth; values range from 120 to 140 million km². Estimates of forest land cover are even more variable, ranging from 30 to 75 million km² (Townshend *et al.*, 1991). Some of this uncertainty results merely from disagreements among scientists on definitions of what is forested land, particularly along transitional zones where forests grade into grasslands in arid environments, or into tundra in boreal regions. Most of the uncertainty arises, however, from our inability to make accurate measurement over millions of square kilometers. Before the era of satellites, global estimates of forestland were derived from exhaustive compilations of national mapping surveys (Matthews, 1983; Olson *et al.*, 1983). Obtaining consistent, accurate mapping of global forests is only now becoming possible with new satellite systems. New models incorporate satellite and ancillary data sources to distinguish forests and other types of vegetation (Prentice *et al.*, 1992; Neilson, 1995; Woodward *et al.*, 1995) and to estimate regional and global scale biogeochemical activity (Melillo *et al.*, 1993; Schimel *et al.*, 1994; Field *et al.*, 1995; Hunt *et al.*, 1996).

This chapter summarizes current analyses of the distribution of global forests and their interactions with the biosphere. It also previews technical advances likely to increase our understanding of the role that forests play globally. To expand our analysis to a global perspective we continue on the logical path set in Chapters 7 and 8, where we seek further to simplify the kinds of questions asked, reduce the detail of data required, and select appropriate tools for the new endeavor. From a functional viewpoint, we emphasize the role of forests in global biogeochemistry and climate dynamics, but we will also consider the importance of forests in protecting global biodiversity and in sustaining a natural resource base for supporting cultural and economic goals.

II. GLOBAL FOREST DISTRIBUTION

A. Paleoecological Evidence of Past Forest Distributions

Paleoecological studies of tree pollen and fossilized wood in peat bog cores have documented that global forest cover changed dramatically as climate has changed, with particularly strong evidence from the last 20,000 years. Earth's climate has undergone a transition from glacial to interglacial conditions in the last 20,000 years that is as large as any climate fluctuation in the last 3,000,000 years (Webb and Bartlein, 1992). This recent and rapid climatic change has allowed pollen counts, dendrochronology, and carbon isotope analyses on ancient samples of wood to provide an accurate picture of the rate at which forests change in their distribution in response to climatic variations. From these paleoecological sources, for example, we know that the geographic distribution of tree species of the eastern North American forest has moved hundreds of kilometers in the last

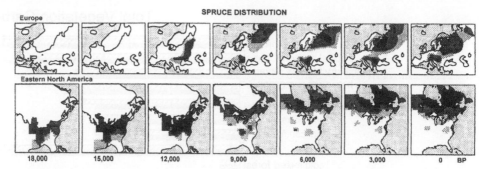

FIGURE 9.1. Changing distribution of spruce (*Picea*) during the past 18,000 years (BP, before present) in Europe and eastern North America. These maps are derived from radiocarbon-dated surface and fossil pollen records interpolated with climatic data and graphed to 100-km^2 cells. Shading refers to pollen abundance: black, >20%; medium, 5–20%; light, <5%. (Redrawn from Prentice *et al.,* 1991, and Webb and Bartlein, 1992.)

18,000 years as temperature and precipitation patterns changed. As the last major ice age ended, temperatures have warmed by at least 5°C and caused continental glaciers to melt, with species like spruce expanding their range slowly into areas once covered by ice (Fig. 9.1; Prentice *et al.,* 1991). Arctic timberline in northwestern Canada was 350 km further north during the warm mid-Holocene period 8500 to 5500 years ago relative to the current forest–tundra boundary (Ritchie *et al.,* 1983). However, there is a lag between the time of climatic change and the response of forest distribution because of the slow pace of tree mortality, seed dispersal, and regeneration. Gear and Huntley (1991) estimated a maximum migration rate of Scots pine in Scotland of about 0.5 km year^{-1} as Scots pine first moved north 70–80 km, then retreated back south during the last 4000 years in response to regional temperature changes. Prentice *et al.* (1991) estimated that it took 1000–1500 years for the distribution of spruce in Fig. 9.1 to equilibrate to a new climate.

Altitude acts as a surrogate for latitude in providing a climatic gradient for forests to move across. There is convincing evidence that alpine timberline in the Rocky Mountains and European Alps was 50–200 m higher during the mid-Holocene period than it is today. Movement of the timberline boundary is a complex response to both episodic trends and occasional extremes in a host of climate factors, making simple predictions of forest responses to future changes difficult (Graumlich and Brubaker, 1995).

These historical reconstructions give a sense of how climate controls the long-term distribution of forests over thousands of years, and they allow us to explore with models how responsive the terrestrial biosphere might be to future climatic change. Foley (1994) used a GCM (Global Circulation Model, often called a Global Climate Model) to simulate the mid-Holocene climate of 6000 years before present (BP) and then contrasted simulated global terrestrial carbon balances with present-day climate using a biospheric model. He concluded that a 20% decrease in carbon storage in boreal forests was offset by an increase of 20% in the area occupied by arid forest savanna, resulting in almost no change in total global terrestrial carbon storage (Fig. 9.2).

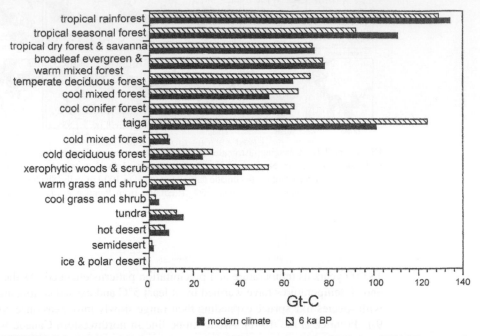

FIGURE 9.2. Simulation of global terrestrial carbon storage in the mid-Holocene (6000 BP) compared to the present. Comparative simulations were run with a global biosphere model using estimated 6000 BP climate and a present-day climate. Results suggest that tropical forests have gained carbon storage from 6000 BP to the present as continental monsoon rains strengthened. Temperate and boreal forests have lost carbon during the last 6000 years, primarily because the boreal forest boundary has retreated 300–500 km south of its northern limits in the mid-Holocene. (From J. A. Foley, *Global Biogeochemical Cycles* **8,** 505–525, 1994, published by the American Geophysical Union.)

B. Bioclimatic Definition of Potential Forest Distribution

Detailed analysis of specific climatic variables combined with additional knowledge of general tree physiology has allowed biogeography models to quantify the critical environmental limits for vegetation distribution. Trees, being perennial organisms must first endure the extremes of their environment in order to persist. Prentice *et al.* (1992) and Haxeltine and Prentice (1996) used absolute minimum air temperatures to discriminate boreal, temperate, and tropical forest ranges in their BIOME2 and BIOME3 biogeography models. Neilson (1995) used a combination of annual growing degree-days and monthly minimum temperatures to define boundaries between thermal zones (Tables 9.1 and 9.2). Woodward (1987) summarized the cellular mechanisms of cold tolerance physiology that are the basis for interpreting how minimum temperatures control global forest distribution, building on the original work of Sakai and Weiser (1973). Tropical trees are unable to withstand ice crystal formation within living cells. Temperate trees are able to withstand temperatures down to −45°C (where unbound water freezes spontaneously) by expelling

TABLE 9.1

Lethal Minimum Temperatures (°C) That Limit the Geographic Distribution of Various Woody Plant Forms[a]

Plant form	Lethal minimum temperature (°C)
Broad-leaved tropical deciduous (rain-sensitive)	0 to 10
Broad-leaved evergreen (frost-sensitive)	0
Broad-leaved evergreen (frost-resistant)	−15
Broad-leaved temperate deciduous	−40
Broad-leaved boreal deciduous (e.g., *Betula, Populus* spp.)	No limit
Needle-leaved evergreen (e.g., *Agathis, Araucaria*)	−15
Needle-leaved evergreen (temperate taxa)	−45
Needle-leaved evergreen (boreal taxa)	−60
Needle-leaved boreal deciduous	No limit

[a]From data summarized in Woodward (1987).

free water from cells during development of cold hardiness in autumn, and by binding the remaining water to cell walls during winter severe freezing conditions. Boreal species exhibit special protective mechanisms that allow them to withstand freezing in liquid nitrogen at −196°C when in their full cold hardened state, and to repair cellular damage in the spring, which are fascinating physiological feats (Havranek and Tranquillini, 1995).

Beyond adapting to different temperature limits, trees require a certain length of growing season to complete their morphological development and an adequate supply of water to support sufficient leaf area required to maintain a positive annual carbon balance. The summed number of growing degree-days above a minimum temperature level, usually 0°C, is used to quantify the overall "heat requirements" for tree development and, more significantly, to define the transition zone (*ecotone*) between tundra and boreal or alpine forests (Table 9.2). Although 25 cm is estimated as an approximate lower limit of annual precipitation required to support forests, a more precise definition is obtained by computing a water balance between precipitation and potential losses in evapotranspiration

TABLE 9.2

Definition of Global Biome Boundaries with a Combination of Limiting Minimum Temperatures and Temperature Thresholds for Morphological Development[a]

Boundary	Calibrated temperature threshold	Physiological interpretation
Tundra–taiga/tundra	735 growing degree-days	Short growing season (frost desiccation)
Taiga/tundra–boreal	1330 growing degree-days	Short growing season (no reproduction)
Boreal–temperate	−16°C (average monthly)	Supercooled freezing point (−45°C)
Temperate–subtropical	1.25°C (average monthly)	Annual hard frost (24 hr < 0°C)
Subtropical–tropical	13°C (average monthly)	No annual hard frost

[a]From Neilson (1995).

(Stephenson, 1990; Neilson, 1995). The transition from arid grasslands and shrubs to forest is often gradual; savannas support a variable mix of trees, grasses, and shrubs. An increase in the water availability favors more trees in savannas, but the seasonal distribution of precipitation and ET are also important. Haxeltine and Prentice (1996) were able to delineate savanna from grassland by defining which of the two types had the highest modeled NPP in a particular area. Once sufficient water is available to support some kind of forest, additional water provides greater development, particularly in the amount of leaves (LAI) and the duration of their seasonal display, as discussed in Chapter 2 (Grier and Running, 1977; Woodward, 1987).

It is important to recognize that climatic constraints alone only help to define *potential* forest cover. The actual distribution of forests is heavily influenced by both natural and human-induced disturbances. In Chapter 6 we considered that forest development over time rarely progresses to a hypothetical "steady-state" or "climax" condition because of frequent disturbances. These same mechanisms operate at continental scales to influence the geographic distribution of forests. The tall grass prairie of the North American Great Plains normally receives sufficient precipitation to support forests, yet periodic droughts, frequent fires, and historical grazing by bison herds maintained a grassland until European immigrants arrived and established what is now the wheat belt of the continent (Neilson *et al.,* 1992). Until the advent of satellites, of course, we had no means of obtaining a consistent measure of the *actual distribution* of vegetation on Earth.

C. Satellite Estimates of Current Forest Distribution

Since the launch of the Landsat series of satellites in the early 1970s, numerous remote sensing studies have quantified local land cover and provided quantitative estimates of the conversion of forestland to other uses (Peterson and Running, 1989; Loveland *et al.,* 1991). D. P. Turner *et al.* (1993) calculated that 45% of the potential forest cover of the continental United States has been converted to other land cover types. There has been much interest in defining tropical deforestation rates by satellite (Skole and Tucker, 1993; Skole *et al.,* 1994) to compare with temperate forest cutting rates (Chapter 8). Such analyses have to date been largely restricted, however, because of the difficulties in dealing with the huge volume of data required and the lack of an accurate means to distinguish forest cutting from large-scale disturbances such as fire or the conversion of forestland to other uses. Nemani and Running (1995) produced a satellite-derived estimate of present global forest cover, 52.6 million km^2, which lies near the middle of the range of 30–75 million km^2 cited by Townshend *et al.* (1991). Their analysis of global forest distribution was 49% tropical, 29% temperate, and 22% boreal (see Plate 12).

New processing of AVHRR data using visible, near-infrared, and thermal channels has also allowed the first mapping of global land cover *change.* The vast majority of global land cover change has been attributed predominantly to forest conversion to agricultural fields, pastures, and urban areas (Meyer and Turner, 1994). Nemani and Running (1995) first computed the potential land cover from climatic data with a simple biogeography model based on principles similar to those in models developed by Prentice *et al.* (1992) and Neilson (1995). Next, an annual sequence of biweekly AVHRR data was processed to define existing global land cover based on procedures outlined in Running *et al.* (1995).

By comparing the two maps a third could be derived that distinguished potential distribution of vegetation from the existing pattern, which emphasized regions of major land cover conversion (see Plate 13).

The surprising results were that only 26% of potential tropical forests have been deforested, substantially less than the estimates of nearly 50% given by Myers (1993). The analysis in Fig. 9.3 suggests that 40% of temperate forests have been altered, which is close to the 45% estimated by Turner and Koerper (1995) for U.S. forestlands. In boreal forests, the estimated reduction in area from the potential was only 20%, which probably reflects the remoteness and generally low adaptation of these ecosystems for other uses. In contrast, much of the once vast forests of Europe have been converted to agriculture and urban use. In this analysis, young, regenerating forests also are interpreted by the satellite as a land cover change, even if the change is temporary, which illustrates the importance of repeating land cover change analyses at regular intervals. The most concentrated regional change has occurred in Southeast Asia and India, where dense tropical and subtropical forests have been cleared for extensive agriculture, a conclusion verified by Dale (1994) from ground-based data. New EOS satellites launched in 2000 provide more precise and regular mapping of global forest land cover (Plate 14).

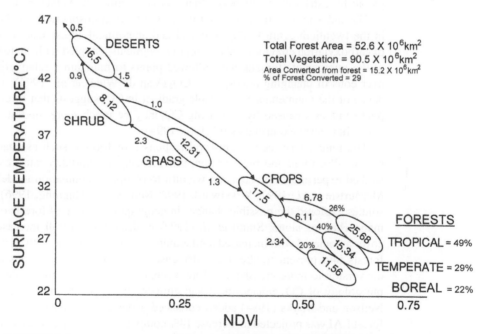

FIGURE 9.3. Areal changes in land surface area for each biome type from the land cover change map in Plate 13. The change trajectory is plotted on the NDVI versus surface temperature axes used by Nemani and Running (1995) to define biome types with satellite data. The numbers indicate areas in $10^6 km^2$, except where percentages are noted. Numbers in the ovals are current land area in each biome type, and labeled arrows are the change in biome area from the maximum potential.

D. Future Forest Responses to Climate Change

The future distribution and activity of forests will be controlled predominantly by changes in climate, forest response to elevated CO_2, and land cover changes associated directly with human activities. Although changes in forest land cover are controlled substantially by socioeconomic factors that cannot be easily predicted, forest responses to potential climate change and CO_2 concentration can be assessed with various computer simulation models. As we covered in Chapter 3, the first response of increasing atmospheric (CO_2) will be increased photosynthetic rates and partial closing of stomata producing higher water use efficiency. An extensive literature review by Ceulemans and Mousseau (1994) of studies done with exposures of less than 2 years found that a doubling of CO_2 from 340 to 680 ppm enhanced photosynthesis rates by 40% in conifers and by 61% in broadleaf trees; however, the range of responses is large. The mean estimated increase in growth was 38% for conifers and 63% for broadleaf trees.

Evidence suggests that the long-term response of trees to rising CO_2 may not be nearly so large as the short-term response. In the last 200 years, atmospheric CO_2 concentrations have risen from about 280 to >360 ppm (Keeling *et al.*, 1995) as a result of land clearing and combustion of fossil fuels. The isotopic composition of C in the atmosphere has changed over recent decades and continues to change. Lloyd and Farquhar (1994) estimate $\delta^{13}C$ discrimination of photosynthesis to be 17.8‰ for C_3 plants that produce 79% of global terrestrial photosynthesis. Plant tissues, with $\delta^{13}C$ values predominantly between −27‰ and −32.5‰, have lowered the $\delta^{13}C$ of the atmosphere from about −6‰ to −8‰ in the twentieth century. Physiological and anatomical adjustments in the concentration of carboxylation enzymes and chlorophyll pigments, coupled with anatomical adjustments of the number of stomata, have allowed plants to maintain a relatively stable c_i/c_a ratio over eons of changing atmospheric CO_2 (Van de Water *et al.*, 1994). The historical evidence of the maintenance of a stable gradient in CO_2 suggests that future increases in net photosynthesis caused by increasing CO_2 may be much more modest than that inferred from short-term experiments (Gifford, 1994).

The longer term ecosystem-level responses of forests, such as changes in leaf area, carbon allocation, and regeneration, senescence, and mortality rates, cannot be as easily studied experimentally, so again we turn to computer simulation models (Bazzaz, 1990; McMurtrie *et al.*, 1992; Woodward, 1992; Neilson and Running, 1996). Initial modeling studies explored the possible changes in geographical range of forests on the basis of climatic changes alone. Smith *et al.* (1992) projected an overall increase in global forest cover associated with increased temperatures and precipitation, with an extension of boreal forests into present tundra and a decrease in the area of dry forests as drought causes conversion to more grasslands. More mechanistic biogeographic models now include the physiology of CO_2 enhancement and climate change in the biome redistribution. When Neilson and Marks (1994) added enhanced water use efficiency to their model analysis, forest LAI was projected to increase 14% compared to current climate conditions. McGuire *et al.* (1993) also illustrated that forest response to climatic changes may be highly variable in different parts of the world. High latitude forests, where development is limited by incident radiation, day length, and temperature, may accelerate in growth, while tropical forests may endure higher water stress and respiration losses.

Longer term simulations typically combine both direct CO_2 effects on tree physiology with climatic change scenarios from GCMs. A more comprehensive model analysis by VEMAP (Vegetation/Ecosystem Modeling and Analysis Project) developed climate chance scenarios from GCM simulations that were coupled with biogeographical models to compute shifts in biome distribution, which then were used in biogeochemical models to project final changes in NPP (VEMAP, 1995). Forest cover in the United States under the present climate was simulated with three biogeographic models to fall within a uniform range of 42–46% of the total land area. Simulations of future forest cover were much more variable, however, ranging from a reduction to 38% of the present area to an increase up to 53% of the area (see Plate 15). Likewise, although current NPP of the United States was calculated to be between 3.1 and 3.8 Pg year^{-1} with the three biogeochemical models, when simulations were made that incorporated changed climate, CO_2, and biome redistribution, predictions of increases in NPP ranged from 2 to 35% (Plate 15). These differences in model predictions primarily result from our inadequate understanding of the balance between water and nitrogen limitations in controlling carbon allocation.

Although the predictions from the VEMAP analysis should be interpreted cautiously, there were some important general conclusions derived from the project. First, the levels of uncertainty regarding future climates, future biome distribution, and future ecosystem biogeochemistry were roughly equal, suggesting the scientific understanding of each of these subjects is similar. This analysis probably brackets the possibilities of change in NPP, and all model combinations suggest at least a modest increase, although less than the NPP enhancements derived from short-term physiological experiments. Finally, because NPP is a conservative ecosystem property, important alterations in forest composition may still occur and not be noticed until all representatives of a functional group are lost.

The replacement of one species with one better adapted to a changed climate has historically taken many centuries, and, in the process, an increase in wildfires and insect outbreaks may cause much damage to surviving forests (Clark, 1988; Cannell *et al.*, 1989). Flannigan and Van Wagner (1991) projected that the area of Canadian boreal forest burned annually may increase 40% based on current GCM projections of climatic change. Price and Rind (1994) used GCM results to estimate that the number of lightning-caused fires could increase by 44% and the area burned could increase by 78% for the United States in a doubled CO_2 climate. Overpeck *et al.* (1991) projected that plant distribution could shift by 500–1000 km in the next 200–500 years based on current GCM projections of the magnitude and rapidity of changing climate. Overpeck *et al.* (1990) hypothesized that these increases in disturbance frequency could provide opportunities for accelerated movement of forest distributions in a highly unpredictable way.

E. Monitoring Future Changes in Forest Dynamics

The impact of global change on forests, whether induced by climatic or sociopolitical forces, will require an accurate, regular monitoring of forest cover and productivity. Because forest redistribution is so hard to predict and will take many centuries, possibly the earliest evidence of change may come from following the interannual variability in phenology, the seasonal timing of leaf emergence in the spring and senescence in the autumn. Spring leaf onset of native trees was observed to vary by more than 90 days in a unique natural history record from England that spans 211 years (Sparks and Carey, 1995). In temperate forests, cold, rainy springs result in delayed budburst in trees relative to sunny, warm spring conditions, and interannual variability of 30–40 days is common (Lechowicz, 1995). Phenological observations were commonly recorded at field research stations earlier in the twentieth century, but most reports were discontinued in the 1960s and 1970s in favor of different types of research, interrupting what could have been a priceless record of global change (Hari and Hakkinen, 1991). In the future, both spring and autumn phenology of forest canopies will be monitored by combining satellite and surface meteorological data to produce maps that cover each continent (see Plate 16). Any trend toward earlier spring leaf emergence and/or later autumn senescence will provide sound evidence of warming climates at mid- to high latitudes.

New satellite systems will allow accurate regular mapping of global forest cover at 1 km scales (Nemani and Running, 1996). Documenting changes in observable ecotones, such as alpine timberline, forest—grassland, and boreal forest—tundra borders, should provide a sensitive measure of climatic influence on the distribution of various types of forests over the coming decades. More locally, tree-ring dendrochronology will continue to offer a valuable index of tree-growth responses to climate fluctuations (Jacoby and D'Arrigo, 1995). Where CO_2 and H_2O flux networks are permanently established, continuous measures of forest ecosystem gas exchange provide detailed insights into how net ecosystem carbon balances and other ecosystem properties respond to changing climate (Baldocchi *et al.*, 1996).

In all of these cases, systematic measurements following standard procedures must be sustained over decades to provide adequate documentation of the effects of global change on forest distribution and function. Global scale monitoring will provide the best assessment of changes in forest cover. Regional scale monitoring will best define shifts in phenology and the displacement of ecotonal boundaries. Local monitoring will provide insights into shifts in carbon balance between photosynthesis, changes in biodiversity, and other important properties. There should, however, also be modifications in climate-forecasting models so that they better represent the actual distributions of vegetation and incorporate the interactions associated with changes in phenology (Henderson-Sellers and McGuffie, 1995; Chase *et al.*, 1996).

III. FOREST–CLIMATE INTERACTIONS

A. Evidence of Climatic Warming

Although projections based on GCM simulations forecast rising global temperatures of around 2°–4°C as atmospheric CO_2 concentrations double in the coming centuries, the measured changes in air temperatures seem to bear out only a modest warming trend to date (IPCC, 1996). Pollution-producing aerosols cause haze to develop over urban areas, reducing incident solar radiation, cooling the land surface, and partially offsetting greenhouse gas-induced warming (Mitchell *et al.*, 1995). Air temperatures are inherently highly variable on daily, seasonal, and interannual basis. In addition, lack of standards in instruments, sensor calibration, and shifts in weather station locations and reporting times contribute to inconsistencies in the global climate database (Karl *et al.*, 1995). Intensive analyses in selected regions of air temperature extremes and their seasonal variations, however, show clear evidence of regional climate warming trends (Oechel *et al.*, 1993). For example, Karl *et al.* (1993) report that the winter minimum temperatures in Alaska have increased 3.5°C since 1951, although the annual average temperatures have changed only slightly. Groisman *et al.* (1994) estimate that the maximum extent of seasonal snow cover in the northern hemisphere has retreated 10% from 1972 to 1992.

There is some clear evidence of warming at high latitudes based on simple but important observations that date back well over a century. Oerlemans (1994) plotted the long-term change in length of 48 glaciers worldwide, some with records going back to 1850 (Fig. 9.4). Every glacier in the study has retreated, and for the period 1884–1978 the average

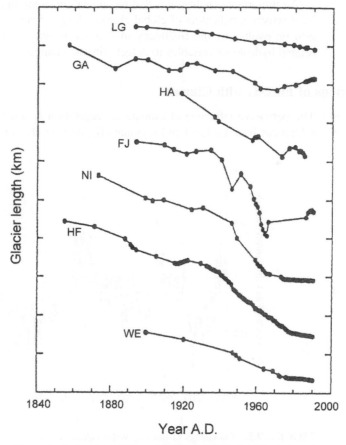

FIGURE 9.4.   Fluctuation in length of glaciers distributed globally that have been monitored for at least 100 years. The locations of glaciers shown are as follows: LG, Kenya; GA, France; HA, Spitzbergen; FJ, New Zealand; NI, Norway; HF, Austria; WE, Canada. All 48 glaciers in the survey have retreated, on average more than 1.2 km, a significant indication of global warming. (Reprinted with permission from J. Oerlemans, "Quantifying global warming from the retreat of glaciers," *Science* **264,** 243–245. Copyright 1994 American Association for the Advancement of Science.)

retreat was 1.23 km, correlated with a 0.62°C increase in temperatures through this period. Walsh (1995) reported data of Kuusisto (1993) showing that duration of ice cover in three lakes in Finland has decreased by 3 weeks since records began in 1830. A record of bud-burst dates from 1907 to 1950 for birch in central Finland suggests that leaf onset now averages about 1 week earlier than at the turn of the twentieth century (Hari and Hakkinen, 1991). Jacoby *et al.* (1996) analyzed 450-year tree-ring chronologies for Siberian pine at timberline (2450 m) at 48°N latitude in Mongolia. Cores averaged from 25 trees showed an extended period of enhanced growth since 1940 that corresponded with warmer regional temperatures (Fig. 9.5). In this severely energy-limited environment, warmer temperatures in conjunction with increased atmospheric CO_2 should result in longer growing seasons, higher photosynthetic rates, and increased primary production, decomposition, and mineral cycling rates.

An important conclusion is that the annual mean air temperature of Earth is among the least sensitive indicators of global climate change. Scientists now are placing more atten-tion on evaluating the frequency of extreme events, trends in regional phenology, and related hydrologic variables to detect climatic changes (Chapman and Walsh, 1993).

B. Interactions of Forests with Climate

The pervasive influence of climate on vegetation is obvious, but vegetation also exerts influences back on local and regional climate (see also Fig. 8.5). Understanding the role

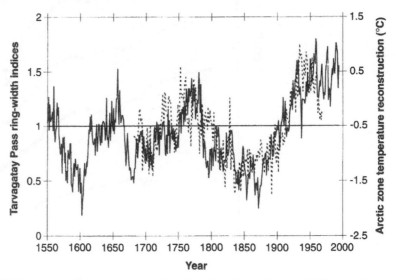

FIGURE 9.5. Chronology of tree-ring width indices for 300- to 500-year-old Siberian pine in Mongolia. The dashed line represents a reconstruction of annual air temperature departures from average for the region. Although isolated years early in the record showed high growth, the 10 widest annual growth rings have all been produced since 1920. Climate models predict that the greatest warming and biospheric models predict that the greatest increases in growth will occur at high latitudes. (Reprinted with permission from G. C. Jacoby, R. D. D'Arrigo, and T. Davaajamts, "Mongolian tree rings and 20th-century warming," *Science* **273**, 771–773. Copyright 1996 American Association for the Advancement of Science.)

of forests in directly influencing climate began with simple sensitivity studies of global climate models. Dickinson and Henderson-Sellers (1988) first ran their GCM with a standard vegetation, then replaced all of the tropical forests of the Amazon basin with a degraded grassland and concluded that evapotranspiration would be reduced substantially and result in an increase in surface temperature of 3°–5°C. A more accurate analysis used the satellite-derived estimates of partial Amazon deforestation reported by Skole and Tucker (1993) and found that evapotranspiration was reduced by 18%, leading to a decrease of precipitation in the basin of 8%, roughly 1.2 mm/day (Walker *et al.*, 1995).

Bonan *et al.* (1992, 1995) simulated the influence of a major deforestation in the boreal forest zone on climate and concluded that removal of forest cover would expose the snowpack, increasing the surface albedo from January through April, and result in a predicted temperature *decrease* of 2°–5°C at high latitudes. Sellers *et al.* (1996b) reported that differences in the assumptions regarding stomatal response also affect the climate predicted over a continent. When more realistic stomatal regulation by humidity deficits and CO_2 was added to a GCM, canopy conductance was reduced by 34%. The associated reduction in transpiration increased sensible heat exchange and resulted in a predicted air temperature increase of 2.6°C above the original simulations (Sellers *et al.*, 1996b). Schwartz and Karl (1990) showed dramatic and direct evidence of the control vegetation produces on current local climates. They documented that the regular increase in surface air temperatures during springtime is interrupted temporarily when leaves emerge on vegetation, in response to the cooling effects of additional transpiration (Fig. 9.6).

IV. FORESTS IN THE GLOBAL CARBON CYCLE

A. Elements of the Global Carbon Cycle

There are two important reasons for us to be concerned about the global carbon cycle. First, the CO_2 exchange from the terrestrial surface modulates the atmospheric CO_2 balance, with significant consequences on the climate. Second, and ultimately of more immediate importance, net primary production of vegetation is the renewable source of food, fuel, and fiber that supports our daily lives.

A simple box model shows the major components of the global carbon cycle (Fig. 9.7). Terrestrial vegetation contains about 550 Pg of carbon, soils 1200 Pg, the oceans 36,000 Pg, and the atmosphere 755 Pg. The fluxes of carbon (primarily as CO_2) between the terrestrial surface and the atmosphere are estimated as follows: gross photosynthesis, 110 Pg; autotrophic respiration of plants, 50 Pg; and soil respiration, 60 Pg (Moore and Braswell, 1994). Scientists face a challenge in determining the exchange rates among carbon pools because the annual fluxes are only a fraction of the total pool sizes. The most easily measured flux of carbon is reflected in changes of annual atmospheric CO_2 concentration. The interannual dynamics of atmospheric CO_2 concentrations have been monitored carefully since 1957 (Keeling, 1958), and longer term changes can be inferred from gases trapped in ice-core bubbles since before A.D. 1700 (Moore and Braswell, 1994). Fossil fuel emissions are also known fairly accurately and currently add 5.5 Pg C year^{-1} of CO_2 to the atmosphere. The

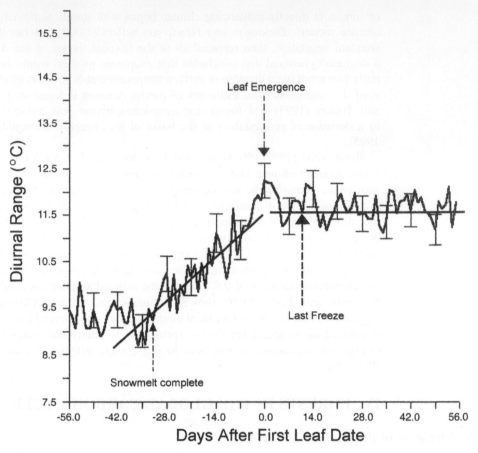

FIGURE 9.6.   Average daily surface temperature amplitude (maximum – minimum), measured before and after spring leaf emergence of local vegetation, for 12 sites across the north-central United States. The rapidly warming spring temperatures generate progressively larger temperature amplitudes during the 2–3 weeks prior to leafing out, which are, however, interrupted by leaf emergence as the partitioning of incoming solar energy is shifted from sensible to latent heat in response to transpiring foliage. This is one of the most obvious illustrations of how forests directly influence climate (Schwartz and Karl, 1990). (From Schwartz, 1996.)

oceans are thought to absorb 2.0 Pg C year^{-1} of CO_2 from the atmosphere, and from the annual increase in atmospheric CO_2 we can compute an increase of 3.2 Pg C year^{-1} in the atmospheric carbon pool size (Schimel, 1995). The remainder, the terrestrial net CO_2 flux, amounts to only 0.3 Pg year^{-1} of carbon uptake from the atmosphere, a very small sink.

Determining the role of forests in the global carbon cycle is even more difficult. Studies estimate that forest ecosystems contain about 80% of all global aboveground carbon. The amount of carbon sequestered in forest biomass is not well established. Estimates range from 380 Pg in biomass, with 770 Pg in forest soils (Dixon *et al.*, 1994), to 458 Pg in biomass and 1200 Pg in the soil (Hunt *et al.*, 1996).

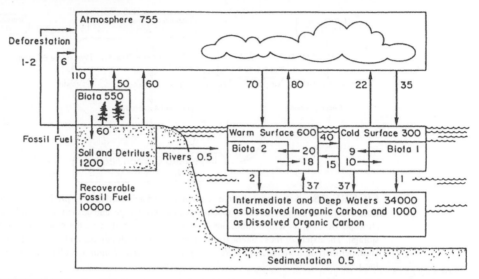

FIGURE 9.7.   The global carbon cycle in units of petagrams ($1\,Pg = 10^{15}\,g$). The annual fluxes of carbon, primarily as CO_2, are a very small percentage of the total carbon in each compartment. (From Moore and Braswell, *Global Biogeochemical Cycles* **8,** 23–28, 1994, copyright by the American Geophysical Union.)

B. Source/Sink Dynamics of the Global Carbon Cycle

An additional complication arises in that some forested regions such as the tropics are thought to be net sources of CO_2, while others, notably temperate forests, are thought to be net sinks (Grace *et al.*, 1995). Field *et al.* (1992) suggested three possible ecosystem mechanisms that could account for net accumulation of carbon in the terrestrial biosphere (Fig. 9.8). Dixon *et al.* (1994) estimated from ground-survey data that mid- and high latitude forests sequestered a net of 0.7 Pg of carbon annually, while deforestation in tropical forests transferred a net of 1.6 Pg of carbon to the atmosphere. Carbon isotope analyses of atmospheres over oceanic and terrestrial stations by Tans *et al.* (1990) projected that temperate, northern hemisphere forests were important sinks, absorbing more carbon in photosynthesis than they release through respiration. In contrast, the isotope analyses support the contention that tropical forests are net contributors of CO_2 to the atmosphere. One interpretation of these data is that temperate forests are regrowing after extensive harvesting in earlier decades, while tropical forests are now being harvested at accelerating rates. Ciais *et al.* (1995) used $^{13}C/^{12}C$ ratios in atmospheric samples from 43 globally distributed sites to estimate that half of the CO_2 produced annually by fossil fuel combustion is currently absorbed by northern temperate forests (3.5 Pg). These isotopic analyses, however, give only approximations of carbon cycling rates at the global scale. Some of the uncertainty in the global carbon budget estimates is reflected in the differences obtained from isotopic studies and the values presented in Fig. 9.7.

Studying subtle details of the seasonal changes in atmospheric CO_2 from stations around the world allows a more refined measure of the balance between terrestrial photosynthesis and respiration. Natural variability in global climate can change the annual terrestrial

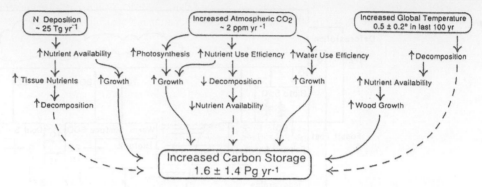

FIGURE 9.8.   Schematic of three ecosystem mechanisms that may produce a terrestrial biospheric carbon sink. Solid arrows depict processes that increase carbon storage, and dashed arrows indicate the reverse. Increased atmospheric CO_2, air temperatures, and nitrogen deposition all have the potential to increase terrestrial carbon storage. The extent of these interactive responses varies with climate and biome type. (From Field *et al.*, 1992. With permission, from the *Annual Review of Ecology and Systematics*, Volume 23, © 1992, by Annual Reviews Inc.)

biospheric exchange by 10 Pg as documented by the atmospheric CO_2 monitoring network (Fig. 9.9; Keeling *et al.*, 1995). For example, periodic disruptions of the global climate, such as the eruption of Mt. Pinatubo in 1991, increased stratospheric aerosols and lowered the global average air temperature slightly for the following 2–3 years. The biosphere quickly responded to this rapid climatic variation as evidenced by a reduced rate in the expected annual increase in atmospheric CO_2. Scientists hypothesize that the slightly lower temperatures generally enhanced photosynthesis in water-limited regions and reduced autotrophic and heterotrophic respiration (Keeling *et al.*, 1995, 1996). The seasonal variation in atmospheric CO_2 concentrations at higher latitudes provides a regional scale impression of the seasonal shift in photosynthetic activity (Fig. 9.10; Hunt *et al.*, 1996).

C. Global Net Primary Production and the Contribution of Forests

The first geographically explicit estimates of global net primary production were based on correlations established between field measurements of NPP and simple climatic indices, because daily surface meteorology was the first accessible global database available to ecologists. Leith (1975) correlated NPP against an estimate of evapotranspiration derived from temperature and precipitation data. However, NPP is controlled by more than evapotranspiration, and climatic indices do not well represent existing biome distributions or actual vegetation cover. Biospheric process models now define the distribution of global biomes, represent biome-specific physiological processes, and incorporate global climate databases to produce considerably improved estimates of global NPP. Melillo *et al.* (1993) produced the first estimates of global NPP derived with an ecosystem model of 53.2 Pg C, 75% coming from forested areas. These global models represent ecosystem processes similarly to the regional simulations presented in Chapter 8 with the RHESSys modeling package. The underlying consistency in logic from regional to global scales provides some confidence that validations confirmed at a smaller scale support extrapolations at the next larger scale.

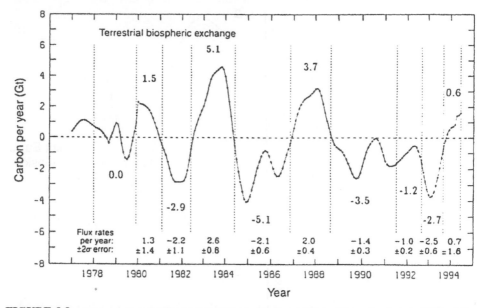

FIGURE 9.9. Interannual variability in terrestrial biospheric net CO_2 exchange, in gigatons (1 Gt = 1 Pg = 10^{15} g). Vertical lines separate periods of persistent positive (CO_2 loss) or negative (CO_2 uptake) fluxes that correspond to warm and cold phase of El Niño events. Annual flux rates are shown with standard errors. In arid areas, cooler temperatures are thought to alleviate water stress, favor photosynthesis, and reduce decomposition rates, leading to a net gain in ecosystem carbon uptake. (From Keeling *et al.,* 1995. Reprinted with permission from *Nature.* Copyright 1995 Macmillan Magazines Limited.)

These biospheric models represent the global land surface rather coarsely, at best currently with about sixty thousand 50 × 50 km cells. The basic principles summarized in Chapters 2–4 for water, carbon, and nutrient cycling are incorporated in these global models, although in less detail than in the stand models described in Chapter 5. As was true with regional modeling, biospheric models are limited mainly by a lack of global data necessary to provide the critical initializing variables. Variables that can be remotely sensed, such as LAI, can be incorporated into a global database with some confidence of the error in estimates involved (see Plate 17). Other variables, such as soil depth, however, cannot be directly estimated and must be inferred from geomorphic relationships. In spite of the obvious limitations, these global simulation models allow scientists to obtain the first estimates of the magnitude and spatial distribution of belowground processes such as decomposition and nitrogen mineralization (see Plate 18). As a result of the coarse spatial resolution, however, the primary value of these model predictions remains at the global scale for studies of biogeochemistry, climate forecasting, and broad policy analyses of natural resources.

For more practical applications, satellite-driven models can now be executed at a $1 km^2$ scale resolution if desired over the entire land surface ($150,000,000 km^2$). Global NPP, for example, can be computed with Light Use Efficiency (LUE) models similar in structure to that presented for stand-level analyses in Table 3.5. Sellers (1985, 1987) demonstrated the nearly linear relation between FPAR and satellite-derived NDVI values. By combining

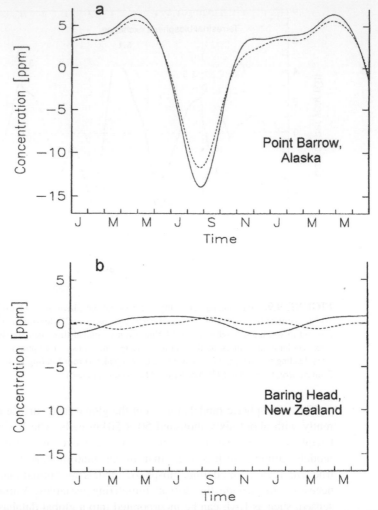

FIGURE 9.10.   Seasonal variability of atmospheric CO_2 concentration that contrasts a boreal location with large seasonal differences in photosynthesis with a more maritime, low latitude site with less seasonal variation in the rates of processes affecting the carbon balance. BIOME-BGC simulations of daily terrestrial net CO_2 flux were entered into an atmospheric transport model to predict atmospheric concentrations, which were compared against measurements from the global sampling network of C. D. Keeling (Law *et al.*, 1996; Piper and Stewart, 1996). The difference in seasonal amplitude of [CO_2], 20 ppm at Point Barrow, Alaska, but only 2 ppm at Baring Head, New Zealand, reflects differences in the magnitude of winter–summer ecosystem processes, which are extreme at high latitudes. Photosynthetic activity draws [CO_2] down in the midsummer at the height of the growing season, but respiration and decomposition produce a net release of CO_2 during the leafless winter season. (From Hunt *et al.*, *Global Biogeochemical Cycles* **10**, 431–456, 1996, copyright by the American Geophysical Union.)

estimates of monthly incoming PAR with FPAR, the upper limits to photosynthesis were established throughout the year. On the basis of climatic restrictions in life forms, a conversion efficiency specific to each biome has been applied to convert APAR to NPP (see Plate 19; Ruimy *et al.*, 1994; Field *et al.*, 1995; Prince and Goward, 1995). Because light conversion efficiency changes with climate and with the relative age and structure of forests, as discussed in Chapter 3, appropriate light use conversion factors (ε) for global scale extrapolations are generally attained from more detailed stand/seasonal level simulation models and then incorporated into global scale biospheric projections (Plate 20).

Validating any global estimate of ecosystem behavior is improbable, but wide scale sampling within regions of some variables such as aboveground forest growth shows good agreement with model predictions derived from satellite and weather station data (Coops *et al.*, 1998). These satellite-based estimates of NPP, however, are seriously limited by continuous cloud cover in some tropical areas, and by the effects of low sun angles on remote sensing at higher latitudes (Prince and Goward, 1995). Still, a quantitative assessment of regional and interannual variability of terrestrial productivity is possible within regions with particularly strong seasonal dynamics (Myneni *et al.*, 1995). Myneni *et al.* (1997b) showed from a record of AVHRR NDVI data for 1979–1990 that the growing season of boreal forests in Canada has increased by 12 days, an estimate consistent with the atmospheric CO_2 record from Keeling *et al.* (1996) for that region. As yet, unfortunately, no other important ecosystem processes have theoretical logic similar to the LUE–NPP relationship, allowing global extrapolation.

D. Other Trace Gas Emissions from Forests

In addition to CO_2, forests exchange other chemical compounds with the atmosphere that contribute to the global carbon cycle, primarily methane and a group of non-methane hydrocarbons such as isoprenes and terpenes. The atmospheric concentration of methane is only 1.7 ppm; however, it is increasing at 1% per year, and each CH_4 molecule is 20 times more active than CO_2 as a greenhouse gas. Fung *el al.* (1991) found intact forests to be a minor contributor to global methane sources compared to wetlands and rice fields. However, biomass burning may contribute 10% to the global atmospheric methane budget. Isoprenes and terpenes are chemicals trees emit when under stress, predominantly during the daytime in conjunction with photosynthesis, as discussed in Chapter 6. These chemicals are rapidly oxidized to carbon monoxide, so are never at high concentrations in the troposphere (Muller and Brasseur, 1995). Some regions, such as the Great Smoky Mountains of the eastern United States and the Amazon tropical forests, are high isoprene/terpene sources (Muller, 1992). Guenther *et al.* (1995) estimated that tropical woodlands at present contribute 50% of global natural nonmethane hydrocarbon emissions. Nonmethane hydrocarbons are thought to contribute 10% of the global CO production annually, although uncertainty in global estimates is large (Muller, 1992).

Forests produce only minor amounts of other atmospherically important gases. Gaseous sulfur in various forms is primarily produced by human industrial pollution, oceanic aerosols, and volcanic emissions. The biotic sources are mostly derived from microbial activity in anaerobic wetlands and from the release of gases during biomass burning, which together may contribute 20% of the annual global atmospheric loading (Schlesinger, 1991,

1997). Nitrous oxide is a trace constituent of only 0.3 ppm atmospheric concentration. However, each N_2O molecule has 200 times the greenhouse effect of a CO_2 molecule in the atmosphere, and because N_2O is increasing 0.3% per year, it may exert a climate warming potential similar to CO_2. As with methane, nitrous oxides are emitted during forest fires (Schlesinger, 1991, 1997; Muller and Brasseur, 1995).

V. FORESTS AND BIODIVERSITY

A. Measures of Biodiversity

The loss of global biodiversity is one of the ecological themes most commonly discussed in the popular press. The finality and irreversibility of a species becoming extinct are discomforting to any biologist. While on a conceptual level most people agree that biodiversity is generally desirable, building a firm quantitative basis for analysis of biodiversity has proved illusive. Various definitions of biodiversity have evolved and now include genetic, species, and ecosystem measures (WCMC, 1992). *Genetic diversity* represents the variability within a species of biochemical attributes, many of which are not externally expressed. Genetic diversity may explain why some tree species have much broader geographic ranges than others, or superior disease resistance.

Species diversity is the most common level of analysis because simple field observations are the only required data source. *Alpha species diversity,* or species richness, is a simple count of the number of species identified in a unit area. The highest published alpha species diversity was found in Amazonian Ecuador, where a single hectare of forest supported 473 tree species (Valencia *et al.,* 1994). The plot-to-plot difference in species count in a local area is defined by the *beta diversity*. Endemism is a term used to describe a species that is restricted to a limited geographic area or ecological habitat. Finally, *gamma species diversity* quantifies variation in the species list found across the landscape, so is a reflection of landscape and microclimate variability as well as evolutionary limitations. For example, the entire 420 million ha northern hemisphere temperate forests contain only 1166 tree species, illustrating rather low gamma diversity (Latham and Ricklefs, 1993).

Ecosystem biodiversity represents changes in the temporal, structural, and functional activity of forests. *Temporal biodiversity* is primarily the species dynamics observed in forest development and replacement over time, and it has a long history of study (Chapter 5). *Structural diversity* is an analysis of the spatial heterogeneity of age, height, and density of forest stands across a landscape. Bird habitat studies illustrated in Chapter 8 evaluated landscape structural patterns and arrangement.

B. Functional Attributes Related to Biodiversity

Functional biodiversity evaluates the role of each species in ecosystem biogeochemical cycles, as well as the redundancy present in the system to assure "normal" rates of carbon and nutrient cycling. In regard to biogeochemical cycles covered in Chapters 2–4, the loss of any one species probably has little effect on the basic carbon cycle of a forest, because another species rapidly fills the void in space and resource utilization (Shaver *et al.,* 1997). In this sense, the basic biogeochemical cycles are conservative properties of ecosystems that proceed in a fairly predictable way with a wide variety of species mixtures. In fact this simplifying assumption is essential in calculating global carbon cycle processes independently of species compositions (Plates 17 and 18). Indeed, critical physiological attributes such as maximum leaf conductance show very little interspecies variability. Kelliher *et al.* (1995) and Körner (1994) found high consistency in reported maximum leaf conductances for a wide variety of global forest types (Table 9.3). However, other ecosystem functions may be much more species sensitive. A critical life cycle activity such as pollination could be disrupted if an insect host were lost from the ecosystem. Susceptibility of trees to insect/disease attack is clearly species specific. These species-specific factors may determine which individuals are present to perform the conservative bigeochemical processes in the forest.

It is not completely clear what biophysical and ecosystem attributes provide optimum opportunities for biodiversity. Suggestions include high topographic and microclimate diversity, high vegetation structural diversity, high primary productivity, and low disturbance frequency (Johnson *et al.,* 1996). Various hypotheses have been advanced to relate species richness to ecosystem stability. Tilman (1996) in an 11-year study of grasslands in Minnesota measured year-to-year variability in total biomass of grasses and found that it was lower in plots with higher species richness. High numbers of species present in a single plot led to increased competition and eventual exclusion of some species. Tilman concluded that the underlying processes controlling biomass production were conservative, and that the dominance of individual species can fluctuate greatly without affecting the overall accumulation of biomass.

At global scales, species richness appears to be related to general rates of biogeochemical activity in association with climatic factors (Fig. 9.11). Ecosystems with favorable climates throughout the year, such as tropical wet forests, support greater numbers of species than seasonally unfavorable boreal or desert environments. At a continental scale, Adams and Woodward (1989) found a high correlation between annual NPP and species

TABLE 9.3

Global Average Maximum Leaf Conductance (g_{smax}) and Maximum Canopy Conductance (G_{smax}) to H$_2$O^a

Superclass	Vegetation type	g_{smax} (mm s^{-1})	G_{smax} (mm s^{-1})
Natural herbaceous	Temperate grassland	8.0 ± 4.0	17.0 ± 4.7
Woody	Conifer forest	5.7 ± 2.4	21.2 ± 7.1
Woody	Eucalypt forest	5.3 ± 3.0	17.0
Woody	Temperate deciduous forest	4.6 ± 1.7	20.7 ± 6.5
Woody	Tropical rain forest	6.1 ± 3.2	13.0
Agricultural crop	Cereals	11.0	32.5 ± 10.9
Agricultural crop	Broad-leaved herbaceous crops	12.2	30.8 ± 10.2

^{a}This summary suggests that canopy gas exchange processes, and more generally biogeochemical cycling, are conservative properties of an ecosystem with limited species differentiation. After *Agricultural and Forest Meteorology,* Volume 73, F. M. Kelliher, R. Leuning, M. R. Raupach, and E. D. Schulze, "Maximum conductances for evaporation from global vegetation types," pp. 1–16, 1995, with kind permission of Elsevier Science–NL, Sara Burgerhartstraat 25, 1055 KV Amsterdam, The Netherlands.

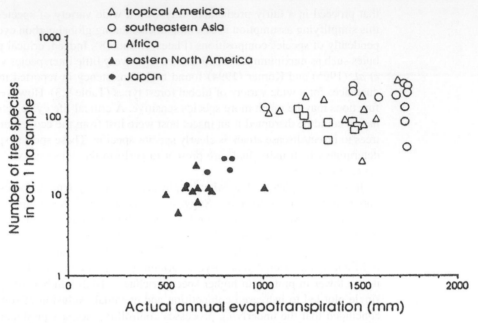

FIGURE 9.11.   Relationship of tree species diversity in a globally distributed set of small forest plots and actual ET. A general trend is for warmer, wetter climates to support more species. (From Latham and Ricklefs, 1993. © 1993 *Oikos.*)

richness of northern hemisphere forests (Fig. 9.12). This suggests that NPP as we have presented at local (Plate 8), regional (Plate 10), and global (Plate 18) scales may provide a general index of biodiversity. Climate maps, as presented in Chapter 7 (see Plate 3), may also be useful for assessing biodiversity. Analysis of climate variability may be more valuable than the analysis of long-term mean conditions. The environment today, of course, may not reflect earlier conditions, so historical environmental and floristic (and zoological) analyses are warranted in areas where species richness differs significantly from general model predictions.

At present, climatic and NPP maps obtained from satellite-derived data offer a first approximation of spatial variation in biodiversity at the global scale. Estimates of future changes in biodiversity over large areas may be made by relating species richness to more observable biophysical characteristics, and by quantifying land cover changes over time. Tuomisto *et al.* (1995) used Landsat TM imagery to extrapolate their estimates of biodiversity of the 500,000-km² Peruvian lowland of Amazonia. Satellite imagery played an important initial role by providing an assessment of landscape-level heterogeneity for the establishment of field transects. Sisk *et al.* (1994) located areas with threatened loss of biodiversity by assessing the rates of land cover change in a region with large-scale AVHRR data sets. The satellite-based analysis indicated that 18.1 million km² of native vegetation has been converted since the mid-1980s to agriculture, the land cover type assumed to have the lowest species diversity. Our estimate from Fig. 9.3 of 17.5 million

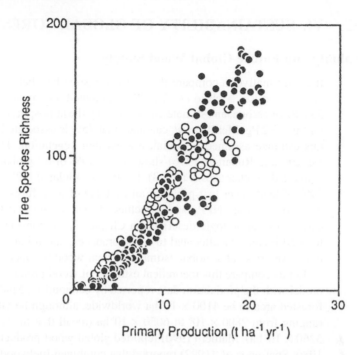

FIGURE 9.12.   Relationship between tree species richness in European and North American forests and NPP. Figures 9.11 and 9.12 suggest that more productive sites exhibit higher species diversity. Harsher boreal and temperate or arid regions (○) support lower species richness than do tropical rain forests (●). (From Adams and Woodward, 1989. Reprinted with permission from *Nature*. Copyright 1989 Macmillan Magazines Limited.)

km² of forestland converted to cropland illustrates the improvement in precision that satellite-based estimates have brought to the land cover calculations that underlie the assessment of global biodiversity.

Related to biodiversity are questions of how forest ecosystems can be maintained through time. What are the critical components and functional activities of a forest ecosystem? As we know, maintaining a certain species composition is much more difficult than maintaining the rates of many conservative processes. Sustaining basic productivity of a forest is probably the easiest task, as many species can fulfill this role, although they will have different commercial value. Exotic species often show higher primary productivity than natives, which is the reason why many of the *Eucalyptus* forests of Australia and *Nothofagus* forests of New Zealand have been replaced with radiata pine (Benson *et al.*, 1992). In some cases, the introduced species have adaptations the native flora lack, and they may also be immune, at least temporarily, to local pathogens and insects. The reduction in native vegetation, of course, has a major impact on the dependent animal populations. Also, if some elements in the flora provided special adaptations to accessing water, fixing nitrogen, or withstanding fire, wind, floods, or other disturbances, their loss may not be immediately appreciated but will have long-term implications.

VI. SUSTAINABILITY OF GLOBAL FORESTS

A. Sustainability and Future Global Wood Supply

It is informative to compare the global forest carbon balance as computed by ecologists and economists. Melillo *et al.* (1993) computed an annual global NPP for all biomes of 53.2 Pg of carbon, an estimate constrained by the atmospheric CO_2 data (Ciais *el al.*, 1995). Of this 53.2 Pg, Melillo *et al.* calculate that 75% is associated with forested land; however, this estimate assumed potential, not existing vegetation. The satellite-based analysis of Nemani and Running (1996) showed that 29% of all global forestland has already been converted to other uses (Fig. 9.3). Reducing simulated NPP of forests (53.2×0.75) by the 29% of land cover lost provides an estimate of total forest NPP of 28.3 Pg year^{-1}. If we next convert total NPP (which includes roots and leaves) to usable fiber, based on the carbon allocation logic described in Chapter 3, we consider that as much as 70% of the total NPP could be allocated to roots, branches, and foliage. As a result of these assumptions, we arrive at a global estimate of stem wood production of 8.5 PgC year^{-1}.

Let us compare this theoretical estimate of wood production derived from a biospheric model with the production data reported by economists. Mather (1990) estimated the total forested area to be 4100×10^6 ha worldwide, although he cites other published estimates ranging from 2800×10^6 to 6050×10^6 ha (recall that the satellite estimate in Fig. 9.3 is 5260×10^6 ha). Mather (1990) reported global wood production of 3.2×10^9 m^3 year^{-1} for 1986. Sharma *et al.* (1992) reported that combined fuelwood and industrial wood production was 4.3×10^9 m^3 year^{-1} harvested from growing stock with a standing live timber volume of 340×10^9 m^3. Using an average specific gravity for wood of 500 kg m^{-3}, and assuming that biomass is 50% carbon, gives global wood production of 1.1 PgC year^{-1}, comparable to the estimates given by Vitousek *et al.* (1986) and Dixon *et al.* (1994). Sharma *et al.* (1992) forecasts future wood fiber demand of 1.8 PgC year^{-1} by the year 2025.

It appears from this simple analysis that current global wood consumption is <15% of annual forest stem wood production and that opportunities could exist for a considerable increase in the rate of harvesting. This conclusion, however, ignores some important facts. First, Houghton (1994) estimated that tropical forests are being harvested at a rate of more than 3% year^{-1}. If this land is not soon returned to forest production, within less than a century the growing global demand for wood will certainly exceed the shrinking land base for production. Second, much of the world's forests is inaccessible and/or serves other critical functions, including protection of watersheds, recreation, and cultural values. Third, even with present rates of wood production, regional shortages have developed that require imports or substitutions of other products. Viewed with these constraints, there is ample reason to support efforts to reduce the rate of forest conversion to other uses, to return degraded lands to production, and to support intensive management of plantations in some regions to offset pressures on native forestlands. As recognized by Vitousek *et al.* (1986), maintaining the production of forests will be a major challenge as population levels and standards of living continue to rise, and the demand for forest products with competition for food and living space further limit our options. For better, or for worse, the extent that future generations succeed in balancing demand for resources with the capacity of the landscape to maintain its functional integrity will be documented in changes in the distribution of forests and estimates of their performance derived from global satellite observations.

B. Forests Planting for Climate Stabilization and Fuel Replacement

It has been suggested that much of the potential for global climate warming could be averted by planting millions of hectares of new forests to sequester more of the 5.5 Pg of C annually transferred by combustion of fossil fuels into the atmosphere (Dixon *et al.*, 1994). Careful analysis of these optimistic scenarios points out two major problems. First, Sedjo (1989) estimated that to sequester just 1.8 Pg of atmospheric carbon annually in forest biomass would be equivalent to (1) planting over 400 million ha of fast growing plantations, which would cost at least $300 billion (US$), (2) increasing the annual production of current forests by about 70%, or (3) expanding the boreal forests northward to cover an additional 3 billion ha. The most productive areas, which provide a good investment, already support forests. Planting forests in marginal areas will not sequester much carbon nor provide as large of an economic return on the investment (Sedjo, 1992). Moreover, as we have discussed, current global trends show a net *loss* in forestland. Second, forests do not permanently sequester carbon, because as the forest is harvested, or dies naturally, the biomass is eventually decomposed or combusted and most of the carbon released back to the atmosphere (Harmon *et al.*, 1990). Although we support the establishment of forests for a myriad of benefits, it is unrealistic to view massive forest planting as a serious alternative to reducing the combustion of fossil fuels.

Wood fiber has also been suggested as an alternative to fossil fuels for energy production. Graham *et al.* (1992) estimated that, if short rotation woody crops were planted on 14 million ha of land in the United States and harvested for commercial electric power production, 10% of national electrical power needs could be supplied. The present cost of fossil fuels, however, makes this option uneconomical; crops with higher values can be grown on the most productive agricultural land, and marginal land cannot produce enough biomass to meet national needs. As costs of fossil fuels rise significantly, however, biomass farms may have a more significant role as a national energy supply, and wood will also be an important local fuel source in many regions.

VII. SUMMARY

According to satellite monitoring, forests currently cover about 53 million km^2 of Earth's surface. In comparison to the area to which they are climatically suited, tropical, temperate, and boreal forests have been reduced through human activities to about 26, 40, and 20%, respectively, of their potential distributions. The geographic distribution of forests has changed continuously over the last few million years. Future changes in forest distribution are difficult to predict without better projections of the rate at which forests are converted to other land uses and the extent to which projected changes in climate occur. Although the general influence of climate on forest distribution and productivity is well established, increasing evidence shows that, on regional scales, the presence of forests also alters the climate through modifications in the rates at which water and heat are exchanged from terrestrial surfaces. At higher latitudes, seasonal and interannual dynamics of atmospheric CO_2 mirror the net metabolic activities of forests in a very direct way. Forests are thought to account for about 28 Pg C year^{-1} of terrestrial NPP, of which we estimate 8.5 Pg year^{-1} is in the form of stem wood biomass. Current global wood consumption is approximately 1.1 Pg year^{-1}, implying that global forest production should be sustainable, although much is inaccessible to direct utilization. The most fundamental concern is the rate at which forestland continues to be converted to other uses. Planting additional land into forests may offset some atmospheric CO_2 increases and replace some fossil fuel uses, but it should not be considered a major alternative to reducing dependence on fossil fuels.

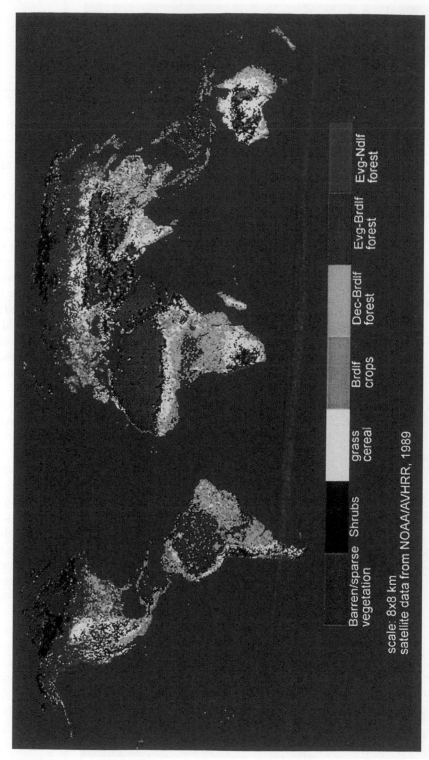

PLATE 12. From 8 × 8 km AVHRR satellite data acquired in 1989, a total of seven major land cover types could be recognized and mapped, which included shrubs, grass/cereal, broadleaf (Brdlf) crops, and three types of forests: deciduous broadleaf (Dec-Brdlf), evergreen broadleaf (evergreen broadleaf (Evg-Brdlf) and evergreen needleleaf (Evg-Ndlf). These types were distinguished with a combination of threshold limits and seasonal amplitude of NDVI, plus maximum temperature estimates acquired from the satellite throughout the year. (From Nemani and Running, 1995).

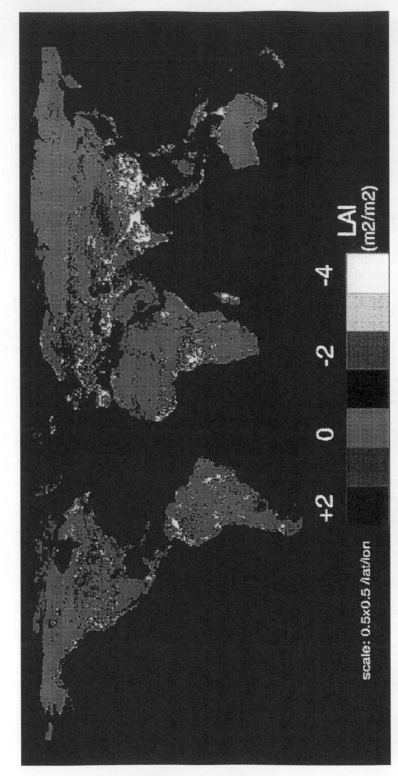

PLATE 13. Global land cover change analysis determined by comparing leaf area index from the satellite-derived existing land cover in Plate 12 with LAI from an ecosystem model that computes potential land cover from bioclimatic principles. (From Nemani *et al.*, 1996).

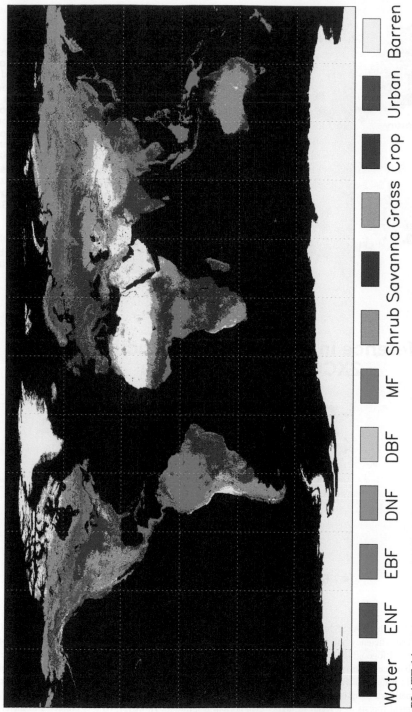

PLATE 14. From 1×1 km MODIS satellite data acquired since 2000, the number of definable cover types increased to twelve. Abbreviations: ENF (evergreen needle-leafed forest), EBF (evergreen broad-leafed forest), DNF (deciduous needle-leafed forest), DBF (deciduous broad-leafed forest), MF (Mixed forest). (Figure provided by Numerical Terradynamic Simulation Group, University of Montana).

Water ENF EBF DNF DBF MF Shrub Savanna Grass Crop Urban Barren

Biome Redistribution with 2XCO2 and Climate Change

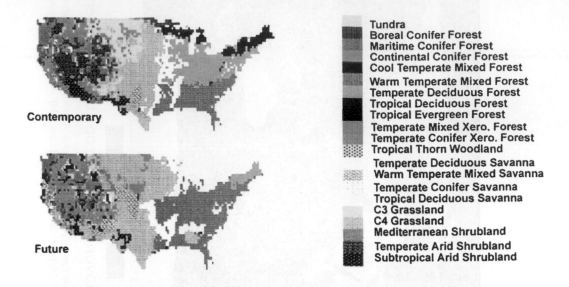

Tundra
Boreal Conifer Forest
Maritime Conifer Forest
Continental Conifer Forest
Cool Temperate Mixed Forest
Warm Temperate Mixed Forest
Temperate Deciduous Forest
Tropical Deciduous Forest
Tropical Evergreen Forest
Temperate Mixed Xero. Forest
Temperate Conifer Xero. Forest
Tropical Thorn Woodland
Temperate Deciduous Savanna
Warm Temperate Mixed Savanna
Temperate Conifer Savanna
Tropical Deciduous Savanna
C3 Grassland
C4 Grassland
Mediterranean Shrubland
Temperate Arid Shrubland
Subtropical Arid Shrubland

Contemporary

Future

Difference in Net Primary Productivity after 2XCO2 Climate Change

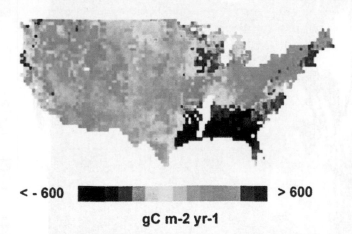

< - 600   > 600

gC m-2 yr-1

PLATE 15. The MAPSS biogeography model of Neilson (1995) evaluated the interactions of increased CO_2 on vegetation water use efficiency and projected climate warming scenarios on the future geographic distribution of forests in the United States. Biogeographic models incorporate various complex measures of temperature limits (Tables 9.1 and 9.2) and water balance controls to define the distribution of biomes. Climate warming scenarios tend to result in expanded grasslands in water-limited regions and expansion northward of boreal forests (VEMAP, 1995). The response of forest NPP to climatic warming is uneven. These BIOME-BGC simulations suggest that climatic warming could result in reduced NPP in tropical forests as a result of higher than normal respiration. However, NPP is projected to increase in semiarid forests associated with enhanced water use efficiency and in boreal forests as the growing season is lengthened and decomposition releases more nitrogen (VEMAP, 1995).

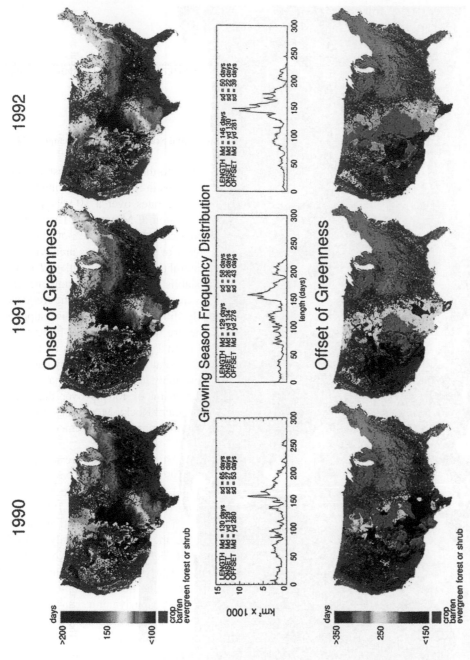

PLATE 16. Interannual variability of spring leaf emergence and autumn leaf senescence for natural deciduous forests and grasslands of the United States for 1990–1992. Traces in the boxes show the aggregate land area with the length of growing season indicated. This continental scale phenological analysis used a combination of daily AVHRR NDVI data and surface meteorological records of air temperature and precipitation. Consistent phenological monitoring provides an important means of quantifying the biospheric response to climate change. (From M. A. White *et al.*, *Global Biogeochemical Cycles* **11**, 217–234, 1998, copyright by the American Geophysical Union.)

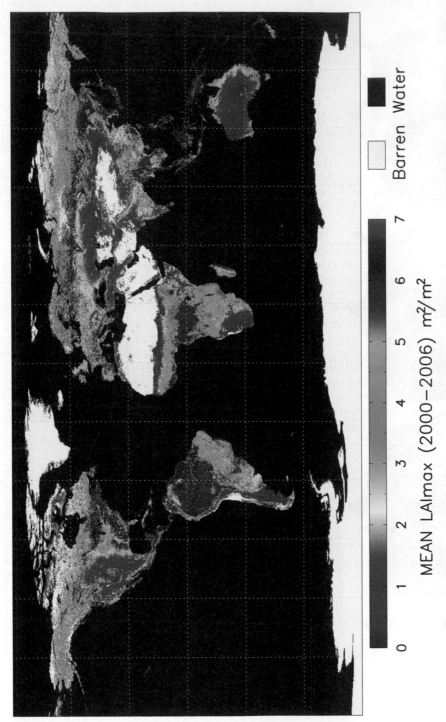

MEAN LAImax (2000–2006) m²/m²

Barren Water

PLATE 17. The annual maximum leaf area index at 1 × 1 km spatial resolution from MODIS data averaged for 2000–2006, with improved spectral and radio-metric resolution over the AVHRR sensor. (Figure provided by Numerical Terradynamic Simulation Group, University of Montana).

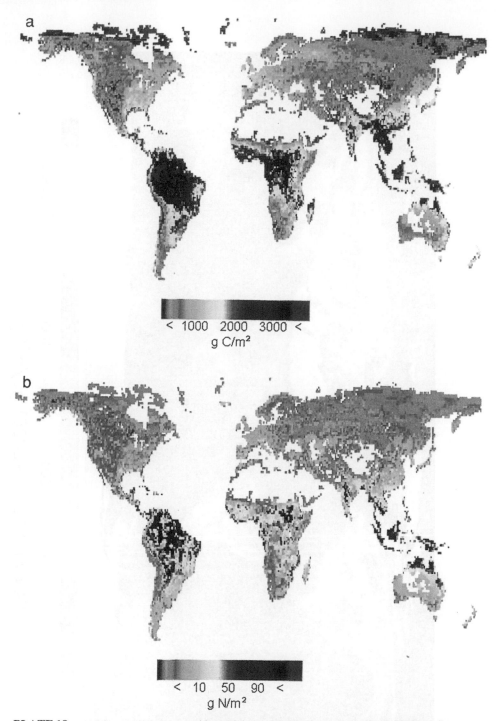

PLATE 18. Global annual (a) decomposition and (b) N mineralization computed by BIOME-BGC operating within the GESSys (Global Ecosystem Simulation System), a global version of the RHESSys model discussed in Chapters 7 and 8. A 1-year simulation was done with 1989 AVHRR data to define biome cover and LAI, with daily climatic data acquired from Piper and Stewart (1996). (From Galina Churkina, University of Montana, unpublished, 1997.)

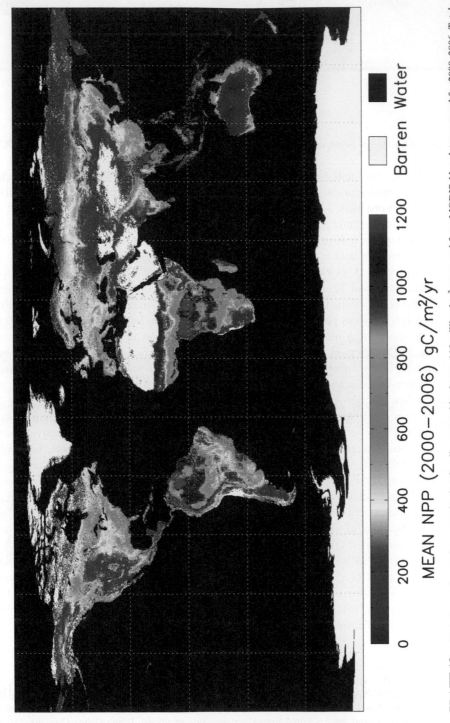

PLATE 19. Global annual net primary production for all vegetated land area, 110 million km² computed from MODIS 1 km data averaged for 2000–2006. Total annual global NPP averaged 821 gC/m²/yr. Total annual global NPP averaged 51.9 Peta grams of carbon, of which 47% was from forests. See Running *et al.* 2004 (Figure provided by Numerical Terradynamic Simulation Group at the University of Montana).

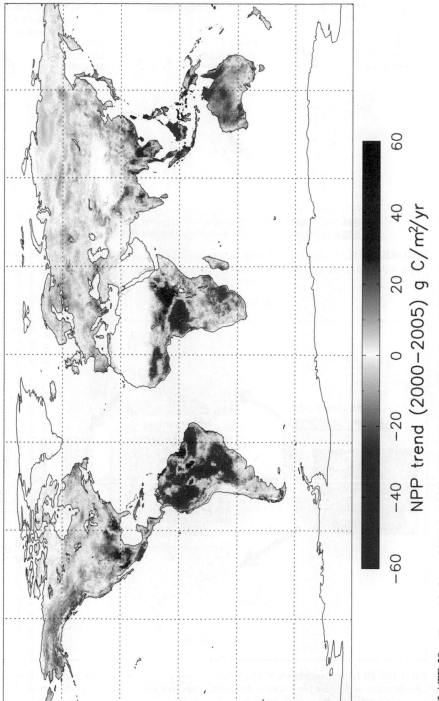

PLATE 20. The trend of annual NPP found from 2000–2005 in the MODIS dataset from Plate 19. Anomalies, or departures from some defined mean value, can be mapped from varying timescales, as an annual departure from a multi-year average productivity, or monthly as a departure from normal growing season seasonality. (After Running *et al.*, 2004).

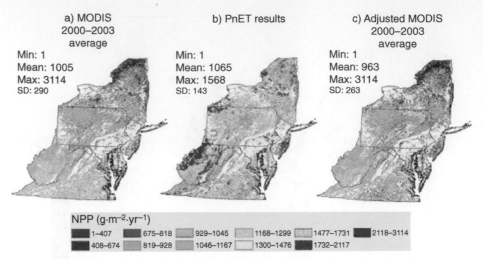

FIGURE 10.10. Calibration of MODIS satellite derived annual net primary production for the New England region with field inventory data to improve local accuracy (Pan *et al.*, 2006).

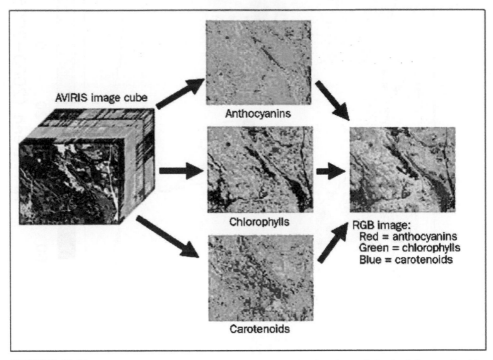

FIGURE 10.13. Classification of a Canadian boreal forest region based on relative leaf-pigment densities corresponding to different forest species derived from spectral mixture analysis of an Airborne Visible/Infrared Spectrometer dataset (Ustin *et al.*, 2004).

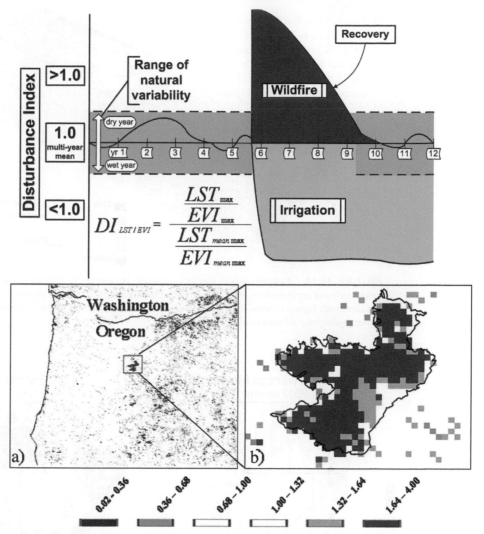

FIGURE 10.15. (a) A disturbance index (DI) uses annual maximum values of MODIS land surface temperature (LST) and Enhanced Vegetation Index (EVI) to detect yearly changes in forest cover and energy partitioning. (b) The DI detected the area burnt by a wildfire in Oregon closely matched that defined with more precise Landsat imagery (black line). (After Mildrexler *et al.*, 2007).

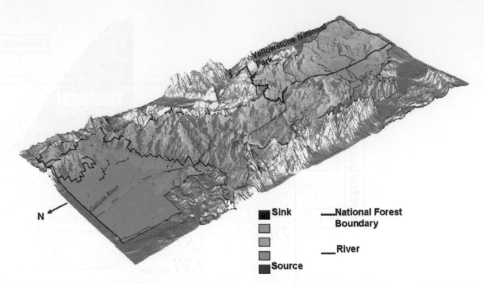

FIGURE 10.16. Richness in bird species vary across the greater Yellowstone area. Areas with high bird species richness are depicted in red. Green and tan colors indicate habitats with progressively fewer bird species. The areas with highest richness for the most part lay outside the National Park boundaries on private lands with higher productivity (Hansen *et al.*, 2002).

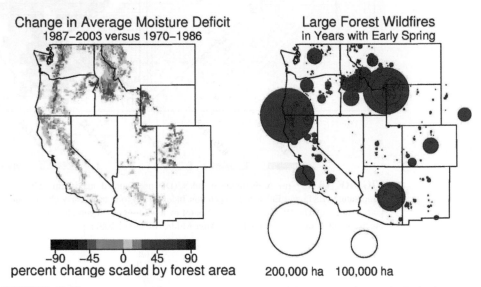

FIGURE 10.19. From 1970 to 2003, summertime moisture deficits increased dramatically in the western United States, and with a concurrent rise in wildfires >400 ha associated with an earlier melting of the snowpack in most years (Running, 2006; Westerling *et al.*, 2006).

CHAPTER 10

Advances in Eddy-Flux Analyses, Remote Sensing, and Evidence of Climate Change

I. INTRODUCTION

Throughout the book, we emphasize the need to identify underlying biophysical principles that can be tested at fine spatial and temporal scales, and then simplified to apply across landscapes and continents. In the first six chapters, we quantify and model the ways that carbon, water, and nutrients move in, through, and out of ecosystems. In chapters seven through nine, we introduce scaling principles and give examples of their use over increasingly larger geographic units. This new chapter does not cover the full spectrum of temporal and spatial scales outlined in Figure 1.1, rather it concentrates on new technologies applied to current conditions.

Since 1998, when we completed the 2nd edition, three major advances have significantly increased the power of multi-scale forest ecosystem analysis. First, the establishment of *regional and global networks of eddy-covariance flux towers* now provides

---

[1] Although NEP includes additional losses of carbon from an ecosystem through leaching, we assume such losses are well within the error estimates of eddy-flux measurements.

continuous measures of carbon, water and energy fluxes from an array of forests and other types of ecosystems. This new set of data, available in comparable format from regional and global data centers, quantifies measurements that increase our confidence in predicting ecosystem responses across landscapes, and provides a basis for generalizations.

Second, *the launch of a new generation of earth observation satellites* has lead to the production of an array of standardized and near-real time delivery of a host of important land data products. These satellite-derived products have spawned a new generation of regional scale modeling of carbon and water cycles, biogeochemistry, biodiversity, and habitat-change analyses.

Third, *clear detection of the effects of global climatic change* has been observed on forested landscapes. We provide evidence that climate change has affected not only forest growth but also the frequency and extent of disturbance and expansion of forests northward. All of these responses have important societal implications for the future.

By 2005, the number of eddy-flux sites grew to exceed 200. These instrumented sites, as described in Chapter 3, offer not only a continuous record of net ecosystem exchange (NEE), or net ecosystem production (NEP),[1] but opportunities to evaluate the underlying processes that affect CO_2 and water vapor exchange, trace gas emissions, and the general health and productivity of forests. The eddy-flux sites cover a sufficient range in productivity and environments to permit testing a host of detailed process-based models. In addition, because the area of the flux measurements must be relatively large ($\sim 1 \, km^2$), eddy-flux sites serve as calibration points for a variety of remote sensing techniques (Section III). In this section we illustrate the extent that an expanding network of flux-measurement sites has helped improve ecosystem models, and thereby our abilities to extrapolate flux estimates with increasing confidence across landscapes and continents.

II. EDDY-COVARIANCE FLUXES

A. Gross Primary Production

Gross primary production (GPP), under near equilibrium conditions, is estimated at eddy-flux sites from the sum of net carbon exchange (NEE) measured during the day plus ecosystem respiration (R_e). Respiration is measured at night when photosynthesis is nil, and then adjusted for increases in temperature during the day.

An annual summary of carbon and water balances is presented in Table 10.1 for 17 forested sites with wide ranges in climate, stand age, LAI, and disturbance. GPP is shown to increase generally from boreal ($700-1000 \, g \, C \, m^2 \, yr^{-1}$) through temperate ($1000-2200 \, g \, C \, m^2 \, yr^{-1}$) to tropical and semitropical forests ($2700-3200 \, g \, C \, m^2 \, yr^{-1}$).

Seasonal patterns in GPP differ significantly depending on the type of forest and degree of disturbance. Wet tropical forests show the highest values with little seasonal variation (Fig. 10.1). Other types show seasonality that parallels trends in solar radiation, with notable reduction in GPP where forests experience drought (e.g., Mediterranean broadleaf forests).

Because GPP and transpiration are well-understood processes, they can be modeled at a 30-minute resolution (Law *et al.*, 2002; Tuzet *et al.*, 2003). Such high temporal resolu-

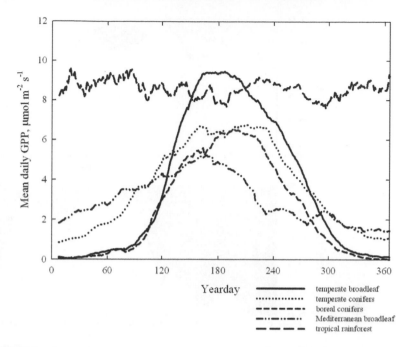

FIGURE 10.1.   Seasonal patterns in gross primary production differ significantly among major forest types. Data provided courtesy of Dr. Eva Falge, Max Planck Institute, Germany. See Falge *et al.* (2002) for additional information.

tion is valuable to derive procedures to fill missing data and to account for residual errors in measurements for a wide variety of ecosystem components (Williams *et al.*, 2005). To drive models with such high temporal resolution, however, requires more precise information than is generally available across landscapes. One solution is to integrate measurements over longer time intervals (Goulden *et al.*, 1996; Law *et al.*, 2000, 2002; Turner *et al.*, 2005).

With continuous monitoring of CO_2 exchange at eddy-flux tower sites, our knowledge of the constraints on forest productivity has greatly increased. Some models focus on predicting the relative constraints on photosynthetic activity throughout the year and between years. For example, Jolly *et al.* (2005) constructed an index that identifies the extent that variation in day length, air temperature, and vapor pressure deficit limits photosynthesis. Figure 10.2 contrasts the limitations of these three variables on photosynthesis for a tropical broadleaf forest in Australia with that of cool temperate evergreen forest in Montana.

B. Ecosystem Respiration

Detailed analyses at 18 forest sites in Europe indicate that on average >70% of the respired CO_2 originates below ground (Janssens *et al.*, 2001). The efflux of CO_2 from soil represents

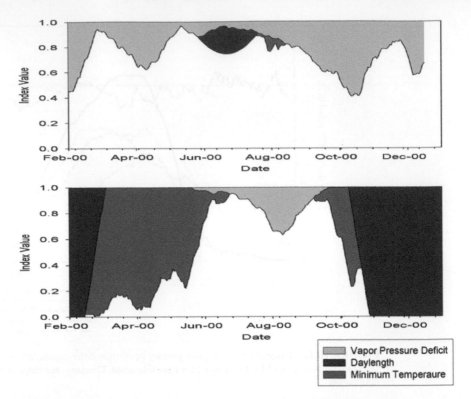

FIGURE 10.2.   The relative photosynthetic activity of a tropical wet-dry rainforest in Australia (upper panel) is generally limited by high evaporative demand. Day length limits GPP for about a month and low temperatures for only a few weeks. In contrast, a temperate conifer forest in the western United States (lower panel) is limited by day length in winter, low temperatures in the spring and fall, and high evaporative demand throughout the summer. (After Jolly *et al.*, 2005.)

both autotrophic (R_a) and heterotrophic (R_h) respiration. Högberg *et al.* (2001) and Bhupinderpal-Singh *et al.* (2003) discovered from tree girdling experiments that more than 50% of the soil CO_2 efflux is associated with the export of current photosynthate to roots and mycorrhizae. The fraction of GPP allocated below ground is known to decrease with increases in soil fertility (Hobbie and Colpaert, 2003; Hobbie, 2006; Johnsen *et al.*, 2007), and to result in commensurate changes in above-ground properties (e.g., increases in wood growth, foliar N content, LAI, and the ratio of actual to potential GPP).

Comparisons of R_a and R_h indicate a generally close relationship, not surprisingly, considering that litter production is associated with both NPP and GPP (Bond-Lamberty *et al.*, 2004). Although GPP can accurately be estimated at hourly and daily time steps, it remains a challenge to document what fraction of current GPP is allocated below ground at weekly and monthly intervals. Some estimate of soil fertility, directly or through its influence on canopy quantum efficiency, is essential to set limits on GPP and the fraction partitioned below ground (Chapter 3; Hobbie and Colpaert, 2003; Hobbie, 2006).

Recent models, supported from data acquired at many sites, predict soil CO_2 efflux as a function of an exponential temperature term, seasonal variation in soil water content, pore space, and maximum LAI (Reichstein *et al.,* 2003). Such models might be improved through a closer link with seasonal variation in GPP and the proportion allocated below ground (Irvine *et al.*, 2005).

C. Net Ecosystem Production (NEP)

The net amount of carbon exchanged by an ecosystem goes from positive to negative following disturbance. Undisturbed forests exhibit a ratio of NEP to GPP between ~0.1 to 0.4 (last column in Table 10.1). With disturbance, the ratio falls below zero as indicated for entries representing regenerating forests after clear-felling operations. Negative values may continue for several decades, depending on the amount of detritus present and the growth rates of planted (Gainesville, Florida) or sprouting trees (Castelporziano, Italy). Although the presence of other vegetation may contribute to a rapid increase in LAI and GPP within a few years following disturbance (e.g., Hyytiala, Finland; Bios, France), competition with tree regeneration may delay full canopy development and a return to a positive carbon balance (Gower, 2003).

Unusual climatic events, as recorded in the third year of data at Harvard Forest, can decrease the NEP/GPP ratio significantly (from >0.2 to 0.11) by increasing R_e without reducing GPP, but the ratio still remains positive. Older stands composed of a single dominant species, such as *Picea mariana* in Canada, may vacillate between being a carbon sink or source from one year to the next depending on spring and summer temperatures (Arain *et al.*, 2002). Undisturbed old-growth forests with a mixture of understory tree species, however, can be expected to maintain a positive NEP and NEP/GPP ratio (e.g., Wind River, Washington, Table 1A).

The most extreme interannual variation in NEP and NEP/GPP reported is for an old-growth *Populus tremuloides* forest in the southern boreal region of Canada, where NEP ranged from 80 to $290\,g\,C\,m^2\,yr^{-1}$ and NEP/GPP from 0.07 to 0.20 over a five-year period (Arain *et al.*, 2002). The interannual variation was attributed to spring conditions that delayed or enhanced budbreak (Arain *et al.*, 2002). On the other hand, young, fast-growing plantations, as represented by the Duke and Gaineville pine forests in Table 10.1, may produce positive ratios of NEP/GPP quickly following disturbance, particularly if soil fertility is inherently high or nutrient supplements are added (Sampson *et al.*, 2006).

D. Net Primary Production (NPP)

Net primary production (NPP) is the residual after autotrophic respiration is subtracted from GPP. To predict variation in NPP across geographic units of increasing size requires progressive simplifications in models, as emphasized in Chapter 7 and the preceding discussion. Recently, a number of simplifying features have been incorporated into predictive models of forest growth that have been widely tested on natural forests and plantations (Landsberg *et al.*, 2003). These simplifying features include expanding from daily to monthly time steps (Coops *et al.*, 2000), assuming that NPP represents an approximately constant proportion of gross photosynthesis (Fig. 3.9; Gifford, 2003; but see Cannell and Thornley, 2000), and that the fraction of NPP allocated aboveground increases with soil

TABLE 10.1

Site Characteristics of Selected Eddy-Flux Sites with Estimates of Annual Carbon and Water Balances. Negative NEP Values Represent Recently Disturbed Sites Losing Carbon to the Atmosphere

Vegetation Place name	Location Latitude & Longitude	Elev., m	Precip. (P), mm yr⁻¹	Water Balance, mm yr⁻¹ P-ET	Stand age, yrs	LAI m²m⁻²	NEP g C m⁻² yr⁻¹	Re g C m⁻² yr⁻¹	GPP[1] g C m⁻² yr⁻¹	NEP/GPP
Evergreen conifers										
Aberfeldy, Scotland	56°N, 4°W	340	1242	355	15	8.0	487	−1270	1757	0.28
Bordeaux, France[1]	56°N, 0°E	60	995	682	30	2.8	525	−1638	2163	0.24
Duke, NC, USA	36°N, 79°W	163	748	505	17	5.2	538	−941	1479	0.36
Flakaliden, Sweden[1]	64°N, 19°E	225	520	298	31	2.0	173	−526	699	0.25
Loobos, Netherlands[1]	52°N, 6°E	25	758	419	80	3.0	323	−1114	1439	0.22
Metolius, Oregon USA[1]	44°N, 121°W	915	867	628	45/250	2.1	287	−885	1172	0.25
Wind River, Wash., USA[2]	46°N, 122°W	355	2528	NA	500	11.0	150	−1420	1570	0.13
Hyytiala, Finland[1,8]	62°N, 24°E	170	540	317	1–2	1.8	−240	−602	364	−0.66
					35	3.0	228	−720	948	0.24
Bilos, France[8]	45°N, 52°E	60	930	NA	1–2	1.9	−225	−878	602	−0.37
					54	4.0	222	−1415	1600	0.14
Gainesville, Florida, USA[3]	30°N, 82°W	46	1332	NA	1–2	0.1–3.0	−1076	−1988	1104	−0.97
					10–11	3.1–5.1	590	−2174	2764	0.21
					24–25	4.0–6.5	675	−1932	2606	0.26

Deciduous Broadleaf & mixed conifers									
Hesse, France[4]	49°N, 7°E	300	335	30	6.0	238	−912	1150	0.21
Walker Branch, Tennessee, USA[2]	36°N, 84°W	375	249	60–90	6.0	470	−1038	1508	0.31
Park Fall, Wisconsin USA[5]	46°N, 90°W	485	378	60–80	4.0	334	−817	1165	0.29
Harvard Forest, Massachusetts, USA[6]	43°N, 72°W	−335	416	90	5.5	280	−960	1210	0.23
						220	−930	1110	0.20
						140	−1140	1270	0.11
						210	−970	1170	0.18
						270	−810	1070	0.25
Willow Creek, Wisconsin, USA	46°N, 90°W	480	591	35–70	4.2	180	−769	949	0.19
Evergreen Broadleaf									
Castelporziano Italy[8]	42°N, 12°E	3	NA	1–2	0.7–2.0	−427	−2220	1420	−0.30
				50	3.5	381	−1160	1600	0.24
Tropical rainforest									
Manaus, Brazil[7]	3°S, 60°W	90	957	mature	5.5	608	−2641	3249	0.19

[1]Law *et al.* (2002) carbon fluxes and water balances for 1996 or 1997.
[2]Paw *et al.* (2004) carbon fluxes for 1998–1999.
[3]Clark *et al.* (2004) carbon fluxes for 1996–1997 for recent clearcut and 10-year-old stand of slash pine; 1998–1999 for 24-year-old stand.
[4]Granier *et al.* (2000) carbon fluxes averaged for 1996 and 1997.
[5]Cook *et al.* (2004) carbon fluxes for 2000.
[6]Goulden *et al.* (1996) carbon fluxes for 1990 through 1994.
[7]Falge *et al.* (2002) carbon fluxes for 1995–1996; Malhi *et al.* (2002) water balance for 1995–1996.
[8]Kowalski *et al.* (2004) carbon fluxes, averaged for two years between 2000 and 2002.

fertility (Fig. 3.15) while that allocated to fine roots and mycorrhizae decreases proportionally (Hobbie, 2006).

Where model predictions deviate from direct measurement of NPP, the relative importance of climatic variation, soil fertility, and soil water storage can be assessed through sensitivity analyses (Rodriguez et al., 2002). Such analyses indicate where additional field measurements might improve model predictions. Stand age is a variable that also must be recognized because older forests generally grow more slowly than younger ones on similar sites (Ryan et al., 1997; Law et al., 2004), but older forests may also have access to deeper soil resources through better developed root systems. Although young forests may exhibit high NPP, the correlation with NEP is contingent, as emphasized earlier, on recovery following disturbance (Fig. 10.3).

E. Trace Gas Emissions

A number of trace gases are produced directly as byproducts of photosynthesis or through the process of decomposition, as discussed in earlier chapters. Through field chamber measurements and continuous monitoring of trace gas emissions at eddy-flux tower sites, our understanding of the controls on trace gas emission has increased significantly in recent years.

In the tropics, the capacity of some tree species to produce isoprenes varies a hundredfold throughout the year (Kuhn et al., 2004). The production of the carbon-based compounds is not related to LAI or canopy phenology but to seasonal changes in canopy photosynthetic activity (Fig. 10.4) and ambient air temperature (Chapter 6). It is reasonable that GPP should be more closely related to isoprene emissions in tropical forests than structural features such as LAI.

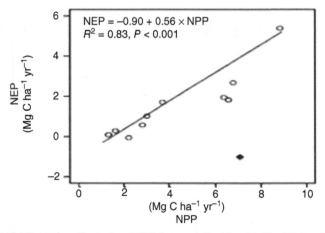

FIGURE 10.3. Net Ecosystem Production (NEP) increases linearly with Net Primary Production (NPP) except when forests are disturbed (black diamond). In the latter case, soil respiration is much enhanced. (After Pregitzer and Euskirchen, 2004.)

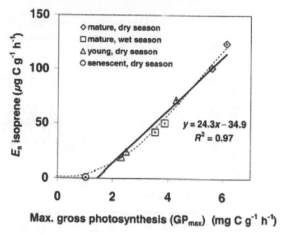

FIGURE 10.4.   Seasonal variation in maximum GPP is closely correlated with isoprene production. (After Kuhn *et al.*, 2004.)

To estimate emissions of NO_x, CO_2, and CH_4 from forested wetlands in both the boreal and subtropical region three models must be integrated: (1) a process-based forest growth model to calculate photosynthesis and its seasonal variation as a function of the environment; (2) a hydrologic model to predict water levels, O_2 concentrations, and water temperature; and (3) a biogeochemical model to compute the degree to which gas diffusion is limited by pore space and the availability of readily decomposable substrate in the soil (Cui *et al.*, 2005). With such integrated models the variation in solar radiation, soil water and oxygen levels, atmospheric N deposition, and substrate quality can be incorporated to predict seasonal differences in GPP, R_{a+h}, and CH_4 with considerable accuracy (Fig. 10.5).

F. Hydrologic and Energy Partitioning

By combining eddy-flux data from a wide range of evergreen and deciduous forests, as well as other types, Law *et al.* (2002) compared measurements of evapotranspiration (ET) when the canopy was dry against measurements of GPP at hourly, daily, weekly, and monthly time steps. Although considerable variation was observed hourly and daily, a linear correlation emerged when data were integrated and averaged at weekly or monthly intervals, showing how detailed analyses help identify appropriate periods for integration (Fig. 10.6).

Important hydrologic insight came from making measurements of volumetric water content, soil water potential, and root distribution at two drought-prone sites in the Pacific Northwest of the United States (Warren *et al.*, 2005). A comparison of diurnal and season changes in water content showed, with reference to eddy-flux measurements, that more than half the total water extracted by trees during the summer drought period came from a few roots that extended below a depth of 2m. When the surface soil dries below water potentials experienced by nontranspiring trees (predawn values), water acquired by deep roots in zones of high potential is accessed. Some of this water is redistributed to surface

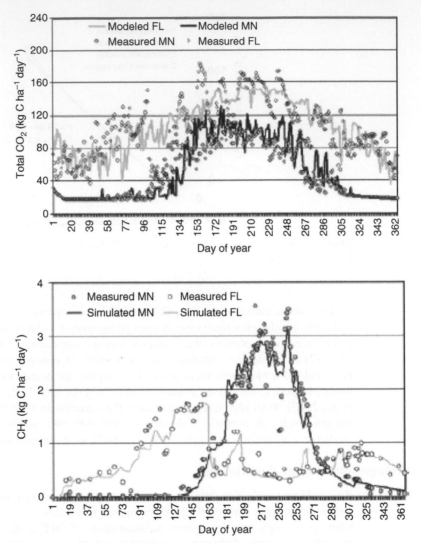

FIGURE 10.5.   Through integration of three models, predictions of CO_2 and CH_4 efflux were compared for bogs in Minnesota (MN) and in Florida (FL). (After Cui *et al.,* 2005.)

roots and leaks out to the surrounding soil (Chapter 2; Fig. 2.15). Evidence of significant water transfer from deep to shallow soils during periods of drought makes it difficult to apply sophisticated soil-plant-atmosphere models that ignore this redistribution.

Alternatively, by monitoring seasonal variation in tree predawn water potential together with modeling (or measuring) daily water use via sap flux monitoring, a reasonable estimate of the total amount of available water in the rooting can be attained. With such knowledge it is not difficult to predict predawn tree water potential values and related seasonal changes in maximum canopy stomatal conductance (Fig. 2.13). Simplified models

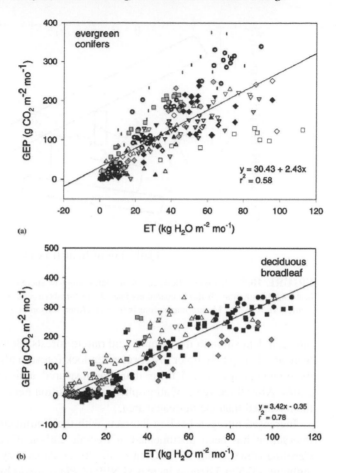

FIGURE 10.6.   Where eddy-flux data were acquired at evergreen and deciduous forest sites over a number of years, a general relationship emerged between ET and gross ecosystem production (GEP) when comparisons were made at weekly (not shown) and monthly time steps. (After Law *et al.*, 2002.)

such as Forest-BGC and 3-PG lend themselves readily to this approach (Coops *et al.*, 2001).

In Chapter 3 we recognized that diurnal and seasonal changes in ratio of sensible to latent heat losses (the Bowen ratio) offered a defining property of different types of vegetation that might be assessed by remote sensing and other means. The value of such Bowen ratio comparisons is demonstrated in Figure 10.7, based on energy balance analyses made at 20 eddy-flux tower sites throughout summer months.

G. Biogeochemistry

Although eddy-flux measurements now extend across a wide range of vegetation and environments, only one experiment to our knowledge has attempted to compare the impact of

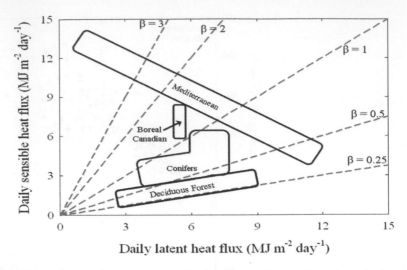

FIGURE 10.7.   Distinct differences in the partitioning of energy between latent (vapor loss) and sensible heat (the Bowen ratio, β) allow separation of deciduous forests (n = 5), boreal coniferous forest (n = 5), temperate conifers (n = 6), and Mediterranean forests (n = 4). (After Wilson *et al.*, 2002.)

nitrogen deposition in the form of acid rain to forests. Beginning in 2001, $18\,kg\,N\,ha^{-1}\,yr^{-1}$ were applied by helicopter to a spruce-hemlock forest in Maine. Initially, NEP measured at the treated area was ~5% lower than that measured at an untreated stand (Hollinger *et al.*, 2004). After three years of atmospheric N deposition net carbon uptake was ~8% higher at the fertilized than the untreated area.

Increased nitrogen loading interacts with the continued rise in atmospheric CO_2. This interaction has been documented at free-air carbon dioxide enrichment sites in North Carolina (Finzi *et al.*, 2002; Oren *et al.*, 2001). At canopy closure, CO_2 enrichment from ambient ~375 to 570 ppm increased NPP of *Pinus taeda* by only 14% on N deficient soils in contrast to 22% where supplements of N were added. These differences in growth responses may be the result of excess photosynthate acquired under elevated CO_2 being shifted to extract more nitrogen from the soil through the growth of fine roots and production of carbon-rich exudates. Sustained additions of nitrogen, through atmospheric deposition, commercial application of fertilizer, or through N-fixation can result in acidification of the soil and leaching of base cations (Chapter 6). In time, increased mortality may result, directly through nutrient imbalance, or indirectly by making trees more susceptible to insect and disease attack. This sequence of events appears to have occurred in unpolluted Douglas-fir forests of western Oregon as a result of conversion of stands of *Alnus*, a native nitrogen-fixing species, to extensive plantations of conifers that became infected with a needle cast disease when N/Ca ratios exceeded a critical threshold (Perakis *et al.*, 2006).

III. NEW REMOTE SENSING OF FORESTS

The latest generation of Earth-observing satellites provides significant improvements in radiometric sensitivity, geolocation accuracy, and spectral calibration over older sensors.

In addition, specific products are being generated for distribution in near-real time, thanks to the efforts of teams of scientists supported by sponsoring agencies such as NASA. The implications of these two advances are discussed in this section.

A. Canopy Fluxes

The accuracy of these new, more quantitative satellite data sets is being assessed in a variety of ways. For example, the global network of eddy-flux monitoring sites is being utilized to compare a range of variables predicted from satellite-derived information. There is a spatial analysis problem to scale from the approximately 1 km radius around a flux-tower that eddy covariance data represents (Schmid, 2002) to landscapes and large geographic areas. Figure 10.8 illustrates a conceptual framework to relate flux tower measurements to data acquired with satellites. For MODIS data, a project called "Bigfoot" developed a sampling protocol to estimate land cover and LAI around each flux tower (Cohen *et al.*, 2003). This project was repeated at many forested flux tower sites with promising results (Turner *et al.*, 2005). Some changes, however, were required in the initial algorithms to predict peak LAI estimates and land cover accurately (Cohen *et al.*, 2003).

1. Gross Primary Production

Remote sensing allows us to expand predictions of GPP from individual stands to regions and continents. Since 2000, imagery from NASA's Moderate Resolution Imaging Spec-

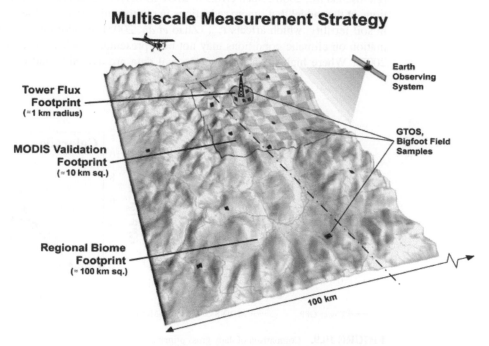

FIGURE 10.8.   Integration of field and remote sensing measurements for landscape scaling of ecosystem fluxes. (After Running *et al.*, 1999.)

troradiometer (MODIS), combined with extrapolations of coarse resolution weather data (1° latitude by 1.25° longitude), has yielded global scale estimates of GPP, averaged over eight-day intervals at 1 km resolution (Running *et al.*, 2004). The MODIS GPP product is calculated:

$$GPP = PAR \times FPAR \times \varepsilon_{max} (S_{Tmin} \times S_{vpd}) \tag{10.1}$$

where PAR represents incident photosynthetically active radiation, FPAR is the fraction of PAR absorbed by the plant canopy, ε_{max} is maximum quantum efficiency, and S_{Tmin} and S_{vpd} are, respectively, scalars for temperature minimum and vapor pressure deficits that are set to vary from 0 (shut down) to 1 (optimum).

The MODIS production efficiency model is simplified. It does not require information on soil water holding capacity, soil fertility, or daily precipitation. Where drought is important the assumption is made that sustained high vapor pressure deficits occur that suppress photosynthesis and ultimately reduce LAI and FPAR to a similar amount predicted by models that include a water balance. Rather than vary quantum efficiency as a function of soil fertility, ε_{max} is defined for representative types of vegetation within a biome.

Figure 10.9 illustrates how weekly flux tower measurements of GPP compare with MODIS satellite-derived estimates for a mature deciduous forest in upper Michigan, U.S.A., and for a Canadian boreal spruce forest. At both sites the seasonal patterns in GPP corresponded well, but MODIS satellite-derived values averaged 20 to 30% higher (Heinsch *et al.*, 2006). Such errors in MODIS-derived estimates of GPP reflect simplifications in the model that ignore reductions in photosynthesis as trees age and local variation in soil fertility, which affects ε_{max} (Zhao *et al.*, 2005). In addition, satellite-derived information on climatic conditions may not be representative when extrapolated (Zhao *et al.*, 2005). Where high quality meteorological data are available and canopy quantum effi-

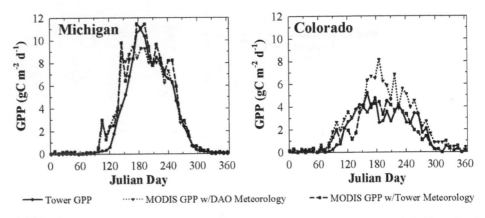

FIGURE 10.9. Comparison of daily gross primary production measured at an old growth deciduous broad-leaf forest in Michigan and a subalpine fir forest in Colorado, U.S.A. with GPP derived from MODIS satellite data (Heinsch *et al.*, 2006).

ciency is known, MODIS-derived estimates of GPP are close to those measured at tower sites (Turner *et al.*, 2006).

Another MODIS product, the enhanced vegetation index (EVI), provides improved estimates of FPAR and ε_{max} compared with the normalized difference vegetation index (NDVI) through addition of a blue spectral band in addition to near-infrared (NIR) and red (R) (Huete *et al.*, 2002). As a result, during the growing season, EVI has been linearly correlated with GPP measured at 10 widely dispersed eddy-flux tower sites (Rahman *et al.*, 2005).

2. Net Primary Production (NPP)

To predict variation in NPP across landscapes of increasing size requires progressive simplifications in models, as emphasized in Chapter 7 and discussed earlier. New global maps of MODIS-derived landcover, LAI, and annual NPP now are produced every year since 2000. Plates 14, 17, and 19 illustrate these new global datasets, which are available at http://edcimswww.cr.usgs.gov/pub/imswelcome/.

With increasing confidence in simplified formulations of stand growth models, their application has been expanded to estimate NPP across landscapes (Tickle *et al.*, 2001) and regions (Whitehead *et al.*, 2002; Swenson *et al.*, 2005). As the spatial scale increases, model accuracy at any specific point tends to decrease, reflecting deficiencies in the reliability of climatic extrapolations, soil maps, and the ability to register details about the vegetation. Nonetheless, spatial patterns predicted in growth potential tend to follow those recorded in reference to scattered point measurements (Swenson *et al.*, 2005).

Pan *et al.* (2006) used an innovative approach to map forest NPP for all of the northeastern United States (Fig. 10.10). A forest biogeochemistry model, PnET, was first run to identify local forest types, and to evaluate harvest impacts and soil water and nutrient

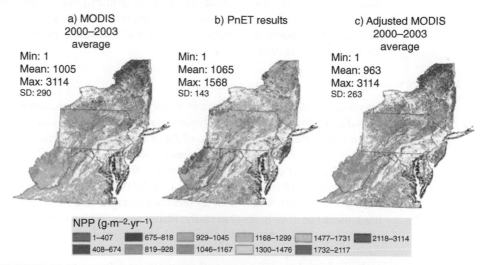

FIGURE 10.10.   Calibration of MODIS satellite derived annual net primary production for the New England region with field inventory data to improve local accuracy (Pan *et al.*, 2006). See Color Plate.

limitations on productivity. Global MODIS NPP products then were recalibrated to improve regional estimates, while retaining the comprehensive geographic coverage the satellite data provide.

Recently established forests are recognizable by their temporarily low leaf area indices. Once the canopy closes, differences exist between the greenness and wetness indices of reflected properties as forests age (Cohen et al., 2002). The standing biomass of forests is not a direct indicator of growth but a variable that new remote sensing tools in combination show increasing promise to estimate (Treuhaft et al., 2004; Lefsky et al., 2005a). If canopy height can be assessed at an accuracy of <1 m, periodic measurements at decadal intervals will provide a direct measure of growth (Lefsky et al., 2005b).

Throughout much of the world, forests are fairly young; thus estimates of leaf area index and related canopy properties are adequate measures from which to generate estimates of gross photosynthesis and autotrophic respiration, the latter a function of surface air temperature (Rahman et al., 2005). As a result, satellite-derived products of GPP and R_a together yield reasonable seasonal estimates of NPP at 8×8 km resolution globally (Running et al., 2004; Rahman et al., 2005). As with GPP, the extent that NPP can be modeled accurately independent of information on soils, precipitation, and canopy quantum efficiency is not fully known but is a recognized concern (Zhao et al., 2005).

3. Evapotranspiration

Daily evapotranspiration from flux tower data also has allowed development of a MODIS-driven ET for continental applications. Cleugh et al. (2007) compared daily ET data from two flux towers in Australia to evaluate how well various equations calculated seasonal drought constraints on ET. They ultimately chose the Penman-Monteith equation (Chapter 2) and, using MODIS landcover to define aerodynamic resistance and LAI, scaled surface resistance to compute daily ET (Fig. 10.11). Mu et al. (2007) produced the first global dataset from 2000 to 2005 of terrestrial daily evapotranspiration with this procedure. Even diurnal cycles of ET can be estimated at landscape scales by incorporating surface reflectance and radiometric temperature data at 15 to 30 minute intervals from geostationary satellites (Norman et al., 2003).

4. Phenology

Improved satellite-derived estimates of large-scale forest phenology are now becoming available. Delbart et al. (2005) compared three spectral indices from the SPOT-VGT satellite to define the growing seasons of *Larix* and *Populus* forests in Siberia from 1999 to 2002. A NDSI = Normalized Difference Snow Index $(0.45\mu - 1.67\mu)/(0.45\mu + 1.67\mu)$ identified the end of snowmelt, whereas the NDVI = Normalized Difference Vegetation Index $(0.83\mu - 0.65\mu)/(0.83\mu + 0.65\mu)$ and NDWI = Normalized Difference Water Index $(0.83\mu - 1.67\mu)/(0.83\mu + 1.67\mu)$ recorded changes in canopy development throughout the growing season. In another study, Fisher et al. (2006) merged Landsat data with field observations to construct landscape-level phenological maps of deciduous forests for U.S. southern New England.

In the boreal zone, the seasonal transition from frozen to thawed conditions is a hydro-ecological trigger clearly detectable in the return backscatter of radar wavelengths. Kimball

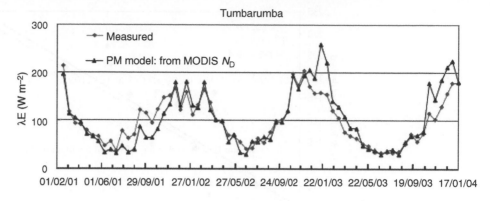

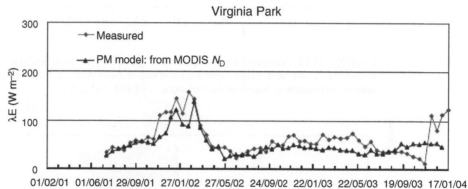

FIGURE 10.11. Comparison of daily evapotranspiration measured at a flux tower and derived from MODIS satellite data for a temperate (Tumbaramba) and savannah (Virginia Park) forests in Australia (Cleugh *et al.*, 2007).

et al. (2004) employed an active radar sensor to record high-latitude seasonal freeze/thaw cycles and thus define the growing seasons for ten predominantly evergreen forests across western North America. While the radar sensor provides spatially coarse coverage (37 × 25 km), radar wavelengths penetrate clouds and offer temporal consistency unavailable from optical sensors (Fig. 10.12).

5. Canopy Chemistry

Canopy biochemical concentrations of Rubisco define the CO_2 uptake capacity of leaves, and are approximated by leaf N concentration. Green *et al.* (2003) found that 85% of the variation in ε_{max} in Equation 10.1 can be explained by leaf N concentration and specific leaf area. Hence, remote sensing of leaf N or chlorophyll would be valuable, but is not possible with broadband visible/IR sensors. New hyperspectral sensors typically have 200 to 300 spectral channels between 0.4 and 2.5 μ of high radiometric sensitivity, and currently are deployed from aircraft, providing much smaller 2 to 10 m, more homogeneous pixels than satellite sensors (Fig. 10.13). Zarco-Tejada *et al.* (2004) used narrow channels at 710 and 750 nm to estimate needle chlorophyll in *Pinus banksiana* near Ontario, Canada.

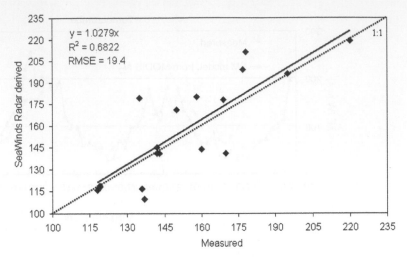

FIGURE 10.12. Estimates of growing season length in days yr^{-1} from SeaWinds radar data compared to measurements made at 10 boreal evergreen forest sites in North America. Radar wavelengths are particularly sensitive to freeze/thaw changes of the land surface (Kimball *et al.*, 2004).

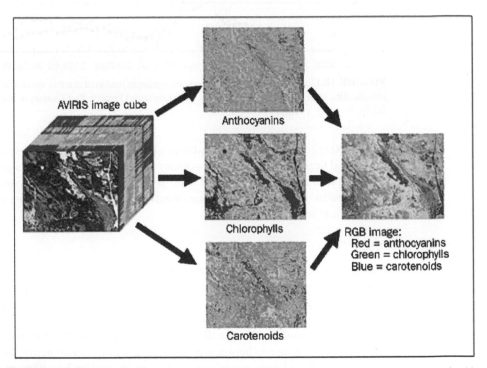

FIGURE 10.13. Classification of a Canadian boreal forest region based on relative leaf-pigment densities corresponding to different forest species derived from spectral mixture analysis of an Airborne Visible/Infrared Spectrometer dataset (Ustin *et al.*, 2004). See Color Plate.

Asner and Vitousek (2005) also demonstrated with hyperspectral data the possibility of distinguishing an overstory tree species from an N-fixing invading tree with two-fold differences in N-concentrations. Roberts *et al.* (2004) were able to identify six canopy species in the evergreen forests of western Washington based on a computed equivalent water thickness (EWT) using a selection of narrow spectral bands from AVIRIS.

Isolation of "red-edge," and SWIR reflectances from hyperspectral imagers has been shown to improve forest LAI estimation over broadband sensors such as MODIS (Lee *et al.*, 2004). However, hyperspectral data become difficult to interpret at landscape scales as multiple plant species and variations in illumination confound the signal.

B. Forest Structure

1. Lidar Measures of Structure and Biomass

Forest structure is important both for biogeochemical cycling, and to quantify other attributes such as wildlife habitat and fuel accumulations. Recent advances in lidar (light detection and ranging) show promise in being able to quantify canopy heights, stand density, and biomass. Airborne canopy lidar sensors fire a 1064 nm laser at over 1000 pulses per second at a canopy; the reflected signal is used to compute a vertical profile of structure (Fig. 10.14).

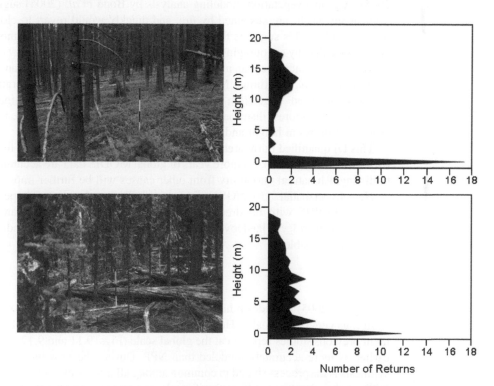

FIGURE 10.14.   The vertical canopy profile of two *Pinus contorta* forests of differing density and structure in Montana measured by airborne scanning lidar (Seielstad *et al.*, 2005).

Lefsky *et al.* (2005a, 2005b) illustrated that airborne canopy lidar was able to estimate tree heights of forests in the Pacific Northwest ranging from 4 to 80 m and stand biomass from 10 to over 1000 Mg ha^{-1}. Airborne canopy lidars currently have only 2 km swath width, so profiling large areas is not yet feasible.

2. Forest Landcover and Disturbance

As we indicated in Chapter 9, launch of the EOS platforms and improved reprocessing of historical satellite datasets provide a new generation of global landcover datasets that contribute more realism to climate models and permit calculation of national carbon and hydrologic budgets (See and Fritz, 2006). The most valuable contribution of this new technology, however, may be the ability to quantify rates of disturbance and recovery. Lepers *et al.* (2005) report that beginning with the global satellite era from 1981, the Amazon represents the highest area of deforestation, and southeast Asia shows the largest areas of forest conversion to cropland. Forests in Siberia are recognized as being rapidly harvested whereas reforestation is occurring on areas that were previously in agricultural use in the southeastern United States and eastern China.

Major disturbances, primarily fire, have impacted an estimated 4 to 6 million km^2 yr^{-1}, with highest frequencies in tropical savannahs (Potter *et al.*, 2002; Mouillot and Field, 2005). A global vegetation modeling analysis by Bond *et al.* (2005) suggests that humid tropical savannahs are sustained by fire, and quickly would revert to closed forest if fire were excluded. This suggests that climate change will impact forests more by increasing disturbance than by encouraging shifts in species distribution (Neilson *et al.*, 2005).

Figure 10.15 introduces an automated analysis of forest disturbance and recovery, using the ratio of *annual* maximum MODIS radiometric land surface temperature (LST) to the MODIS Enhanced Vegetation Index (EVI). Annual maximum LST increases and EVI decreases after a forest disturbance, resulting in an increase in DI value that returns to 1.0 as a forest grows in height and LAI recovers.

This DI quantified burnt areas in the western United Stated after the fire season of 2003 (Mildrexler *et al.*, 2007). As more years of global MODIS data are accumulated, our ability to distinguish natural variability from other causes will be further improved.

New hyperspatial (IKONOS at <1 m resolution) and airborne hyperspectral sensors such as AVIRIS with hundreds of spectral channels between 0.3 and 2.5 μ are now available to map local landcover change, although identification of individual species is not usually possible.

3. Forest Biodiversity

Sarr *et al.* (2005) provides a hierarchical framework for the analysis of species diversity at various geographic scales. Here we report on analyses at the regional level that complement previous studies reported at the global scale (Figs. 9.11 and 9.12). GPP, as explained earlier, is more accurately modeled than NPP. During the growing season, gross photosynthesis is the process shared in common among all tree species, deciduous and evergreen. NPP, on the other hand, differs significantly among species in the fraction allocated above- and belowground.

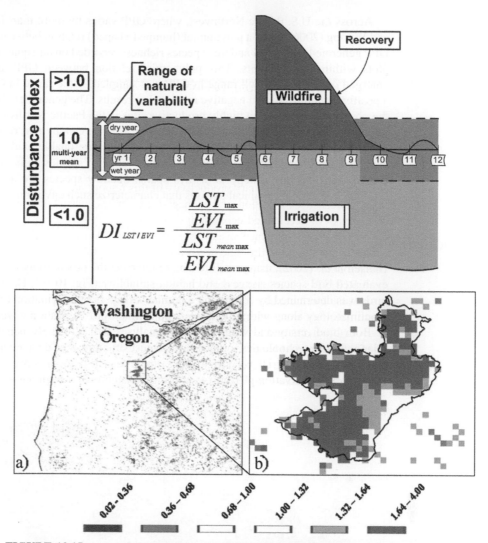

FIGURE 10.15.   (a) A disturbance index (DI) uses annual maximum values of MODIS land surface temperature (LST) and Enhanced Vegetation Index (EVI) to detect yearly changes in forest cover and energy partitioning. (b) The DI detected the area burnt by a wildfire in Oregon closely matched that defined with more precise Landsat imagery (black line). (After Mildrexler *et al.*, 2007). See Color Plate.

Across the U.S. Pacific Northwest, where GPP varies by more than 10-fold, Swenson and Waring (2006) report a polynomial (humped-shaped) relation between growing season GPP generated with 3-PG and tree species richness recorded on an equal number of survey plots within 100 km² units. This polynomial relation between GPP and tree richness emerged only when the full range in GPP was sampled. If the range in GPP is restricted, a positive, neutral, or even negative relationship results. The general explanation given for the observed pattern is that highly productive sites in the Pacific Northwest region contain a few fast-growing species that quickly develop a dense canopy that restricts the number of species present. Sites of intermediate productivity are unable to develop such dense canopies, and thus provide a haven for a wider range of species native to the region. On the harsher sites, the canopy is always sparse but few species are adapted to extreme drought, cold, and mechanical stresses that characterize such environments (Waring *et al.*, 2002).

4. Wildlife Habitat

Hansen *et al.* (2002), using satellite data to interpret the productivity of forest land cover, evaluated bird species richness and habitat suitability (Fig. 10.16). Habitat suitability for birds was determined by combining a number of layers of information on topography and bioclimatology along with satellite-derived estimates of vegetation cover.

Both biodiversity and wildlife habitat analyses will ultimately benefit from the new lidar and radar technologies for measuring forest structure once data are available for large areas. Unmanned aerial vehicles (UAVs) may soon provide a more cost effective and controllable observation platform than orbiting satellites for regional analysis.

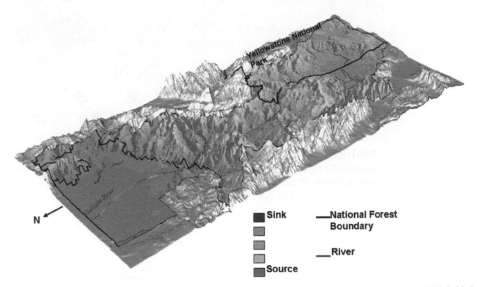

FIGURE 10.16. Richness in bird species vary across the greater Yellowstone area. Areas with high bird species richness are depicted in red. Green and tan colors indicate habitats with progressively fewer bird species. The areas with highest richness for the most part lay outside the National Park boundaries on private lands with higher productivity (Hansen *et al.*, 2002). See Color Plate.

IV. CLIMATE CHANGE AND FORESTS

A. Climatic Trends

The average global air temperature has increased by 0.8°C ± 0.2°C over the last century, with the period since 1970 showing an acceleration to around 0.35°C/decade for the northern hemisphere and 0.16°C/decade for the southern hemisphere (IPCC AR4). Warming trends are highest in the northern boreal latitudes, 1–2°C, since 1970. Most of the warming has occurred in spring and winter, and minimum daily temperatures have increased more rapidly than maximums (Easterling *et al.*, 1997; Boisvenue and Running, 2006; Bonsal *et al.*, 2001).

Global precipitation trends are less consistent but overall show an increase of 3 to 5% in the last century (Groisman *et al.*, 2004; Boisvenue and Running, 2006). This increase in precipitation does not necessarily mean more water available to forests. Higher air temperatures produce more evaporative water loss, and an increased fraction of annual precipitation falls as rain that immediately runs off rather than accumulates in snowpack (Knowles *et al.*, 2006). Snowpack has decreased significantly over the last 30 to 50 years in the western United States as well as in Canada, and spring runoff is occurring one to four weeks earlier (Stewart *et al.*, 2005). Also, there is increasing evidence that wet climates are getting wetter and dry climates drier, with a resulting intensification of extremes in the hydrologic cycle. For example, streamflow has decreased in the western United States by about 20% over the last century (Rood *et al.*, 2005), while in the eastern United States it has increased by 25% over the last 60 years (Groisman *et al.*, 2004).

B. Impacts on Forests

Forests are responding to the previously mentioned climatic trends in numerous ways, the more obvious being through a change in phenology, tree growth rates, and the frequency and extent of disturbance. We summarize recent findings on these topics in this section.

1. Phenology

A comprehensive review of papers on global trends in phenology documents that growing seasons have lengthened by 10 to 20 days over the last 30 to 50 years (Linderholm, 2006). A phenology model developed from extensive field observations of lilac (*Syringa vulgaris*) in reference to meteorological records for the northern hemisphere indicate that the growing season increased by 1.7 days/decade from 1955 to 2000 (Fig. 10.17) (Schwartz *et al.*, 2006). A different analysis using global daily satellite data confirms that the onset of spring "greenness" is 10 to 14 days earlier than it was two decades ago across temperate latitudes in the northern hemisphere (Zhou *et al.*, 2001; Lucht *et al.*, 2002). Flowering dates of honeysuckle (*Lonicera japonica*) in the western United States have advanced by 3.8 days per decade (Cayan, 2001). The first leaves on aspen (*Populus tremuloides*) trees in Edmonton, Canada appear on average 26 days earlier than in 1901 (Beaubien and Freeland, 2000). This extension of the growing season should accelerate growth in temperature-limited forests where water is adequate.

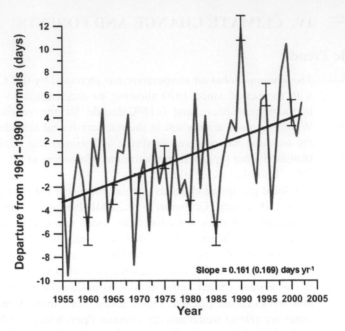

FIGURE 10.17.   A lengthening growing season of the Northern Hemisphere is observed over the last 40 years. Calculations were based on daily temperature data and phenological observations following procedures described in Schwartz *et al.* (2006).

2. Forest Growth

Globally, total terrestrial net primary production on average has increased 6% between 1982 and 1999 based on satellite-derived assessments (Nemani, 2003; Cao *et al.*, 2002). However, this satellite-derived information, coupled with regional atmospheric CO_2 data acquired between 1982 and 2002, indicate a counterbalance between enhancement of photosynthetic activity in the spring and a depression of activity with increased summer drought in forests of the northern hemisphere (Angert *et al.*, 2005).

Field observations of forest growth support similar interpretation. A literature review of climatic influences on forest growth trends of the last 50 years found 37 of the 49 papers cited showed acceleration of forest growth (Boisvenue and Running, 2006). In North America, tree diameter growth has increased by about 1% per decade over the last century in areas with unfavorably low growing season temperatures (McKenzie et al., 2001; Joos, 2002; Caspersen, 2000). In the maritime part of eastern Canada, black spruce (*Picea mariana*) at the forest-tundra transition exhibit accelerated height growth since 1970 (Gamache, 2004). In Alaska, however, warming has adversely affected the water balance on south-facing slopes, causing white spruce (*Picea glauca*) to show reduced growth rates over the last 90 years (Barber et al., 2000).

Drought-prone forests in the southwestern United States also show decreasing growth since 1895, correlated with increasing evaporative demand attributed to rising air temperatures (McKenzie et al., 2001). Over the last century, tree growth at high elevations in the

Pacific Northwest has increased as the depth of the snowpack has dropped. At the same time, tree growth at lower elevations has been reduced, suggesting water limitations associated with increasing evaporative demand (Peterson and Peterson, 2001; Peterson *et al.*, 2002).

For Europe as a whole, forest growth trends are generally positive. Vetter *et al.* (2005) attributed the increase in productivity of high elevation temperate conifer forests of Central Europe to an increase in N deposition between 1982 and 2001 and a similar response by conifer forests at mid- to low-elevations to the continued increase in atmospheric CO_2. The Swedish National Forest Inventory showed a highly significant annual increase in both height and basal area growth (0.5–0.8%) for the period 1953 to 1992 (Elfving *et al.* 1996) and site indices (SI), a measure of height attained at comparable ages, have increased for both Scots pine and Norway spruce during the last decades by 0.05 to 0.11 m/year (Ericksson and Karlsson, 1996). Site index of beech forests in Denmark also have shown an increase between 1920 and 1990 of 3.6 m (Skovsgaard and Henriksen, 1996).

Dendrochronological studies in France record an increasing growth trend over the past 150 years of 50% to 160% (Badeau *et al.*, 1996). Forests in Switzerland also showed an improved growth trend from 4 to 49% since the beginning of the twentieth century (Bräker, 1996). In Austria, studies of Norway spruce show current annual increments have increased 17% since 1961 (Hasenauer *et al.*, 1999; Schadauer, 1996). Beech forest in Slovenia increased current annual growth from 3.1 m³/ha in 1947 to 5.3 m³/ha in 1990 (Kotar, 1996). Analyses of carbon sequestration trends showed higher than expected rates of accumulation in 110-year old beech forests in Europe (Bascietto *et al.*, 2004). On the other hand, growth analysis in Portugal indicates that Maritime pine, eucalyptus, and poplar made no measurable responses between 1970 and 1990, primarily because of water limitations (Tomé *et al.*, 1996).

Across South America, an analysis of 50 long-term monitoring plots spanning years from 1971 to 2002 confirmed increases in stand basal area of 0.1 ± 0.04 m²/ha/year (Lewis *et al.*, 2004). All of these responses are not related to more intensive management or the use of commercial fertilizers. Forests appear to be responding to CO^2 fertilization, N-deposition, and longer growing seasons.

3. Net Forest Carbon Balances

Estimates of net ecosystem exchange, which represents the carbon balance of the land, can be derived from atmospheric inversion, carbon bookkeeping, and biogeochemical process models, augmented with satellite, field inventory, and flux tower data (House *et al.*, 2003). Goodale *et al.* (2002) estimated a forest-sector carbon sink of 0.28 Pg C yr^{-1} for the contiguous United States. Canada, however, because of its generally lower rates of forest growth and increasing wildfires represented a source of 0.04 Pg C yr^{-1}. In Europe, the terrestrial carbon sink during the 1990s was estimated at 0.1 to 0.2 Pg C yr^{-1} (Janssens *et al.*, 2001); forests are the major contributor to this sink (Janssens *et al.*, 2004). These C-sequestration rates have increased significantly since the 1950s when the values were near 0.03 Pg C yr^{-1}. Since the 1990s, they have increased to 0.14 Pg C yr^{-1} based on extensive inventories of European forests. The strong carbon sink in European forests is attributed to an increase in growing season length and N-deposition (Fig. 10.18) (De Vries *et al.*, 2006).

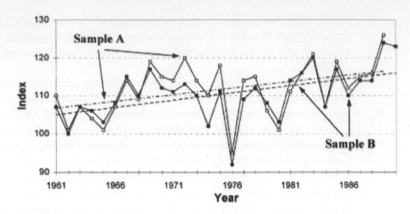

FIGURE 10.18.   Mean annual increment growth of *Picea abies* in Austria measured on 614 (Sample A) and 565 (Sample B) growth plots increased on average by 10% between 1961 and 1990. An increase in average annual air temperature of 0.72°C best accounts for a lengthening of the growing season by 11 days over this period (Hasenauer *et al.*, 1999).

4. Forest Disturbances

Climate change is altering the frequency, intensity, and duration of three major types of forest disturbances: wildfire, epidemics of pathogens, and windstorms (Dale *et al.*, 2001).

In recent decades, the area of forest burned has increased substantially. From 1920 to 1980, wildfires in the United States covered about 13,000 km²/yr. Since 1980, the area has almost doubled to 22,000 km²/yr, and for three of those years fires consumed 30,000 km² (Schoennagel *et al.*, 2004). The forested area burned between 1987 and 2003 is 6.7 times the area burned from 1970 to 1986, with a larger fraction of fires at higher elevations (Westerling *et al.*, 2006). In Canada, the area burned since 1990 averaged 30,000 km²/yr, with three extreme years with between 60 and 76,000 km² consumed (Stocks *et al.*, 2002). Gillett *et al.* (2004) reported a correlation ($r = 0.77$) between warming summer temperatures of 0.8°C and the increase in the area burned by wildfires since 1970 in Canada.

A warming climate fosters wildfires by providing drier fuel and by creating meteorological conditions conducive to rapid spread (Westerling *et al.*, 2006). Earlier snowmelt, longer growing seasons, and higher summer temperatures combine to increase the frequency of wildfires (Fig. 10.19). In addition, in those parts of the world where fire suppression activities were most intensive, an uncharacteristically dense understory has developed along with an accumulation of dead woody material. These contribute, but do not override the effects of climatic change.

Insects and diseases are a natural part of all ecosystems. In forests, periodic insect epidemics kill millions of hectare of trees, providing heavy loads of potential tinder. The dynamics of these outbreaks are related to insect lifecycles that are tightly coupled to climate fluctuations (Williams and Liebhold, 2002). Many insects adapted to cold climates have a two-year lifecycle, so warmer winter temperatures increase larvae survival. As a result, spruce budworm in Alaska now completes its lifecycle in one year (Volney and

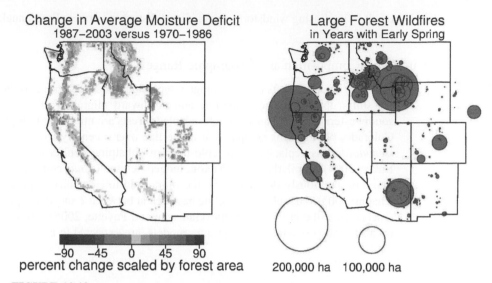

FIGURE 10.19. From 1970 to 2003, summertime moisture deficits increased dramatically in the western United States, and with a concurrent rise in wildfires >400 ha associated with an earlier melting of the snowpack in most years (Running, 2006; Westerling *et al.*, 2006).

Fleming, 2000). Mountain pine beetles have expanded their range in British Columbia into areas previously too cold to support their survival (Carroll *et al.*, 2003). Multiyear droughts also reduce the available carbohydrates, limiting the ability of trees to generate defensive chemicals to repel insect attack (Logan *et al.*, 2003). Hogg *et al.* (2002) found recent dieback of aspen stands in Alberta to be caused by a combination of light snowpacks and drought in the 1980s that triggered defoliation by tent caterpillars, followed by damage from wood-boring insects and fungal pathogens.

Candau and Fleming (2005) studied the bioclimatic control of major spruce budworm defoliations in Ontario, Canada over an area of 400,000 km². Severe defoliations were associated with dry Junes when larvae are feeding, and mild winter temperatures that reduce winter mortality. Woods *et al.* (2005), however, attribute an unprecedented outbreak of *Dothisroma* needle blight on lodgepole pine in central British Columbia to increased summer precipitation. Even in mild temperate climates such as that of the southeast United States, higher winter and spring temperatures intensify southern pine beetle attacks (Gan, 2004). Although the relationships between changing climate and insect dynamics are complicated, there is a consensus that certain thresholds exist, that once exceeded, lead to dramatic increases in pests and pathogen outbreaks.

Although wind damage is common in forests worldwide, hurricanes (or typhoons as they are called in Asia) can in one event cause substantial disturbance over thousands of square kilometers. The hurricane season of 2005 damaged an estimated $5 billion worth of forests in the southeastern United States. McNulty (2002) estimates that a single large hurricane can reduce the annual carbon sequestration in all U.S. forests by 10%. New evidence suggests that climate change is increasing the power of hurricanes, which may result in increasing windstorm-driven forest disturbance rates (Emmanuel, 2005; Webster *et al.*, 2005).

5. Changes in Forest Composition and Geographic Range

The most obvious shifts of forest biogeographical boundaries with current warming trends are shifts upward in the timberline and northward migration of boreal forests. In the southwestern United States, drought-adapted forests are migrating to higher elevations. In Colorado, photographs compared at timberline over a century show that aspen (*Populus tremuloides*) is replacing more cold-tolerant subalpine coniferous species (Elliott and Baker, 2004). Similarly, in the Yukon, lodgepole pine (*Pinus contorta*) is advancing into the zone previously dominated by the more cold-tolerant black spruce (Johnstone and Chapin, 2003). The subarctic treeline has risen at between 2 and 10 cm per year in northern Quebec over the last half-century (Gamache and Payette, 2005). These trends, predicted by nearly all ecosystem biogeography models, are expected to continue.

EPILOGUE

Forests always have been important resources for humans. In ancient times, wars were fought over wood rather than oil. Foley *et al.* (2005) estimate 7 to 11 million square kilometres of forested land have been converted in the last 300 years, mostly to agriculture. On a global basis, production of fiber and fuel wood from forests, although high, is not evenly distributed for the needs of human populations. Moreover, the extent that forests will continue to provide a carbon sink for CO_2-generated from consumption of fossil fuels is questionable. Although international policies such as the Kyoto Protocol encourage delay in harvesting forests through financial incentives, countering efforts to produce cellulosic ethanol would lead to shorter rotations. The challenge to forest managers and policy makers promises to be greater than ever, with unprecedented demand to increase growth rates and harvest activities in the face of climate change and a desire to maintain biodiversity and carbon stocks.

To keep up with such rapid changes, continued monitoring of global resources will be essential. For this reason, we do not provide an update of the CD provided with the 1998 edition; rather we will post various models and updates on global conditions at the web site http://www.ntsg.umt.edu/textbooks/.

R. H. W.
S. W. R.

BIBLIOGRAPHY

Aber, J. D. (1992). Nitrogen cycling and nitrogen saturation in temperate forest ecosystems. *Trends in Ecology and Evolution* **7,** 220–223.

Aber, J. D., and Federer, A. (1992). A generalized, lumped-parameter model of photosynthesis, evapotranspiration and net primary production in temperate and boreal forest ecosystems. *Oecologia* **92,** 463–474.

Aber, J. D., and Melillo, J. M. (1991). "Terrestrial Ecosystems." Saunders College Publishing of Holt, Rinehart and Winston, Orlando, Florida.

Aber, J. D., Driscoll, C., Federer, C. A., Lathrop, R., Lovett, G., Melillo, J. M., Steudler, P., and Vogelmann, J. (1993). A strategy for the regional analysis of the effects of physical and chemical climate change on biogeochemical cycles in northeastern (U.S.) forests. *Ecological Modelling* **67,** 37–47.

Aber, J. D., Melillo, J. M., Nadelhoffer, K. J., Pastor, J., and Boone, R. D. (1991). Factors controlling nitrogen cycling and nitrogen saturation in northern temperate forest ecosystems. *Ecological Applications* **1,** 303–315.

Aber, J. D., Nadelhoffer, K. J., Steudler, P., and Melillo, J. (1989). Nitrogen saturation in northern forest ecosystems. *BioScience* **39,** 378–386.

Aber, J. D., Ollinger, S. V., Federer, C. A., Reich, P. B., Goulden, M. L., Kicklighter, D. W., Melillo, J. M., and Lathrop, R. G., Jr. (1995). Predicting the effects of climate change on water yield and forest production in the northeastern United States. *Climate Research* **5,** 207–222.

Abrams, M. D., Orwig, D. A., and Demeo, T. E. (1995). Dendroecological analysis of successional analysis of successional dynamics for a presettlement-origin white-pine–mixed-oak forest in the southern Appalachians, USA. *Journal of Ecology* **83,** 123–133.

Adams, J. M., and Woodward, F. I. (1989). Patterns in tree species richness as a test of the glacial extinction hypothesis. *Nature* **339,** 699–701.

Ae, N., Arihar, J., Okada, K., Yoshihara, T., and Johanssen, C. (1990). Phosphorus uptake by pigeon pea and its role in cropping systems of the Indian subcontinent. *Science* **248,** 477–480.

Aerts, R., Bakker, C., and DeCaluwe, H. (1992). Root turnover as determinant of the cycling of C, N, and P in dry heathland ecosystem. *Biogeochemistry* **15,** 175–190.

Agee, J. K., and Huff, M. H. (1987). Fuel succession in a western hemlock/Douglas-fir forest. *Canadian Journal of Forest Research* **17,** 697–704.

Ågren, G. I., and Ingestad, T. (1987). Root:shoot ratio as a balance between nitrogen productivity and photosynthesis. *Plant, Cell and Environment* **10,** 579–586.

Ågren, G. I., McMurtrie, R. E., Parton, W. J., Pastor, J., and Shugart, H. H. (1991). State-of-the-art of models of production-decomposition linkages in conifer and grassland ecosystems. *Ecological Applications* **1,** 118–138.

Alexander, I. J., and Fairly, R. I. (1983). Effects on N-fertilization on populations of fine roots and mycorrhizas in spruce humus. *Plant and Soil* **71,** 49–53.

Allen, R. B., Clinton, P. W., and Davis, M. R. (1997). Cation storage and availability along a *Nothofagus* forest development sequence in New Zealand. *Canadian Journal of Forest Research* **27,** 323–330.

Ambrose, S. H., and DeNiro, M. J. (1986). The isotopic ecology of East African mammals. *Oecologia* **69,** 395–406.

Amthor, J. S. (1984). The role of maintenance respiration in plant growth. *Plant, Cell and Environment* **7,** 561–569.

Anderson, E. A. (1968). Development and testing of snowpack energy balance equations. *Water Resources Research* **4,** 19–37.

Angerbjörn, A., Hersteinsson, P., Liden, K., and Nelson, E. (1994). Dietary variation in arctic foxes (*Alopex lagopus*)—An analysis of stable carbon isotopes. *Oecologia* **99,** 226–232.

Angert, A., Biraud, S., Bonfils, C., Henning, C. C., Buermann, W., Pinzon, J., Tucker, C. J., and Fung, I. (2005). Drier summers cancel out the CO_2 uptake enhancement induced by warmer springs. *PNAS* **102,** 10823–10827.

April, K., and Newton, K. (1992). Mineralogy and mineral weathering. *In* "Atmospheric Deposition and Forest Nutrient Cycling" (D. W. Johnson and S. E. Lindberg, eds.), pp. 378–425. Springer-Verlag, New York.

Arain, M. A., Black, T. A., Barr, A. G., Jarvis, P. G., Massheder, J. M., Verseghy, D. L., and Nesic, Z. (2002). *Canadian Journal Forest Research* **32,** 878–891.

Arthur, M. A., and Fahey, I. J. (1993). Controls on soil solution chemistry in a subalpine forest in north-central Colorado. *Soil Science Society of America Journal* **57,** 1122–1130.

Ascaso, C., Galvan, J., and Rodrequez-Oascual, C. (1982). The weathering of calcareous rocks by lichens. *Pedobiologia* **24,** 219–229.

Asner, G. P., and Vitousek, P. M. (2005). Remote analysis of biological invasion and biogeochemical change. *PNAS* **102,** 4383–4386.

Asrar, G., ed. (1989). "Theory and Applications of Optical Remote Sensing." Wiley, New York.

Aston, A. R. (1979). Rainfall interception by eight small trees. *Journal of Hydrology* **42,** 383–396.

Attiwill, P. M. (1994). The disturbance of forest ecosystems: The ecological basis for conservative management. *Forest Ecology and Management* **63,** 247–300.

Attiwill, P. M., and Adams, M. A. (1993). Tansley review No. 50: Nutrient cycling in forests. *New Phytologist* **124,** 561–582.

Axelsson, E., and Axelsson, B. (1986). Changes in carbon allocation patterns in spruce and pine trees following irrigation and fertilization. *Tree Physiology* **2,** 189–204.

Badeau, V., Becker, M., Bert, D., Dupouey, J.-L., Lebourgeois, F., and Picard, J.-F. (1996). Long-term growth trends of trees: Ten years of dendrochronological studies in France. *In* "Growth Trends in European Forests" (H. Spiecker, K. Mielikäinen, K. Köhl, and J. P. Skovsgaard, eds.), pp. 167–181. Springer, Berlin.

Bailey, S. W., Hornbeck, J. W., Driscoll, C. T., and Gaudette, H. E. (1996). Calcium inputs and transport in a base-poor forest ecosystem as interpreted by Sr isotopes. *Water Resources Research* **32,** 707–719.

Baker, F. S. (1944). Mountain climates of the western United States. *Ecological Monographs* **14,** 225–254.

Baker, T. G., and Attiwill, P. M. (1985). Above-ground nutrient distribution and cycling in *Pinus radiata* D. Don and *Eucalyptus obliqua* L'Herit. forests in southeastern Australia. *Forest Ecology and Management* **13,** 41–52.

Baker, W. L. (1989). A review of models of landscape change. *Landscape Ecology* **2,** 111–133.

Baldocchi, D. D. (1988). A multi-layer model for estimating sulfur dioxide deposition to a deciduous oak forest canopy. *Atmospheric Environment* **22,** 869–884.

Baldocchi, D. D., and Harley, P. C. (1995). Scaling carbon dioxide and water vapour exchange from leaf to canopy in a deciduous forest. II. Model testing and application. *Plant, Cell and Environment* **18,** 1157–1173.

Baldocchi, D. D., Hicks, B. B., and Meyers, T. P. (1988). Measuring biosphere–atmosphere exchanges of biologically related gases with micrometeorological methods. *Ecology* **69,** 1331–1340.

Baldocchi, D., Valentini, R., Running, S., Oechel, W., and Dahlman, R. (1996). Strategies for measuring and modelling carbon dioxide and water vapor fluxes over terrestrial ecosystems. *Global Change Biology* **2,** 159–168.

Ball, J. T., Woodrow, I. E., and Berry, J. A. (1987). A model predicting stomatal conductance and its contribution to the control of photosynthesis under different environmental conditions. *In* "Progress in Photosynthetic Research" (J. Biggins, ed.), Vol. 4, pp. 221–224. Nijhoff, Dordrecht, The Netherlands.

Band, L. E. (1989a). A terrain based watershed information system. *Hydrological Processes* **3,** 151–163.

Band, L. E. (1989b). Spatial aggregation of complex terrain. *Geographical Analysis* **21,** 279–293.

Band, L. E. (1993). Effect of land surface representation on forest water and carbon budgets. *Journal of Hydrology* **150,** 749–772.

Band, L. E., and Moore, I. D. (1995). Scale: Landscape attributes and geographical information systems. *Hydrological Processes* **9,** 401–422.

Band, L. E., Patterson, P., Nemani, R. R., and Running, S. W. (1993). Forest ecosystem processes at the watershed scale: Incorporating hillslope hydrology. *Agricultural and Forest Meteorology* **63,** 93–126.

Band, L. E., Peterson, D. L., Running, S. W., Coughlan, J., Lammers, R., Dungan, J., and Nemani, R. (1991). Forest ecosystem processes at the watershed scale: Basis for distributed simulation. *Ecological Modelling* **56,** 171–196.

Barber, V. A., Juday, G. P., and Finney, B. P. (2000). Reduced growth of Alaskan white spruce in the twentieth century from temperature-induced drought stress. *Nature* **405,** 668–673.

Barry, R. G. (1992). "Mountain Weather and Climate," 2nd Ed. Routledge, London.

Bartell, S. M., Cale, W. G., O'Neill, R. V., and Gardner, R. H. (1988). Aggregation error: Research objectives and relevant model structure. *Ecological Modelling* **41,** 157–168.

Bascietto, M., Cherubini, P., and Scarascia-Mugnozza, G. (2004). Tree rings from a European beech forest chronosequence are useful for detecting growth trends and carbon sequestration. *Canadian Journal of Forest Research* **34,** 481–492.

Bassow, S. L., McConnaughay, K. D. M., and Bazazz, F. A. (1994). The response of temperate tree seedlings grown in elevated CO_2 to extreme temperature events. *Ecological Applications* **4,** 593–603.

Bazazz, F. A. (1990). The response of natural ecosystems to the rising of global CO_2 levels. *Annual Review of Ecology and Systematics* **21,** 167–196.

Beaubien, E. G., and Freeland, H. J. (2000). Spring phenology trends in Alberta, Canada: Links to ocean temperature. *International Journal of Biometeorology* **44,** 53–59.

Beaufils, E. R. (1973). Diagnosis and Recommendation Integrated System (DRIS). Department of Soil Science and Agro-Meteorology, University of Natal, Pietermariteburg, South Africa.

Beckwith, R. C., and Burnell, D. G. (1982). Spring larval dispersal of the western spruce budworm (Lepidoptera: Tortriciole) in north central Washington. *Environmental Entomology* **11,** 828–832.

Beerling, D. J., and Chaloner, W. G. (1993). Evolutionary responses of stomatal density to global CO_2 change. *Biological Journal Linnaean Society* **48,** 343–353.

Beets, P. N., and Madgwick, H. A. I. (1988). Above-ground dry matter and nutrient content of *Pinus radiata* as affected by lupins, fertiliser, thinning and stand age. *New Zealand Journal of Forestry* **18,** 43–64.

Beets, P. N., and Whitehead, D. (1996). Carbon partitioning in *Pinus radiata* in relation to foliage nitrogen status. *Tree Physiology* **16,** 131–138.

Benecke, U., and Evans, G. (1987). Growth and water use in *Nothofagus truncata* (hard beech) in temperate hill country, Nelson, New Zealand. *In* "The Temperate Forest Ecosystem" (Y. Hanzi,

W. Zhan, J. N. R. Jeffers, and P. A. Ward, eds.), pp. 131–140. Institute of Terrestrial Ecology Symposium No. 20. Lavenham Press, Lavenham, Suffolk, U.K.

Benecke, U., and Nordmeyer, A. H. (1982). Carbon uptake and allocation by *Nothofagus solandri* var. *cliffortiodes* (Hook. f.) Poole and *Pinus contorta* Douglas ex Loudon ssp. *contorta* at montane and sub-alpine altitudes. *In* "Carbon Uptake and Allocation in Subalpine Ecosystems as a Key to Management" (R. H. Waring, ed.), pp. 9–21. Oregon State University, Forest Research Laboratory, Corvallis, Oregon.

Benner, R., Fogel, M. L., Sprague, E. K., and Hodson, R. E. (1987). Depletion of ^{13}C in lignin and its implications for stable isotope studies. *Nature* **329,** 708–710.

Benson, M. L., Landsberg, J. J., and Borough, C. J. (1992). The biology of forest growth experiment: An introduction. *Forest Ecology and Management* **52,** 1–16.

Berg, B., and Staaf, H. (1981). Leaching, accumulation and release of nitrogen in decomposing forest litter. *In* "Terrestrial Nitrogen Cycles" (F. E. Clark and T. Rosswall, eds.), pp. 163–178. Swedish National Science Research Council, Stockholm.

Berg, B., Berg, M. P., Bottner, P., Box, E., Breymeyer, A., Calvo de Anta, R., Couteaux, M., Escudero, A., Gallardo, A., Kratz, W., Madeira, M., Mälkönen, E., McClaugherty, C., Meentemeyer, V., Muñoz, F., Piussi, P., Remacle, J., and Virzo de Santo, A. (1993). Litter mass loss rates in pine forests of Europe and eastern United States: Some relationships with climate and litter quality. *Biogeochemistry* **20,** 127–159.

Bernays, E. A., and Woodhead, S. (1982). Plant phenols utilized by a phytophagous insect. *Science* **216,** 201–203.

Berry, J. A., and Downton, W. J. S. (1982). Environmental regulation of photosynthesis. *In* "Photosynthesis, Development, Carbon Metabolism, and Plant Productivity" (R. Govindjee, ed.), pp. 263–343. Academic Press, New York.

Beven, K. J. (1989). Changing ideas in hydrology: The case of physically based models. *Journal of Hydrology* **105,** 157–172.

Beven, K. J., and Kirkby, M. J. (1979). A physically based, variable contributing model of basin hydrology. *Hydrology Science Bulletin* **24,** 43–69.

Bhupinderpal-Singh, Nordgren, A., Ottosson Löfvenius, M., Hogberg, M. N., Mellander, P.-E., and Högberg, P. (2003). Tree root and soil heterotrophic respiration as revealed by girdling of boreal Scots pine forest: Extending observations beyond the first year. *Plant, Cell, and Environment* **26,** 1287–1296.

Biederbeck, V. O., and Campbell, C. A. (1973). Soil microbial activity as influenced by temperature trends and fluctuations. *Canadian Journal of Forest Research* **53,** 363–376.

Billings, W. D. (1978). "Plants and the Ecosystem," 3rd Ed. Wadsworth, Belmont, California.

Binkley, D. (1986). "Forest Nutrition Management." Wiley, New York.

Binkley, D., and Brown, T. C. (1993). "Management Impacts on Water Quality of Forests and Rangelands." *USDA Forest Service, General Technical. Report* **RM-239.** Fort Collins, Colorado.

Binkley, D., and Valentine, D. (1991). Fifty-year biogeochemical effects of green ash, white pine, and Norway spruce in a replicated experiment. *Forest Ecology and Management* **40,** 13–25.

Binkley, D., Cromack, K., Jr., and Baker, D. D. (1994). Nitrogen fixation by red alder: Biology and rates of control. *In* "The Biology and Management of Red Alder" (D. Hibbs, D. DeBell, and R. Tarrant, eds.), pp. 57–72. Oregon State Univ. Press, Corvallis, Oregon.

Blaise, T., and Garbaye, J. (1983). Effects de la fertilisation minerale sur les ectomycorhizes d'une hetraie. *Acta Oecologia Plantarum* **4,** 165–169.

Bockheim, J. G. (1980). Solution and use of chronofunctions in studying soil development. *Geoderma* **24,** 71–85.

Boisvenue, C., and Running, S. W. (2006). Impacts of climate change on natural forest productivity—Evidence since the middle of the 20th century. *Global Change Biology* **12,** 862–882.

Bolan, N. S., Robson, A. D., and Barrow, N. J. (1987). Effects of vesicular–arbuscular mycorrhizae on the availability of iron phosphates in plants. *Plants and Soil* **99,** 401–410.

Bonan, G. B. (1989). Environmental factors and ecological processes controlling vegetation patterns in boreal forests. *Landscape Ecology* **3,** 111–130.

Bonan, G. B. (1993). Importance of leaf area index and forest type when estimating photosynthesis in boreal forests. *Remote Sensing of the Environment* **43,** 303–314.

Bonan, G. B. (1995). Land atmosphere CO_2 exchange simulated by a land-surface process model coupled to an atmospheric general circulation model. *Journal of Geophysical Review* **100,** 2817–2831.

Bonan, G. B., Chapin III, F. S., and Thompson, S. L. (1995). Boreal forest and tundra ecosystems as components of the climate system. *Climatic Change* **29,** 147–167.

Bonan, G. B., Pollard, D., and Thompson, S. L. (1992). Effects of boreal forest vegetation on global climate. *Nature* **359,** 716–718.

Bond, W. J., Woodward, F. I., and Midgley, G. F. (2005). The global distribution of ecosystems in a world without fire. New Phytologist **165,** 525–538.

Bond-Lamberty, B., Wang, C., and Gower, S. T. (2004). A global relationship between heterotrophic and autotrophic components of soil respiration? *Global Change Biology* **10,** 1756–1766.

Bonsal, B. R., Zhang, X., Vincent, L. A., and Hood, W. D. (2001). Characteristics of daily and extreme temperatures over Canada. *Journal of Climate* **14,** 1959–1976.

Borchert, R. (1973). Simulation of rhythmic tree growth under constant conditions. *Physiologia Plantarum* **29,** 173–180.

Boring, L. R., Swank, W. T., Waide, J. B., and Henderson, G. S. (1988). Sources, fates, and impacts of nitrogen inputs to terrestrial ecosystems: Review and synthesis. *Biogeochemistry* **6,** 119–159.

Bormann, B. T., Bormann, F. H., Bowden, W. B., Pierce, R. S., Wang, D., Snyder, M. C., Li, C. Y., and Ingersoll, R. C. (1993). Rapid N_2 fixation in pines, alder, and locust: Evidence from the sandbox ecosystem study. *Ecology* **74,** 583–598.

Bormann, B. T., Spaltenstein, H., McClellan, M. H., Ugolini, F. C., Cromack, K., Jr., and Nay, S. M. (1995). Rapid soil development after windthrow disturbance in pristine forests. *Journal of Ecology* **83,** 1–20.

Bormann, F. H. (1985). Air pollution and forests: An ecosystem perspective. *BioScience* **35,** 434–441.

Bormann, F. H., and Likens, G. E. (1979). "Pattern and Process in a Forested Ecosystem." Springer-Verlag, New York.

Bosch, J. M., and Hewlett, J. D. (1982). A review of catchment experiments to determine the effect of vegetation changes on water yield and evapotranspiration. *Journal of Hydrology* **55,** 3–23.

Bossel, H. (1991). Modelling forest dynamics: Moving from description to explanation. *Forest Ecology and Management* **42,** 129–142.

Bossel, H. (1996). TREEDYN3 forest simulation model. *Ecological Modeling* **90,** 187–227.

Botkin, D. B., and Simpson, L. G. (1990). Biomass of the North American boreal forest: A step toward accurate global measures. *Biogeochemistry* **9,** 161–174.

Botkin, D. B., Janak, J. F., and Wallis, J. R. (1972). Some ecological consequences of a computer model of forest growth. *Journal of Ecology* **60,** 849–872.

Boucher, J. F., Munson, A. D., and Bernier, P. Y. (1995). Foliar absorption of dew influences shoot water potential and root growth in *Pinus strobus* seedlings. *Tree Physiology* **15,** 819–824.

Bouten, W., and Jansson, P.-E. (1995). Evaluation of model behaviour with respect to the hydrology at the Solling spruce site. *Ecological Modelling* **83,** 245–253.

Boutton, T. W. (1991). Stable carbon isotope ratios of natural materials: II. Atmospheric, terrestrial, marine, and freshwater environments. *In* "Carbon Isotope Techniques" (D. C. Coleman and B. Fry, eds.), pp. 173–185. Academic Press, San Diego.

Bradshaw, R. H. W., and Zackrisson, O. (1990). A two thousand year history of a northern Swedish boreal forest stand. *Journal of Ecology* **83,** 123–133.

Bräker, O. U. (1996). Growth Trends of Swiss Forests: Tree-Ring Data. Case Study Toppwald. *In* "Growth trends in European Forests" (H. Spiecker, K. Mielikäinen, K. Köhl, and J. P. Skovsgaard, eds.), pp. 199–217. Springer, Berlin.

Braun, E. L. (1950). "Deciduous Forests of Eastern North America." Blakiston, Philadelphia, Pennsylvania.

Bridgham, S. D., Pastor, J., McClaugherty, C. A., and Richardson, C. J. (1995). Nutrient-use efficiency: A litterfall index, a model, and a test along a nutrient-availability gradient in North Carolina peatlands. *American Naturalist* **145,** 1–21.

Briggs, G. E., Kidd, F., and West, C. (1920). A quantitative analysis of plant growth. *Annals of Applied Biology* **7,** 202–223.

Bristow, K. L., and Campbell, G. S. (1984). On the relationship between incoming solar radiation and daily maximum and minimum temperature. *Agricultural and Forest Meteorology* **31,** 159–166.

Brown, S., Gillespie, A. J. R., and Lugo, A. E. (1991). Biomass of tropical forests of south and southeast Asia. *Canadian Journal of Forest Research* **21,** 111–117.

Bryant, J. P., Chapin III, F. S., Reichardt, P., and Clausen, T. (1986). Chemical mediated interactions between plants and other organisms. *Recent Advances in Phytochemistry* **19,** 219–237.

Bryant, J. P., Provenza, F. D., Pastor, J., Reichardt, P. B., Clausen, T. P., and du Toit, J. T. (1991). Interactions between woody plants and browsing mammals mediated by secondary metabolites. *Annual Review of Ecology and Systematics* **22,** 431–446.

Buchmann, N., Gebauer, G., and Schulze, E.-D. (1996). Partitioning of ^{15}N-labeled ammonium and nitrate among soil, litter, below- and above-ground biomass of trees and understory in a 15–year-old *Picea abies* plantation. *Biogeochemistry* **33,** 1–23.

Buchmann, N., Guehl, J.-M., Barigah, T. S., and Ehleringer, J. R. (1997). Interseasonal comparison of CO_2 concentrations, isotopic composition and carbon dynamics in an Amazonian rainforest (French Guiana). *Oecologia* **110,** 120–131.

Burgan, R. E. (1988). 1988 revisions to the 1978 National Fire-Danger Rating System. *USDA Forest Service Research Paper* **SE-273.** Asheville, North Carolina.

Burger, H. (1929). Holz, Blattmenge und Zuwachs. *Mitteilungen Schweizerische Zentralanstalt für das Forstliche Versuchswesen* **15,** 243–292.

Burgman, M. A. (1996). Characterisation and delineation of the eucalypt old-growth forest estate in Australia: A review. *Forest Ecology and Management* **83,** 149–161.

Burke, I. C., and Lauenroth, W. K. (1993). What do LTER results mean? Extrapolating from site to region and decade to century. *Ecological Modelling* **67,** 19–35.

Burke, I. C., Kittel, T. G. F., Lauenroth, W. K., Snook, P., Yonker, C. M., and Parton, W. J. (1991). Regional analysis of the central great plains. *BioScience* **41,** 685–692.

Burns, R. G. (1982). Enzyme activity in soil: Location and a possible role in microbial ecology. *Soil Biology and Biochemistry* **14,** 423–427.

Burns, R. M., and Honkala, B. H. (1990). "Silvics of North America." Agriculture Handbook 654, United States Dept. of Agriculture, Forest Service, Washington, D.C.

Burton, A. J., Pregitzer, K. S., and Reed, D. D. (1991). Leaf area and foliar biomass relationships in northern hardwood forest located along an 800km acid deposition gradient. *Forest Science* **37,** 1041–1052.

Burton, A. J., Zogg, G. P., Pregitzer, K. S., and Zak, D. R. (1997). Effect of measurement CO_2 concentration on sugar maple root respiration. *Tree Physiology* **17,** 421–427.

Cadogan, B. L., Nealis, V. G., and van Frankenhuyzen, K. (1995). Control of spruce budworm (Lepidoptera: Tortricidae) with *Bacillus thuringiensis* applications timed to conserve a larval parasitoid. *Crop Protection* **14,** 31–36.

Canadell, J., Jackson, R. B., Ehleringer, J. R., Mooney, H. A., Sala, O. E., and Schulze, E.-D. (1996). Maximum rooting depth of vegetation types at the global scale. *Oecologia* **108,** 583–595.

Candau, J.-N., and Fleming, R. A. (2005). Landscape-scale spatial distribution of spruce budworm defoliation in relation to bioclimatic conditions. *Canadian Journal of Forest Research* **35,** 2218–2232.

Cannell, M. G. R. (1989). Physiological basis of wood production: A review. *Scandinavian Journal of Forest Research* **4,** 459–490.

Cannell, M. G. R., and Dewar, R. C. (1994). Carbon allocation in trees: A review of concepts for modelling. *Advances in Ecological Research* **25,** 60–140.

Cannell, M. G. R., and Thornley, J. H. M. (2000). Modeling the components of plant respiration: Some guiding principles. *Annals of Botany* **85,** 45–54.

Cannell, M. G. R., Grace, J., and Booth, A. (1989). Possible impacts of climatic warming on trees and forests in the United Kingdom: A review. *Forestry* **62,** 337–364.

Cao, M., Prince, S. D., and Shugart, H. H. (2002). Increasing terrestrial carbon uptake from the 1980s to the 1990s with changes in climate and atmospheric CO_2. *Global Biogeochemical Cycles* **16,** 1069–1079.

Carlyle, J. C., and Malcolm, D. C. (1986). Nitrogen availability beneath pure spruce and mixed larch + spruce stands growing on a deep peat. II. A comparison of N availability as measured by plant uptake and long-term laboratory incubations. *Plant and Soil* **93,** 115–122.

Carroll, A. L., Taylor, S. W., Régnière, J., and Safranyik, L. (2003). Effects of climate change on range expansion by the mountain pine beetle in British Columbia. *In* "Information Report BC-X-399 Mountain Pine Beetle Symposium: Challenges and Solutions" (T. L. Shore, J. E. Brookes, and J. E. Stone, eds.), pp. 223–232. Natural Resources Canada, Victoria, British Columbia.

Carter, G. A. (1994). Ratios of leaf reflectances in narrow wavebands as indicators of plant stress. *International Journal of Remote Sensing* **15,** 697–703.

Caspersen, J., Pacala, S., Jenkins, J., Hurtt, G., Moorcroft, P., and Birdsey, R. (2000). Contributions of land-use history to carbon accumulation in U.S. forests. *Science* **290,** 1148–1151.

Cattelino, P. J., Noble, I. R., Slatyer, R. O., and Kessell, S. R. (1979). Predicting the multiple pathways of plant succession. *Environmental Management* **3,** 41–50.

Cayan, D. R., Kammerdiener, S. A., Dettinger, M. D., Caprio, J. M., and Peterson, D. H. (2001). Changes in the onset of spring in the western United States. *Bulletin of the American Meteorological Society* **82,** 399–425.

Ceulemans, R., and Mousseau, M. (1994). Effects of elevated atmospheric CO_2 on woody plants. *New Phytologist* **127,** 425–446.

Ceulemans, R. J., and Saugier, B. (1991). Photosynthesis. *In* "Physiology of Trees" (A. S. Raghavendra, ed.), pp. 21–50. Wiley, New York.

Chapin, F. S. (1980). The mineral nutrition of wild plants. *Annual Review of Ecology and Systematics* **11,** 233–260.

Chapin, F. S., and Kedrowski, R. A. (1983). Seasonal changes in nitrogen and phosphorus fractions and autumn retranslocation in evergreen and deciduous taiga trees. *Ecology* **64,** 376–391.

Chapin III, F. S., Vitousek, P. M., and Van Cleve, K. (1986). The nature of nutrient limitations in plant communities. *American Naturalist* **127,** 48–58.

Chapman, W. L., and Walsh, J. E. (1993). Recent variations of sea ice and air temperature in high latitudes. *Bulletin of the American Meteorological Society* **74,** 33–47.

Chase, T. N., Pielke, R. A., Kittel, T. G. F., Nemani, R., and Running, S. W. (1996). Sensitivity of a general circulation model to global changes in leaf area index. *Journal of Geophysical Research* **101**(D3), 7393–7408.

Chen, J. M., and Cihlar, J. (1996). Retrieving leaf area index of boreal conifer forests using Landsat TM images. *Remote Sensing of the Environment* **55,** 153–162.

Chen, J., and Franklin, J. F. (1992). Vegetation responses to edge environments in old-growth Douglas-fir environments. *Ecological Applications* **2,** 387–396.

Christensen, N. K., Agee, J. K., Brussard, P. F., Hughes, J., Knight, D. H., Minshall, G. W., Peek, J. M., Pyne, S. J., Swanson, F. J., Thomas, J. W., Wells, S., Williams, S. E., and Wright, H. A. (1989). Interpreting the Yellowstone fires of 1988. *BioScience* **39,** 678–685.

Christensen, N. L., and Peet, R. K. (1981). Secondary forest succession on the North Carolina Piedmont. *In* "Forest Succession: Concepts and Applications" (D. C. West, H. H. Shugart, and D. B. Botkin, eds.), pp. 230–245. Springer-Verlag, New York.

Christensen, N. L., Bartuska, A. M., Brown, J. H., Carpenter, S., DeAntonio, C., Francis, R., Franklin, J. F., MacMahon, J. A., Noss, R. F., Parsons, D. J., Peterson, C. H., Turner, M. G., and Woodmansee, R. G. (1996). The report of the Ecological Society of America committee on the scientific basis for ecosystem management. *Ecological Applications* **6,** 665–691.

Christiansen, E., and Ericsson, A. (1986). Starch reserves in *Picea abies* in relation to defense reaction against a bark beetle transmitted blue-stain fungus, *Ceratocystis polonica. Canadian Journal of Forest Research* **16,** 78–83.

Christiansen, E., Waring, R. H., and Berryman, A. A. (1987). Resistance of conifers to bark beetle attack: Searching for general relationships. *Forest Ecology and Management* **22,** 89–106.

Chung, H. H., and Barnes, R. L. (1977). Photosynthate allocation in *Pinus taeda.* I. Substrate requirements for synthesis of shoot biomass. *Canadian Journal of Forest Research* **7,** 106–111.

Chuvieco, E., and Martin, M. P. (1994). A simple method for fire growth mapping using AVHRR channel 3 data. *International Journal of Remote Sensing* **15,** 3141–3146.

Ciais, P., Tans, P. P., Trolier, M., White, J. W. C., and Francey, R. J. (1995). A large northern hemisphere terrestrial CO_2 sink indicated by the $^{13}C/^{12}C$ ratio of atmospheric CO_2. *Science* **269,** 1098–1102.

Cienciala, E., Running, S. W., Lindroth, A., Grelle, A., and Ryan, M. G. (1998). Analysis of carbon and water fluxes from the NOPEX boreal forest: Comparison of measurements with FOREST-BGC simulations. *Journal of Hydrology* **212–213,** 62–78.

Clapp, R. B., and Hornberger, G. M. (1978). Empirical equations for some soil hydraulic properties. *Water Resources Research* **14,** 601–604.

Clark, J. S. (1988). Effect of climate change on fire regimes in northwestern Minnesota. *Nature* **334,** 233–235.

Clark, K. L., Gholz, H. L., and Castro, M. S. (2004). Carbon dynamics along a chronosequence of slash pine plantations in north Florida. *Ecological Applications* **14,** 1154–1171.

Cleary, B. D., and Waring, R. H. (1969). Temperature collection of data and its analysis for the interpretation of plant growth and distribution. *Canadian Journal of Botany* **47,** 167–173.

Clements, F. E. (1936). Nature and structure of the climax. *Journal of Ecology* **24,** 252–284.

Cleugh, H. A., Leuning, R., Mu, Q., and Running, S. W. (2007). Regional evaporation estimates from flux tower and MODIS satellite data. *Remote Sensing of Environment.* **106,** 285–304.

Clinton, P. W., Allen, R. B., and Davis, M. R. (2002). Nitrogen storage and availability during stand development in a New Zealand *Nothofagus* forest. *Canadian Journal of Forest Research* **32,** 344–352.

Cohen, W. B., and Spies, T. A. (1992). Estimating structural attributes of Douglas-fir/western hemlock forest stands from Landsat and SPOT imagery. *Remote Sensing of the Environment* **41,** 1–17.

Cohen, W. B., Harmon, M. E., Wallin, D. O., and Fiorella, M. (1996). Two decades of carbon flux from forests of the Pacific Northwest. *BioScience* **46,** 836–844.

Cohen, W. B., Maiersperger, T. K., Yang, Z., Gower, S. T., Turner, D. P., Ritts, W. D., Berterretche, M., and Running, S. W. (2003). Comparisons of land cover and LAI estimates derived from

ETM+ and MODIS for four sites in North America: A quality assessment of 2000/2001 provisional MODIS products. *Remote Sensing of Environment* **88,** 233–255.

Cohen, W. B., Spies, T. A., Alig, R. J., Oetter, D. R., Maiersperger, T. K., and Fiorella, M. (2002). Characterizing 23 years (1972–95) of stand replacement disturbance in western Oregon forests with Landsat imagery. *Ecosystems* **5,** 122–137.

Cohen, W. B., Spies, T. A., and Fiorella, M. (1995). Estimating the age and structure of forests in a multiownership landscape of western Oregon, U.S.A. *International Journal of Remote Sensing* **16,** 721–746.

Cole, D. W., and Rapp, M. (1981). Elemental cycling in forest ecosystems. *In* "Dynamic Properties of Forest Ecosystems" (D. E. Reichle, ed.), pp. 341–409. International Biological Program 23. Cambridge Univ. Press, Cambridge.

Coleman, J. D., Gillman, A., and Green, W. Q. (1980). Forest patterns and possum densities within podocarp/mixed hardwood forests on Mt. Bryan O'Lynn, Westland, New Zealand. *New Zealand Journal of Ecology* **3,** 69–84.

Coley, P. D. (1988). Effects of plant growth rate and leaf lifetime on the amount and type of antiherbivore defense. *Oecologia* **74,** 531–536.

Comerford, N. B., and Skinner, M. F. (1989). Residual phosphorus solubility for an acid, clayey, forested soil in the presence of oxalate and citrate. *Canadian Journal of Soil Science* **69,** 111–117.

Comins, H. N., and McMurtrie, R. E. (1993). Long-term response of nutrient-limited forests to CO_2 enrichment: Equilibrium behaviour of plant–soil models. *Ecological Applications* **3,** 666–681.

Comstock, G. L. (1965). Longitudinal permeability of green eastern hemlock. *Forest Products Journal* **15,** 441–449.

Connell, J. H., and Slatyer, R. O. (1977). Mechanisms of succession in natural communities and their role in community stability and organization. *American Naturalist* **111,** 1119–1144.

Cook, B. D., Davis, K. J., Wang, W., Desai, A., Berger, B. W., Teclaw, R. M., Martin, J. G., Bolstad, P. V., Bakwin, P. S., Yi, C., and Heilman, W. (2004). Carbon exchange and venting anomalies in an upland deciduous forest in northern Wisconsin. *Agricultural and Forest Meteorology* **126,** 271–295.

Cook, E., Bird, T., Peterson, M., Barbetti, M., Buckley, B., D'Arrigo, R., Francey, R., and Tans, P. (1991). Climatic change in Tasmania inferred from a 1089–year tree-ring chronology of Huon pine. *Science* **253,** 1266–1268.

Cook, J. E. (1996). Implications of modern successional theory for habitat typing: A review. *Forest Science* **42,** 67–75.

Coops, N. C., Waring, R. H., and Landsberg, J. J. (1998). Assessing forest productivity in Australia and New Zealand using a physiologically-based model driven with averaged monthly weather data and satellite derived estimates of canopy photosynthetic capacity. *Forest Ecology and Management* **104,** 113–127.

Coops, N. C., Waring, R. H., and Moncreiff, J. B. (2000). Estimating mean monthly incident solar radiation on horizontal and inclined slopes from mean monthly temperature extremes. *International Journal of Biometeorology* **44,** 204–211.

Coops, N. C., Waring, R. H., Brown, S. R., and Running, S. W. (2001). Comparison of predictions of net primary production and seasonal patterns in water use derived with two forest growth models in southwestern Oregon. *Ecological Modeling* **142,** 61–81.

Cosby, B. J., Hornberger, G. M., Galloway, J. N., and Wright, R. F. (1985). Modeling the effects of acid deposition: Assessment of a lumped parameter model of soil water and streamwater chemistry. *Water Resources Research* **21,** 51–63.

Cote, B., and Camire, C. (1995). Application of leaf, soil, and tree ring chemistry to determine the nutritional status of sugar maple on sites of different levels of decline. *Water, Air and Soil Pollution* **83,** 363–373.

Coughlan, J. C., and Running, S. W. (1997a). Regional ecosystem simulation: A general model for simulating snow accumulation and melt in mountainous terrain. *Landscape Ecology* **12,** 119–136.

Coughlan, J. C., and Running, S. W. (1997b). Biophysical aggregations of forested landscape using an ecological diagnostic system. *Transactions in GIS* **1,** 25–39.

Covington, W. W., and Moore, M. M. (1994). Southwestern ponderosa pine forest structure and resource conditions: Changes since Euro-American settlement. *Journal of Forestry* **92,** 39–47.

Cowles, H. C. (1899). The ecological relations of the vegetation of the sand dunes of Lake Michigan. *Botanical Gazette* **27,** 95–117, 167–202, 281–308, and 361–391.

Coyea, M. R., and Margolis, H. A. (1992). Factors affecting the relationship between sapwood and leaf area of balsam fir. *Canadian Journal of Forest Research* **22,** 1684–1693.

Coyea, M. R., and Margolis, H. A. (1994). The historical reconstruction of growth efficiency and its relationship to tree mortality in balsam fir ecosystems affected by spruce budworm. *Canadian Journal of Forest Research* **24,** 2208–2221.

Coyea, M. R., Margolis, H. A., and Gagnon, R. R. (1990). A method for reconstructing the development of the sapwood area of balsam fir. *Tree Physiology* **6,** 283–291.

Crane, W. J. B., and Banks, J. C. G. (1992). Accumulation and retranslocation of foliar nitrogen in fertilised and irrigated *Pinus radiata. Forest Ecology and Management* **52,** 201–223.

Crawley, M. J. (1983). "Herbivory: The Dynamics of Animal–Plant Interactions." Univ. of California Press, Berkeley.

Creed, I. F., Band, L. E., Foster, N. W., Morrison, I. K., Nicolson, J. A., Semkin, R. S., and Jeffries, D. S. (1996). Regulation of nitrate-N release from temperate forests: A test of the N flushing hypothesis. *Water Resources Research* **23,** 3337–3354.

Crill, P. M. (1991). Seasonal patterns of methane uptake and carbon dioxide release by a temperate woodland soil. *Global Biogeochemical Cycles* **5,** 319–334.

Cromer, R. N., Cameron, D. M., Rance, S. J., Ryan, P. A., and Brown, M. (1993a). Response to nutrients in *Eucalyptus grandis.* 2. Nitrogen accumulation. *Forest Ecology and Management* **62,** 231–243.

Cromer, R. N., Cameron, D. M., Rance, S. J., Ryan, P. A., and Brown, M. (1993b). Response to nutrients in *Eucalyptus grandis.* 1. Biomass accumulation. *Forest Ecology and Management* **62,** 211–230.

Cropper, W. P., Jr., and Gholz, H. L. (1993). Simulation of the carbon dynamics of Florida slash pine plantation. *Ecological Modelling* **66,** 231–249.

Cropper, W., Jr., and Gholz, H. L. (1994). Evaluating potential response mechanisms of a forest stand to fertilization and night temperature: A case study using *Pinus elliottii. Ecological Bulletins* **43,** 154–160.

Cui, J., Li, C., and Trettin, C. (2005). Analyzing the ecosystem carbon and hydrologic characteristics of forested wetland using a biogeochemical process model. *Global Change Biology* **11,** 278–289.

Curran, P. J. (1994). Imaging spectrometry. *Progress in Physical Geography* **18,** 247–266.

Currie, W. S., Aber, J. A., McDowell, W. H., Boone, R. D., and Magill, A. H. (1996). Vertical transport of dissolved organic C and N under long-term N amendments in pine and hardwood forests. *Biogeochemistry* **35,** 471–505.

Dahlgren, R. A. (1994). Soil acidification and nitrogen saturation from weathering of ammonium-bearing rock. *Nature* **368,** 838–841.

Dale, V. H., and Rauscher, H. M. (1994). Assessing impacts of climate change on forests: The state of biological modeling. *Climatic Change* **28,** 65–90.

Dale, V. H., Doyle, T. W., and Shugart, H. H. (1985). A comparison of tree growth models. *Ecological Modelling* **29,** 145–169.

Dale, V. H., ed. (1994). "Effects of Land-Use Change on Atmospheric CO_2 Concentrations: South and Southeast Asia as a Case Study." Springer-Verlag, New York.

Dale, V. H., Gardner, R. H., DeAngelis, D. L., Eagar, C., and Webb, J. W. (1991). Elevation-mediated effects of balsam wooly adeligid on southern Appalachian spruce–fir forests. *Canadian Journal of Forest Research* **21,** 1639–1648.

Dale, V. H., Joyce, L. A., McNulty, S., Neilson, R. P., Ayres, M. P., Flannigan, M. D., Hanson, P. J., Irland, L. C., Lugo, A. E., Peterson, C. J., Simberloff, D., Swanson, F. J., Stocks, B. J., Wotton, M. (2001). Climate change and forest disturbances. *BioScience* **51,** 723–734.

Daly, C., Neilson, R. P., and Phillips, D. L. (1994). A statistical-topographic model for mapping climatological precipitation over mountainous terrain. *Journal of Applied Meteorology* **33,** 140–158.

Dang, Q.-L., Margolis, H. A., Sy, M., Coyea, M. R., Collartz, G. J., and Wathall, C. L. (1997). Profiles of photosynthetically active radiation, nitrogen, and photosynthetic capacity in the boreal forest: Implications for scaling for leaf to canopy. *Journal of Geophysical Research* **102(D24),** 28,845–828,860.

Darley-Hill, S., and Johnson, W. C. (1981). Acorn dispersal by the blue jay (*Cyanocitta cristata*). *Oecologia* **50,** 231–232.

Davidson, E. A., Hart, S. C., and Firestone, M. K. (1992). Internal cycling of nitrate in soils of a mature coniferous forest. *Ecology* **73,** 1148–1156.

Davidson, R. L. (1969). Effect of root–leaf temperature differentials on root–shoot ratios in some pasture grasses and clover. *Annals of Botany* **33,** 561–569.

Dawson, T. E. (1993). Water sources of plants as determined from xylem-water isotopic composition: Perspectives on plant competition, distribution, and water relations. *In* "Stable Isotopes and Plant Carbon–Water Relations" (J. R. Ehleringer, A. E. Hall, and G. D. Farquhar, eds.), pp. 465–496. Academic Press, San Diego.

De Vries, W., Reinds, G. J., Gundersen, P., and Sterba, H. (2006). The impact of nitrogen deposition on carbon sequestration in European forests and forest soils. *Global Change Biology* **12,** 1151–1173.

Deans, J, D. (1981). Dynamics of coarse root production in a young plantation of *Picea sitchensis*. *Forestry* **54,** 139–155.

DeBano, L. F., and Rice, R. M. (1973). Water repellent soils: Their implications in forestry. *Journal of Forestry* **71,** 220–223.

Delbart, N., Kergoat, L., Toan, T. L., Lhermitte, J., and Picard, G. (2005). Determination of pheno-logical dates in boreal regions using normalized difference water index. *Remote Sensing of Environment* **97,** 26–38.

Delcourt, H. R., and Delcourt, P. A. (1985). Comparison of taxon calibrations, modern analogue techniques, and forest-stand simulation models for the quantitative reconstruction of past vegeta-tion. *Earth Surface Processes and Landforms* **10,** 293–304.

Delwiche, C. C. (1970). The nitrogen cycle. *Scientific American* **223,** 136–146.

Dement, W. A., and Mooney, H. A. (1974). Seasonal variation in the production of tannins and cyanogenic glucosides in the chaparral shrub, *Heteromeles arbutifolia. Oecologia* **15,** 65–76.

Desjardins, R. L., Hart, R. L., MacPherson, J. I., Schuepp, P. H., and Verma, S. B. (1992a). Aircraft-and tower-based fluxes of carbon dioxide, latent, and sensible heat. *Journal of Geophysical Research* **97,** 18477–18485.

Desjardins, R. L., Schuepp, P. H., MacPherson, J. I., and Buckley, D. J. (1992b). Spatial and tem-poral variations of the fluxes of carbon dioxide and sensible and latent heat over the FIFE site. *Journal of Geophysical Research* **97,** 18467–18475.

DeSteven, D., Kline, J., and Matthiae, P. E. (1991). Long-term changes in a Wisconsin *Fagus–Acer* forest in relation to glaze storm distribution. *Journal of Vegetation Science* **2,** 201–208.

Dewar, R. C. (1993). A root–shoot partitioning model based on carbon–nitrogen–water interactions and Munch phloem flow. *Functional Ecology* **7,** 356–368.

Dewar, R. C. (1995). Interpretation of an empirical model for stomatal conductance in terms of guard cell function: Theoretical paper. *Plant, Cell and Environment* **18,** 365–372.

Dewar, R. C., Ludlow, A. R., and Dougherty, P. M. (1994). Environmental influences on carbon allocation in pines. *Ecological Bulletins* **43,** 92–101.

Diaz-Ravina, M., Acea, M. J., and Carballas, T. (1995). Seasonal changes in microbial biomass and nutrient flush in forest soils. *Biology and Fertility of Soils* **19,** 220–226.

Dickinson, R. E., and Henderson-Sellers, A. (1988). Modelling tropical deforestation: A study of GCM land-surface parameterizations. *Quarterly Journal of the Royal Meteorology Society* **114,** 439–462.

Dickson, B. A., and Crocker, R. L. (1953). A chronosequence of soils and vegetation near Mt. Shasta, California. II. The development of the forest floors and the carbon and nitrogen profiles of the soils. *Journal of Soil Science* **4,** 142–154.

Dickson, R. E., and Broyer, T. C. (1972). Effects of aeration, water supply, and nitrogen source on growth and development of tupelo gum and bald cypress. *Ecology* **53,** 626–634.

Dise, N. B., and Wright, R. F. (1995). Nitrogen leaching from European forests in relation to nitrogen deposition. *Forest Ecology and Management* **71,** 153–161.

Dixon, R. K., Brown, S., Houghton, R. A., Solomon, A. M., Trexler, M. C., and Wisniewski, J. (1994). Carbon pools and flux of global forest ecosystems. *Science* **263,** 185–190.

Doane, C. C., and McManus, M. L., eds. (1981). The gypsy moth: Research toward integrated pest management. *United States Department of Agriculture, Technical Bulletin* **1584,** 65–86.

Dobson, M. C., Ulaby, F. T., and Pierce, L. E. (1995). Land-cover classification and estimation of terrain attributes using synthetic aperture radar. *Remote Sensing of the Environment* **51,** 199–214.

Doley, D. (1981). Tropical and subtropical forests and woodlands. *In* "Water Deficits and Plant Growth" (T. T. Kozlowski, ed.), pp. 209–223. Academic Press, New York.

Dougherty, P. M., Whitehead, D., and Vose, J. M. (1994). Environmental influences on the phenology of pine. *Ecological Bulletins* **43,** 64–75.

Drew, T. J., and Flewelling, J. W. (1977). Some recent Japanese theories of yield–density relationships and their application to Monterey pine plantations. *Forest Science* **23,** 517–534.

Driscoll, C. T., Likens, G. E., Hedin, L. O., Eaton, J. S., and Bormann, F. H. (1989). Changes in the chemistry of surface water. *Environmental Science and Technology* **23,** 137–143.

Duncan, R. P., and Stewart, G. H. (1991). The temporal and spatial analysis of tree age distributions. *Canadian Journal of Forest Research* **21,** 1703–1710.

Dunin, F. X., McIlroy, I. C., and O'Loughlin, E. M. (1985). A lysimeter characterization of evaporation by eucalypt forest and its representatitiveness for the local environment. *In* "The Forest–Atmosphere Interaction" (B. A. Hutchison and B. B. Hicks, eds.), pp. 271–291. Reidel, Dordrecht, The Netherlands.

Dunning, J. B., Stewart, D. J., Danielson, B. J., Noon, B. R., Root, T. L., Lamberson, R. H., and Stevens, E. E. (1995). Spatially explicit population models: Current forms and future uses. *Ecological Applications* **5,** 3–11.

Durka, W., Schulze, E.-D., Gebauer, G., and Voerkelius, S. (1994). Effects of forest decline on uptake and leaching of deposited nitrate determined from ^{15}N and ^{18}O measurements. *Nature* **372,** 765–767.

Dye, D. G., and Shibasaki, R. (1995). Intercomparison of global PAR data sets. *Geophysical Research Letters* **38,** 2013–2016.

Easterling, D. R., Horton, B., Jones, P. D., Peterson, T. C., Karl, T. R., Parker, D. E., Salinger, M. J., Razuvayev, V., Plummer, N., Jamason, P., and Folland, C. K. (1997). Maximum and minimum temperature trends for the globe. *Science* **277,** 364–367.

Eaton, J. S., Likens, G. E., and Bormann, F. H. (1978). The input of gaseous and particulate sulfur to a forest ecosystem. *Tellus* **30,** 546–551.

Edmonds, R. L., ed. (1982). "Analysis of Coniferous Forest Ecosystems in the Western United States." Hutchinson and Ross, Stroudsburg, Pennsylvania.

Edwards, D. R., and Dixon, M. A. (1995). Mechanisms of drought response in *Thuja occidentalis* L. I. Water stress conditioning and osmotic adjustment. *Tree Physiology* **15,** 121–127.

Edwards, N. T., and Hanson, P. J. (1996). Stem respiration in a closed-canopy upland oak forest. *Tree Physiology* **16,** 433–439.

Edwards, N. T., and Harris, W. F. (1977). Carbon cycling in a mixed deciduous forest floor. *Ecology* **58,** 430–437.

Edwards, T. W. D., and Fritz, P. (1986). Assessing meteoric water composition and relative humidity from $\delta^{18}O$ and δ^2H in wood cellulose: Paleoclimatic implications for southern Ontario, Canada. *Applied Geochemistry* **1,** 715–723.

Edwards, W. R. N., Jarvis, P. G., Grace, J., and Moncrieff, J. B. (1994). Reversing cavitation in tracheids of *Pinus sylvestris* L. under negative water potentials. *Plant, Cell and Environment* **17,** 389–397.

Ehleringer, J. R., and Dawson, T. E. (1992). Water uptake by plants: Perspectives from stable isotope composition. *Plant, Cell and Environment* **15,** 1073–1082.

Eidenshink, J. C., and Foundeen, J. L. (1994). The 1 km AVHRR global land data set: First stages in implementation. *International Journal of Remote Sensing* **15,** 3443–3462.

Eis, S., Garman, H., and Ebell, L. F. (1965). Relation between cone production and diameter increment of Douglas-fir [*Pseudotsuga menziesii* (Mirb.) Franco], grand fir [*Abies grandis* (Dougl.) Lindl.], and western white pine (*Pinus moniticola* Dougl.). *Canadian Journal of Botany* **43,** 1553–1559.

Eissenstat, D. M., and Van Rees, K. C. J. (1994). The growth and function of pine roots. *Ecological Bulletins* **43,** 76–91.

Ekblak, A., Lindquist, P.-O., Sjöström, M., and Huss-Danell, K. (1994). Day-to-day variation in nitrogenase activity of *Alnus incana* explained by weather variables: A multivariate time series analysis. *Plant, Cell and Environment* **17,** 319–325.

Eklund, M. (1995). Cadmium and lead deposition around a Swedish battery plant as recorded in oak tree rings. *Journal of Environmental Quality* **24,** 126–131.

Elfving, B., Tegnhammar, L., and Tveite, B. (1996). Studies on growth trends of forests in Sweden and Norway. *In* "Growth Trends in European Forests" (H. Spiecker, K. Mielikäinen, K. Köhl, and J. P. Skovsgaard, eds.), pp. 61–70. Springer, Berlin.

Elliott, G. P., and Baker, W. L. (2004). Quaking aspen (*Populus tremuloides* Michx.) at treeline: A century of change in the San Juan Mountains, Colorado, USA. *Journal of Biogeography* **31,** 733–745.

Ellsworth, D., and Reich, P. B. (1993). Canopy structure and vertical patterns of photosynthesis and related leaf traits in a deciduous forest. *Oecologia* **96,** 169–178.

Emmanuel, K. A. (2005). Increasing destructiveness of tropical cyclones over the past 30 years. *Nature* **436,** 686–688.

Emmingham, W. H. (1977). Comparison of selected Douglas-fir seed sources for cambial and leader growth patterns in four western environments. *Canadian Journal of Forest Research* **7,** 154–164.

Engman, E. T., and Chauhan, N. (1995). Status of microwave soil moisture measurements with remote sensing. *Remote Sensing of the Environment* **51,** 189–198.

Entry, J. A., Cromack, K., Jr., Hansen, E., and Waring, R. (1991a). Response of western coniferous seedlings to infection by *Armillaria ostoyae* under limited light and nitrogen. *Physiology and Biochemistry* **81,** 89–94.

Entry, J. A., Cromack, K., Jr., Kelsey, R. G., and Martin, N. E. (1991b). Response of Douglas-fir to infection by *Armillaria ostoyae* after thinning or thinning plus fertilization. *Phytopathology* **81,** 682–689.

Entry, J. A., Martin, N. E., Cromack, K., Jr., and Stafford, S. (1986). Light and nutrient limitations in *Pinus moniticola:* Seedling susceptibility to *Armillaria mellea* infection. *Forest Ecology and Management* **17,** 189–198.

Ericksson, H., and Karlsson, K. (1996). Long-term changes in site index in growth and yield experiments with Norway spruce (*Picea abies*, [L.] Karst) and Scots pine (*Pinus sylvestris*, L.). *In* "Growth trends in European Forests" (H. Spiecker, K. Mielikäinen, K. Köhl, and J. P. Skovsgaard, eds.), pp. 80–88. Springer, Berlin.

Ericsson, A., Nordén, L.-G., Näsholm, T., and Walheim, M. (1993). Mineral nutrient imbalances and arginine concentrations in needles of *Picea abies* (L.) Karst. from two areas with different levels of airborne deposition. *Trees* **8,** 67–74.

Ericsson, T. (1994). Nutrient dynamics and requirements of forest crops. *New Zealand Journal of Forestry Science* **24,** 133–168.

Escarré, A., Gracia, C., and Terradas, J. (1984). Nutrient cycling and vertical structure in evergreen-oak forest. *In* "State and Change of Forest Ecosystems—indicators in Current Research" (G. I. Ågren, ed.), pp. 181–191. Department of Ecology and Environmental Research, Report No. 13, Swedish Univ. of Agric. Sciences, Uppsala, Sweden.

Evans, F. C. (1956). Ecosystem as the basic unit in ecology. *Science* **123,** 1127–1128.

Evans, R. D., and Ehleringer, J. R. (1993). A break in the nitrogen cycle in arid lands? Evidence from $d^{15}N$ of soils. *Oecologia* **94,** 314–317.

Everett, R. L., Java-Sharpe, B. J., Scherer, G. R., Wilt, F. M., and Ottmar, R. D. (1995). Co-occurrence of hydrophobicity and allelopathy in sand pits under burned slash. *Soil Science Society of America Journal* **59,** 1176–1183.

Evison, H. C. (1993). "Te Wai Pounamu: The Greenstone Island: A History of the Southern Maori during the European Colonization of New Zealand." Aoraki Press, Wellington, New Zealand.

Ewel, J., Berish, C., Brown, B., Price, N., and Raich, J. (1981). Slash and burn impacts on a Costa Rican wet forest site. *Ecology* **62,** 816–829.

Ewel, J. J., Mazzarino, M. J., and Berish, C. W. (1991). Tropical soil fertility changes under monocultures and successional communities of different structure. *Ecological Applications* **1,** 289–302.

Ewel, K. C., and Gholz, H. L. (1991). A simulation model of the role of below-ground dynamics in a slash pine plantation. *Forest Science* **37,** 397–438.

Fahey, T. J., and Young, D. B. (1984). Soil and xylem water potential and soil water content in contrasting *Pinus contorta* ecosystems, southeastern Wyoming, U.S.A. *Oecologia* **61,** 346–351.

Fain, J. J., Volk, T. A., and Fahey, T. J. (1994). Fifty years of change in an upland forest in south-central New York: General patterns. *Bulletin of the Torrey Botanical Club* **12,** 130–139.

Fairfield, J., and Leymarie, P. (1991). Drainage networks from grid digital elevation models. *Water Resources Research* **27,** 709–717.

Falge, E., Baldocchi, D., Tenhunen, J., Aubinet, M., Bakwin, P., Berbigier, P., Bernhofer, C., Burba, G., Clement, R., Davis, K. J., Elbers, J. A., Goldstein, A. H., Grelle, A., Granier, A., Gudmundsson, J., Hollinger, D., Kowalski, A. S., Katul, G., Law, B. E., Malhi, Y., Meyers, T., Monson, R. K., Munger, J. W., Oechel, W., Paw U., K. T., Pilegaard, K., Rannik, Ü., Rebmann, C., Suyker, A., Valentini, R., Wilson, K., and Wofsy, S. (2002). Seasonality of ecosystem respiration and gross primary production as derived from FLUXNET measurements. *Agricultural and Forest Meteorology* **113,** 53–74.

Farquhar, G. D., and Sharkey, T. D. (1982). Stomatal conductance and photosynthesis. *Annual Review of Plant Physiology* **33,** 317–345.

Farquhar, G. D., Caemmerer, S. von, and Berry, J. A. (1980). A biochemical model of photosynthetic CO_2 assimilation in leaves of C_3 species. *Planta* **149,** 78–90.

Farquhar, G. D., O'Leary, M. H., and Berry, J. A. (1982). On the relationship between carbon isotope discrimination and the intercellular carbon dioxide concentration in leaves. *Australian Journal of Plant Physiology* **9,** 121–137.

Fassnacht, K. S., Gower, S. T., Norman, J. M., and McMurtrie, R. E. (1994). A comparison of optical and direct methods for estimating foliage surface area index in forests. *Agricultural and Forest Meteorology* **71,** 183–207.

Fellin, D. G. (1980). A review of some interactions between insects and disease. *USDA Forest Service, General Technical Report* **INT-90,** 335–414. Ogden, Utah.

Fenn, M. (1991). Increased site fertility and litter decomposition rate in high-pollution sites in the San Bernardino Mountains. *Forest Science* **37,** 1163–1181.

Field, C., and Mooney, H. A. (1986). The photosynthesis–nitrogen relationship in wild plants. *In* "On the Economy of Plant Form and Function" (T. J. Givnish, ed.), pp. 25–55. Cambridge Univ. Press, Cambridge.

Field, C. B., Chapin III, F. S., Matson, P. A., and Mooney, H. A. (1992). Responses of terrestrial ecosystems to the changing atmosphere: A resource-based approach. *Annual Review of Ecology and Systematics* **23,** 201–235.

Field, C. B., Gamon, J. A., and Penuelas, J. (1994). Remote sensing of terrestrial photosynthesis. *In* "Ecophysiology of Photosynthesis" (E.-D. Schulze and M. M. Caldwell, eds.), pp. 511–527. Springer-Verlag, Berlin.

Field, C. B., Randerson, J. T., and Malmstrom, C. M. (1995). Global net primary production: Combining ecology and remote sensing. *Remote Sensing of the Environment* **51,** 74–88.

Finzi, A. C., DeLucia, E. H., Hamilton, J. G., Richter, D. D., and Schlesinger, W. H. (2002). The nitrogen budget of a pine forest under free air CO_2 enrichment. *Oecologia* **132,** 567–578.

Fisher, J. I., Mustard, J. F., and Vadeboncoeur, M. A. (2006). Green leaf phenology at Landsat resolution: Scaling from the field to the satellite. *Remote Sensing of Environment* **100,** 265–279.

Flanagan, L. B., Brooks, J. R., Varney, G. T., and Ehleringer, J. R. (1997). Discrimination against $C^{18}O^{16}O$ during photosynthesis and the oxygen isotope ratio of respired CO_2 in boreal ecosystems. *Global Biogeochemical Cycles* **11,** 83–98.

Flanagan, L. B., Brooks, J. R., Varney, G. T., Berry, S. C., and Ehleringer, J. R. (1996). Carbon isotope discrimination during photosynthesis and the isotope ratio of respired CO_2 in boreal forest ecosystems. *Global Biogeochemical Cycles* **10,** 629–640.

Flannigan, M. D., and Van Wagner, C. E. (1991). Climate change and wildfire in Canada. *Canadian Journal of Forest Research* **21,** 61–72.

Foley, J. A. (1994). The sensitivity of the terrestrial biosphere to climatic change: A simulation of the middle Holocene. *Global Biogeochemical Cycles* **8,** 505–525.

Foley, J. A., DeFries, R., Asner, G. P., Barford, C., Bonan, G., Carpenter, S. R., Chapin, F. S., Coe, M. T., Daily, G. C., Gibbs, H. K., Helkowski, J. H., Holloway, T., Howard, E. A., Kucharik, C. J., Monfreda, C., Patz, J. A., Prentice, I. C., Ramankutty, N., Snyder, P. K. (2005). Global consequences of land use. *Science* **309,** 570–574.

Ford, E. D., and Kiester, R. (1990). Modeling the effects of pollutants on the process of tree growth. *In* "Process Modeling of Forest Growth Responses to Environmental Stress" (R. K. Dixon, R. S. Meldahl, G. A. Ruark, and W. G. Warren, eds.), pp. 324–337. Timber Press, Portland, Oregon.

Forman, R. T. T. (1995). "Land Mosaics: The Ecology of Landscapes and Regions." Cambridge Univ. Press, Cambridge.

Foster, D. R., and Zebryk, T. M. (1993). Long-term vegetation dynamics and disturbance history of a *Tsuga*-dominated forest in New England. *Ecology* **74,** 982–998.

Fowells, H. A. (1965). "Silvics of Forest Trees of the United States." Agriculture Handbook No. 271. U.S. Dept. of Agriculture, Forest Service, Washington, D.C.

Franchini, M., Wendling, J., Obled, C., and Todini, E. (1996). Physical interpretation and sensitivity analysis of the TOPMODEL. *Journal of Hydrology* **175,** 293–338.

Francy, R. J. (1981). Tasmanian tree rings belie suggested anthropogenic $^{13}C/^{12}C$ trends. *Nature* **290,** 232–235.

Franklin, J. F., and Forman, R. T. T. (1987). Creating landscape patterns by forest cutting: Ecological consequences and principles. *Landscape Ecology* **1,** 5–18.

Franklin, J. F., MacMahon, J. A., Swanson, F. J., and Sedell, J. R. (1985). Ecosystem responses to the eruption of Mount St. Helens. *National Geographic Research* **Spring,** 197–216.

Frelich, L. E., and Reich, P. B. (1995). Spatial patterns and succession in a Minnesota southern-boreal forest. *Ecological Monographs* **65,** 325–346.

Freyer, H. D., and Belacy, N. (1983). $^{13}C/^{12}C$ records in northern hemispheric trees during the last 150 years. *Tellus* **31,** 124–137.

Friend, A. D. (1995). PGEN: An integrated model of leaf photosynthesis, transpiration, and conductance. *Ecological Modelling* **77,** 233–255.

Friend, A. D., and Woodward, F. I. (1990). Evolutionary and ecophysiological responses of mountain plants to the growing season environment. *Advances in Ecological Research* **20,** 59–124.

Friend, A. D., Shugart, H. H., and Running, S. W. (1993). A physiology-based gap model of forest dynamics. *Ecology.* **74,** 792–797.

Fritts, H. C., and Swetnam, T. W. (1989). Dendrochronology: A tool for evaluating variations in past and present forest environments. *Advances in Ecological Research* **19,** 111–188.

Frouin, R., and Pinker, R. T. (1995). Estimating photosynthetically active radiation (PAR) at the earth's surface from satellite observations. *Remote Sensing of the Environment* **51,** 98–107.

Fung, I., John, J., Lerner, J., Matthews, E., Prather, M., Steele, L. P., and Fraser, P. J. (1991). Three-dimensional model synthesis of the global methane cycle. *Journal of Geophysical Research* **96,** 13033–13065.

Gamache, I., and Payette, S. (2004). Height growth response of tree line black spruce to recent climate warming across the forest-tundra of eastern Canada. *Journal of Ecology* **92,** 835–845.

Gamache, I., and Payette, S. (2005). Latitudinal response of subarctic tree lines to recent climate change in eastern Canada. *Journal of Biogeography* **32,** 849–862.

Gamon, J. A., Penuelas, J., and Field, C. B. (1992). A narrow-waveband spectral index that tracks diurnal changes in photosynthetic efficiency. *Remote Sensing of the Environment* **41,** 35–44.

Gan, J. (2004). Risk and damage of southern pine beetle outbreaks under global climate change. *Forest Ecology and Management* **191,** 61–71.

Garnier, B. J., and Ohmura, A. (1968). A method of calculating the direct shortwave radiation income of slopes. *Journal of Applied Meteorology* **7,** 796–800.

Garten, C. T., Jr., and Van Miegroet, H. (1994). Relationships between soil nitrogen dynamics and natural ^{15}N abundance in plant foliage from Great Smoky Mountains National Park. *Canadian Journal of Forest Research* **24,** 1636–1645.

Gartlan, J. S., McKey, D. B., Waterman, P. G., Mbi, C. N., and Stuhsaker, T. T. (1980). A comparative study of the phytochemistry of two African rain forests. *Biol. Chem. Syst. Ecol.* **8,** 401–422.

Gash, J. H. C. (1979). An analytical model of rainfall interception by forests. *Quarterly Journal of the Royal Meteorology Society* **105,** 43–55.

Gates, D. M. (1980). "Biophysical Ecology." Springer-Verlag, Berlin and New York.

Gear, A. J., and Huntley, B. (1991). Rapid changes in the range limits of Scots pine 4000 years ago. *Science* **251,** 544–547.

Gemmell, F. M. (1995). Effects of forest cover, terrain, and scale on timber volume estimation with thematic mapper data in a rocky mountain site. *Remote Sensing of the Environment* **51**, 291–305.

Gersper, P. L., and Holowaychuk, N. (1971). Some effects of stem flow from forest canopy trees on chemical properties of soils. *Ecology* **52**, 691–702.

Gherini, S. A., Mok, L., Hudson, R. J. M., Davis, G. F., Chen, C. W., and Goldstein, R. A. (1985). The ILWAS model: Formulation and application. *Water Resources Research* **26**, 425–459.

Gholz, H. L. (1982). Environmental limits on aboveground net primary production, leaf area, and biomass in vegetation zones of the Pacific Northwest. *Ecology* **63**, 469–481.

Gholz, H. L., Grier, C. C., Campbell, A. G., and Brown, A. T. (1979). "Equations and Their Use for Estimating Biomass and Leaf Area of Pacific Northwest Plants," Research Paper No. 41. Forest Research Laboratory, Oregon State University, Corvallis.

Giblin, A. E., Laundre, J. A., Nadelhoffer, K. J., and Shaver, G. R. (1994). Measuring nutrient availability in arctic soils using ion exchange resins: A field test. *Soil Science Society of America Journal* **58**, 1154–1162.

Gifford, R. M. (1994). The global carbon cycle: A viewpoint on the missing sink. *Australian Journal of Plant Physiology* **21**, 1–15.

Gifford, R. M. (2003). Plant respiration in productivity models: Conceptualisation, representation and issues for global terrestrial carbon-cycle research. *Functional Plant Biology* **30**, 171–186.

Gilbert, G. S., Hubbell, S. P., and Foster, R. B. (1994). Density and distance-to-adult effects of a canker disease of trees in a moist tropical forest. *Oecologia* **98**, 100–108.

Gillett, N. P., Weaver, A. J., Zwiers, F. W., and Flannigan, M. D. (2004). Detecting the effect of climate change on Canadian forest fires. *Geophysical Research Letters* **31**, L18211–L18214.

Gilmour, D. A., Bonell, M., and Cassells, D. S. (1987). The effects of forestation on soil hydraulic properties in the Middle Hills of Nepal: A preliminary assessment. *Mountain Research and Development* **7**, 239–249.

Glassy, J., and Running, S. W. (1994). Validating diurnal climatology logic of the MT-CLIM logic across a climatic gradient of Oregon. *Ecological Applications* **4**, 248–257.

Glatzel, G. (1991). The impact of historic land use and modern forestry on nutrient relations of central European forest ecosystems. *Fertilizer Research* **27**, 1–8.

Gleason, H. A. (1926). The individualistic concept of the plant association. *Bulletin of the Torrey Botanical Club* **53**, 7–26.

Goldammer, J. G. (1983). "Sicherung des sudbrasilianischen Kiefernanbaues durch Kontrolliertes Brennen," Forstwirtsch. Vol. 4. Hochschul-Verlag, Freiburg, Germany.

Golden, J. M. (1980). Percolation theory and model of unsaturated porous media. *Water Resources Research* **16**, 201–209.

Goodale C. L., Apps, M. J., Birdsey, R. A., Field, C. B., Heath, L. S., Houghton, R. A., Jenkins, J. C., Kohlmaier, G. H., Kurz, W., Liu, S., Nabuurs, G.-J., Nilsson, S., and Shvidenko, A. Z. (2002). Forest carbon sinks in the northern hemisphere. *Ecological Applications* **12**, 891–899.

Gosz, J. R., Likens, G. E., and Bormann, F. H. (1972). Nutrient content of litterfall on the Hubbard Brook Experimental Forest, New Hampshire. *Ecology* **53**, 769–784.

Goulden, M. L., Munger, J. W., Fan, S.-M., Daube, B. C., and Wofsy, S. C. (1996). Exchange of carbon dioxide by a deciduous forest: Response to interannual climate variability. *Science* **271**, 1576–1578.

Goward, S. N., Markham, B., Dye, D. G., Dulaney, W., and Yang, J. (1991). Normalized difference vegetation index measurements from the advanced very high resolution radiometer. *Remote Sensing of the Environment* **35**, 257–277.

Goward, S. N., Waring, R. H., Dye, D. G., and Yang, J. (1994). Ecological remote sensing at Otter: Satellite macroscale observations. *Ecological Applications* **4**, 322–343.

Gower, S. T. (2003). Patterns and mechanisms of the forest carbon cycle. *Annual Review of Environmental Resources* **28,** 169–204.

Gower, S. T., and Norman, J. M. (1991). Rapid estimation of leaf area index in conifer and broad-leaf plantations using the LI-COH LAI-2000. *Ecology* **72,** 1896–1900.

Gower, S. T., and Son, Y. (1992). Differences in soil and leaf litterfall nitrogen dynamics for five forest plantations. *Soil Science Society of America Journal* **55,** 1959–1966.

Gower, S. T., Haynes, B. E., Fassnacht, K. S., Running, S. W., Hunt, E. R., Jr. (1993). Influence of fertilization on the allometnic relations for two pines in contrasting environments. *Canadian Journal of Forest Research* **23,** 1104–1111.

Gower, S. T., McMurtrie, R. E., and Murty, D. (1996). Aboveground net primary productivity decline with stand age: Potential causes. *Trees* **11,** 378–382.

Grace, J. (1981). Some effects of wind on plants. *In* "Plants and their Atmospheric Environment" (J. Grace, E. D. Ford, and P. G. Jarvis, eds.), pp. 31–56. Blackwell, Oxford.

Grace, J. C. (1987). Theoretical ratio between "one-sided" and total surface area for pine needles. *New Zealand Journal of Forest Research* **17,** 292–296.

Grace, J., Lloyd, J., McIntyre, J., Miranda, A. C., Meir, P., Miranda, H. S., Nobre, C., Moncrieff, J., Massheder, J., and Malhi, Y. (1995). Carbon dioxide uptake by an undisturbed tropical rain forest in southwest Amazonia, 1992 to 1993. *Science* **270,** 778–780.

Graham, R. L., Wright, L. L., and Turhollow, A. F. (1992). The potential for short-rotation woody crops to reduce U.S. CO_2 emissions. *Climatic Change* **22,** 223–238.

Granhall, U. (1981). Biological nitrogen fixation in relation to environmental factors and functioning of natural ecosystems. *Ecological Bulletins (Stockholm)* **33,** 131–144.

Granier, A., Biron, P., Bréda, N., Pontailler, J.-Y., and Saugier, B. (1996). Transpiration of trees and forest stands: Short- and long-term monitoring using sapflow methods. *Global Change Biology* **2,** 265–274.

Granier, A., Ceschia, E., Dames, C., Dufrenet, E., Epront, D., Gross, P., Lebaube, S., Le Dantec, V., Le Goff, N., Lucott, E., Ottorin, J. M., Pontaillert, J. Y., and Saugier, B. (2000). The carbon balance of a young beech forest. *Functional Ecology* **14,** 312–325.

Graumlich, L. J. and Brubaker, L. B. (1995). Long-term records of growth and distribution of conifers: Integration of paleoecology and physiological ecology. *In* "Ecophysiology of Coniferous Forest" (W. K. Smith and T. M. Hinckley, eds.), pp. 37–62. Academic Press, San Diego.

Graumlich, L. J., Brubaker, L. B., and Grier, C. C. (1989). Long-term trends in forest net primary productivity: Cascade Mountains, Washington. *Ecology* **70,** 405–410.

Graustein, W. C., Cromack, K., and Sollins, P. (1977). Calcium oxalate: Occurrence in soils and effect on nutrient and geochemical cycles. *Science* **198,** 1252–1254.

Green, D. S., Erickson, J. E., and Kruger, E. L. (2003). Foliar morphology and canopy nitrogen as predictors of light-use efficiency in terrestrial vegetation. *Agric. and Forest Meteorology* **115,** 163–171.

Gregory, R. A. (1971). Cambial activity in Alaskan white spruce. *American Journal of Botany* **58,** 160–171.

Grier, C. C., and Running, S. W. (1977). Leaf area of mature northwestern coniferous forests: Relation to site water balance. *Ecology* **58,** 893–899.

Grier, C. C., and Waring, R. H. (1974). Conifer foliage mass related to sapwood area. *Forest Science* **20,** 205–206.

Griffin, K. L. (1994). Calorimetric estimates of construction cost and their use in ecological studies. *Functional Ecology* **8,** 551–562.

Grime, J. P. (1977). Evidence for the existence of three primary strategies in plants and its relevance to ecological and evolutionary theory. *American Naturalist* **111,** 1169–1194.

Groisman, P. Y., Karl, T. R., Knight, R. W., and Stenchikov, G. L. (1994). Changes of snow cover, temperature and radiative heat balance over the Northern Hemisphere. *Journal of Climate* **7,** 1633–1656.

Groisman, P. Y., Knight, R. W., Kari, T. R., Easterling, D. R., Sun, B., and Lawrimore, J. H. (2004). Contemporary changes of the hydrological cycle of the contiguous United States: Trends derived from in situ observations. *Journal of Hydrometeorology* **5,** 64–85.

Gruell, G. E. (1983). Fire and vegetative trends in the Rockies: Interpretations from 1871–1982 photographs. *USDA General Technical Report* **INT-158**.

Grupe, G., and Krueger, H. H. (1990). Feeding habitats of free ranging martens (*Martes marines* and *Marines foina*) evaluated by means of trace element and table carbon isotopes of their skeletons. *Science Total Environment* **90,** 227–240.

Guenther, A., Hewitt, C. N., Erickson, D., Fall, R., Geron, C., Graedel, T., Harley, P., Klinger, L., Lerdaue, M. (1995). A global model of natural volatile organic compound emissions. *Journal of Geophysical Research* **100**(D5), 8873–8892.

Gullion, G. W. (1990). Ruffed grouse use of conifer plantations. *Wildlife Society Bulletin* **18,** 183–187.

Gustafson, E. J., and Crow, T. R. (1994). Modeling the effects of forest harvesting on landscape structure and the spatial distribution of cowbird brood parasitism. *Landscape Ecology* **9,** 237–248.

Hagihara, A., and Hozumi, K. (1991). Respiration. *In* "Physiology of Trees" (A. S. Raghavendra, ed.), pp. 87–110. Wiley, New York.

Hall, F. G., Botkin, D. B., Strebel, D. E., Woods, K. D., and Goetz, S. J. (1991). Large-scale patterns of forest succession as determined by remote sensing. *Ecology* **72,** 628–640.

Hall, F. G., Shimabukuro, Y. E., and Huemmrich, K. F. (1995). Remote sensing of forest biophysical structure using mixture decomposition and geometric reflectance models. *Ecological Applications* **5,** 993–1013.

Hällgren, J.-D., Strand, M., and Lundmark, T. (1991). Temperature stress. *In* "Physiology of Trees" (A. S. Raghavendra, ed.), pp. 301–355. Wiley, New York.

Handley, L. L., and Raven, J. A. (1992). The use of natural abundance of nitrogen isotopes in plant physiology and ecology. *Plant, Cell and Environment* **15,** 965–985.

Handley, L. L., Odee, D., and Scrimgeour, C. M. (1994). $\delta^{15}N$ and $\delta^{13}C$ patterns in savanna vegetation: Dependence on water availability and disturbance. *Functional Ecology* **8,** 306–314.

Hanley, T. A. (1993). Balancing economic development, biological conservation, and human cultures: The Sitka black-tailed deer *Odocoileus heminonus sitkensis* as an ecological indicator. *Biological Conservation* **66,** 61–67.

Hanley, T. A., and McKendrick, J. D. (1985). Potential nutritional limitations for black-tailed deer in a spruce–hemlock forest, southeastern Alaska. *Journal of Wildlife Management* **49,** 103–114.

Hanley, T. A., Robbins, C. T., Hagerman, A. E., and McArthur, C. (1992). Predicting digestible protein and digestible dry matter in tannin-containing forages consumed by ruminants. *Ecology* **73,** 537–541.

Hansen, A. J., Garman, S. L., Marks, B., and Urban, D. L. (1993). An approach for managing vertebrate diversity across multiple-use landscapes. *Ecological Applications* **3,** 481–496.

Hansen, A. J., Rasker, R., Maxwell, B., Rotella, J. J., Johnson, J. D., Parmenter, A. W., Langner, U., Cohen, W. B., Lawrence, R. L., and Kraska, M. P. V. (2002). Ecological causes and consequences of demographic change in the New West. *BioScience* **52,** 151–162.

Hansen, H. L., Krefting, L. W., and Kurmis, V. (1973). The forest of Isle Royale in relation to fire history and wildlife. *Technical Bulletin, Agricultural Experimental Station, University of Minnesota, St. Paul,* **294**.

Hanson, A. D., and Hitz, W. D. (1982). Metabolic responses of mesophytes to plant water deficits. *Annual Review of Plant Physiology* **33,** 163–203.

Hanson, P. J., Wullschleger, S. D., Bohlman, S. A., and Todd, D. E. (1993). Seasonal and topographic patterns of forest floor CO_2 efflux from an upland oak forest. *Tree Physiology* **13,** 1–15.

Harborne, J. B. (1982). "Introduction to Ecological Biochemistry." Academic Press, New York.

Harcombe, P. A., Harmon, M. E., and Greene, S. E. (1990). Changes in biomass and production over 53 years in a coastal *Picea sitchensis–Tsuga heterophylla* forest approaching maturity. *Canadian Journal of Forest Research* **20,** 1602–1610.

Hari, P., and Hakkinen, R. (1991). The utilization of old phenological time series of budburst to compare models describing annual cycles of plants. *Tree Physiology* **8,** 281–287.

Harley, P. C., Thomas, R. B., Reynolds, J. F., and Strain, B. R. (1992). Modelling photosynthesis of cotton grown in elevated CO_2. *Plant, Cell and Environment* **15,** 271–282.

Harmon, M. E., Ferrell, W. K., and Franklin, J. F. (1990). Effects on carbon storage of conversion of old-growth forests to young forests. *Science* **247,** 699–702.

Harmon, M. E., Franklin, J. F., Swanson, F. J., Sollins, P., Gregory, S. V., Lattin, J. D., Anderson, N. H., Cline, S. P., Aumen, N. G., Sedell, J. R., Lienkaemper, G. W., Cromack, K., Jr., and Cummins, K. W. (1986). Ecology of coarse woody debris in temperate ecosystems. *Advances in Ecological Research* **15,** 133–302.

Harr, R. D., Fredrickson, R. L., and Rothacher, J. (1979). Changes in streamflow following timber harvest in southwestern Oregon. *Pacific Northwest Forest and Range Experiment Station, USDA, Forest Service, Research Paper* **PNW-249.** Portland, Oregon.

Harris, W. F., Sollins, P., Edwards, N. T., Dinger, B. E., and Shugart, H. H. (1975). Analysis of carbon flow and productivity in a temperate deciduous forest ecosystem. *In* "Productivity of World Ecosystems" (D. E. Reichle, J. F. Franklin, and D. W. Goodall, eds.), pp. 116–122. Special Committee for Int. Biol. Prog., Natl. Acad. Sci., Washington, D.C.

Hartshorn, G. S. (1978). Treefalls and tropical forest dynamics. *In* "Tropical Trees as Living Systems" (P. B. Tomlinson and M. H. Zimmerman, eds.), pp. 617–638. Cambridge Univ. Press, London and New York.

Harvey, P. J., Schoemaker, H. E., and Palmer, J. M. (1987). Lignin degradation by white rot fungi. *Plant, Cell and Environment* **10,** 709–714.

Hasenauer, H., Nemani, R. R., Schadauer, K., and Running, S. W. (1999). Forest growth response to changing climate between 1961 and 1990 in Austria. *Forest Ecology and Management* **122,** 209–219.

Haukioja, E., and Niemelä, P. (1979). Birch leaves as a resource for herbivores: Seasonal occurrence of increased resistance in foliage after mechanical damage of adjacent leaves. *Oecologia* **39,** 151–159.

Havranek, W. M., and Tranquillini, W. (1995). Physiological processes during winter dormancy and their ecological significance. *In* "Ecophysiology of Coniferous Forests" (W. K. Smith and T. M. Hinckley, eds.), pp. 95–124. Academic Press, San Diego.

Hawkins, B. J., and Sweet, G. B. (1989). Photosynthesis and growth of present New Zealand forest trees relate to ancient climates. *Annales des Sciences Forestières* **46,** 512–514.

Haxeltine, A., and Prentice, I. C. (1996). BIOME 3: An equilibrium terrestrial biosphere model based on ecophysiological constraints, resource availability, and competition among plant functional types. *Global Biogeochemical Cycles* **10,** 693–709.

Hedin, L. O., and Likens, G. E. (1996). Atmospheric dust and acid rain. *Scientific American* **248,** 87–92.

Hedin, L. O., Armesto, J. J., and Johnson, A. H. (1995). Patterns of nutrient loss from unpolluted old-growth temperate forests: Evaluation of biogeochemical theory. *Ecology* **76,** 493–509.

Heilman, P. (1981). Root penetration of Douglas-fir seedlings into compacted soil. *Forest Science* **27,** 660–666.

Heinsch, F. A., Zhao, M., Running, S. W., Kimball, J. S., Nemani, R. R., Davis, K. J., Bolstad, P. V., Cook, B. D., Desai, A. R., Ricciuto, D. M., Law, B. E., Oechel, W. C., Kwon, H., Luo, H., Wofsy, S. C., Dunn, A. L., Munger, J. W., Baldocchi, D. D., Xu, L., Hollinger, D. Y., Richardson, A. D., Stoy, P. C., Siqueira, M. B. S., Monson, R. K., Burns, S., and Flanagan, L. B. (2006). Evaluation of remote sensing based terrestrial productivity from MODIS using AmeriFlux tower eddy flux network observations. *IEEE Transactions on Geoscience and Remote Sensing* **44,** 1908–1925.

Heinselman, M. L. (1973). Fire in the virgin forests of the Boundary Waters Canoe Area. *Minnesota Quaternary Review* **3,** 329–382.

Helvey, J. D. (1971). A summary of rainfall interception by certain conifers of North America. *In* "Proceedings of the Third International Symposium for Hydrology Professors. Biological Effects in the Hydrological Cycle" (E. J. Monke, ed.), pp. 103–113. Purdue University, West Lafayette, Indiana.

Helvey, J. D., and Patric, J. H. (1965). Canopy and litter interception by hardwoods of eastern United States. *Water Resources Research* **1,** 193–206.

Henderson-Sellers, A., and McGuffie, K. (1995). Global climate models and "dynamic" vegetation changes. *Global Change Biology* **1,** 63–75.

Hendrick, R. L., and Pregitzer, K. S. (1993). Patterns of fine root mortality in two sugar maple forests. *Nature* **361,** 59–61.

Henry, J. D., and Swan, J. M. A. (1974). Reconstructing forest history from live and dead plant material—An approach to the study of forest succession in southwest New Hampshire. *Ecology* **55,** 772–783.

Hinckley, T. M., Duhme, F., Hinckley, A. R., and Richter, H. (1983). Drought relations of shrub species: Assessment of the mechanisms of drought resistance. *Oecologia* **59,** 344–350.

Hinckley, T. M., Teskey, R. O., Duhme, F., and Richter, H. (1981). Temperate hardwood forests. *In* "Water Deficits and Plant Growth" (T. T. Kozlowski, ed.). Vol. 6, pp. 153–208. Academic Press, New York.

Hobbie, E. A. (2006). Carbon allocation to ectomycorrhizal fungi correlates with belowground allocation in culture studies. *Ecology* **87,** 563–569.

Hobbie, E. A., and Colpaert, J. V. (2003). Nitrogen availability and colonization by mycorrhizal fungi correlate with nitrogen isotope patterns in plants. *New Phytologist* **157,** 115–126.

Högberg, P., Nordgren, A., Buchmann, N., Taylor, A. F. S., Ekblad, A., Högberg, M. N., Nyberg, G., Ottosson-Löfvenius, M., and Read, D. J. (2001). Large-scale forest girdling shows that current photosynthesis drives soil respiration. *Nature* **411,** 789–792.

Hogg, E. H., Brandt, J. P., and Kochtubajda, B. (2002). Growth and dieback of aspen forests in northwestern Alberta, Canada, in relation to climate and insects. *Canadian Journal of Forest Research* **32,** 823–832.

Hole, F. D. (1981). Effects of animals on soil. *Geoderma* **25,** 75–112.

Hollinger, D. Y. (1989). Canopy organization and foliage photosynthetic capacity in a broad-leaved evergreen montane forest. *Functional Ecology* **3,** 53–62.

Hollinger, D. Y., Aber, J., Dail, B., Davidson, E. A., Goltz, S. M., Hughes, H., Leclerc, M. Y., Lee, J. T., Richardson, A. D., Rodrigues, C., Scott, N. A., Achuatavarier, D., and Walsh, J. (2004). Spatial and temporal variability in forest-atmosphere CO_2 exchange. *Global Change Biology* **10,** 1689–1706.

Holmes, R. T., Schultz, J. C., and Nothnage, P. (1979). Bird predation on forest insects: An exclosure experiment. *Science* **206,** 462–463.

Hook, D. D., Langdon, O. G., Stubbs, J., and Brown, C. L. (1970). Effect of water regimes on the survival, growth, and morphology of tupelo seedlings. *Forest Science* **16,** 304–311.

Hornbeck, J. W., and Kropelin, W. (1982). Nutrient removal and leaching from a whole-tree harvest of northern hardwoods. *Journal of Environmental Quality* **11,** 309–316.

Houghton, R. A. (1994). The worldwide extent of land-use change. *BioScience* **44,** 305–313.

House, J. I., Prentice, I. C., Ramankutty, N., Houghton, R. A., and Heimann, M. (2003). Reconciling apparent inconsistencies in estimates of terrestrial CO_2 sources and sinks. *Tellus* **55B,** 345–363.

Howard, W. E. (1964). Introduced browsing mammals and habitat stability in New Zealand. *Journal of Wildlife Management* **28,** 421–429.

Huete, A., Didan, K., Miura, T., Rodriguez, E. P., Gao, X., and Ferreira, L. G. (2002). Overview of the radiometric and biophysical performance of the MODIS vegetation indices. *Remote Sensing of Environment* **83,** 195–213.

Hungerford, R. D., Nemani, R. R., Running, S. W., Coughlan, J. C. (1989). MT-CLIM: A mountain microclimate simulation model. *USDA Forest Service Research Paper* **INT-414.** Ogden, Utah.

Hunsaker, C. T., and Levine, D. A. (1995). Hierarchical approaches to the study of water quality in rivers. *BioScience* **45,** 193–203.

Hunt, E. R., Jr., Martin, F. C., and Running, S. W. (1991). Simulating the effect of climatic variation on stem carbon accumulation of a ponderosa pine stand: Comparison with annual growth increment data. *Tree Physiology* **9,** 161–172.

Hunt, E. R., Jr., Piper, S. C., Nemani, R., Keeling, C. D., Otto, R. D., and Running, S. W. (1996). Global net carbon exchange and intra-annual atmospheric CO_2 concentrations predicted by an ecosystem process model and three-dimensional atmospheric transport model. *Global Biogeochemical Cycles* **10,** 431–456.

Hunter, I. R., Nicholson, G., and Thorn, A. J. (1985). Chemical analysis of pine litter: An alternative to foliage analysis? *New Zealand Journal of Forestry Science* **15,** 101–110.

Huston, M. A. (1991). Use of individual-based forest succession models to link physiological whole-tree models to landscape-scale ecosystem models. *Tree Physiology* **9,** 293–306.

Huston, M., and Smith, T. (1987). Plant succession: Life history and competition. *American Naturalist* **130,** 168–198.

Ingestad, T. (1979). Nitrogen stress in birch seedlings. II. N, K, P, Ca and Mg nutrition. *Physiologia Plantarum* **45,** 149–157.

Ingestad, T. (1982). Relative addition rate and external concentration; driving variables used in plant nutrition research. *Plant, Cell and Environment* **5,** 443–453.

IPCC. (1996). "Climate Change 1995, The Science of Climate Change" (J. T. Houghton, L. G. Meira Filho, B. A. Callender, N. Harris, A. Kattenberg, and K. Maskell, eds.). Cambridge Univ. Press, Cambridge.

Irons, J. R., Hanson, K. J., Williams, D. L., Irish, R. R., and Huegel, F. G. (1991). An off-nadir-pointing imaging spectroradiometer for terrestrial ecosystem studies. *IEEE Transactions on Geoscience and Remote Sensing* **29,** 66–74.

Irvine, J., Law, B. E., and Kurpius, M. R. (2005). Coupling of canopy gas exchange with root and rhizosphere respiration in a semi-arid forest. *Biogeochemistry* **73,** 271–282.

Iverson, L. R., Cook, E. A., and Graham, R. L. (1994). Regional forest cover estimation via remote sensing: The calibration center concept. *Landscape Ecology* **4,** 5–19.

Jackson, G. E., Irvine, J., and Grace, J. (1995). Xylem cavitation in Scots pine and Sitka spruce saplings during water stress. *Tree Physiology* **15,** 783–790.

Jackson, M. B. (1994). Root-to-shoot communication in flooded plants: Involvement of abscisic acid, ethylene, and 1-aminocyclopropane-1-carboxylic acid. *Agronomy Journal* **86,** 775–782.

Jacoby, G. C., and D'Arrigo, R. D. (1995). Tree ring width and density evidence of climatic and potential forest change in Alaska. *Global Biogeochemical Cycles* **9,** 227–234.

Jacoby, G. C., D'Arrigo, R. D., and Davaajamts, T. (1996). Mongolian tree rings and 20th-century warming. *Science* **273,** 771–773.

Jakubas, W. J., Gullion, G. W., and Clausen, T. P. (1989). Ruffed grouse feeding behavior and its relationship to secondary metabolites of quaking aspen flower buds. *Journal of Chemical Ecology* **15,** 1899–1917.

Janssens, I. A., Freibauer, A., Ciais, P., Smith, P., Nabuurs, G.-J., Folberth, G., Schlamadinger, B., Hutjes, R. W. A., Ceulemans, R., Schulze, E.-D., Valentini, R., and Dolman, A. J. (2003). Europe's Terrestrial Biosphere absorbs 7 to 12% of European anthropogenic CO_2 emissions. *Science* **300,** 1538–1542.

Janssens, I. A., Freibauer, A., Schlamadinger, B., Ceulemans, R., Ciais, P., Dolman, A. J., Helmann, M., Nabuurs, G.-J., Smith, P., Valentini, R., and Schulze, E.-D. (2004). The carbon budget of terrestrial ecosystems at country-scale—A European case study. *Biogeosciences Discussions* **1,** 167–193.

Janssens, I. A., Landkreijer, H., Matteucci, G., Kowalski, A. S., Buchmann, N., Epron, D., Pilegaard, K., Kutsch, W., Longdoz, B., Grünwald, T., Montagnani, L., Dore, S., Rebmann, C., Moors, E. J., Grelle, A., Rannik, Ü., Morgenstern, K., Oltchev, S., Clement, R., Gudmundsson, J., Minerbi, S., Berbigier, P., Ibrom, A., Moncrieff, J., Aubinet, M., Bernhofer, C., Jensen, N. O., Vesala, T., Granier, A., Schultze, E.-D., Lindroth, A., Dolman, A. J., Jarvis, P. G., Ceulemans, R., and Valentini, R. (2001). Productivity overshadows temperature in determining soil and ecosystem respiration across European forests. *Global Change Biology* **7,** 269–278.

Jansson, P. E., and Berg, B. (1985). Temporal variation of litter decomposition in relation to simulated soil climate. Long-term decomposition in a Scots pine forest. V. *Canadian Journal of Forest Research* **63,** 1008–1016.

Janzen, D. H. (1979). New horizons in the biology of plant defenses. *In* "Herbivores: Their Interactions with Secondary Plant Metabolites" (G. A. Rosenthal and D. H. Janzen, eds.), pp. 331–348. Academic Press, New York.

Janzen, D. H., and Martin, P. S. (1982). Neotropical anachronisms: The fruits the Gomphotheres ate. *Science* **215,** 19–27.

Janzen, R. A., Cook, F. D., and McGill, W. B. (1995). Compost extract added to microcosms may simulate community-level control on soil microorganisms involved in element cycling. *Soil Biology and Biochemistry* **27,** 181–188.

Jarvis, P. G. (1976). The interpretation of the variations in leaf water potential and stomatal conductance found in canopies in the field. *Philosophical Transactions of the Royal Society of London, Series B* **273,** 593–610.

Jarvis, P. G., and Leverenz, J. W. (1983). Productivity of temperate, deciduous, and evergreen forests. *In* "Encyclopedia of Plant Physiology" (O. L. Lange, P. S. Nobel, C. B. Osmond, and H. Ziegler, eds.), pp. 233–280. Springer-Verlag, New York.

Jayachandran, K., Schwab, A. P., and Hetrick, B. A. D. (1992). Mineralization of organic phosphorus by vesicular–arbuscular mycorrhizal fungi. *Soil Biology and Biochemistry* **24,** 897–903.

Jenny, H. (1980). "The Soil Resource." Springer-Verlag, Berlin and New York.

Jensen, J. R. (1996). "Introductory Digital Image Processing: A Remote Sensing Perspective." Prentice-Hall, Upper Saddle River, New Jersey.

Johnsen, K., Maier, C., Butnor, J., Anderson, P., Waring, R. H., and Linder, S. (2007). Physiological girdling of pine trees via phloem chilling: Proof of concept. *Plant, Cell and Environment* **30,** 128–134.

Johnson, A. H., Friedland, A. J., Miller, E. K., and Siccama, T. G. (1994). Acid rain and soils of the Adirondacks. III. Rates of soil acidification in a montane spruce–fir forest at Whiteface Mountain, New York. *Canadian Journal of Forest Research* **24,** 663–669.

Johnson, D. W. (1992). Nitrogen retention in forest soils. *Journal of Environmental Quality* **21,** 1–12.

Johnson, D. W., and Cole, D. W. (1980). Anion mobility in soils: Relevance to nutrient transport from forest ecosystems. *Environment International* **3,** 79–90.

Johnson, D. W., and Lindberg, S. E., eds. (1992). "Atmospheric Deposition and Forest Nutrient Cycling: A Synthesis of the Integrated Forest Study." Springer-Verlag, New York.

Johnson, D. W., Cole, D. W., and Gessel, S. P. (1979). Acid precipitation and soil sulfate adsorption properties in a tropical and in a temperate forest soil. *Biotropica* **11,** 38–42.

Johnson, D. W., Cole, D. W., Van Miegroet, H., and Horng, F. W. (1986). Factors affecting anion movement and retention in four forest soils. *Soil Science Society of America Journal* **50,** 776–783.

Johnson, D. W., Cresser, M. S., Nilsson, S. I., Turner, J., Ulrich, B., Binkley, D., and Cole, D. W. (1991). Soil changes in forest ecosystems: Evidence for and probable causes. *Proceedings of the Royal Society of Edinburgh* **97B,** 81–116.

Johnson, D. W., West, D. C., Todd, D. E., and Mann, L. K. (1982). Effects of sawlog vs. whole-tree harvesting on the nitrogen, phosphorus, potassium and calcium budgets of an upland mixed oak forest. *Soil Science Society of America Journal* **46,** 1353–1363.

Johnson, I. R., and Thornley, J. H. M. (1987). A model of shoot:root partitioning with optimal growth. *Annals of Botany* **60,** 133–142.

Johnson, K. H., Vogt, K. A., Clark, H. J., Schmitz, O. J., and Vogt, D. J. (1996). Biodiversity and the productivity and stability of ecosystems. *Tree* **11,** 372–377.

Johnson, P. C., and Denton, R. E. (1975). Outbreaks of western spruce budworm in the American Northern Rocky Mountain area from 1922 through 1971. *USDA Forest Service, General Technical Report* **INT-20,** 1–144.

Johnston, C. A., and Naiman, R. J. (1990). The use of a geographic information system to analyze long-term landscape alteration by beaver. *Landscape Ecology* **4,** 5–19.

Johnstone, J. F., and Chapin, F. S. (2003). Non-equilibrium succession dynamics indicate continued northern migration of lodgepole pine. *Global Change Biology* **9,** 1401–1409.

Jolly, W. M., Nemani, R., and Running, S. W. (2005). A generalized, bioclimatic index to predict foliar phenology in response to climate. *Global Change Biology* **11,** 619–632.

Jones, H. G. (1992). "Plants and Microclimate." Cambridge Univ. Press, Cambridge.

Jones, J. A., and Grant, G. E. (1996). Peak flow responses to clearcutting and roads in small and large basins, western Cascades, Oregon. *Water Resources Research* **32,** 959–974.

Joos, F., Prentice, I. C., and House, J. I. (2002). Growth enhancement due to global atmospheric change as predicted by terrestrial ecosystem models: Consistent with US forest inventory data. *Global Change Biology* **8,** 299–303.

Jordon, D. N., Wright, L. M., and Lockaby, B. G. (1990). Relationship between xylem trace metals and radial growth of loblolly pine in rural Alabama. *Journal of Environmental Quality* **19,** 504–508.

Jusoff, K., and Majid, N. M. (1987). Effect of crawler tractor logging on soil compaction in central Pahang Malaysia. *Malaysian Forester* **50,** 274–280.

Kafkafi, U., Bar-Yosef, B., Rosenberg, R., and Sposito, G. (1988). Phosphorus adsorption by kaolinite and montomorillonite. II. Organic anion competition. *Soil Science Society of America Journal* **52,** 1585–1589.

Karban, R., and Myers, J. H. (1989). Induced plant responses to herbivory. *Annual Review of Ecology and Systematics* **20,** 331–348.

Karieiva, P. M., and Andersen, M. (1988). Spatial aspects of species interactions: The wedding of models and experiments. *In* "Lecture Notes in Biomathematics" (A. Hastings, ed.), Vol. 77, pp. 35–50. Springer-Verlag, Berlin.

Karl, T. R., Derr, V. E., Fastening, D. R., Folland, C. K., Hofmann, D. J., Levitus, S., Nicholls, N., Parker, D. E., and Withee, G. W. (1995). Critical issues for long-term climate monitoring. *Climatic Change* **31,** 185–221.

Karl, T. R., Jones, P. D., Knight, R. W., Kukla, G., Plummer, N., Razuvayev, V., Gallo, K. P., Lindseay, J., Charlson, R. J., and Peterson, T. C. (1993). Asymmetric trends of daily maximum and minimum temperature. *Bulletin of the American Meteorological Society* **74,** 1007–1023.

Kasischke, E. S., and French, N. H. F. (1995). Locating and estimating the areal extent of wildfires in Alaskan boreal forests using multiple-season AVHRH NDVI composite data. *Remote Sensing of the Environment* **51,** 263–275.

Kaufmann, J. B., Cummings, D. L., and Ward, D. E. (1994). Relationships of fire, biomass and nutrient dynamics along a vegetation gradient in the Brazilian cerrado. *Journal of Ecology* **82,** 519–531.

Kaufmann, J. B., Sanford, R. L., Jr., Cummings, D. L., Salcedo, I. H., and Sampaio, E. V. S. B. (1993). Biomass and nutrient dynamics associated with slash fires in neotropical dry forests. *Ecology* **74,** 140–151.

Kaufmann, M. R. (1977). Soil temperature and drying cycle effects on water relations of *Pinus radiata. Canadian Journal of Botany* **55,** 2413–2418.

Kaufmann, M. R. (1996). To live fast or not: Growth, vigor and longevity of old-growth ponderosa pine and lodgepole pine trees. *Tree Physiology* **16,** 139–144.

Kaufmann, M. R., and Troendle, C. A. (1981). The relationship of leaf area and foliage biomass to sapwood conducting area in four subalpine forest tree species. *Forest Science* **27,** 477–482.

Keane, R. E., Hardy, C., Ryan, K., and Finney, M. (1997). Simulating effects of fire management on gaseous emissions from future landscape of Glacier National Park, Montana, USA. *World Resource Review* **9,** 177–205.

Keane, R. E., Morgan, P., and Running, S. W. (1996a). FIRE-BGC—A mechanistic ecological process model for simulating fire succession on coniferous forest landscapes of the northern Rocky Mountains. *USDA Research Paper* **INT-RP-484.** Ogden, Utah.

Keane, R. E., Ryan, K. C., and Running, S. W. (1996b). Simulating effects of fire on northern Rocky Mountain landscapes with the ecological process model FIRE-BGC. *Tree Physiology* **16,** 319–331.

Keeley, J. E. (1979). Population differentiation along a flood frequency gradient: Physiological adaptations to flooding in *Nyssa sylvatica. Ecological Monographs* **49,** 89–108.

Keeling, C, D., Chin, J. F. S., and Whorf, T. P. (1996). Increased activity of northern vegetation inferred from atmospheric CO_2 measurements. *Nature* **382,** 146–149.

Keeling, C. D. (1958). The concentration and isotopic abundances of atmospheric carbon dioxide in rural areas. *Geochimica et Cosmochimica Acta* **13,** 322–334.

Keeling, C. D., Whorf, T. P., Wahlen, M., and Van der Plicht, J. (1995). Interannual extremes in the rate of rise of atmospheric carbon dioxide since 1980. *Nature* **375,** 666–670.

Keller, M., and Reiners, W. A. (1994). Soil-atmosphere exchange of nitrous oxide, nitric oxide, and methane under secondary succession of pasture to forest in the Atlantic lowlands of Costa Rica. *Global Biogeochemical Cycles* **8,** 399–409.

Kelliher, F. M., and Scotter, D. R. (1992). Evaporation, soil, and water. *In* "Waters of New Zealand," pp. 136–146. New Zealand Hydrological Society, Wellington, New Zealand.

Kelliher, F. M., Leuning, R., and Schulze, E.-D. (1993). Evaporation and canopy characteristics of coniferous forests and grasslands. *Oecologia* **95,** 153–163.

Kelliher, F. M., Leuning, R., Raupach, M. R., and Schulze, E. D. (1995). Maximum conductances for evaporation from global vegetation types. *Agricultural and Forest Meteorology* **73**, 1–16.

Kercher, J. R., and Axelrod, M. C. (1984). Analysis of SILVA: A model for forecasting the effects of SO$_2$ [sulfur dioxide] pollution and fire on western coniferous forest [*Pinus ponderosa, Pinus lambertiana, Abies concolor,* forest succession simulation, Sierra Nevada, California]. *Ecological Modelling* **23**, 165–184.

Kern, J. S. (1995a). Evaluation of soil water retention models based on basic soil physical properties. *Soil Science Society of America Journal* **59**, 1134–1141.

Kern, J. S. (1995b). Geographic patterns of soil water-holding capacity in the contiguous United States. *Soil Science Society of America Journal* **59**, 1126–1133.

Kicklighter, D. W., Melillo, J. M., Peterjohn, W. T., Rastetter, E. B., McGuire, A. D., and Steudler, P. A. (1994). Aspects of spatial and temporal aggregation in estimating regional carbon dioxide fluxes from temperate forest soils. *Journal of Geophysical Research* **99**, 1303–1315.

Kilgore, B. M., and Taylor, D. (1979). Fire history of a sequoia–mixed conifer forest. *Ecology* **60**, 129–142.

Kimball, J. S., McDonald, K. C., Running, S. W., and Frolking, S. E. (2004). Satellite radar remote sensing of seasonal growing seasons for boreal and subalpine evergreen forests. *Remote Sensing of Environment* **90**, 243–258.

Kimball, J. S., White, M. A., and Running, S. W. (1997a). Biome-BGC simulations of stand hydrologic processes for BOREAS. *Journal of Geophysical Research* **102(D24)**, 29,043–29,051.

Kimball, J. S., Running, S. W., and Nemani, R. (1997b). An improved method for estimating surface humidity from daily minimum temperature. *Agricultural and Forest Meteorology* **85**, 87–98.

Kira, T., and Ogawa, H. (1971). Assessment of primary production in tropical and equatorial forests. *Ecology and Conservation* **4**, 309–321.

Kitzberger, T., Veblen, T. T., and Villalba, R. (1995). Tectonic influences on tree growth in northern Patagonia, Argentina: The roles of substrate stability and climatic variation. *Canadian Journal of Forest Research* **25**, 1684–1696.

Kloft, W. (1957). Further investigations concerning the inter-relationship between bark conditions of *Abias alba* and infestation by *Adelges piceae typica* and *A. nusslini schneideri. Zeitschrift für angewante Entomologie* **41**, 438–442.

Knight, D. H., Fahey, T. J., and Running, S. W. (1985). Factors affecting water and nutrient outflow from lodgepole pine forests in Wyoming. *Ecological Monographs* **55**, 29–78.

Knowles, N., Dettinger, M. D., and Cayan, D. R. (2006). Trends in snowfall versus rainfall for the western United States, 1949–2004. *Journal of Climate* **19**, 4545–4559.

Kohyama, T. (1980). Growth pattern of *Abies mariessi* saplings under conditions of open-growth and suppression. *Botanical Magazine (Tokyo)* **93**, 13–24.

Körner, C. (1985). Humidity responses in forest trees: Precautions in thermal scanning surveys. *Archives for Meteorology, Geophysics, and Bioclimatology, Series B* **36**, 83–98.

Körner, C. (1994). Leaf diffusive conductances in the major vegetation types of the globe. *In* "Ecophysiology of Photosynthesis" (E.-D. Schulze and M. M. Caldwell, eds.), pp. 463–490. Springer-Verlag, Heidelberg, Germany.

Korol, R. L., Milner, K. S., and Running, S. W. (1996). Testing a mechanistic model for predicting stand and tree growth. *Forest Science* **42**, 139–153.

Korol, R. L., Running, S. W., and Milner, K. S. (1995). Incorporating intertree competition into an ecosystem model. *Canadian Journal of Forest Research* **25**, 413–424.

Korol, R. L., Running, S. W., Milner, K. S., and Hunt, E. R., Jr. (1991). Testing a mechanistic carbon balance model against observed tree growth. *Canadian Journal of Forest Research* **21**, 1098–1105.

Kotar, M. (1996). Volume and height growth of fully stocked mature beech stands in Slovenia during the past three decades. *In* "Growth Trends in European Forests" (H. Spiecker, K. Mielikäinen, K. Köhl, and J. P. Skovsgaard, eds.), pp. 291–312. Springer, Berlin.

Kowalski, A. S., Loustau, D., Berbigier, P., Manca, G., Tedeschi, V., Borghetti, M., Balentini, R., Kolari, P., Berninger, F., Rannik, U., Hari, P., Rayment, M., Mencuccini, M., Moncrieff, J., and Grace, J. (2004). Paired comparisons of carbon exchange between disturbed and regenerating stands in four managed forests in Europe. *Global Change Biology* **10,** 1707–1723.

Kozlowski, T. T., and Keller, T. (1966). Food relations of woody plants. *Botanical Review* **32,** 293–382.

Kozlowski, T. T., and Pallardy, S. G. (1979). Stomata response of *Fraxinus pennsylvanica* seedlings during and after flooding. *Physiologia Plantarum* **46,** 155–158.

Kros, H., and Warfvinge, P. (1995). Evaluation of model behaviour with respect to the biogeo-chemistry at the Solling spruce site. *Ecological Modelling* **83,** 255–262.

Kuhn, U., Rottenberger, S., Biesenthal, T., Wolf, A., Schebeske, G., Ciccioli, P., and Kesselmeier, J. (2004). Strong correlation between isoprene emission and gross photosynthetic capacity during leaf phenology of the tropical tree species *Hymenaea courbaril* with fundamental changes in volatile organic compounds emission composition during early leaf development. *Plant, Cell and Environment* **27,** 1469–1485.

Kull, O., and Jarvis, P. G. (1995). The role of nitrogen in a simple scheme to scale up photosynthesis from leaf to canopy. *Plant, Cell and Environment* **18,** 1174–1182.

Kurz, W. A., and Kimmins, J. P. (1987). Analysis of some sources of error in methods used to determine fine root production in forest ecosystems: A simulation approach. *Canadian Journal of Forest Research* **17,** 901–912.

Kuusisto, E. (1993). Lake ice observations in Finland in the 19th and 20th century: Any message for the 21st? *In* "Snow Watch '92: Detection Strategies for Snow and Ice, Glaciological Data" (R. G. Barry, B. E. Goodison, and E. F. LeDrew, eds.), pp. 57–65. Glaciological Data Report GD-25.

Lahde, E., Nieminen, J., Etholen, K., and Suolahti, P. (1982). Older lodgepole pine stands in southern Finland. *Folia Forestalia* No. 533. Institute of Finnish Forestry, Helsinki, Finland.

LaMarche, V. C., and Pittock, A. B. (1982). Preliminary temperature reconstruction for Tasmania. *In* "Climate from Tree Rings" (M. K. Hughes, P. M. Kelly, J. R. Pilcher, and V. C. LaMarche, eds.), pp. 177–185. Cambridge Univ. Press, London and New York.

Lamberson, R. H., McKelvey, R., Noon, B. R., and Voss, C. (1992). A dynamic analysis of northern spotted owl viability in a fragmented forest landscape. *Conservation Biology* **6,** 1–8.

Landsberg, J. J. (1986a). "Physiological Ecology of Forest Production." Academic Press, London.

Landsberg, J. J. (1986b). Experimental approaches to the study of the effects of nutrients and water on carbon assimilation by trees. *Tree Physiology* **2,** 427–444.

Landsberg, J. J., and Fowkes, N. D. (1978). Water movement through plant roots. *Annals of Botany* **42,** 493–508.

Landsberg, J. J., and Gower, S. T. (1997). "Applications of Physiological Ecology to Forest Management." Academic Press, San Diego.

Landsberg, J. J., and McMurtrie, R. (1984). Water use by isolated trees. *Agriculture and Water Management* **8,** 223–242.

Landsberg, J. J., and Waring, R. H. (1997). A generalised model of forest productivity using simpli-fied concepts of radiation-use efficiency, carbon balance and partitioning. *Forest Ecology and Management* **95,** 209–228.

Landsberg, J. J., Prince, S. G. D., Jarvis, P. G., McMurtrie, R. E., Luxmoore, R., and Medlyn, B. E. (1996). Energy conversion and use in forests: The analysis of forest production in terms of radiation utilisation efficiency (ε). *In* "The Use of Remote Sensing in the Modeling of Forest

Productivity at Scales from the Stand to the Globe" (H. L. Gholz, K. Nakane, and H. Shimoda, eds.), pp. 273–298. Kluwer, Dordrecht, The Netherlands.

Landsberg, J. J., Waring, R. H., and Coops, N. C. (2003). Performance of the forest productivity model 3-PG applied to a wide range of forest types. *Forest Ecology and Management* **172,** 199–214.

Lang, A. R. G. (1991). Application of some of Cauchy's theorems to estimation of surface areas of leaves, needles and branches of plants, and light transmittance. *Agricultural and Forest Meteorology* **55,** 191–212.

Larsson, S., Oren, R., Waring, R. H., and Barrett, J. W. (1983). Attacks of mountain pine beetle as related to tree vigor of ponderosa pine. *Forest Science* **29,** 395–402.

Latham, R. E., and Ricklefs, R. E. (1993). Global patterns of tree species richness in moist forests: Energy-diversity theory does not account for variation in tree species richness. *Oikos* **67,** 325–333.

Lavelle, P., Blanchart, E., Martin, A., Martin, S., Spain, A., Toutain, F., Barois, I., and Schaefer, R. (1993). A hierarchical model for decomposition in terrestrial ecosystems: Application to soils of the humid tropics. *Biotropica* **25,** 130–150.

Law, B. E. (1995). Estimation of leaf area index and light interception by shrubs from digital videography. *Remote Sensing of Environment* **51,** 276–280.

Law, B. E., and Waring, R. H. (1994). Combining remote sensing and climatic data to estimate net primary production. *Ecological Applications* **4,** 717–728.

Law, B. E., Falge, E., Gu, L., Baldocchi, D. D., Bakwin, P., Berbigier, P., Davis, K., Dolman, A. J., Falk, M., Fuentes, J. D., Goldstein, A., Granier, A., Grelle, A., Hollinger, D., Janssens, I. A., Jarvis, P., Jensen, N. O., Katul, G., Mahli, Y., Matteucci, G., Meyers, T., Monson, R., Munger, W., Oechel, W., Olson, R., Pilegaard, K., Paw U. K. T., Thorgeirsson, H., Valentini, R., Verma, S., Vesala, T., Wilson, K., and Wofsy, S. (2002). Environmental controls over carbon dioxide and water vapor exchange of terrestrial vegetation. *Agricultural and Forest Meteorology* **113,** 97–120.

Law, B. E., Turner, D., Campbell, J., Sun, O. J., Van Tuyl, S., Ritts, W. D., and Cohen, W. B. (2004). Disturbance and climate effects on carbon stocks and fluxes across Western Oregon USA. *Global Change Biology* **10,** 1429–1444.

Law, B. E., Waring, R. H., Anthoni, P. M., and Aber, J. D. (2000). Measurements of gross and net ecosystem productivity and water vapour exchange of a *Pinus ponderosa* ecosystem, and an evaluation of two general models. *Global Change Biology* **5,** 155–168.

Law, R. M., Rayner, P. J., Denning, A. S., Erickson, D. Fung, I. Y., Heimann, M., Piper, S. C., Ramonet, M., Taguchi, S., Taylor, J. A., Trudinger, C. M., and Watterson, I. G. (1996). Variation in modeled atmospheric transport of carbon dioxide and the consequences for CO_2 inversions. *Global Biogeochemical Cycles* **10,** 783–796.

Leahey, A. (1947). Characteristics of soils adjacent to the MacKenzie River in the Northwest Territories of Canada. *Soil Science Society of America Proceedings* **12,** 458–461.

Leak, W. B. (1987). Fifty years of compositional change in deciduous and coniferous forest types in New Hampshire. *Canadian Journal of Forest Research* **17,** 388–393.

Leavitt, S. W., and Long, A. (1989). The atmospheric $\delta^{13}C$ record as derived from 56 pinyon trees at 14 sites in the southwest United States. *Radiocarbon* **31,** 469–474.

Lechowicz, M. J. (1995). Seasonality of flowering and fruiting in temperate forest trees. *Canadian Journal of Botany* **73,** 175–182.

Lee, K.-S., Cohen, W. B., Kennedy, R. E., Maiersperger, T. K., and Gower, S. T. (2004). Hyperspectral versus multispectral data for estimating leaf area index in four different biomes. *Remote Sensing of Environment* **91,** 508–520.

Lee, R., and Sypolt, C. R. (1974). Toward a biophysical evaluation of forest site quality. *Forest Science* **20,** 145–154.

Leech, R. H., and Kim, Y. T. (1981). Foliar analysis and DRIS as a guide to fertilizer amendments in poplar plantations. *Forestry Chronicle* **57,** 17–21.

Lefsky, M. A., Hudak, A. T., Cohen, W. B., and Acker, S. A. (2005a). Geographic variability in lidar predictions of forest stand structure in the Pacific Northwest. *Remote Sensing of Environment* **95,** 532–548.

Lefsky, M. A., Turner, D. P., Guzy, M., and Cohen, W. B. (2005b). Combining lidar estimates of biomass and Landsat estimates of stand age for spatially extensive validation of modeled forest productivity. *Remote Sensing of Environment* **95,** 549–558.

Leith, H. (1975). Modeling the primary productivity of the world. *In* "Primary Productivity of the Biosphere" (H. Leith and R. H. Whittaker, eds.), pp. 237–263. Springer-Verlag, New York.

Lemon, P. C. (1961). Forest ecology of ice storms. *Bulletin of Torrey Botanical Club* **88,** 21–29.

Lepers, E., Lambin, E. F., Janetos, A. C., DeFries, R., Achard, F., Ramankutty, N., and Scholes, R. J. (2005). A synthesis of information on rapid land-cover change for the period 1981–2000. *BioScience* **55,** 115–124.

Leuning, R. (1995). A critical appraisal of a combined stomatal-photosynthesis model for C_3 plants. *Plant, Cell and Environment* **18,** 339–355.

Leuning, R., Kelliher, F. M., DePury, D. G. G., and Schulze, E.-D. (1995). Leaf nitrogen, photosynthesis, conductance and transpiration: Scaling from leaves to canopies. *Plant, Cell and Environment* **18,** 1183–1200.

Levin, S. A. (1992). The problem of pattern and scale in ecology. *Ecology* **73,** 1943–1967.

Levine, E. R., Knox, R. G., and Lawrence, W. T. (1994). Relationships between soil properties and vegetation at the northern experimental forest, Howland, Maine. *Remote Sensing of Environment* **47,** 231–241.

Levitt, J. (1980). "Responses of Plants to Environmental Stresses." Academic Press, New York.

Lewin, R. (1984). Parks: How big is big is enough? *Science* **225,** 611–612.

Lewis, S. L., Phillips, O. L., Baker, T. R. *et al.* (2004). Concerted changes in tropical forest structure and dynamics: Evidence from 50 South American long-term plots. *Philosophical Transcripts of the Royal Society of London, Series B* **359,** 421–436.

Li, H., Franklin, J. F., Swanson, F. J., and Spies, T. A. (1993). Developing alternative forest cutting patterns: A simulation approach. *Landscape Ecology* **8,** 63–75.

Likens, G. E., and Bormann, F. H. (1995). "Biogeochemistry of a Forest Ecosystem," 2nd Ed. Springer-Verlag, New York.

Likens, G. E., Bormann, F. H., Pierce, R. S., Eaton, J. S., and Johnson, N. M. (1977). "Biogeochemistry of a Forested Ecosystem." Springer-Verlag, New York.

Lillesand, T. M., and Kiefer, R. W. (1994). "Remote Sensing and Image Interpretation," 3rd ed. Wiley, New York.

Lindberg, S. E., Lovett, G. M., Richter, D. D., and Johnson, D. W. (1986). Atmospheric deposition and canopy interactions of major ions in a forest. *Science* **231,** 141–146.

Linder, S. (1986). Responses to water and nutrients in coniferous ecosystems. *In* "Potentials and Limitations in Ecosystem Analysis" (E.-D. Schulze and H. Zwöllfer, eds.), pp. 180–202. Springer-Verlag, Berlin.

Linder, S. (1995). Foliar analysis for detecting and correcting nutrient imbalances in Norway spruce. *Ecological Bulletins* **44,** 178–190.

Linder, S., and Rook, D. A. (1984). Effects of mineral nutrition on carbon dioxide exchange and partitioning carbon in trees. *In* "Nutrition of Plantation Forests" (G. D. Bowen and E. K. S. Nambiar, eds.), pp. 211–236. Academic Press, Orlando, Florida.

Linder, S., and Troeng, E. (1981). The seasonal variation in stem and coarse root respiration in a 20-year old Scots pine (*Pinus sylvestris* L.). *Mitteilungender Forstlichen Bundes-Versuchsanstalt (Wein)* **142,** 125–139.

Linder, S., Benson, M. L., Myers, B. J., and Raison, R. J. (1987). Canopy dynamics and growth of *Pinus radiata*. Effects of irrigation and fertilization during a drought. *Canadian Journal of Forest Research* **17,** 117–126.

Linderholm, H. W. (2006). Growing season changes in the last century. *Agricultural and Forest Meteorology* **137,** 1–14.

Lines, R., and Howell, R. S. (1963). The use of flags to estimate the relative exposure of trial plantations. *Forestry Commission, Forest Records* **51,** 1–31.

Liu, G. F., and Deng, T. X. (1991). Mathematical model of the relationship between nitrogen-fixation by black locust and soil conditions. *Soil Biology and Biochemistry* **23,** 1–7.

Liu, S., Munson, R., Johnson, D., Gherini, S., Summers, K., Hudson, R., Wilkinson, K., and Pitelka, L. (1991). Applications of a nutrient cycling model (NuCM) to a northern mixed hardwood and southern coniferous forest. *Tree Physiology* **9,** 173–184.

Lloyd, J., and Farquhar, G. D. (1994). ^{13}C discrimination during CO_2 assimilation by the terrestrial biosphere. *Oecologia* **99,** 201–215.

Lloyd, J., Grace, J., Miranda, A. C., Meir, P., Wong, S. C., Miranda, H. S., Wright, I. R., Gash, J. H. C., and McIntyre, J. (1995). A simple calibrated model of Amazon rainforest productivity based on leaf biochemical properties. *Plant, Cell and Environment* **18,** 1129–1145.

Lo Gullo, M. A., Salleo, S., Piaceri, E. C., and Rosso, R. (1995). Relations between vulnerability to xylem embolism and xylem conduit dimensions in young trees of *Quercus cerris*. *Plant, Cell and Environment* **18,** 661–669.

Logan, J. A., Régnière, J., and Powell, J. A. (2003). Assessing the impacts of global warming on forest pest dynamics. *Frontiers in Ecology and the Environment* **1,** 130–137.

Lorio, P. L., Jr. (1986). Growth–differentiation balance: A basis for understanding southern pine beetle–tree interactions. *Forest Ecology and Management* **14,** 259–273.

Loveland, T. R., Merchant, J. W., Ohlen, D. O., and Brown, J. F. (1991). Development of a land cover characteristics database for the conterminous U.S. *Photogrammetric Engineering and Remote Sensing* **57,** 1453–1463.

Lovett, G. M. (1994). Atmospheric deposition of nutrients and pollutants in North America: An ecological perspective. *Ecological Applications* **4,** 629–650.

Lovett, G. M., Reiners, W. A., and Olson, R. K. (1982). Cloud droplet deposition in subalpine balsam fir forests: Hydrologic and chemical inputs. *Science* **218,** 1303–1304.

Lowry, W. P. (1969). "Weather and Life: An Introduction to Biometeorology." Academic Press, New York.

Luan, J., Muetzefeldt, R. L., and Grace, J. (1996). Hierarchical approach to forest ecosystem simulation. *Ecological Modelling* **86,** 37–50.

Luce, C. H., and Cundy, T. W. (1994). Parameter identification for a runoff model for forest road. *Water Resources Research* **30,** 1057–1069.

Lucht, W., Prentice, C. I., Myneni, R. B., Sitch, S., Friedlingstein, P., Cramer, W., Bousquet, P., Buermann, W., and Smith, B. (2002). Climatic control of the high-latitude vegetation greening trend and Pinatubo effect. *Science* **296,** 1687–1689.

Lugo, A. E., Applefield, M., Pool, D. J., and McDonald, R. B. (1983). The impact of Hurricane David on the forest of Dominica. *Canadian Journal of Forest Research* **13,** 201–211.

Luizao, F., Matson, P., Livingston, G., Luizao, R., and Vitousek, P. (1989). Nitrous oxide flux following tropical land clearing. *Global Biogeochemical Cycles* **3,** 281–285.

Luxmoore, R. J. (1983). Infiltration and runoff prediction for a grassland watershed. *Journal of Hydrology* **65,** 271–278.

Luxmoore, R. J. (1991). A source–sink framework for coupling water, carbon and nutrient dynamics of vegetation. *Tree Physiology* **9,** 267–280.

MacKney, D. (1961). A podzol development sequence in oakwoods and heath in central England. *Journal of Soil Science* **12,** 22–40.

Macko, A., and Estep, M. F. L. (1984). Microbial alteration of stable nitrogen and carbon isotopic composition of organic matter. *Organic Geochemistry* **6,** 787–790.

MacLean, D. A., and Wein, R. W. (1977). Nutrient accumulation for post fire jack pine and hardwood successional patterns in New Brunswick. *Canadian Journal of Forest Research* **7,** 562–578.

Magnani, F., and Borghetti, M. (1995). Interpretation of seasonal changes in xylem embolism and plant hydraulic resistance in *Fagus sylvatica. Plant, Cell and Environment* **18,** 689–696.

Magnuson, J. J. (1990). Long-term ecological research and the invisible present. *BioScience* **40,** 495–501.

Malmer, A., and Grip, H. (1990). Soil disturbance and loss of infiltrability caused by mechanized and manual extraction of tropical rainforest in Sabah, Malaysia. *Forest Ecology and Management* **38,** 1–12.

Mandzak, J. M., and Moore, J. A. (1994). The role of nutrition in the health of inland western forests. *Journal of Sustainable Forestry* **2,** 191–210.

Mariotti, A., German, J. C., Hubert, P., Kaiser, P., Letolle, R., Tardieus, A., and Tardieus, P. (1981). Experimental determination of nitrogen kinetic isotope fractionation: Some principles; illustration for the denitrification and nitrification processes. *Plant and Soil* **62,** 453–430.

Marks, P. L. (1974). The role of pin cherry (*Prunus pennsylvanica* L.) in the maintenance of stability in northern hardwood ecosystems. *Ecological Monographs* **44,** 73–88.

Marks, P. L., and Bormann, F. H. (1972). Revegetation following forest cutting: Mechanisms for return to steady state nutrient cycling. *Science* **176,** 914–915.

Marschner, H. (1995). "Mineral Nutrition of Higher Plants." Academic Press, San Diego.

Marshall, J. D., and Monserud, R. A. (1996). Homeostatic gas-exchange parameters inferred from $^{13}C/^{12}C$ in tree rings of conifers. *Oecologia* **105,** 13–21.

Marshall, J. D., and Waring, R. H. (1985). Predicting fine root production and turnover by monitoring root starch and soil temperature. *Canadian Journal of Forest Research* **15,** 791–800.

Marshall, J. D., and Waring, R. H. (1986). Comparison of methods of estimating leaf-area index in old-growth Douglas-fir. *Ecology* **67,** 975–979.

Martikainen, P. J., and Palojarvi, A. (1990). Evaluation of the fumigation–extraction method for the determination of microbial C and N in a range of forest soils. *Soil Biology and Biochemistry* **22,** 797–802.

Martin, B., and Sutherland, E. K. (1990). Air pollution in the past recorded in width and stable carbon isotope composition of annual growth rings of Douglas-fir. *Plant, Cell and Environment* **13,** 839–844.

Martin, F., Canet, D., Rolin, D., Marchal, J. P., and Larher, F. (1983). Phosphorus-31 nuclear magnetic resonance study of polyphosphate metabolism in intact mycorrhizal fungi. *Plant and Soil* **71,** 469–476.

Martin, M. E., and Aber, J. D. (1997). High spectral resolution remote sensing of forest canopy lignin, nitrogen, and ecosystem processes. *Ecological Applications* **7,** 431–443.

Marumoto, T., Anderson, J. P. E., and Domsch, K. H. (1982). Mineralization of nutrients from soil microbial biomass. *Soil Biology and Biochemistry* **14,** 469–475.

Marx, D. H., Hatch, A. B., and Mendicino, J. F. (1977). High soil fertility decreases sucrose content and susceptibility of loblolly pine roots to ectomycorrhizal infection by *Pisolithus tinctorius. Canadian Journal of Botany* **55,** 1569–1574.

Maser, C., Trappe, J. M., and Nussbaum, R. A. (1978). Fungal–small mammal interrelationships with emphasis on Oregon coniferous forests. *Ecology* **59,** 799–809.

Mason, R. R., Wickman, B. E., Beckwith, R. C., and Paul, H. G. (1992). Thinning and nitrogen fertilization in a grand fir stand infested with western spruce budworm. Part I. Insect response. *Forest Science* **38,** 235–251.

Masscheleyn, P. H., Pardue, J. H., DeLaune, R. D., and Patrick, W. H., Jr. (1992). Phosphorus release and assimilatory capacity of two lower Mississippi Valley freshwater wetland soils. *Water Resources Research* **28,** 763–773.

Mather, A. S. (1990). "Global Forest Resources." Timber Press, Portland, Oregon.

Matson, P. A., and Harriss, R. C. (1988). Prospects for aircraft-based gas exchange measurements in ecosystem studies. *Ecology* **69,** 1318–1325.

Matson, P. A., and Waring, R. H. (1984). Effects of nutrient and light limitation on mountain hemlock: Susceptibility to laminated root rot. *Ecology* **65,** 1517–1524.

Matson, P. A., Johnson, L., Billow, C., Miller, J., and Pu, R. (1994). Seasonal patterns and remote spectral estimation of canopy chemistry across the Oregon transect. *Ecological Applications* **4,** 280–298.

Matson, P. A., Vitousek, P. M., Ewel, J. J., Mazzarino, M. J., and Robertson, G. P. (1987). Nitrogen transformations following tropical forest felling and burning on a volcanic soil. *Ecology* **68,** 491–502.

Matthews, E. (1983). Global vegetation and land use: New high-resolution data bases for climate studies. *Journal of Climatic Applied Meteorology* **22,** 474–487.

Matyssek, R., Reich, P., Oren, R., and Winner, W. E. (1995). Response mechanisms of conifers to air pollutants. *In* "Ecophysiology of Coniferous Forests" (W. K. Smith and T. M. Hinckley, eds.), pp. 255–308. Academic Press, San Diego.

Matyssek, R., Günthardt-Goerg, M. S., Saurer, M., and Keller, T. (1992). Seasonal growth, $\delta^{13}C$ in leaves and stem, and phloem structure of birch (*Betula pendula*) under low ozone concentrations. *Tree Physiology* **6,** 69–76.

McCauley, K. J., and Cook, S. A. (1980). *Phellinus weirii* infestation of two mountain hemlock forests in the Oregon Cascades. *Forest Science* **26,** 201–211.

McClaugherty, C. A., Aber, J. D., and Melillo, J. M. (1982). The role of fine roots in the organic matter and nitrogen budgets of two forested ecosystems. *Ecology* **63,** 1481–1490.

McClelland, B. R. (1973). Autumn concentration of bald eagles in Glacier National Park. *Condor* **75,** 121–123.

McComb, W. C., Bonney, S. A., Sheffield, R. M., and Cost, N. D. (1986). Den tree characteristics and abundance in Florida and North Carolina. *Journal of Wildlife Management* **50,** 584–591.

McCune, B. (1983). Fire frequency reduced by two orders of magnitude in the Bitterroot Canyons, Montana. *Canadian Journal of Forest Research* **13,** 212–218.

McDermitt, D. K., and Loomis, R. S. (1981). Elemental composition of biomass and its relation to energy content, growth efficiency, and growth yield. *Annals of Botany* **48,** 275–290.

McGarigal, K., and Marks, G. J. (1995). FRAGSTATS: Spatial pattern analysis program for quantifying landscape structure. *USDA Forest Service, General Technical Report* **PNW–351.** Portland, Oregon.

McGinty, D. T. (1976). Comparative root and soil dynamics on a white pine watershed and in the hardwood forest in the Coweeta basin. Ph.D. Dissertation, University of Georgia, Athens.

McGuire, A. D., Joyce, L. A., Kicklighter, D. W., Melillo, J. M., Esser, G., and Vorosmarty, C. J. (1993). Productivity response of climax temperate forests to elevated temperature and carbon dioxide: A North American comparison between two global models. *Climatic Change* **24,** 287–310.

McKay, D. S., and Band, L. E. (1997). Forest ecosystem processes at the watershed scale: Dynamic coupling of distributed hydrology and canopy growth. *Hydrological Processes* **11,** 1197–1217.

McKenzie, D., Hessl, A. E., and Peterson, D. L. (2001). Recent growth of conifer species of western North American: Assessing spatial patterns of radial growth trends. *Canadian Journal of Forest Research* **31,** 526–538.

McKey, D., Waterman, P. G., Mbi, C. N., Cartlan, J. S., and Struhsaker, T. T. (1978). Phenolic content of vegetation in two African rain forests: Ecological implications. *Science* **202,** 61–64.

McLaughlin, S. B. (1985). Effects of air pollution on forests, a critical review. *Journal of Air Pollution Control Association* **35,** 513–534.

McLaughlin, S. B., and Shriner, D. S. (1980). Allocation of resources to defense and repair. *Plant Diseases* **5,** 22–40.

McLaughlin, S. B., McConathy, R. K., Barnes, R. L., and Edwards, N. T. (1980). Seasonal changes in energy allocation by white oak (*Quercus alba*). *Canadian Journal of Forest Research* **10,** 379–388.

McLeod, S., and Running, S. W. (1988). Comparing site quality indices and productivity of ponderosa pine stands in western Montana. *Canadian Journal of Forest Research* **18,** 346–352.

McMurtrie, R. E. (1993). Modelling canopy carbon and water balance. *In* "Photosynthesis and Production in a Changing Environment: A Field and Laboratory Manual" (D. O. Hall, J. M. O. Scurlock, H. R. Bolhar-Nordenkampf, R. C. Leegood, and S. P. Long, eds.), pp. 220–231. Chapman & Hall, London.

McMurtrie, R. E., and Landsberg, J. J. (1992). Using a simulation model to evaluate the effects of water and nutrients on the growth and carbon partitioning of *Pinus radiata*. *Forest Ecology and Management* **52,** 243–260.

McMurtrie, R. E., Comins, H. N., Kirschbaum, M. U. F., and Wang, Y. P. (1992). Modifying existing forest growth models to take account of effects of elevated CO_2. *Australian Journal of Botany* **40,** 657–677.

McMurtrie, R. E., Gholz, H. L., Linder, S., and Gower, S. T. (1994). Climatic factors controlling the productivity of pine stands: A model-based analysis. *Ecological Bulletins* **43,** 173–188.

McMurtrie, R., and Wolf, L. (1983). A model of competition between trees and grass for radiation, water, and nutrients. *Annals of Botany* **52,** 449–458.

McNaughton, K. G. (1976). Evaporation and advection I: evaporation from extensive homogeneous surfaces. *Quarterly Journal of the Royal Meteorology Society* **102,** 181–191.

McNaughton, K. G., and Jarvis, P. G. (1983). Predicting effects of vegetation changes on transpiration and evaporation, *In* "Water Deficits and Plant Growth" (T. T. Kozlowski, ed.), Vol. 7, pp. 1–47. Academic Press, London.

McNeill, A. M., Hood, R. C., and Wood, M. (1994). Direct measurement of nitrogen fixation by *Trifolium repens* L. and *Alnus glutinosa* using $^{15}N_2$. *Journal of Experimental Botany* **45,** 749–755.

McNulty, S. G. (2002). Hurricane impacts on US forest carbon sequestration. *Environmental Pollution* **116,** S17–S24.

Mead, D. J., and Preston, C. M. (1994). Distribution and retranslocation of ^{15}N in lodgepole pine oven eight growing seasons. *Tree Physiology* **14,** 389–402.

Meinzer, F. C., and Grantz, D. A. (1991). Coordination of stomatal, hydraulic, and canopy boundary layer properties: Do stomata balance conductances by measuring transpiration? *Physiologia Plantarum* **83,** 324–329.

Melillo, J. M., Aber, J. D., and Muratore, J. F. (1982). Nitrogen and lignin control of hardwood leaf litter decomposition dynamics. *Ecology* **63,** 621–626.

Melillo, J. M., McGuire, A. D., Kicklighten, A. D., Moore III, B., Vorsmarty, C. J., and Schloss, A. L. (1993). Global climate change and terrestrial net primary production. *Nature* **363,** 234–240.

Mencuccini, M., and Grace, J. (1995). Climate influences the leaf area–sapwood area relationship in Scots pine *(Pinus sylvestris* L.). *Plant, Cell and Environment* **15,** 1–10.

Mencuccini, M., and Grace, J. (1996). Hydraulic conductance, light interception, and needle nutrient concentration in Scots pine stands (Thetford, U.K.) and their relation with net primary production. *Tree Physiology* **16,** 459–469.

Meyer, W. B., and Turner, B. L. (1994). "Changes in Land Use and Land Cover: A Global Perspective." Cambridge Univ. Press, New York.

Michaud, J., and Sorooshian, S. (1994). Comparison of simple versus complex distributed runoff models on a midsized semiarid watershed. *Water Resources Research* **30,** 593–605.

Mildrexler, D. J., Zhao, M., Heinsch, F. A., and Running, S. W. (2007). A new satellite based methodology for continental scale disturbance detection. *Ecological Applications.* **17,** 235–250.

Miles, J. (1985). The pedogenic effects of different species and vegetation types and the implications on succession. *Journal of Soil Science* **36,** 571–584.

Miller, E. K., Blum, J. D., and Friedland, A. J. (1993). Determination of soil exchangeable-cation loss and weathering using Sr isotopes. *Nature* **362,** 438–441.

Miller, H. G., and Miller, J. D. (1976). Analysis of needle fall as a means of assessing nitrogen status in pine. *Forestry* **49,** 57–61.

Milne, B. T. (1992). Spatial aggregation and neutral models in fractal landscapes. *The American Naturalist* **139,** 32–57.

Milner, K. S., Running, S. W., and Coble, D. W. (1996). A biophysical soil/site model for estimating potential productivity of forested landscapes. *Canadian Journal of Forest Research* **26,** 1174–1186.

Mitchell, J. F. B., Johns, T. C., Gregory, J. M., and Tett, S. F. B. (1995). Climate response to increasing levels of greenhouse gases and sulphate aerosols. *Nature* **376,** 501–504.

Mitchell, M. J., Driscoll, C. T., Kahl, J. S., Likens, G. E., Murdoch, P. S., and Pardo, L. H. (1996). Climatic control of nitrate loss from forested watersheds in the Northeast United States. *Environmental Science and Technology* **30,** 2609–2612.

Mitchell, R. G., Waring, R. H., and Pitman, G. B. (1983). Thinning lodgepole pine increases tree vigor and resistance to mountain pine beetle. *Forest Science* **29,** 204–211.

Moffat, A. S. (1996). Biodiversity is a boon to ecosystems, not species. *Science* **271,** 1497.

Mohren, G. M. J., and Ilvesniemi, H. (1995). Modelling effects of soil acidification on tree growth and nutrient status. *Ecological Modelling* **83,** 263–272.

Monson, R. K., Harley, P. C., Litvak, M. E., Wildermuth, M., Guenther, A. B., Zimmerman, P. R., and Fall, R. (1994). Environmental and developmental controls over the seasonal pattern of isoprene emission from aspen leaves. *Oecologia* **99,** 260–270.

Monteith, J. L. (1965). Evaporation and environment. *Symposia of the Society for Experimental Biology* **29,** 205–234.

Monteith, J. L. (1972). Solar radiation and productivity in tropical ecosystems. *Journal of Applied Ecology* **9,** 747–766.

Monteith, J. L. (1977). Climate and the efficiency of crop production in Britain. *Philosophical Transactions of the Royal Society of London, Series B* **281,** 277–294.

Monteith, J. L., and Unsworth, M. H. (1990). "Principles of Environmental Physics," 2nd Ed. Arnold, London.

Mooney, H. A. (1972). The carbon balance of plants. *Annual Review of Ecology and Systematics* **3,** 315–346.

Mooney, H. A., and Chu, C. (1974). Seasonal carbon allocation in *Heteromeles arbutifolia* a California evergreen shrub. *Oecologia* **14,** 295–306.

Moore III, B., and Braswell, B. H. (1994). The lifetime of excess atmospheric carbon dioxide. *Global Biogeochemical Cycles* **8,** 23–28.

Moore, P. D. (1994). Trials in bad taste. *Nature* **372,** 410.

Moore, R. D., and Thompson, J. C. (1996). Are water table variations in a shallow forest soil consistent with the TOPMODEL concept? *Water Resources Research* **32,** 663–669.

Mott, K. A., and Parkhurst, D. F. (1991). Stomatal responses to humidity in air and helox. *Plant, Cell and Environment* **14,** 509–515.

Mouillot, F., and Field, C. B. (2005). Fire history and the global carbon budget: A $1° \times 1°$ fire history reconstruction for the 20th century. *Global Change Biology* **11,** 398–420.

Mu, Q, Heinsch, F. A., Zhao, M., and Running, S. W. (2007). Development of a global evapotranspiration algorithm based on MODIS and global meteorology data. *Remote Sensing of Environment*. In press.

Muller, J.-F. (1992). Geographic distribution and seasonal variation of surface emissions and deposition velocities of atmospheric trace gases. *Journal of Geophysical Research* **97**(D4), 3787–3804.

Muller, J.-F., and Brasseur, G. (1995). IMAGES: A three-dimensional chemical transport model of the global troposphere. *Journal of Geophysical Research* **11**(D8), 16445–16490.

Myers, N. (1993). Tropical forests: The main deforestation fronts. *Environmental Conservation* **20,** 9–16.

Myneni, R. B. (2001). Variations in northern vegetation activity inferred from satellite data of vegetation index during 1981–1999. *Journal of Geophysical Research* **106,** 20069–20084.

Myneni, R. B., Nemani, R. R., and Running, S. W. (1997a). Algorithm for the estimation of global land cover, LAI, and FPAR based on radiative transfer models. *IEEE Transactions in Geoscience and Remote Sensing* **35,** 302–314.

Myneni, R. B., Keeling, C. D., Tucker, C. J., Asrar, G., and Nemani, R. R. (1997b). Increased plant growth in the northern high latitudes from 1981–1991. *Nature* **386,** 698–702.

Myneni, R. B., Los, S. O., and Asrar, G. (1995). Potential gross primary productivity of terrestrial vegetation from 1982–1990. *Geophysical Research Letters* **22,** 2617–2620.

Myrold, D. D. (1987). Relationship between microbial biomass nitrogen and a nitrogen availability index. *Soil Society of America Journal* **51,** 1047–1049.

Näsholm, T., and Ericsson, A. (1990). Seasonal changes in amino acids, protein, and total nitrogen in needles of fertilized Scots pine. *Tree Physiology* **6,** 267–282.

Nadelhoffer, K. J., Aber, J. D., and Melillo, J. M. (1983). Leaf-litter production and soil organic matter dynamics along a nitrogen-availability gradient in Southern Wisconsin (U.S.A.). *Canadian Journal of Forest Research* **13,** 12–21.

Nambiar, E. K. S., and Fife, D. N. (1991). Nutrient retranslocation in temperate conifers. *Tree Physiology* **9,** 185–207.

Nason, G. E., and Myrold, D. D. (1992). Nitrogen fertilizers: Fates and environmental effects in forest. *In* "Forest Fertilization: Sustaining and Improving Nutrition and Growth of Western Forests" (H. N. Chappell, G. F. Weetman, and R. E. Miller, eds.), pp. 67–81. Institute of Forest Resources, Contrib. No. 73. University of Washington, Seattle, Washington.

Naveh, Z., and Lieberman, A. (1994). "Landscape Ecology: Theory and Application." Springer-Verlag, New York.

Neilson, R. P. (1995). A model for predicting continental-scale vegetation distribution and water balance. *Ecological Applications* **5,** 362–385.

Neilson, R. P., and Marks, D. (1994). A global perspective of regional vegetation and hydrologic sensitivities from climatic change. *Journal of Vegetation Science* **5,** 715–730.

Neilson, R. P., Pitelka, L. F., Solomon, A. M., Nathan, R., Migley, G. F., Fragoso, J. M. V., Lischke, H., and Thompson. (2005). Forecasting regional to global plant migration in response to climate change. *BioScience* **55,** 749–759.

Neilson, R. P., and Running, S. W. (1996). Global dynamic vegetation modelling: Coupling biogeochemistry and biogeography models. *In* "Global Change and Terrestrial Ecosystems: The

First GCTE Core Project of the IGBP" (B. Walker and W. Steffen, eds.), pp. 451–465. Cambridge Univ. Press, Cambridge.

Neilson, R. P., King, G. A., and Koerper, G. (1992). Toward a rule-based biome model. *Landscape Ecology* **7**, 27–43.

Nel, E. M., and Wessman, C. A. (1993). Canopy transmittance models for estimating forest leaf area index. *Canadian Journal of Forest Research* **23**, 2579–2568.

Nemani, R. R., and Running, S. W. (1989a). Testing a theoretical climate–sail–leaf area hydrologic equilibrium of forests using satellite data and ecosystem simulation. *Agricultural and Forest Meteorology* **44**, 245–260.

Nemani, R., and Running, S. W. (1989b). Estimation of regional surface resistance to evapotranspiration from NDVI and thermal infrared AVHRR data. *Journal of Applied Meteorology* **28**, 276–284.

Nemani, R. R., and Running, S. W. (1995). Satellite monitoring of global land cover changes and their impact on climate. *Climatic Change* **31**, 395–413.

Nemani, R. R., Keeling, C. D., Hashimoto, H., Jolly, W. M., Piper, S. C., Tucker, C. J., Myneni, R. B., and Running, S. W. (2003). Climate-driven increases in global terrestrial net primary production from 1982 to 1999. *Science* **300**, 1560–1563.

Nemani, R. R., Running, S. W., Pierce, L. L., and Goward, S. (1993a). Developing satellite derived estimates of regional surface resistance to evapotranspiration. *Journal of Applied Meteorology* **32**, 548–557.

Nemani, R. R., Running, S. W., Band, L. E., and Peterson, D. L. (1993b). Regional hydro-ecological simulation system: An illustration of the integration of ecosystem modeling with a GIS. *In* "Environmental Modeling with GIS" (M. F. Goodchild, B. O. Parks, and L. T. Steyart, eds.), pp. 296–304. Oxford Univ. Press, London.

Nemani, R., and Running, S. W. (1996). Implementation of a hierarchical global vegetation classification in ecosystem function models. *Journal of Vegetation Science* **7**, 337–346.

Nemani, R., Pierce, L., Running, S., and Band, L. (1993c). Forest ecosystem processes at the watershed scale: Sensitivity to remotely-sensed leaf area index estimates. *International Journal of Remote Sensing* **14**, 2519–2534.

Nemani, R., Running, S. W., Pielke, R. A., and Chase, T. N. (1996). Global vegetation cover changes from coarse resolution satellite data. *Journal of Geophysical Research* **101**(D3), 7157–7162.

Nepstad, D. C., de Carvalho, C. R., Davidson, E. A., Jipp, P. H., Lefebvre, P. A., Negreiros, G. H., Stone, T. A., Trumbore, S. E., and Vieira, S. (1994). The role of deep roots in hydrological and carbon cycles of Amazonian forests and pastures. *Nature* **372**, 666–669.

Newnham, R. M. (1992). A 30,000 year pollen, vegetation and climate record from Otakairangi (Hikurangi), Northland, New Zealand. *Journal of Biogeography* **19**, 541–554.

Ni, B.-R., and Pallardy, S. G. (1991). Response of gas exchange to water stress in seedlings of woody angiosperms. *Tree Physiology* **8**, 1–10.

Nicholas, N. S., and Zedaker, S. M. (1989). Ice damage in spruce–fir forests of the Black Mountains, North Carolina. *Canadian Journal of Forest Research* **19**, 1487–1491.

Nnyamah, J. U., and Black, T. A. (1977). Rates and patterns of water uptake in a Douglas-fir forest. *Soil Science Society of America Journal* **41**, 972–979.

Nobel, P. S. (1991). "Physiochemical and Environmental Plant Physiology." Academic Press, San Diego.

Noon, B. R., and McKelvey, K. S. (1996). Management of the spotted owl: A case history in conservation biology. *Annual Review of Ecology and Systematics* **27**, 135–162.

Norby, R. J. (1996). Forest canopy productivity index. *Nature* **381**, 56.

Nordmeyer, A. H., Kelland, C. M., Evans, G. R., and Ledgard, H. J. (1987). Nutrients and the restoration of mountain forests in New Zealand. *In* "Human Impacts and Management of Mountain Forests" (T. Fujimori and M. Kimura, eds.), pp. 145–154. Forestry and Forest Products Research Institute, Ibaraki, Japan.

Norman, J. M., Anderson, M. C., Kustas, W. P., French, A. N., Mecikalski, J., Tom, R., Diak, G. R., Schmugge, T. J., and Tanner, B. C. W. (2003). Remote sensing of surface energy fluxes at 101-m pixel resolutions. *Water Resources Research* **39,** 1221–1238.

Norman, J. M., Divakarla, M., and Goel, N. S. (1995). Algorithms for extracting information from remote thermal-IR observations of the earth's surface. *Remote Sensing of the Environment* **51,** 157–168.

Northrup, R. R., Yu, Z., Dahgren, R. A., and Vogt, K. A. (1995). Polyphenol control of nitrogen release from pine litter. *Nature* **377,** 227–229.

Norton, S. A., Wright, R. F., Kahl, J. S., and Schofield, J. P. (1992). The MAGIC simulation of surface water acidification at, and first year results from, the Bear Brook watershed manipulation, Maine, USA. *Environmental Pollution* **77,** 279–286.

O'Connell, A. M. (1990). Microbial decomposition (respiration) of litter in eucalypt forests of southwestern Australia: An empirical model based on laboratory incubations. *Soil Biology and Biochemistry* **22,** 153–160.

O'Hara, K. L. (1988). Stand structure and growing space efficiency following thinning in an even-aged Douglas-fir stand. *Canadian Journal of Forest Research* **18,** 859–866.

O'Hara, K. L. (1996). Dynamics and stocking-level relationships of multi-aged ponderosa pine stands. *Forest Science Monographs* **33,** 1–34.

O'Neill, R. V., Ausmus, B. S., Jackson, D. R., Van Hook, R. I., Van Voris, P., Washburne, C., and Watson, A. P. (1977). Monitoring terrestrial ecosystems by analysis of nutrient export. *Water, Air and Soil Pollution* **8,** 271–277.

O'Neill, R. V., DeAngelis, D. L., Waide, J. B., and Allen, T. F. H. (1986). "A Hierarchical Concept of Ecosystems." Princeton Univ. Press, Princeton, New Jersey.

Odum, H. T. (1983). "Systems Ecology: An Introduction." Wiley, New York.

Oechel, W. C., Hastings, S. J., Vourlitis, G., Jenkins, M., Riechers, G., and Grulke, N. (1993). Recent change of Arctic tundra ecosystems from a net carbon dioxide sink to a source. *Nature* **361,** 520–523.

Oerlemans, J. (1994). Quantifying global warming from the retreat of glaciers. *Science* **264,** 243–245.

Ohno, T., Sucoff, E. I., Drich, M. S., Bloom, P. R., Buschena, C. A., and Dixon, R. K. (1988). Growth and nutrient content of red spruce seedlings in soil amended with aluminum. *Journal of Environmental Quality* **17,** 666–672.

Oliver, C. D. (1981). Forest development in North America following major disturbances. *Forest Ecology and Management* **3,** 153–168.

Oliver, C. D., and Larson, B. C. (1996). "Forest Stand Dynamics." Wiley, New York.

Oliver, C. D., and Stephens, E. P. (1977). Reconstruction of a mixed species forest in central New England. *Ecology* **58,** 562–572.

Ollinger, S. V., Aber, J. D., Lovett, G. M., Millham, S. E., Lathrop, R. G., and Ellis, J. M. (1993). A spatial model of atmospheric deposition for the northeastern U.S. *Ecological Applications* **3,** 459–472.

Olson, J. S. (1963). Energy storage and balance of producers and decomposers in ecological systems. *Ecology* **44,** 322–331.

Olson, J. S., Watts, J. A., and Allison, L. J. (1983). "Carbon in Life Vegetation of Major World Ecosystems." TR004, U.S. Dept. of Energy, Washington, D.C.

Olsson, H. (1994). Changes in satellite-measured reflectances caused by thinning cuttings in boreal forest. *Remote Sensing of the Environment* **50,** 221–230.

Olsthroon, A. F. M., and Tiktak, A. (1991). Fine root density and root biomass of two Douglas-fir stands on sandy soil in the Netherlands. 2. Periodicity of fine root growth and estimation of below-ground carbon allocation. *Netherlands Journal of Agricultural Science* **39,** 61–77.

Orchard, V. A., Cook, F. J., and Corderoy, D. M. (1992). Field and laboratory studies on the relationships between respiration and moisture for two sails of contrasting fertility status. *Pedobiologia* **36,** 21–33.

Oren, R., Ellsworth, D. S., Johnsen, K. H., Phillips, N., Ewers, B. E., Maier, C., Schäfer, K. V. R., McCarthy, H., Hendrey, G., McNulty, S. G., and Katul, G. G. (2001). Soil fertility limits carbon sequestration by forest ecosystems in a CO_2-enriched atmosphere. *Nature* **411,** 469–472.

Oren, R., Schulze, E.-D., Werk, K. S., Meyer, J., Schneider, B. U., and Heilmeier, H. (1988a). Performance of two *Picea abies* (L.) Karst. stands at different stages of decline. I. Carbon relations and stand growth. *Oecologia* **75,** 25–37.

Oren, R., Schulze, E.-D., Werk, K. S., Meyer, J., and Schneider. J. (1988b). Performance of two *Picea abies* (L.) Karst. stands at different stages of decline. VII. Nutrient relations and growth. *Oecologia* **77,** 163–173.

Oren, R., Thies, W. G., and Waring, R. H. (1985). Tree vigor and stand growth of Douglas-fir as influenced by laminated root rot. *Canadian Journal of Forest Research* **15,** 985–988.

Osonubi, O., Oren, R., Werk, K. S., Schulze, E.-D., and Heilmeier, H. (1988). Performance of two *Picea abies* (L.) Karst. stands at different stages of decline. IV. Xylem sap concentrations of magnesium, calcium, potassium and nitrogen. *Oecologia* **77,** 1–6.

Ostrofsky, A., and Shigo, A. L. (1984). Relationship between canker size and wood starch in American chestnut. *European Journal of Forest Pathology* **14,** 65–68.

Overpeck, J. T., Bartlein, P. J., and Webb, T. (1991). Potential magnitude of future vegetation change in Eastern North America: Comparisons with the past. *Science* **254,** 692–695.

Overpeck, J. T., Rind, D., and Goldberg, R. (1990). Climate-induced changes in forest disturbance and vegetation. *Nature* **343,** 51–53.

Pacala, S. W., Canham, C. D., and Silander, J. A., Jr. (1993). Forest models defined by field measurements: I. The design of a northeastern forest simulator. *Canadian Journal of Forest Research* **23,** 1980–1988.

Pacala, S. W., Canham, C. D., Saponara, J., Silander, J. A., Kobe, R. K., and Ribbens, E. (1996). Forest models defined by field measurements: Estimation, error analysis and dynamics. *Ecological Monographs* **66,** 1–43.

Pan, Y., Birdsey, R., Hom, J., McCullough, K., and Clark, K. (2006). Improved estimates of net primary productivity from MODIS satellite data at regional and local scales. *Ecological Applications* **16,** 125–132.

Panek, J. A. (1996). Correlations between stable carbon-isotope abundance and hydraulic conductivity in Douglas-fir across a climate gradient in Oregon, USA. *Tree Physiology* **16,** 747–755.

Pang, P. C. K. (1984). Absorption of gaseous ammonia by Douglas-fir. *Soil Science Society of America Journal* **48,** 1178–1181.

Paniconi, C., and Wood, E. F. (1993). A detailed model for simulation of catchment scale subsurface hydrologic processes. *Water Resources Research* **29,** 1601–1620.

Panshin, A. J., de Zeeuw, C., and Brown, H. P. (1964). "Textbook of Wood Technology." 2nd Ed., Vol. 1. McGraw-Hill, New York.

Parrotta, J. A., Baker, D. D., and Fried, M. (1994). Applications of ^{15}N-enrichment methodologies to estimate nitrogen fixation in *Casuarina equisetifolia*. *Canadian Journal of Forest Research* **24,** 201–207.

Parsons, W. F. J., Mitre, M. E., Keller, M., and Reiners, W. A. (1993). Nitrate limitation of N_2O production and denitrification from tropical pasture and rain forest soils. *Biogeochemistry* **22,** 179–193.

Parton, W. J. (1984). Predicting soil temperatures in a shortgrass steppe. *Soil Science* **138,** 93–101.

Parton, W. J., Mosier, A. R., Ojima, D. S., Valentine, D. W., Schimel, D. S., Weier, K., and Kulmala, A. E. (1996). Generalized model for N_2 and N_2O. production from nitrification and denitrification. *Global Biogeochemical Cycles* **10,** 401–412.

Parton, W. J., Schimel, D. S., Cole, C. V., and Ojima, D. S. (1987). Analysis of factors controlling soil organic matter levels in the Great Plains grasslands. *Soil Society of America Journal* **51,** 1173–1179.

Parton, W. J., Scurlock, J. M. O., Ojima, D. S., Gilmanob, T. G., Scholes, R. J., Schimei, D. S., Kirchner, T., Menaut, J.-C., Seastadt, T., Garcia Moya, E., Kamnalrut, A., and Kinyamario, J. (1993). Observations and modeling of biomass and soil organic matter dynamics for the grassland biome worldwide. *Global Biogeochemical Cycles* **7,** 785–809.

Parton, W. J., Stewart, J. W. B., and Cole, C. V. (1988). Dynamics of C, N, P and S in grassland soils: A model. *Biogeochemistry* **5,** 109–131.

Passioura, J. B. (1979). Accountability, philosophy and plant physiology. *Search* **10,** 347–350.

Pastor, J., Aber, J. D., and Melillo, J. M. (1984). Biomass prediction using generalized allometric regressions for some Northeast tree species. *Forest Ecology and Management* **7,** 265–274.

Pastor, J., and Bockheim, J. G. (1984). Distribution and cycling of nutrients in an aspen–mixed-hardwood–spodosol ecosystem in northern Wisconsin. *Ecology* **65,** 339–353.

Pastor, J., Dewey, B., Naiman, R. J., McInnes, P. F., and Cohen, Y. (1993). Moose browsing and soil fertility in the boreal forests of Isle Royale National Park. *Ecology* **74,** 467–480.

Paul, E. A., and Clark, F. E. (1996). "Soil Microbiology and Biochemistry," 2nd Ed. Academic Press, New York.

Paw, U. K. T., Falk, M., Suchanek, T. H., Ustin, S. L., Chen, J., Park, Y.-S., Winner, W. E., Thomas, S. C., Hsiao, T. C., Shaw, R. H., King, T. S., Pyles, R. D., Schroeder, M., and Matista, A. A. (2004). Carbon dioxide exchange between an old-growth forest and the atmosphere. *Ecosystems* **7,** 513–524.

Pearson, S. M., Turner, M. G., Wallace, L. L., and Romme, W. H. (1995). Winter habitat use by large ungulates following fire in northern Yellowstone National Park. *Ecological Applications* **5,** 744–755.

Pedro, G., Jamagne, M., and Began, J. C. (1978). Two routes in genesis of strongly differentiated acid soils under humid, coal-temperate conditions. *Geoderma* **20,** 173–189.

Penman, H. C. (1948). Natural evaporation from open water, bare soil and grass. *Proceedings of the Royal Society of London, Series A* **194,** 120–145.

Penning de Vries, F. W. T. (1975). The cost of maintenance processes in plant cells. *Annals of Botany* **39,** 77–92.

Perakis, S. S., Maguire, D. A., Bullen, T. D., Cromack, K., Waring, R. H., and Boyle, J. R. (2006). Coupled nitrogen and calcium cycles in forests of the Oregon Coast Range. *Ecosystems* **9,** 63–74.

Pereira, J. S., and Kozlowski, T. T. (1977). Influence of light intensity, temperature, and leaf area on stomatal aperture and water potential of woody plants. *Canadian Journal of Forest Research* **7,** 145–153.

Perkins, T. D., and Adams, G. T. (1995). Rapid freezing induces winter injury symptomatology in red spruce foliage. *Tree Physiology* **15,** 259–266.

Perry, D. A. (1994). "Forest Ecosystems." Johns Hopkins Univ. Press, Baltimore and London.

Persson, H. (1978). Root dynamics in a young Scots pine stand in central Sweden. *Oikos* **30,** 508–519.

Peterjohn, W. T., Melillo, J. M., Steudler, P. A., Newkirk, K. M., Bowles, F. P., and Aber, J. D. (1994). Responses of trace gas fluxes and N availability to experimentally elevated soil temperatures. *Ecological Applications* **4,** 77–87.

Peters, R. H. (1991). "A Critique for Ecology." Cambridge Univ. Press, Cambridge.

Peterson, B. J., and Fry, B. (1987). Stable isotopes in ecosystem studies. *Annual Review of Ecology and Systematics* **18,** 293–320.

Peterson, D. L., and Running, S. W. (1989). Applications in Forest Science and Management. *In* "Theory and Applications of Optical Remote Sensing" (G. Asrar, ed.), pp. 429–463. Wiley, New York.

Peterson, D. L., and Waring, R. H. (1994). Overview of the Oregon Transect Ecosystem research project. *Ecological Applications* **4,** 211–225.

Peterson, D. W., and Peterson, D. L. (2001). Mountain hemlock growth trends to climatic variability at annual and decadal time scales. *Ecology* **82,** 3330–3345.

Peterson, D. W., Peterson, D. L., and Ettl, G. J. (2002). Growth responses of subalpine fir to climatic variability in the Pacific Northwest. *Canadian Journal of Forest Research* **32,** 1503–1517.

Petty, G. W. (1995). The status of satellite-based rainfall estimation over land. *Remote Sensing of the Environment* **51,** 125–137.

Petty, J. A., and Worrell, R. (1981). Stability of coniferous tree stems in relation to damage by snow. *Forestry* **54,** 115–128.

Phillipson, J., Aber, R., Steel, J., and Woodell, S. R. J. (1978). Earthworm numbers, biomass and respiratory metabolism in a beech woodland—Wytham Woods, Oxford. *Oecologia* **33,** 291–309.

Pickett, S. T. A. (1976). Succession: An evolutionary interpretation. *American Naturalist* **110,** 107–119.

Pielke, R. A., Cotton, W. R., Walko, R. L., Tremback, C. J., Lyons, W. A., Grasso, L. D., Nicholls, M. E., Moran, M. D., Wesley, D. A., Lee, T. J., and Copeland, J. H. (1992). A comprehensive meteorological modeling system—RAMS. *Meteorology and Atmospheric Physics* **49,** 69–91.

Pielke, R. A., Lee, T. J., Copeland, J. H., Eastman, J. L., Ziegler, C. L., and Finley, C. A. (1997). Use of USGS-provided data to improve weather and climate simulations. *Ecological Applications* **7,** 3–21.

Pierce, L. L., and Running, S. W. (1988). Rapid estimation of coniferous forest leaf area index using a portable integrating radiometer. *Ecology* **69,** 1762–1767.

Pierce, L. L., and Running, S. W. (1995). The effects of aggregating sub-grid land surface variation on large-scale estimates of net primary production. *Landscape Ecology* **10,** 239–253.

Pierce, L. L., Running, S. W., and Walker, J. (1994). Regional-scale relationships of leaf area index to specific leaf area and leaf nitrogen content. *Ecological Applications* **4,** 313–321.

Pierce, L. L., Walker, J., Dowling, T. I., McVicar, T. R., Hatton, T. J., Running, S. W., and Caughlan, J. C. (1993). Ecohydrological changes in the Murray-Darling basin. III. A simulation of regional hydrological changes. *Journal of Applied Ecology* **30,** 283–294.

Pinker, R. T., Frouin, R., and Li, A. (1995). A review of satellite methods to derive surface shortwave irradiance. *Remote Sensing of the Environment* **51,** 108–124.

Piper, S. C., and Stewart, E. F. (1996). A gridded global data set of daily temperature and precipitation for terrestrial biospheric modeling. *Global Biogeochemical Cycles* **10,** 757–782.

Placet, M., Battye, R. E., Fehsenfeld, F. C., and Bassett, G. W. (1990). Emissions involved in acidic deposition processes. State of science/technology report 1. National Acid Precipitation Assessment Program, Washington, D.C.

Polglase, P. J., Attiwill, P. M., and Adams, M. A. (1992). Nitrogen and phosphorus cycling in relation to stand age of *Eucalyptus regans* F. Muell. III. Labile inorganic and organic P, phosphatase activity and P availability. *Plant and Soil* **142,** 177–185.

Polley, H. W., Johnson, H. B., Marina, B. D., and Mayeux, H. S. (1993). Increase in C_3 plant water-use efficiency and biomass over glacial to present CO_2 concentrations. *Nature* **361,** 61–64.

Pothier, D., Margolis, H. A., and Waring, R. H. (1989). Patterns of change of saturated sapwood permeability and sapwood conductance with stand development. *Canadian Journal of Forest Research* **19,** 432–439.

Potter, C. S., Matson, P. A., Vitousek, P. M., and Davidson, E. A. (1996). Process modeling of controls on nitrogen trace gas emissions from soils worldwide. *Journal of Geophysical Research—Atmospheres* **101,** 1361–1377.

Potter, C., BrooksGenovese, V., Klooster, S., and Torregrosa, A. (2002). Biomass burning emission of reactive gases estimated from satellite data analysis and ecosystem modeling for the Brazilian Amazon region. *Journal of Geophysical Research* **107**(D20), 8056–8065.

Pregitzer, K. S., and Burton, A. J. (1991). Sugar maple seed production and nitrogen in litterfall. *Canadian Journal of Forest Research* **21,** 1148–1152.

Pregitzer, K. S., and Euskirchen, E. S. (2004). Carbon cycling and storage in world forests: Biome patterns related to forest age. *Global Change Biology* **10,** 2052–2077.

Prentice, I. C., Bartlein, P. J., and Webb III, T. (1991). Vegetation and climate change in eastern North America since the last glacial maximum. *Ecology* **72,** 2038–2056.

Prentice, I. C., Bramer, W., Harrison, S. P., Leemans, R., Monserud, R. A., and Solomon, A. M. (1992). A global biome model based on plant physiology and dominance, soil properties and climate. *Journal of Biogeography* **19,** 117–134.

Prenzel, J., and Meiwes, K. J. (1994). Sulfate sorption in soils under acid deposition: Modeling field data from forest liming. *Journal of Environmental Quality* **23,** 1212–1217.

Preston, C. M., Marshall, V. G., McCullough, K., and Mead, D. J. (1990). Fate of ^{15}N-labelled fertilizer applied on snow at two forest sites in British Columbia. *Canadian Journal of Forest Research* **20,** 1583–1592.

Prevedello, C. L., Libardi, P. L., and Reichardt, K. (1991). Water flow in homogeneous unsaturated soils: An attempt using a diffusional model to describe gravitational flow. *Soil Science* **152,** 184–188.

Price, C., and Rind, D. (1994). The impact of a $2 \times CO_2$ climate on lightning-caused fires. *Journal of Climate* **7,** 1484–1494.

Price, D. T., and Black, T. A. (1990). Effects of short-term variation in weather on diurnal canopy CO_2 flux and evaporation of a juvenile Douglas-fir stand. *Agricultural and Forest Meteorology* **50,** 139–158.

Price, J. C. (1989). Quantitative aspects of remote sensing in the thermal infrared. *In* "Theory and Applications of Optical Remote Sensing" (G. Asrar, ed.), pp. 578–603. Wiley, New York.

Price, P. W., Bouton, C. E., Gross, P., McPheron, B. A., Thompson, J. N., and Weis, A. E. (1980). Interactions among three trophic levels: Influence of plants on interactions between insect herbivores and natural enemies. *Annual Review of Ecology and Systematics* **11,** 41–65.

Prince, S. D., and Goward, S. N. (1995). Global primary production: A remote sensing approach. *Journal of Biogeography* **22,** 815–835.

Proctor, J., and Woodell, S. R. J. (1975). The ecology of serpentine soils. *Advances in Ecological Research* **9,** 255–366.

Prudhomme, T. I. (1983). Carbon allocation to antiherbivore compounds in a deciduous and evergreen subarctic shrub species. *Oikos* **40,** 344–356.

Putz, F. E., and Milton, K. (1982). Tree mortality rates on Barro Colorado Island. *In* "The Ecology of a Tropical Forest" (E. G. Leigh, Jr., A. S. Rand, and D. M. Windsor, eds.), pp. 95–100. Smithsonian Institution, Washington, D.C.

Qi, J., Marshall, J. D., and Mattson, K. G. (1994). High soil carbon dioxide concentrations inhibit root respiration of Douglas-fir. *New Phytologist* **128,** 435–442.

Rafes, P. M. (1971). Pests and the damage which they cause to forests. *In* "Productivity of Forest Ecosystems," pp. 357–367. UNESCO, Rome.

Raffa, K. F., and Berryman, A. A. (1983). Physiological differences between lodgepole pines resistant and susceptible to the mountain pine beetle and associated microorganisms. *Environmental Entomology* **11,** 486–492.

Ragsdale, H. L., and Berish, C. W. (1988). The decline of lead in tree rings of *Carya* spp. in urban Atlanta, GA, USA. *Biogeochemistry* **6,** 21–29.

Rahman, A. F., Sims, D. A., Cordova, V. D., and El-Masri, B. Z. (2005). Potential of MODIS EVI and surface temperature for directly estimating per-pixel ecosystem C fluxes. *Geophysical Research Letters* **32,** L19404–L19407.

Raich, J. W., and Nadelhoffer, K. J. (1989). Belowground carbon allocation in forest ecosystems: Global trends. *Ecology* **70,** 1346–1354.

Raich, J. W., and Schlesinger, W. H. (1992). The global carbon dioxide flux in soil respiration and its relationship to vegetation and climate. *Tellus* **44B,** 81–99.

Raison, R. J. (1980). A review of the role of fire in nutrient cycling in Australian native forests, and of methodology for studying the fire–nutrient interaction. *Australian Journal of Ecology* **5,** 15–21.

Raison, R. J., O'Connell, A. M., Khanna, P. K., and Keith, H. (1993). Effects of repeated fires on nitrogen and phosphorus budgets and cycling processes in forest ecosystems. *In* "Fire in Mediterranean Ecosystems" (L. Trabaud and R. Prodon, eds.), pp. 347–363. Publ. No. EUR 15089 EN. Commission of the European Communities, Brussels–Luxembourg.

Ranson, K. J., Irons, J. R., and Williams, D. L. (1994). Multispectral bidirectional reflectance of northern forest canopies with the advanced solid-state array spectroradiometer (ASAS). *Remote Sensing of the Environment* **47,** 276–289.

Rastetter, E. B. (1996). Validating models of ecosystem response to global change. *Bioscience* **46,** 190–198.

Rastetter, E. B., King, A. W., Crosby, B. J., Hornberger, G. M., O'Neill, R. V., and Hobbie, J. E. (1992). Aggregating fine-scale ecological knowledge to model coarser-scale attributes of ecosystems. *Ecological Applications* **2,** 55–70.

Rastetter, E. B., Ryan, M. G., Shaver, G. R., Melillo, J. M., Nadelhoffer, K. J., Hobbie, J. E., and Aber, J. D. (1991). A general biogeochemical model describing the responses of the C and N cycles in terrestrial ecosystems to changes in CO_2 climate, and N deposition. *Tree Physiology* **9,** 101–126.

Raven, P. H., Evert, R. F., and Curtis, H. (1981). "Biology of Plants." Worth, New York.

Reich, P. B. (1987). Quantifying plant response to ozone: A unifying theory. *Tree Physiology* **3,** 63–91.

Reich, P. B. (1995). Phenology of tropical forests: patterns, causes, and consequences. *Canadian Journal of Botany* **73,** 164–174.

Reich, P. B., and Hinckley, T. M. (1989). Influence of pre-dawn water potential and soil-to-leaf hydraulic conductance on maximum daily leaf diffusive conductance in two oak species. *Functional Ecology* **3,** 719–726.

Reich, P. B., and Walters, M. B. (1994). Photosynthesis–nitrogen relations in Amazonian tree species II. Variation in nitrogen vis-à-vis specific leaf area influences mass- and area-based expressions. *Oecologia* **97,** 73–81.

Reich, P. B., Kaike, T., Gower, S. T., and Schoettle, A. W. (1995a). Causes and consequences of variation in conifer leaf life-span. *In* "Ecophysiology of Coniferous Forests" (W. K. Smith and T. M. Hinckley, eds.), pp. 225–254. Academic Press, San Diego.

Reich, P. B., Kloeppel, B. D., Ellsworth, D. S., and Walters, M. B. (1995b). Different photosynthesis–nitrogen relation in deciduous hardwood and evergreen coniferous tree species. *Oecologia* **104,** 24–30.

Reichle, D. E., ed. (1981). "Dynamic Properties of Forest Ecosystems." Cambridge Univ. Press, Cambridge.

Reichle, D. E., Goldstein, R. A., van Hook, R. I., and Dodson, G. J. (1973). Analysis of insect consumption in a forest canopy. *Ecology* **54**, 1076–1084.

Reichstein, M., Rey, A., Freibauer, A., Tenhunen, J., Valenti, R., Banza, J., Casals, P., Cheng, Y., Grünzweig, J. M., Irvine, J., Joffre, R., Law, B. E., Loustau, D., Miglietta, F., Oechel, W., Ourcival, J.-M., Pereira, J. S., Peressotti, A., Ponti, F., Qi, Y., Rambal, S., Rayment, M., Romanya, J., Rossi, F., Tecdeschi, V., Tirone, G., Xu, M., and Yakir, D. (2003). Modeling temporal and large-scale variability of soil respiration from soil water availability, temperature and vegetation productivity indices. *Global Biogeochemical Cycles* **17**, 1104–1118.

Reid, J. B., and Bowden, M. (1995). Temporal variability and scaling behavior in the fine-root system of kiwifruit *(Actinidia deliciosa)*. *Plant, Cell and Environment* **18**, 555–564.

Reid, L. M., and Dunne, T. (1984). Sediment production from forest road surfaces. *Water Resources Research* **20**, 1753–1761.

Rhoades, D. F., and Cates, R. G. (1976). Toward a general theory of plant antiherbivore chemistry. *Recent Advances in Phytochemisty* **10**, 168–213.

Rice, R. M., and Lewis, J. (1991). Estimating erosion risks associated with logging and forest roads in northwestern California. *Water Resources Bulletin* **27**, 809–817.

Riley, S. J. (1984). Effect of clearing and road operations on the permeability of forest soils, Karuah catchment, New South Wales, Australia. *Forest Ecology & Management* **9**, 283–293.

Rinne, P., Tuominen, H., and Junttila, O. (1992). Arrested leaf abscission in the non-abscising variety of pubescent birch: Developmental, morphological and hormonal aspects. *Journal of Experimental Botany* **43**, 975–982.

Ritchie, G. A., and Hinckley, T. M. (1975). The pressure chamber as an instrument for ecological research. *Advances in Ecological Research* **9**, 165–254.

Ritchie, J. C., Cwynar, L. C., and Spear, R. W. (1983). Evidence from northwest Canada for an early Holocene Milankovitch thermal maximum. *Nature* **305**, 126–128.

Roberts, C. T., Hanley, T. A., Hagerman, A. E., Hjeljord, O., Baker, D. L., Schwarts, C. C., and Mautz, W. W. (1987). Role of tannins in defending plants against ruminants: Reduction in protein availability. *Ecology* **68**, 98–107.

Roberts, D. (1996a). Landscape vegetation modeling with vital attributes and fuzzy systems theory. *Ecological Modelling* **90**, 175–184.

Roberts, D. W. (1996b). Modeling forest dynamics with vital attributes and fuzzy systems theory. *Ecological Modelling* **90**, 161–173.

Roberts, D. A., Ustin, S. L., Ogunjemiyo, S., Greenberg, J., Dobrowski, S. Z., Chen, J., and Hinckley, T. M. (2004). Spectral and structural measures of Northwest forest vegetation at leaf to landscape scales. *Ecosystems* **7**, 545–562.

Robertson, G. P. (1982). Nitrification in forested ecosystems. *Philosophical Transactions of the Royal Society of London, Series B, Biological Sciences* **296**, 445–457.

Robertson, G. P., and Tiedje, J. M. (1984). Denitrification and nitrous oxide production in successional and old-growth Michigan forests. *Soil Science Society of America Journal* **48**, 383–389.

Robinson, J. (1994). Speculations on carbon dioxide starvation, late Tertiary evolution of stomatal regulation and floristic modernization. *Plant, Cell and Environment* **17**, 345–354.

Robinson, J. S., Sivapalan, M., and Snell, J. D. (1995). On the relative roles of hillslope processes, channel routing, and network geomorphology in the hydrologic response of natural catchments. *Water Resources Research* **31**, 3089–3101.

Rodríguez, R., Espinosa, M., Real, P., and Inzunza, J. (2002). Analysis of productivity of radiata pine plantations under different silvicultural regimes using the 3-PG process-based model. *Australian Forestry* **65**, 165–172.

Rogers, R., and Hinckley, T. M. (1979). Foliage weight and area related to current sapwood area in oak. *Forest Science* **25,** 298–303.

Romme, W. H. (1982). Fire and landscape diversity in subalpine forests of Yellowstone National Park. *Ecological Monographs* **52,** 199–221.

Rood, S. B., Samuelson, G. M., Weber, J. K., and Wywrot, K. A. (2005). Twentieth-century decline in streamflows from the hydrographic apex of North America. *Journal of Hydrology* **306,** 215–233.

Rose, C. L. (1990). Application of the carbon/nutrient balance concept to predicting the nutritional qualities of blueberry foliage to deer in southeastern Alaska. Ph.D. Thesis, Oregon State University, Corvallis, Oregon.

Rothermel, R. C. (1972). A mathematical model for predicting fire spread in wildland fuels. *USDA Research Paper* **INT-443.** Ogden, Utah.

Ruel, J. C. (1995). Understanding windthrow: Silvicultural implications. *Forestry Chronicle* **71,** 434–445.

Ruess, R. W., Van Cleve, K., Yarie, J., and Viereck, L. A. (1996). Contributions of fine root production and turnover to the carbon and nitrogen cycling in tiaga forests of the Alaskan interior. *Canadian Journal of Forest Research* **26,** 1326–1336.

Ruimy, A., Saugier, B., and Dedieu, G. (1994). Methodology for the estimation of terrestrial net primary production from remotely sensed data. *Journal of Geophysical Research* **99,** 5263–5283.

Running, S. W. (1976). Environmental control of leaf water conductance in conifers. *Canadian Journal of Forest Research* **6,** 104–112.

Running, S. W. (1984a). Documentation and preliminary validation of H2OTRANS and DAYTRANS, two models for predicting transpiration and water stress in western coniferous forests. *USDA Rocky Mountain Forest and Range Experiment Station Research Paper* **RM-252.** Fort Collins, Colorado.

Running, S. W. (1984b). Microclimate control of forest productivity: Analysis by computer simulation of annual photosynthesis/transpiration balance in different environments. *Agriculture and Forest Meteorology* **32,** 267–288.

Running, S. W. (1994). Testing FOREST-BGC ecosystem process simulations across a climatic gradient in Oregon. *Ecological Applications* **4,** 238–247.

Running, S. W. (2006). Is global warming causing more, larger wildfires? *Science* **313,** 927–928.

Running, S. W., and Gower, S. T. (1991). FOREST-BGC, a general model of forest ecosystem processes for regional applications II. Dynamic carbon allocation and nitrogen budgets. *Tree Physiology* **9,** 147–160.

Running, S. W., and Hunt, E. R., Jr. (1993). Generalization of a forest ecosystem process model for other biomes, BIOME-BGC, and an application for global-scale models, *In* "Scaling Processes between Leaf and Landscape Levels" (J. R. Ehleringer and C. Field, eds.), pp. 141–158. Academic Press, San Diego.

Running, S. W., and Nemani, R. R. (1991). Regional hydrologic and carbon balance responses of forests resulting from potential climate change. *Climatic Change* **19,** 349–368.

Running, S. W., and Reid, C. P. (1980). Soil temperature influences root resistance of *Pinus contorta* seedlings. *Plant Physiology* **65,** 535–640.

Running, S. W., Baldocchi, D. D., Turner, D. P., Gower, S. T., Bakwin, P. S., and Hibbard, K. A. (1999). A global terrestrial monitoring network integrating tower fluxes, flask sampling, ecosystem modeling and EOS satellite data. *Remote Sensing of Environment* **70,** 108–127.

Running, S. W., Justice, C. O., Salomonson, V., Hall, D., Barker, J., Kaufmann, Y. J., Strahler, A. H., Huete, A. R., Muller, J.-P., Vanderbilt, V., Wan, Z. M., Teillet, P., and Carneggie, D.

(1994). Terrestrial remote sensing science and algorithms planned for EOS/MODIS. *International Journal of Remote Sensing* **15,** 3587–3620.

Running, S. W., Nemani, R. R., Heinsch, F. A., Zhao, M., Reeves, M., and Hashimoto, H. (2004). A continuous satellite-derived measure of global terrestrial primary production. *BioScience* **54,** 547–560.

Running, S. W., Nemani, R. R., Peterson, D. L., Band, L. E., Potts, D. F., Pierce, L. L., and Spanner, M. A. (1989). Mapping regional forest evapotranspiration and photosynthesis by coupling satellite data with ecosystem simulation. *Ecology* **70,** 1090–1101.

Running, S. W., and Coughlan, J. C. (1988). A general model of forest ecosystem processes for regional applications. I. Hydrologic balance, canopy gas exchange and primary production processes. *Ecological Modelling* **42,** 125–154.

Running, S. W., Loveland, T. R., Pierce, L. L., and Hunt, E. R., Jr. (1995). A remote sensing based vegetation classification logic for global land cover analysis. *Remote Sensing of Environment* **51,** 39–48.

Running, S. W., Nemani, R., and Hungerford, R. D. (1987). Extrapolation of synoptic meteorological data in mountainous terrain, and its use for simulating forest evapotranspiration and photosynthesis. *Canadian Journal of Forest Research* **17,** 472–483.

Running, S. W., Waring, R. H., and Rydell, R. A. (1975). Physiological control of water flux in conifers: A computer simulation model. *Oecologia* **18,** 1–16.

Runyon, J., Waring, R. H., Goward, S. N., and Welles, J. M. (1994). Environmental limits on net primary production and light-use efficiency across the Oregon transect. *Ecological Applications* **4,** 226–237.

Rutter, A. J. (1963). Studies in water relations of *Pinus sylvestris* in plantation conditions. *Journal of Ecology* **51,** 191–203.

Rutter, A. J. (1975). The hydrological cycle in vegetation. *In* "Vegetation and the Atmosphere" (J. L. Monteith, ed.), Vol. 1, pp. 111–154. Academic Press, London.

Rutter, A. J., Kershaw, K. A., Robins, P. C., and Morton, A. J. (1971). A predictive model of rainfall interception in forests I. Derivation of the model from observations in a plantation of Corsican pine. *Agricultural Meteorology* **9,** 367–383.

Ryan, M. G. (1991a). Effects of climate change on plant respiration. *Ecological Applications* **1,** 157–167.

Ryan, M. G. (1991b). A simple method for estimating gross carbon budgets for vegetation in forest ecosystems. *Tree Physiology* **9,** 255–266.

Ryan, M. G. (1995). Foliar maintenance respiration of subalpine and boreal trees and shrubs in relation to nitrogen content. *Plant, Cell and Environment* **18,** 765–772.

Ryan, M. G., and Yoder, B. J. (1997). Hydraulic limits to tree height and tree growth. *BioScience* **47,** 235–242.

Ryan, M. G., Binkley, D., and Fownes, J. H. (1997). Age-related decline in forest productivity: Pattern and process. *Advances in Ecological Research* **27,** 213–262.

Ryan, M. G., Lavigne, M. B., and Gower, S. T. (1997a). Annual carbon cost of autotrophic respiration in boreal forest ecosystems in relation to species and climate. *Journal of Geophysical Research* **102,** 28871–28884.

Ryan, M. G., Gower, S. T., Hubbard, R. M., Waring, R. H., Gholz, H. L., Cropper, W. P., and Running, S. W. (1995). Stem maintenance respiration of four conifers in contrasting climates. *Oecologia* **101,** 133–140.

Ryan, M. G., Hunt, E. R., Jr., McMurtrie, R. E., Ågren, G. I., Aber, J. D., Friend, A. D., Rastetter, E. B., Pulliam, W. M., Raison, R. J., and Linder, S. (1996a). Comparing models of ecosystem function for temperate conifer forests. I. Model description and validation. *In* "Global Change: Effects on Coniferous Forests and Grasslands" (A. I. Breymeyer, D. O. Hall, J. M. Melillo, and G. I. Ågren, eds.), SCOPE Vol. 56, pp. 313–362. Wiley, New York.

Ryan, M. G., Hubbard, R. M., Pongracic, S., Raison, R. J., and McMurtrie, R. E. (1996b). Foliage, fine-root, woody-tissue and stand respiration in *Pinus radiata* in relation to nitrogen status. *Tree Physiology* **16**, 333–344.

Ryerson, R. A. (1996). Earth observing platforms and sensors. *In* "Manual of Remote Sensing" (S. A. Morain and A. M. Budge, eds.), 3rd Ed., American Society for Photogrammetry and Remote Sensing, Bethesda, Maryland.

Rygiewicz, P. T., and Andersen, C. P. (1994). Mycorrhizae alter quality and quantity of carbon allocated below ground. *Nature* **369**, 58.

Rykiel, E. J., Jr. (1996). Testing ecological models: The meaning of validation. *Ecological Modelling* **90**, 229–244.

Sackett, S. S. (1975). Scheduling prescribed burns for hazard reduction in the southeast. *Journal of Forestry* **73**, 143–147.

Sakai, A., and Weiser, C. J. (1973). Freezing resistance of trees in North America with reference to tree regions. *Ecology* **54**, 118–126.

Sample, V. A., ed. (1994) "Remote Sensing and GIS in Ecosystem Management." Island Press, Washington, D.C.

Sampson, D. A., Waring, R. H., Maier, C. A., Gough, C. M., Ducey, M. J., and Johnsen, K. N. (2006). Fertilization effects on forest carbon storage and exchange, and net primary production; a new hybrid process model for stand management. *Forest Ecology and Management* **211**, 91–109.

Sanchez, P. A., Bandy, D. E., Villachica, J. H., and Nicolaides, J. J. (1982a). Amazon Basin Soils: Management for continuous crop production. *Science* **216**, 821–827.

Sanchez, P. A., Gichuru, M. P., and Katz, L. B. (1982b). Organic matter in major soils of the tropical and temperate regions. *Twelfth International Congress on Soil Science 1982,* **1**, 99–114.

Sandford, A. P., and Grace, J. (1985). The measurement and interpretation of ultrasound from woody stems. *Journal of Experimental Botany* **36**, 1–14.

Sarr, D. A., Hibbs, D. E., and Huston, M. A. (2005). A hierarchical perspective of plant diversity. *Quarterly Review of Biology* **80**, 187–212.

Schadauer, K. (1996). Growth trend in Austria. *In* "Growth Trends in European Forests" (H. Spiecker, K. Mielikäinen, K. Köhl, and J. P. Skovsgaard, eds.), pp. 275–289. Springer, Berlin.

Schimel, D. S. (1993). "Theory and Application of Tracers." Academic Press, San Diego.

Schimel, D. S. (1995). Terrestrial ecosystems and the carbon cycle. *Global Change Biology* **1**, 77–91.

Schimel, D. S., Braswell, B. H., Holland, E. A., McKeon, R., Ojima, D. S., Painter, T. H., Parton, W. J., and Townsend, A. R. (1994). Climatic, edaphic, and biotic controls over storage and turnover of carbon in soils. *Global Biogeochemical Cycles* **8**, 279–293.

Schimel, J. P., Scott, W., and Killham, K. (1989). Changes in cytoplasmic carbon and nitrogen pools in a soil bacterium and a fungus in response to salt stress. *Applied Environmental Microbiology* **55**, 1635–1637.

Schlesinger, W. H. (1991). "Biogeochemistry—An Analysis of Global Change." Academic Press, San Diego.

Schlesinger, W. H. (1997). "Biogeochemistry—An Analysis of Global Change," 2nd Ed. Academic Press, San Diego.

Schlesinger, W. H., and Marks, P. L. (1977). Mineral cycling and the niche of Spanish moss, *Tillandsia usneoides* L. *American Journal of Botany* **64**, 1254–1262.

Schlesinger, W. H., Gray, J. T., and Gilliam, F. S. (1982). Atmospheric deposition processes and their importance as sources of nutrients in a chaparral ecosystem of southern California. *Water Resources Research* **18**, 623–629.

Schmid, H. P. (2002). Footprint modeling for vegetation atmosphere exchange studies: A review and perspective. *Agricultural and Forest Meteorology* **113,** 159–183.

Schnoor, J. L., and Stumm, W. (1984). Acidification of aquatic and terrestrial systems. *In* "Chemical Processes in Lakes" (W. Stumm, ed.), pp. 311–338. Wiley, New York.

Schoennagel, T., Veblen, T. T., and Romme, W. H. (2004). The interaction of fire, fuels, and climate across Rocky Mountain forests. *BioScience* **54,** 661–676.

Schoettle, A. W., and Fahey, T. D. (1994). Foliage and fine root longevity of pines. *Ecological Bulletin* **43,** 136–153.

Scholander, P. F., Hammel, H. T., Bradstreet, E. D., and Hemmingsen, E. A. (1965). Sap pressure in vascular plants. *Science* **148,** 339–346.

Schowalter, T. D. (1989). Canopy arthropod community structure and herbivory in old-growth and regenerating forests in western Oregon. *Canadian Journal of Forest Research* **19,** 318–322.

Schultz, J. C., and Baldwin, I. T. (1982). Oak leaf quality declines in response to defoliation by gypsy moth larvae. *Science* **217,** 149–151.

Schulze, E.-D. (1986). Carbon dioxide and water vapor exchange in response to drought in the atmosphere and in the soil. *Annual Review of Plant Physiology* **37,** 247–274.

Schulze, E.-D. (1989). Air pollution and forest decline in a spruce *(Picea abies)* forest. *Science* **244,** 776–783.

Schulze, E.-D., Chapin III, F. S., and Gebauer, G. (1994b). Nitrogen nutrition and isotope differences among life forms at the northern treeline of Alaska. *Oecologia* **100,** 406–412.

Schulze, E.-D., Kelliher, F. M., Körner, C., Lloyd, J. N., and Leuning, R. (1994a). Relationships among stomatal conductance, ecosystem surface conductance, carbon assimilation rates, and plant nitrogen nutrition: A global ecology scaling exercise. *Annual Review of Ecology and Systematics* **25,** 629–660.

Schulze, E.-D., Robichaux, R. H., Grace, J., Rundel, P. W., and Ehleringer, J. R. (1987). Plant water balance. *BioScience* **37,** 30–37.

Schwartz, M. D. (1996). Examining the spring discontinuity in daily temperature ranges. *Journal of Climate* **9,** 803–808.

Schwartz, M. D., Ahas, R., and Aasa, A. (2006). Onset of spring starting earlier across the Northern Hemisphere. *Global Change Biology* **12,** 343–351.

Schwartz, M. D., and Karl, T. R. (1990). Spring phenology: Nature's experiment to detect the effect of "green-up" on surface maximum temperatures. *Monthly Weather Review* **118,** 883–890.

Scriber, R. A., and Slansky, F. (1981). The nutritional ecology of immature insects. *Annual Review of Entomology* **76,** 183–211.

Seastedt, T. R., and Tate, C. M. (1981). Decomposition rates and nutrient contents of arthropod remains in forest litter. *Ecology* **62,** 13–19.

Sedjo, R. A. (1989). Forests to offset the greenhouse. *Journal of Forestry* **87,** 12–15.

Sedjo, R. A. (1992). Temperate forest ecosystems in the global carbon cycle. *Ambio* **21,** 274–277.

See, L. M., and Fritz, S. (2006). A method to compare and improve land cover datasets: Application to the GLC-2000 and MODIS land cover products. *IEEE Transactions on Geoscience and Remote Sensing* **44,** 1740–1746.

Seielstad, C. A., and Queen, L. P. (2003). Using airborne laser altimetry to determine fuel models for estimating fire behavior. *Journal of Forestry* **101,** 10–15.

Seif el Din, A., and Obeid, M. (1971). Ecological studies of the vegetation of the Sudan. IV. The effect of simulated grazing on the growth of *Acacia senegal* (L.) Willd. seedlings. *Journal of Applied Ecology* **8,** 211–216.

Sellers, P. J. (1985). Canopy reflectance, photosynthesis and transpiration. *International Journal of Remote Sensing* **6,** 1335–1372.

Sellers, P. J. (1987). Canopy reflectance, photosynthesis and transpiration. II. The role of biophysics in the linearity of their interdependence. *Remote Sensing of Environment* **21**, 143–183.

Sellers, P. J., Berry, J. A., Collatz, G. J., Field, C. B., and Hall, F. G. (1992a). Canopy reflectance, photosynthesis, and transpiration. III. A reanalysis using improved leaf models and a new canopy integration scheme. *Remote Sensing of Environment* **42**, 187–216.

Sellers, P. J., Heiser, M. D., and Hall, F. G. (1992b). Relations between surface conductance and spectral vegetation indices at intermediate ($100 \, m^2$ to $15 \, km^2$) length scales. *Journal of Geophysical Research* **97**, 19,033–19,059.

Sellers, P. J., Randall, D. A., Collatz, G. J., Berry, J., Field, C., Dazlich, D. A., and Zhang, C. (1996a). A revised land-surface parameterization (SiB2) for atmospheric GCMs. Part 1: Model formulation. *Journal of Climate* **9**, 738–763.

Sellers, P. J., Bounoua, L., Collatz, G. J., Randall, D. A., Dazlich, D. A., Los, S. O., Berry, J. A., Fung, I., Tucker, C. J., Field, C. B., and Jensen, T. G. (1996b). Comparison of radiative and physiological effects of doubled atmospheric CO_2 on climate. *Science* **271**, 1402–1406.

Sellers, P., Hall, F., Margolis, H., Kelly, B., Baldocchi, D., denHartog, G., Cihlar, J., Ryan, M. G., Goodison, B., Crill, P., Hanson, K. J., Lettenmaier., D., and Wickland, D. E. (1995). The Boreal Ecosystem–Atmosphere Study (BOREAS): An overview and early results from the 1994 field year. *Bulletin of the American Meteorological Society* **76**, 1549–1577.

Sellin, A. (1993). Resistance to water flow in xylem of *Picea abies* (L.) Karst. trees grown under contrasting light conditions. *Trees* **7**, 220–226.

Shahlaee, A. K., Nutter, W. L., Burroughs, E. R., Jr., and Morris, L. A. (1991). Runoff and sediment production from burned forest sites in the Georgia Piedmont. *Water Resources Bulletin* **27**, 485–493.

Sharkey, T. D., and Singsaas, E. L. (1995). Why plants emit isoprene. *Nature* **374**, 769.

Sharkey, T. D., Singsaas, E. L., Vanderveer, P. J., and Geron, C. (1995). Response of isoprene and photosynthesis to temperature and light in a North-Carolina forest. *Plant Physiology* **108**, 60–69.

Sharma, N. P., Rowe, R., Openshaw, K., and Jacobson, M. (1992). World forests in perspective. *In* "Managing the World's Forests" (N. P. Sharma, ed.), pp. 17–31. Kendall/Hunt, Dubuque, Iowa.

Shaver, G. R., and Melillo, J. M. (1984). Nutrient budgets of marsh plants: Efficiency concepts and relation to availability. *Ecology* **65**, 1491–1510.

Shaver, G. R., Giblin, A. E., Nadelhoffer, K. J., and Rastetter, E. B. (1997). Plant functional types and ecosystem change. *In* "Plant Functional Types" (T. Smith, H. H. Shugart, and F. I. Woodward, eds.), Cambridge Univ. Press, Cambridge.

Shearer, G., and Kohl, D. H. (1986). N_2-fixation in field settings: Estimations based on natural $\delta^{15}N$ abundance. *Australian Journal of Plant Physiology* **13**, 699–756.

Shelburne, V. B., and Hedden, R. L. (1996). Effect of stem height, dominance class, and site quality on sapwood permeability in loblolly pine *(Pinus taeda* L.). *Forest Ecology and Management* **83**, 163–169.

Sheriff, D. W., Narmbiar, E. K. S., and Fife, D. N. (1986). Relationships between nutrient status, carbon assimilation and water use efficiency in *Pinus radiata* (D. Don) needles. *Tree Physiology* **2**, 73–88.

Shugart, H. H. (1984). "A Theory of Forest Dynamics: The Ecological Implications of Forest Succession Models." Springer-Verlag, New York.

Shugart, H. H., and Noble, I. R. (1981). A computer model of succession and fire response of the high altitude *Eucalyptus* forests of the Brindabella Range, Australian Capital Territory. *Australian Journal of Ecology* **6**, 149–164.

Shugart, H. H., Smith, T. M., and Post, W. M. (1992). The potential for application of individual-based simulation models for assessing the effects of global change. *Annual Review of Ecology and Systematics* **23,** 15–35.

Shuttleworth, W. J. (1989). Micrometeorology of temperate and tropical forest. *Philosophical Transactions of the Royal Society of London, Series B* **324,** 299–334.

Siau, J. F. (1971). "Flow in Wood," Syracuse Wood Science Series 1. Syracuse Univ. Press, Syracuse, New York.

Silkworth, D. R., and Grigal, D. F. (1982). Determining and evaluating nutrient losses following whole-tree harvesting of aspen. *Soil Science Society of America Journal* **46,** 626–631.

Silvester, W. B. (1989). Molybdenum limitations of asymbiotic nitrogen fixation in forests of Pacific Northwest America. *Soil Biology and Biochemistry* **21,** 283–289.

Silvester, W. B., Carter, D. A., and Sprent, J. I. (1979). Nitrogen input by *Lupinus* and *Coriaria* in *Pinus radiata* forest in New Zealand. *In* "Symbiotic Nitrogen Fixation in the Management of Temperate Forests" (J. C. Gordon, C. T. Wheeler, and D. A. Perry, eds.), pp. 253–265. Forest Research Lab., Oregon State University, Corvallis, Oregon.

Singh, J. S., Lauenroth, W. K., Hunt, W. H., and Swift, D. M. (1984). Bias and random errors in estimators of net root production: A simulation approach. *Ecology* **65,** 1760–1764.

Sirois, L., Bonan, G. B., and Shugart, H. H. (1994). Development of a simulation model of the forest–tundra transition zone of northeastern Canada. *Canadian Journal of Forest Research* **24,** 697–706.

Sisk, T. D., Launer, A. E., Switky, K. R., and Ehrlich, P. R. (1994). Identifying extinction threats. *BioScience* **44,** 592–604.

Skole, D., and Tucker, C. (1993). Tropical deforestation and habitat fragmentation in the Amazon: Satellite data from 1978 to 1988. *Science* **260,** 1905–1910.

Skole, D. L., Chomentowski, W. H., Salas, W. A., and Nobre, A. D. (1994). Physical and human dimensions of deforestation in Amazonia. *BioScience* **44,** 314–322.

Skovsgaard, J. P., and Henriksen, H. A. (1996). Increasing site productivity during consecutive generations of naturally regenerated and planted beech (*Fagus sylvatica* L.) in Denmark. *In* "Growth Trends in European Forests" (H. Spiecker, K. Mielikäinen, K. Köhl, and J. P. Skovsgaard, eds.), pp. 89–97. Springer, Berlin.

Slatyer, R. O., and Morrow, P. A. (1977). Attitudinal variation in photosynthetic characteristics of snow gum, *Eucalyptus pauciflora* Sieb. ex Spreng. I. Seasonal changes under field conditions in the Snowy Mountains area of south-eastern Australia. *Australian Journal of Botany* **25,** 1–20.

Smirnoff, N., and Stewart, G. R. (1985). Nitrate assimilation and translocation by higher plants: Comparative physiology and ecological consequences. *Physiologia Plantarum* **64,** 133–140.

Smirnoff, N., Todd, P., and Stewart, G. R. (1984). The occurrence of nitrate reduction in the leaves of woody plants. *Annals of Botany* **54,** 363–374.

Smith, D. H. (1946). Storm damage in New England Forests. M.S. Thesis, Yale University, New Haven, Connecticut.

Smith, N. J., Chen, J. M., and Black, T. A. (1993). Effects of clumping on estimates of stand leaf area index using the LI-COR LAI-2000. *Canadian Journal of Forest Research* **23,** 1940–1943.

Smith, T. A., Leemans, R., and Shugart, H. H. (1992). Sensitivity of terrestrial carbon storage to CO_2-induced climate change: Comparison of four scenarios based on general circulation models. *Climatic Change* **21,** 367–384.

Smith, W. K., and Hinckley, T. M. (1995a). "Resource Physiology of Conifers—Acquisition, Allocation, and Utilization." Academic Press, San Diego.

Smith, W. K., and Hinckley, T. M. (1995b). "Ecophysiology of Coniferous Forests." Academic Press. San Diego.

Snell, J. K. A., and Brown, J. K. (1978). Comparison of tree biomass estimators—dbh and sapwood area. *Forest Science* **24,** 455–457.

Sobrado, M. A., Grace, J., and Jarvis, P. G. (1992). The limits to xylem embolism recovery in *Pinus sylvestris* L. *Journal of Experimental Botany* **43,** 831–836.

Sohlenius, B. (1980). Abundance, biomass and contribution to energy flow by soil nematodes in terrestrial ecosystems. *Oikos* **34,** 186–194.

Sollins, P., Grier, C. C., McCorison, F. M., Cromack, K., Jr., Fogel, R., and Fredriksen, R. L. (1980). The internal element cycles of an old-growth Douglas-fir ecosystem in western Oregon. *Ecological Monographs* **50,** 261–285.

Sollins, P., Spycher, G., and Glassman, C. A. (1984). Net nitrogen mineralization from light- and heavy-fraction forest soil organic matter. *Soil Biology and Biochemistry* **16,** 31–37.

Solomon, A. M., and Bartlein, P. J. (1992). Past and future climate change: Response by mixed deciduous–coniferous forest ecosystems in northern Michigan. *Canadian Journal of Forest Research* **22,** 1727–1738.

Solomon, A. M., Delcourt, H. R., West, D. C., and Blasing, T. J. (1980). Testing simulation model for reconstruction of prehistoric forest-stand dynamics. *Quaternary Research* **14,** 275–293.

Sørenson, L. H. (1974). Rate of decomposition of organic matter in soil as influenced by repeated air drying–rewetting and repeated additions of organic material. *Soil Biology and Biochemistry* **6,** 287–292.

Spanner, M. A., Pierce, L. L., Peterson, D. L., and Running, S. W. (1990). Remote sensing of temperate coniferous forest leaf area index: The influence of canopy closure, understory vegetation and background reflectance. *International Journal of Remote Sensing* **11,** 95–111.

Spanner, M., Johnson, L., Miller, J., McCreight, R., Freemantle, J., Runyon, J., and Gong, P. (1994). Remote sensing of seasonal leaf area index across the Oregon transect. *Ecological Applications* **4,** 258–271.

Sparks, T. H., and Carey, P. D. (1995). The responses of species to climate over two centuries: An analysis of the Marsham phenological record, 1736–1947. *Journal of Ecology* **83,** 321–329.

Sparling, G. P., and Williams, B. L. (1986). Microbial biomass in organic soils: Estimation of biomass C, and effect of glucose or cellulose amendments on the amounts of N and P released by fumigation. *Soil Biology and Biochemistry* **18,** 507–513.

Specht, R. L., and Specht, A. (1989). Canopy structure in *Eucalyptus*-dominated communities in Australia along climatic gradients. *Acta Oecologica* **10,** 191–213.

Sperry, J. S., and Pockman, W. T. (1993). Limitation of transpiration by hydraulic conductance and xylem cavitation in *Betula occidentalis*. *Plant, Cell and Environment* **16,** 279–287.

Sperry, J. S., Donnelly, J. R., and Tyree, M. T. (1988). Seasonal occurrence of xylem embolism in sugar maple (*Acer saccharum*). *American Journal of Botany* **75,** 831–836.

Spies, T. A., Ripple, W. J., and Bradshaw, G. A. (1994). Dynamics and pattern of a managed coniferous forest landscape in Oregon. *Ecological Applications* **4,** 555–568.

Staaf, H., and Berg, B. (1982). Accumulation and release of plant nutrients in decomposing Scots pine needle litter. Long-term decomposition in a Scots pine forest II. *Canadian Journal of Botany* **60,** 1561–1568.

Stage, A. R. (1977). Forest inventory data and construction of growth models. *Eidgenossiche Anstalt für das Forstliche Versuchswesen Mitteilungen* **171,** 23–27.

Stathers, R. J., Black, T. A., and Novak, M. D. (1985). Modelling soil temperature in forest clearcuts using climate station data. *Agricultural Forest Meteorology* **36,** 153–164.

Steingraeber, D. A., Kascht, L. J., and Franck, D. H. (1979). Variation of shoot morphology and bifurcation ratio in sugar maple (*Acer saccharum*) saplings. *American Journal of Botany* **66,** 441–445.

Stenberg, P. (1996). Correcting LAI-2000 estimates for the clumping of needles in shoots of conifers. *Agricultural and Forest Meteorology* **79,** 1–8.

Stephenson, N. L. (1990). Climatic control of vegetation distribution: The role of the water balance. *The American Naturalist* **135,** 649–670.

Sternberg, L. (1989). A model to estimate carbon dioxide recycling in forests using $^{13}C/^{12}C$ ratios and concentrations of ambient carbon dioxide. *Agricultural and Forest Meteorology* **48,** 163–173.

Steudler, P. A., Bowden, R. D., Melillo, J. M., and Aber, J. D. (1989). Influence of nitrogen fertilization on methane uptake in temperate forest soils. *Nature* **341,** 314–316.

Steudler, P. A., Melillo, J. M., Bowden, R. D., Castro, M. S., and Lugo, A. E. (1991). The effects of natural and human disturbances on soil nitrogen dynamics and trace gas fluxes in a Puerto Rican wet forest. *Biotropica* **23,** 356–363.

Stevenson, F. J. (1986). "Cycles of Soil: Carbon, Nitrogen, Phosphorus, Sulfur, Micronutrients." Wiley, New York.

Stewart, I. T., Cayan, D. R., and Dettinger, M. D. (2005). Changes toward earlier streamflow timing across western North America. *Journal of Climate* **18,** 1136–1155.

Stocks, B. J., Mason, J. A., Todd, J. B., Bosch, E. M., Wotton, B. M., Amiro, B. D., Flannigan, M. D., Hirsch, K. G., Logan, K. A., Martell, D. L., and Skinner, W. R. (2002). Large forest fires in Canada, 1959–1997. *Journal of Geophysical Research* **108,** 8149–8160.

Stone, E. C., and Vasey, R. B. (1968). Preservation of coast redwood on alluvial flats. *Science* **159,** 157–161.

Strain, B. R., Higginbotham, K. O., and Mulroy, J. C. (1976). Temperature preconditioning and photosynthetic capacity of *Pinus taeda* L. *Photosynthetica* **10,** 47–53.

Strand, M., and Öquist, G. (1985). Inhibition of photosynthesis by freezing temperatures and high light levels in cold-acclimated seedlings of Scots pine (*Pinus sylvestris*). I. Effects on the light-limited and light-saturated rates of CO_2 assimilation. *Physiologia Plantarum* **64,** 425–430.

Stroo, H. F., Reich, P. B., Schoettle, A. W., and Amundson, R. G. (1988). Effects of ozone and acid rain on white pine (*Pinus strobus*) seedlings grown in five soils. II. Mycorrhizal infection. *Canadian Journal of Botany* **66,** 1510–1516.

Stuiver, M., and Braziunas, T. F. (1987). Tree cellulose $^{13}C/^{12}C$ isotope ratios and climatic change. *Nature* **328,** 58–60.

Sturtevant, B. R., Bissonette, J. A., Long, J. N., and Roberts, D. W. (1997). Coarse woody debris as a function of age, stand structure, and disturbance in boreal Newfoundland. *Ecological Applications* **7,** 702–712.

Sucoff, E. (1972). Water potential in red pine: Soil moisture, evapotranspiration, crown position. *Ecology* **52,** 681–686.

Sukwong, S., Frayer, W. E., and Morgen, E. W. (1971). Generalized comparisons of the precision of fixed-radius and variable-radius plots for basal-area estimates. *Forest Science* **17,** 263–271.

Sun, O. J., Sweet, G. B., Whitehead, D., and Buchan, G. D. (1995). Physiological responses to water stress and waterlogging in *Nothofagus* species. *Tree Physiology* **15,** 629–638.

Swain, T. (1977). Secondary compounds as protective agents. *Annual Review of Plant Physiology* **28,** 479–501.

Swartzman, G. L. (1979). Simulation modeling of material and energy flow through an ecosystem: Methods and documentation. *Ecological Modelling* **7,** 55–81.

Swenson, J., and Waring, R. H. (2006). Modeled photosynthesis predicts woody plant richness at three geographic scales across the northwestern U.S.A. *Global Ecology & Biogeography* **15,** 470–485.

Swenson, J. J., Waring, R. H., Fan, W., and Coops, N. (2005). Predicting site index with a physiologically based growth model across Oregon, USA. *Canadian Journal of Forest Research* **35,** 1697–1707.

Swift, L. W. (1976). Algorithm for solar radiation on mountain slopes. *Water Resources Research* **12,** 108–112.

Tadaki, Y., Sato, A., Sakurai, S., Takeuchi, I., and Kawahara, T. (1977). Studies on the production structure of forest. XVII. Structure and primary production in subalpine "dead tree strips" *Abies* forest near Mt. A. Sahi. *Japanese Journal of Ecology* **27**, 83–90.

Talsma, T., and Gardner, E. A. (1986). Soil water extraction by a mixed eucalypt forest during a drought period. *Australia Journal of Soil Research* **24**, 25–32.

Tan, K. H. (1982). "Principles of Soil Chemistry." Dekker, New York and Basel.

Tan, K. H., and Troth, P. S. (1982). Silica–sesquioxide ratios as aids in characterization of some temperate region and tropical soil clays. *Soil Science Society of America Journal* **46**, 1109–1114.

Tang, S., Meng, C. H., Meng, F.-R., and Wang, Y. H. (1994). A growth and self-thinning model for pure even-age stands: Theory and applications. *Forest Ecology and Management* **70**, 67–73.

Tans, P. P., Fung, I. Y., and Takahashi, T. (1990). Observational constraints on the global atmospheric CO_2 budget. *Science* **247**, 1431–1438.

Tansley, A. G. (1935). The use and abuse of vegetational terms and concepts. *Ecology* **16**, 284–307.

Teklehaimanot, Z., and Jarvis, P. G. (1991). Direct measurement of evaporation of intercepted water from forest canopies. *Journal of Applied Ecology* **28**, 603–618.

Temple, S. A. (1977). Plant–animal mutualism: Coevolution with dodo leads to near extinction of plant. *Science* **197**, 885–886.

Tesch, S. D. (1981a). The evolution of forest yield determination and site classification. *Forest Ecology and Management* **3**, 169–182.

Tesch, S. D. (1981b). Comparative stand development in an old-growth Douglas-fir (*Pseudotsuga menziesii* var. *glauca*) forest in western Montana. *Canadian Journal of Forest Research* **11**, 82–89.

Thies, W. G. (1984). Laminated root rot—The quest for control. *Journal of Forestry* **82**, 345–356.

Thies, W. G., Neilson, E. E., and Zabowski, D. (1994). Removal of stumps from a *Phellinus weirii* infected site and fertilization affect mortality and growth of planted Douglas-fir. *Canadian Journal of Forest Research* **24**, 234–239.

Thomas, J. W. (1979). "Wildlife Habitats in Managed Forests of the Blue Mountains of Oregon and Washington." Agricultural Handbook No. 553. USDA Forest Service, Washington, D.C.

Thornley, J. H. M. (1991). A transport-resistance model of forest growth and partitioning. *Annals of Botany* **68**, 211–226.

Thornley, J. H. M., and Cannell, M. G. R. (1996). Temperate forest responses to carbon dioxide, temperate and nitrogen: A model analysis. *Plant, Cell and Environment* **19**, 1331–1348.

Thornton, P. E., and Running, S. W. (1996). Generating daily surfaces of temperature and precipitation over complex topography. *In* "GIS and Environmental Modeling: Progress and Research Issues" (M. F. Goodchild, L. T. Steyaert, B. O. Parks, C. Johnston, D. Maidment, M. Crane, and S. Glendinning, eds.), pp. 93–98. GIS World, Fort Collins, Colorado.

Thornton, P. E., Running, S. W., and White, M. A. (1997). Generating surfaces of daily meteorological variables over large regions of complex terrain. *Journal of Hydrology* **190**, 214–251.

Tickle, P. K., Coops, N. C., Hafner, S. D., and The Bago Science Team. (2001). Assessing forest productivity at local scales across a native eucalypt forest using a process model, 3PG-SPATIAL. *Forest Ecology and Management* **152**, 275–291.

Tiedje, J. M., Sextone, A. J., Parkin, T. B., Revsbech, N. P., and Shelton, D. R. (1984). Anaerobic processes in soil. *Plant and Soil* **76**, 197–212.

Tiessen, H., Cuevas, E., and Chacon, P. (1994). The role of soil organic matter in sustaining soil fertility. *Nature* **371**, 783–785.

Tiktak, A., and van Grinsven, H. J. M. (1995). Review of sixteen forest–soil–atmosphere models. *Ecological Modelling* **83**, 35–53.

Tilman, D. (1985). The resource-ratio hypothesis of plant succession. *American Naturalist* **125**, 827–852.

Tilman, D. (1996). Biodiversity: Population versus ecosystem stability. *Ecology* **77**, 350–363.

Tilman, D., and Downing, J. A. (1994). Biodiversity and stability in grasslands. *Nature* **367**, 363–365.

Tissue, D. T., Thomas, R. B., and Strain, B. R. (1993). Long-term effects of elevated CO_2 and nutrients on photosynthesis and Rubisco in loblolly pine seedlings. *Plant, Cell and Environment* **16**, 859–865.

Tognetti, R., Raschi, A., Béres, C., Fenyvesi, A., and Ridder, H.-W. (1996). Comparison of sap flow, cavitation and water status of *Quercus petraea* and *Ouercus cerris* trees with special reference to computer tomography. *Plant, Cell and Environment* **19**, 928–938.

Tomé, M., Ribeiro, F., Páscoa, F., Silva, R., Tavares, M., Palma, A., Joäo, C., and Paulo, M. (1996). Growth trend in Portuguese forests: An exploratory analysis. *In* "Growth Trends in European Forests" (H. Spiecker, K. Mielikäinen, K. Köhl, and J. P. Skovsgaard, eds.), pp. 329–353. Springer, Berlin.

Torgersen, T. R., Mason, R. R., and Campbell, R. W. (1990). Predation by birds and ants on two forest insect pests in the Pacific Northwest. *Studies in Avian Biology* **13**, 14–19.

Townsend, A. R., Vitousek, P. M., and Trumbore, S. E. (1995). Soil organic matter dynamics along gradients in temperate and land use on the island of Hawaii. *Ecology* **76**, 721–733.

Townshend, J., Justice, C., Li, W., Gurney, C., and McManus, J. (1991). Global land cover classification by remote sensing: Present capabilities and future possibilities. *Remote Sensing of the Environment* **35**, 243–255.

Treuhaft, R. N., Law, B. E., and Asner, G. P. (2004). Forest attributes from radar inteferometric structure and its fusion with optical remote sensing. *BioScience* **54**, 561–571.

Trofymow, J. A., Barclay, H. J., and McCullough, K. M. (1991). Annual rates and elemental concentrations of litter fall in thinned and fertilized Douglas-fir. *Canadian Journal of Forest Research* **21**, 1601–1615.

Trumbore, S. E., Chadwick, O. A., and Amundson, R. (1996). Rapid exchange between soil carbon and atmospheric carbon dioxide driven by temperature change. *Science* **272**, 393–396.

Tuomisto, H., Ruokolainen, K., Kalliola, R., Linna, A., Danjoy, W., and Rodriquez, Z. (1995). Dissecting Amazon biodiversity. *Science* **269**, 63–66.

Turnbull, M. H., Goodall, R., and Stewart, G. R. (1995). The impact of mycorrhizal colonization upon nitrogen source utilization and metabolism in seedlings of *Eucalyptus grandis* Hill ex Maiden and *Eucalyptus maculata* Hook. *Plant, Cell and Environment* **18**, 1386–1394.

Turner, D. P., and Koerper, G. J. (1995). A carbon budget for forests of the conterminous United States. *Ecological Applications* **5**, 421–436.

Turner, D. P., Dodson, R., and Marks, D. (1996). Comparison of alternative spatial resolutions in the application of a spatially distributed biogeochemical model over complex terrain. *Ecological Modelling* **90**, 53–67.

Turner, D. P., Koerper, G., Gucinski, H., Peterson, C., and Dixon, R. K. (1993). Monitoring global change: Comparison of forest cover estimates using remote sensing and inventory approaches. *Environmental Monitoring and Assessment* **26**, 295–305.

Turner, D. P., Ritts, W. D., Cohen, W. B., Maeirsperger, T. K., Gower, S. T., Kirschbaum, A. A., Running, S. W., Zhao, M., Wofsy, S. C., Dunn, A. L., Law, B. E., Campbell, J. L., Oechel, W. C., Kwon, H. J., Meyers, T. P., Small, E. E., Kurc, S. A., and Gamon, J. A. (2005). Site-level evaluation of satellite-based global terrestrial gross primary production and net primary production monitoring. *Global Change Biology* **11**, 666–684.

Turner, D. P., Ritts, W. D., Zhao, M., Kurc, S. A., Dunn, A. L., Wofsy, S. C., Small, E. E., and Running, S. W. (2006). Assessing interannual variation in MODIS-based estimates of gross primary production. *IEEE Transactions on Geoscience and Remote Sensing* **44,** 1899–1907.

Turner, J. (1977). Effects of nitrogen availability on nitrogen cycle in a Douglas-fir stand. *Forest Science* **23,** 307–316.

Turner, J., and Gessel, S. P. (1990). Forest productivity in the southern hemisphere with particular emphasis on managed forests. *In* "Sustained Productivity of Forest Soils" (S. P. Gessel, D. S. Lacate, G. F. Weetman, and R. F. Powers, eds.), pp. 23–39. University of British Columbia, Faculty of Forestry. University of British Columbia, Vancouver, B.C., Canada.

Turner, J., and Lambert, M. J. (1983). Nutrient cycling within a 27-year-old *Eucalyptus grandis* plantation in New South Wales. *Forest Ecology and Management* **6,** 155–168.

Turner, J., and Lambert, M. J. (1986). Nutrition and nutritional relationships of *Pinus radiata*. *Annual Review of Ecology and Systematics* **17,** 325–350.

Turner, J., and Lambert, M. J. (1988). Soil properties as affected by *Pinus radiata* plantations. *New Zealand Journal of Forest Science* **18,** 77–91.

Turner, J., and Long, J. N. (1975). Accumulation of organic matter in a series of Douglas-fir stands. *Canadian Journal of Forest Research* **5,** 681–690.

Turner, M. G. (1989). Landscape ecology: The effect of pattern on process. *Annual Review of Ecological Systems* **20,** 171–197.

Turner, M. G., and Gardner, R. H., eds. (1991) "Quantitative Methods in Landscape Ecology: The Analysis and Interpretation of Landscape Heterogeneity." Springer-Verlag, New York.

Turner, M. G., and Romme, W. H. (1994). Landscape dynamics in crown fire ecosystems. *Landscape Ecology* **9,** 59–77.

Turner, M. G., Romme, W. H., Gardner, R. H., O'Neill, R. V., and Kratz, T. K. (1993). A revised concept of landscape equilibrium: Disturbance and stability on scaled landscapes. *Landscape Ecology* **8,** 213–227.

Turner, M. G., Wu, Y., Wallace, L. L., Romme, W. H., and Brenkert, A. (1994). Simulating winter interactions among ungulates, vegetation, and fire in northern Yellowstone Park. *Ecological Applications* **4,** 472–496.

Turner, M. G., Arthaud, G. J., Engstrom, R. T., Hejl, S. J., Liu, J., Loeb, S., and McKelvey, K. (1995). Usefulness of spatially explicit population models in land management. *Ecological Applications* **5,** 12–16.

Tuzet, A., Perrier, A., and Leuning, R. (2003). A coupled model of stomatal conductance, photosynthesis and transpiration. *Plant, Cell and Environment* **26,** 1097–1116.

Tworkoski, T. J., Burger, J. A., and Smith, D. W. (1983). Soil texture and bulk density affect early growth of white oak seedlings. *Tree Planters Notes* **34,** 22–25.

Tyler, G. (1972). Heavy metals pollute nature—may reduce productivity. *Ambio* **1,** 52–59.

Tyree, M. T., and Dixon, M. A. (1986). Water stress induced cavitation and embolism in some woody plants. *Physiologia Plantarum* **66,** 397–405.

Tyree, M. T., and Ewers, F. W. (1991). The hydraulic architecture of trees and other woody plants (Tansley Review 34). *New Phytologist* **119,** 345–360.

Tyree, M. T., Davis, S. D., and Cochard, H. (1994). Biophysical perspectives of xylem evolution: Is there a tradeoff of hydraulic efficiency for vulnerability to dysfunction? *IAWA Journal* **15,** 335–360.

Ugolini, F. C., Dawson, H., and Zachara, J. (1977a). Direct evidence of particle migration in the soil solution of a podzol. *Science* **198,** 603–605.

Ugolini, F. C., Minden, R., Dawson, H., and Zachara, J. (1977b). An example of soil processes in the *Abies amabilis* zone of central Cascades. Washington. *Soil Science* **124,** 291–302.

Ulrich, B. (1983). A concept of forest ecosystem stability and of acid deposition as driving force for destabilization. *In* "Effects of Accumulation of Air Pollutants in Forest Ecosystems" (B. Ulrich and J. Pankrath, eds.), pp. 1–29. Reidel, Dordrecht, The Netherlands.

Ulrich, B., Benecke, P., Harris, W. F., Khanna, P. K., and Mayer, R. (1981). Soil processes. *In* "Dynamic Properties of Forest Ecosystems" (D. E. Reichle, ed.), pp. 265–359. Cambridge Univ. Press, London and New York.

USDA. (1991). State soil geographic database (STATSGO) data user guide. Misc. Pub. USDA No. 1492, Washington, D.C.

Ustin, S. L., Roberts, D. A., Gamon, J. A., Asner, G. P., and Green, R. O. (2004). Using imaging spectroscopy to study ecosystem processes and properties. *BioScience* **54**, 523–534.

Valencia, R., Balsiev, H., and Pazy Mio, G. (1994). High tree alpha diversity in Amazonian Ecuador. *Biodiversity Conservation* **3**, 21–28.

Valentini, R., Mugnozza, G. E. S., De Angelis, P., and Bimbi, R. (1991). An experimental test of the eddy correlation technique over Mediterranean macchia canopy. *Plant, Cell and Environment* **14**, 987–994.

Van Bell, A. J. E. (1990). Xylem-phloem exchange via the rays: The undervalued route of transport. *Journal of Experimental Botany* **41**, 631–644.

Van Cleve, K., Dyrness, C. T., Viereck, L. A., Fox, J., Chapin III, F. S., and Oechel, W. C. (1983). Taiga ecosystems in interior Alaska. *BioScience* **33**, 39–44.

Van de Water, P. K., Leavitt, S. W., and Betancourt, J. L. (1994). Trends in stomatal density and $^{13}C/^{12}C$ ratios of *Pinus flexilis* needles during last glacial–interglacial cycle. *Science* **264**, 239–243.

Van den Driessche, R. (1984). Nutrient storage, retranslocation and relationship of stress to nutrition. *In* "Nutrition of Plantation Forests" (G. D. Bowen and E. K. S. Nambiar, eds.), pp. 181–209. Academic Press, Orlando, Florida.

Van der Merwe, N. J., Lee-Thorp, J. A., Thaderay, J. F., Hull-Martion, A., Kruger, F. J., Coetzees, H., Bell, R. H., and Lindeque, M. (1990). Source-area determination of elephant ivory by isotopic analyses. *Nature* **346**, 744–746.

Van Hook, R. I., Johnson, D. W., West, D. C., and Mann, L. K. (1982). Environmental effects of harvesting forests for energy. *Forest Ecology and Management* **4**, 79–94.

Van Horne, B., Hanley, T. A., and Cates, R. G., McKendrick, J., and Hornet, J. D. (1988). Influence of seral stage and season on leaf chemistry of southeastern Alaska deer forage. *Canadian Journal of Forest Research* **18**, 90–99.

Vance, C. P., and Heichel, G. H. (1991). Carbon in N_2 fixation: Limitation or exquisite adaptation. *Plant Physiology and Plant Molecular Biology* **42**, 373–392.

Vann, D. R., Johnson, A. H., and Casper, B. B. (1994). Effect of elevated temperatures on carbon dioxide exchange in *Picea rubens*. *Tree Physiology* **14**, 1339–1349.

Veblen, T. T., and Alaback, P. B. (1996). A comparative review of forest dynamics and disturbance in the temperate rainforests of North and South America. *In* "High-Latitude Rainforests and Associated Ecosystems of the West Coast of the Americas" (R. G. Lawford, P. B. Alaback, and E. Fuentes, eds.), pp. 105–133. Springer-Verlag, New York.

VEMAP Members. (1995). Vegetation/ecosystem modeling and analysis project (VEMAP): Comparing biogeography and biogeochemistry models in a continental-scale study of terrestrial ecosystem responses to climate change and CO_2 doubling. *Global Biogeochemical Cycles* **9**, 407–437.

Vepraskas, M. J., Jongmans, A. G., Hoover, M. T., and Bouma, J. (1991). Hydraulic conductivity of saprolite as determined by channels and porous groundmass. *Soil Science Society of America Journal* **55**, 923–938.

Verbyla, D. L. (1995). "Satellite Remote Sensing of Natural Resources." CRC Press, Boca Raton, Florida.

Vertessy, R. A., Benyon, R. G., O'Sulllivan, S. K., and Gribben, P. R. (1995). Relationships between stem diameter, sapwood area, leaf area and transpiration in a young mountain ash forest. *Tree Physiology* **15,** 559–568.

Vertregt, N., and Penny de Vries, F. W. T. (1987). A rapid method for determining the efficiency of biosynthesis of plant biomass. *Journal of Theoretical Biology* **128,** 109–119.

Vetter, M., Wirth, C., Böttcher, H., Churkina, G., Schulze, E.-D., Wutzler, T., and Weber, G. (2005). Partitioning direct and indirect human-induced effects on carbon sequestration of managed coniferous forests using model simulations and forest inventories. *Global Change Biology* **11,** 810–827.

Vidal, A., Pinglo, F., Durand, H., Devaux-Ros, C., and Maillet, A. (1994). Evaluation of a temporal fire risk index in Mediterranean forests from NOAA thermal IR. *Remote Sensing of the Environment* **49,** 96–303.

Viereck, L. A., Dyrness, C. T., and Van Cleve, K. (1983). Vegetation, soils, and forest productivity in selected forest types in interior Alaska. *Canadian Journal of Forest Research* **13,** 703–720.

Vitousek, P. (1982). Nutrient cycling and nutrient use efficiency. *American Naturalist* **119,** 553–572.

Vitousek, P. M. (1994). Beyond global warming: Ecology and global change. *Ecology* **75,** 1861–1876.

Vitousek, P. M., and Hooper, D. U. (1993). Biological diversity and terrestrial ecosystem biogeochemistry. *In* "Biodiversity and Ecosystem Function" (E.-D. Schulze and H. A. Mooney, eds.), pp. 3–14. Springer-Verlag, New York.

Vitousek, P. M., and Matson, P. (1993). Agriculture, the global nitrogen cycle, and trace gas flux. *In* "Biogeochemistry of Global Change: Radiatively Active Trace Gases" (R. Oremland, ed.), pp. 193–208. Chapman & Hall, New York.

Vitousek, P. M., Ehrlich, P. R., Ehrlich, A. H., and Matson, P. A. (1986). Human appropriation of the products of photosynthesis. *BioScience* **36,** 368–373.

Vitousek, P. M., Gosz, J. R., Grier, C. C., Melillo, J. M., and Reiners, W. A. (1982). A comparative analysis of potential nitrification and nitrate mobility in forest ecosystem. *Ecological Monographs* **52,** 155–177.

Vitousek, P. (1984). Litterfall, nutrient cycling, and nutrient limitations in tropical forests. *Ecology* **65,** 285–298.

Vogel, J. C., Eglington, B., and Auret, J. M. (1990). Isotope fingerprints in elephant bone and ivory. *Nature* **346,** 747–749.

Vogt, K. A., Grier, C. C., and Vogt, D. J. (1986). Production, turnover, and nutrient dynamics of above- and belowground detritus of world forests. *Advances in Ecological Research* **15,** 303–377.

Volney, W. J. A., and Fleming, R. A. (2000). Climate change and impacts of boreal forest insects. *Agriculture, Ecosystems and Environment* **82,** 283–294.

Von Caemmerer, S., and Farquhar, G. D. (1981). Some relations between the biochemistry of photosynthesis and the gas exchange of leaves. *Planta* **153,** 376–387.

Walcroft, A. S., Silvester, W. B., Whitehead, D., and Kelliher, F. M. (1997). Seasonal changes in stable carbon isotope ratios within annual growth rings of *Pinus radiata* reflect environmental regulation of growth processes. *Australian Journal of Plant Physiology* **24,** 57–68.

Walker, G. K., Sud, Y. C., and Atlas, R. (1995). Impact of the ongoing Amazonian deforestation on local precipitation: A GCM simulation study. *Bulletin of the American Meteorological Society* **76,** 346–361.

Walker, J., Bullen, F., and Williams, B. G. (1993). Ecohydrological changes in the Murray—Darling basin. I. The number of trees cleared over two centuries. *Journal of Applied Ecology* **30**, 265–273.

Walker, J., Raison, R. J., and Khanna, P. K. (1983). Fire. *In* "Australian Soils: The Human Impact" (J. S. Russell and R. F. Isbell, eds.), pp. 185–216. Univ. of Queensland Press, Brisbane, Australia.

Walker, T. W., and Syers, J. K. (1976). The fate of phosphorus during pedogenesis. *Geoderma* **15**, 1–19.

Wallin, D. O., Harmon, M. E., Cohen, W. B., Fiorella, M., and Ferrell, W. K. (1996). Use of remote sensing to model land use effects on carbon flux in forests of the Pacific Northwest, USA. *In* "The Use of Remote Sensing in the Modeling of Forest Productivity" (H. L. Gholz, K. Nakane, and H. Shimoda, eds.), pp. 219–237. Kluwer, Dordrecht, The Netherlands.

Wallin, D. O., Swanson, F. J., and Marks, B. (1994). Landscape pattern response to changes in pattern generation rules: Land-use legacies in forestry. *Ecological Applications* **4**, 569–580.

Walsh, J. E. (1995). Long-term observations for monitoring of the cryosphere. *Climatic Change* **31**, 369–394.

Walter, H. (1954). "Grundlangen der Pflanzenverbreitung. III. Einfuhrung in die Pflanzengeographie." Ulmer, Stuttgart.

Wang, Y. P., and Jarvis, P. G. (1990). Influence of crown structural properties on PAR absorption, photosynthesis, and transpiration in Sitka spruce: Application of a model (MAESTRO). *Tree Physiology* **7**, 297–316.

Wang, Y. P., and Polglase, P. J. (1995). Carbon balance in tundra, boreal forest and humid tropical forest during climate change: Scaling up from leaf physiology and soil carbon dynamics. *Plant, Cell and Environment* **18**, 1226–1244.

Wang, Y. P., McMurtrie, R. E., and Landsberg, J. J. (1992a). Modelling canopy photosynthetic productivity. *In* "Crop Photosynthesis: Spatial and Temporal Determinants" (N. R. Baker and H. Thomas, eds.), pp. 43–67. Elsevier, Amsterdam.

Wang, Y. S., Miller, D. R., Welles, J. M., and Heisler, G. M. (1992b). Spatial variability of canopy foliage in an oak forest estimated with fisheye sensors. *Forest Science* **38**, 854–865.

Wargo, P. M. (1972). Defoliation-induced chemical changes in sugar maple roots stimulate growth of *Armillaria mellea*. *Phytopathology* **62**, 1278–1283.

Wargo, P. M., Parker, J., and Houston, D. R. (1972). Starch content in roots of defoliated sugar maple. *Forest Science* **18**, 203–204.

Waring, R. H. (1969). Forest plants of the eastern Siskiyous: Their environmental and vegetational distribution. *Northwest Science* **43**, 1–17.

Waring, R. H. (1980). Site, leaf area, and phytomass production in trees. *In* "Mountain Environments and Subalpine Tree Growth" (U. Benecke, ed.), pp. 125–135. IUFRO Workshop, Forest Research Institute Tech. Paper No. 70, Christchurch, New Zealand.

Waring, R. H. (1983). Estimating forest growth and efficiency in relation to canopy leaf area. *Advances in Ecological Research* **13**, 327–354.

Waring, R. H. (1987). Characteristics of trees predisposed to die. *BioScience* **37**, 569–574.

Waring, R. H. (1989). Ecosystems: Fluxes of matter and energy. *In* "Ecological Concepts: The Contribution of Ecology to an Understanding of the Natural World" (J. M. Cherrett, A. D. Bradshaw, P. J. Grubb, F. B. Goldsmith, and J. R. Krebs, eds.), pp. 17–41. Blackwell, Oxford.

Waring, R. H., and Franklin, J. F. (1979). Evergreen coniferous forests of the Pacific Northwest. *Science* **204**, 1380–1386.

Waring, R. H., and Pitman, G. B. (1983). Physiological stress in lodgepole pine as a precursor for mountain pine beetle attack. *Zeitschrift für angewandte Entomologie* **96**, 265–270.

Waring, R. H., and Pitman, G. B. (1985). Modifying lodgepole pine stands to change susceptibility to mountain pine beetle attack. *Ecology* **66,** 889–897.

Waring, R. H., and Running, S. W. (1976). Water uptake, storage, and transpiration by conifers: A physiological model. *In* "Water and Plant Life" (O. L. Lange, E.-D. Schulze, and L. Kappen, eds.), pp. 189–202. Springer-Verlag, New York.

Waring, R. H., and Running, S. W. (1978). Sapwood water storage: Its contribution to transpiration and effect upon water conductance through the stems of old-growth Douglas-fir. *Plant, Cell and Environment* **1,** 131–140.

Waring, R. H., and Schlesinger, W. H. (1985). "Forest Ecosystems: Concepts and Management." Academic Press, Orlando, Florida.

Waring, R. H., and Silvester, W. B. (1994). Interpreting variation in carbon isotopes within tree crowns. *Tree Physiology* **14,** 1203–1213.

Waring, R. H., and Winner, W. E. (1996). Constraints on terrestrial primary production in temperate forests along the Pacific Coast of North and South America. *In* "High Latitude Rain Forests and Associated Ecosystems of the West Coast of the Americas: Climate, Hydrology, Ecology and Conservation" (R. G. Lawford, P. Alaback, and E. R. Fuentes, eds.), pp. 89–102. Springer-Verlag, New York.

Waring, R. H., Coops, N. C., Ohmann, J. L., and Sarr, D. A. (2002). Interpreting woody plant richness from seasonal ratios of photosynthesis. *Ecology* **83,** 2964–2970.

Waring, R. H., Cromack, K., Jr., Boone, R. D., and Stafford, S. G. (1987). Responses to pathogen-induced disturbance: Decomposition, nutrient availability, and tree vigour. *Forestry* **60,** 219–227.

Waring, R. H., Landsberg, J. J., and Williams, M. (1998). Net primary production of forests: A constant fraction of gross primary production? *Tree Physiology* **18,** 129–134.

Waring, R. H., Law, B. E., Goulden, M. L., Bassow, S. L., McCreight, R. W., Wofsy, S. C., and Bazzaz, F. A. (1995a). Scaling gross ecosystem production at Harvard Forest with remote sensing: A comparison of estimates from a constrained quantum-use efficiency model and eddy correlation. *Plant, Cell and Environment* **18,** 1201–1213.

Waring, R. H., Way, J., Hunt, E. R., Jr., Morrissey, L., Ranson, K. J., Weishampel, J. F., Oren, R., and Franklin, S. E. (1995b). Imaging radar for ecosystem studies. *BioScience* **45,** 715–723.

Waring, R. H., Savage, T., Cromack, K., Jr., and Rose, C. (1992). Thinning and nitrogen fertilization in a grand fir stand infested with western spruce budworm. Part IV. An ecosystem management perspective. *Forest Science* **38,** 275–286.

Waring, R. H., Schroeder, P. E., and Oren, R. (1982). Application of the pipe model theory to predict canopy leaf area. *Canadian Journal of Forest Research* **12,** 556–560.

Waring, R. H., Whitehead, D., and Jarvis, P. G. (1979). The contribution of stored water to transpiration in Scots pine. *Plant, Cell and Environment* **2,** 309–317.

Waring, R. H., Whitehead, D., and Jarvis, P. G. (1980). Comparison of an isotopic method and the Penman–Monteith equation for estimating transpiration from Scots pine. *Canadian Journal of Forest Research* **10,** 555–558.

Warren, J. M., Meinzer, F. C., Brooks, J. R., and Domec, J. C. (2005). Vertical stratification of soil water storage and release dynamics in Pacific Northwest coniferous forests. *Agricultural and Forest Meteorology* **130,** 39–58.

WCMC (World Conservation Monitoring Center). (1992). "Global Biodiversity: Status of the Earth's Living Resources." Chapman & Hall, London.

Webb III, T., and Bartlein, P. J. (1992). Global changes during the last 3 million years: Climatic controls and biotic responses. *Annual Review of Ecology and Systematics* **23,** 141–173.

Webster, P. J., Holland, G. J., Curry, J. A., and Chang, H.-R. (2005). Changes in tropical cyclone number, duration, and intensity in a warming environment. *Science* **309,** 1844–1846.

Weier, T. E., Stocking, C. R., and Barbour, M. G. (1974). "Botany, An Introduction to Plant Biology." Wiley, New York.

Weinstein, D. A., and Yanai, R. D. (1994). Integrating the effects of simultaneous multiple stresses on plants using the simulation model TREGRO. *Journal of Environmental Quality* **23,** 418–428.

Weinstein, D. A., Beloin, R. M., and Yanai, R. D. (1991). Modeling changes in red spruce carbon balance and allocation in response to interacting ozone and nutrient stresses. *Tree Physiology* **9,** 127–146.

Weishampel, J. F., Ranson, K. J., Harding, D. J., and Blair, J. B. (1996). Remote sensing of forest canopies. *Selbyana* **7,** 6–14.

Weller, D. A. (1987). A reevaluation of the −3/2 power rule of plant self-thinning. *Ecological Monographs* **57,** 23–43.

Wendler, R., and Millard, P. (1996). Impact of water and nitrogen supplies on the physiology, leaf demography and nitrogen dynamics of *Betula pendula. Tree Physiology* **16,** 153–159.

Westerling, A. L., Hidalgo, H. G., Cayan, D. R., and Swetnam, T. W. (2006). Warming and earlier spring increases western U.S. forest wildfire activity. *Science* **313,** 940–943.

Westoby, M. (1977). Self-thinning driven by leaf area not by weight. *Nature* **265,** 330–331.

Westoby, M. (1984). The self-thinning rule. *Advances in Ecological Research* **14,** 167–225.

White, C. S. (1986). Effects of prescribed fire on rates of decomposition and nitrogen mineralization in a ponderosa pine ecosystem. *Biology and Fertility of Soils* **2,** 87–95.

White, C. S., Moore, D. I., Homer, J. D., and Gosz, J. R. (1988). Nitrogen mineralization–immobilization response to field N or C perturbations: An evaluation of a theoretical model. *Soil Biology and Biochemistry* **20,** 101–105.

White, J. (1981). The allometric interpretation of the self-thinning rule. *Journal of Theoretical Biology* **89,** 475–500.

White, J. D., and Running, S. W. (1994). Testing scale dependent assumptions in regional ecosystem simulations. *Journal of Vegetation Science* **5,** 687–702.

White, J. D., Running, S. W., Nemani, R., Keane, R. E., and Ryan, K. C. (1997). Measurement and mapping of LAI in Rocky Mountain montane ecosystems. *Canadian Journal of Forest Research* **8,** 805–823.

White, J. D., Running, S. W., Thornton, P. E., Keane, R. E., Ryan, K. C., Fagre, D. B., and Key, C. H. (1998). Assessing regional ecosystem simulations of carbon and water budgets for climate change research at Glacier National Park, USA. *Ecological Applications* **8,** 805–823.

White, J. W. C., Cook, E. R., Lawrence, J. R., and Broecker, W. S. (1985). The D/H ratios of sap in trees: Implications for water sources and tree ring D/H ratios. *Geochimica et Cosmochimica Acta* **49,** 237–246.

White, M. A., and Mladenoff, D. J. (1994). Old-growth forest landscape transitions from pre-European settlement to present. *Landscape Ecology* **9,** 191–205.

White, M. A., Thornton, P. E., and Running, S. W. (1997). A continental phenology model for monitoring vegetation responses to interannual climatic variability. *Global Biogeochemical Cycles* **11,** 217–234.

Whitehead, D., and Jarvis, P. G. (1981). Coniferous forest and plantations. *In* "Water Deficits and Plant Growth" (T. T. Kozlowski, ed.), pp. 49–152. Academic Press, New York.

Whitehead, D., and Kelliher, F. M. (1991). Modeling the water balance of a small *Pinus radiata* catchment. *Tree Physiology* **9,** 17–33.

Whitehead, D., Hall, G. M. J., Walcroft, A. S., Brown, K. J., Landsberg, J. J., Tissue, D. T., Turnbull, M. H., Griffin, K. L., Schuster, W. S., Carswell, F. E., Trotter, C. M., James, I. L., and Norton, D. A. (2002). Analysis of the growth of rimu (*Dacrydium cupressinum*) in South Westland, New Zealand, using process-based simulation models. *International Journal of Biometeorology* **46,** 66–75.

Whitehead, D., Kelliher, F. M., Frampton, C. M., and Godfrey, M. J. S. (1994). Seasonal develop-ment of leaf area in a young, widely spaced *Pinus radiata* D. Don stand. *Tree Physiology* **14,** 1019–1038.

Whitehead, D., Livingston, N. J., Kelliher, F. M., Hogan, K. P., Pepin, S., McSeveny, T. M., and Byers, J. N. (1996). Response of transpiration and photosynthesis to a transient change in illu-minated foliage area for a *Pinus radiata* D. Don tree. *Plant, Cell and Environment* **19,** 949–957.

Whitehead, D., Okali, D. U. U., and Fasehun, F. E. (1981). Stomatal response to environmental variables in two tropical forest species during the dry season in Nigeria. *Journal of Applied Ecology* **18,** 571–587.

Whittaker, R. H. (1953). A consideration of climax theory: The climax as a population and pattern. *Ecological Monographs* **23,** 41–78.

Whittaker, R. H. (1975). "Communities and Ecosystems." 2nd Ed. Macmillan, New York.

Whittaker, R. H., and Woodwell, G. M. (1968). Dimension and production relations of trees and shrubs it the Brookhaven Forest, New York. *Journal of Ecology* **56,** 1–24.

Wickman, B. E. (1978). Ecological effects. *In* "The Douglas-fir Tussock Moth: A Synthesis" (M. H. Brookes, R. W. Stark, and R. W. Campbell, eds.), pp. 63–75. Forest Service Technical Bulletin 1585. Washington, D.C.

Wickman, B. E., Mason, R. R., and Paul, H. G. (1992). Thinning and nitrogen fertilization in a grand fir stand infested with western spruce budworm. Part II: Tree growth response. *Forest Science* **38,** 252–264.

Wiens, J. A., Stenseth, N. C., Van Borne, B., and Ims, R. A. (1993). Ecological mechanisms and landscape ecology. *Oikos* **66,** 369–380.

Wigmosta, M. S., Vail, L. W., and Lettenmaier, D. P. (1994). A distributed hydrology-vegetation model for complex terrain. *Water Resources Research* **30,** 1665–1679.

Wikström, F., and Ericsson, T. (1995). Allocation of mass in trees subject to nitrogen and magnesium limitation. *Tree Physiology* **15,** 339–344.

Williams, D. W., and Liebhold, A. M. (2002). Climate change and the outbreak ranges of two North American bark beetles. *Agricultural and Forest Entomology* **4,** 87–99.

Williams, K., Percival, F., Merino, J., and Mooney, H. A. (1987). Estimation of tissue construction cost from heat of combustion and organic nitrogen content. *Plant, Cell and Environment* **10,** 725–734.

Williams, M., Rastetter, E. B., Fernandes, D., Goulden, M. L., Wofsy, S. C., Shaver, G. R., Melillo, J. M., Nadelhoffer, K. J., Fan, S.-M., and Monger, D. W. (1996). Modelling the soil–plant–atmosphere continuum in a *Quercus–Acer* stand at Harvard Forest: The regulation of stomatal conductance by light, nitrogen, and soil/plant hydraulic properties. *Plant, Cell and Environment* **19,** 911–927.

Williams, M., Rastetter, E. R., Rernandes, D. N., Goulden, M. L., and Shaver, G. R. (1997). Predicting gross primary productivity in terrestrial ecosystems. *Ecological Applications* **7,** 882–894.

Williams, M., Schwarz, P. A., Law, B. E., Irvine, J., and Kurpius, M. R. (2005). An improved analysis of forest carbon dynamics using data assimilation. *Global Change Biology* **11,** 89–105.

Wilson, K. B., Baldocchi, D. D., Aubinet, M., Berbigier, P., Bernhofer, C., Dolman, H., Falge, E., Field, C., Goldstein, A., Granier, A., Grelle, A., Halldor, T., Hollinger, D., Katul, G., Law, B. E., Lindroth, A., Meyers, T., Moncrieff, J., Monson, R., Oechel, W., Tenhunen, J., Valentini, R., Verma, S., Vesala, T., and Wofsy, S. (2002). Energy partitioning between latent and sensible heat flux during the warm season at FLUXNET sites. *Water Resources Research* **38,** 1294–1304.

Winner, W. M., Bewley, J. D., Krause, H. R., and Brown, H. M. (1978). Stable sulfur isotope analysis of SO_2 pollution impact on vegetation. *Oecologia* **36,** 351–361.

Witkamp, M. (1971). Soils as components of ecosystems. *Annual Review of Ecology and Systematics* **2,** 85–110.

Witkamp, M., and Ausmus, B. S. (1976). Processes in decomposition and nutrient transfer in forest systems. *In* "The Role of Terrestrial and Aquatic Organisms in Decomposition Processes" (J. M. Anderson and A. MacFadyen, eds.), pp. 375–396. Blackwell, Oxford.

Wofsy, S. C., Goulden, M. L., Munger, J. W., Fan, S.-M., Bakwin, P. S., Daube, B. C., Bassow, S. L., and Bazzaz, F. A. (1993). Net exchange of CO_2 in a mid-latitude forest. *Science* **260,** 1314–1317.

Woldendorp, J. W., and Laanbroek, H. J. (1989). Activity of nitrifers in relation to nitrogen nutrition of plants in natural ecosystems. *Plant and Soil* **115,** 217–228.

Wolfe, M. L., and Berg, F. C. (1988). Deer and forestry in Germany—Half a century after Aldo Leopold. *Journal of Forestry* **86,** 25–28.

Wood, T. G. (1976). The role of termites (Isoptera) in decomposition processes. *In* "The Role of Terrestrial and Aquatic Organisms in Decomposition Processes" (J. M. Anderson and A. MacFadyen, eds.), pp. 145–168. Blackwell, Oxford.

Woods, A. S., Coates, K. D., and Hamann, A. (2005). Is an unprecedented Dothistroma needle blight epidemic related to climate change? *BioScience* **55,** 761–769.

Woodward, F. I. (1987). "Climate and Plant Distribution" (R. S. K. Barnes, H. J. B. Birks, E. F. Connor, J. L. Harper, and R. T. Paine, eds.), pp. 11–165. Cambridge Univ. Press, New York.

Woodward, F. I. (1992). Predicting plant responses to global environmental change. *New Phytology* **122,** 239–255.

Woodward, F. I., Smith, T. M., and Emanuel, W. R. (1995). A global land primary productivity and phytogeography model. *Global Biogeochemical Cycles* **9,** 471–490.

Woodward, F. I., Thompson, G. B., and McKee, I. F. (1991). Effects of elevated concentrations of carbon dioxide on individual plants, populations, communities and ecosystems. *Annals of Botany* **67,** 23–38.

Woodwell, G. M. (1974). Variation in the nutrient content of leaves of *Quercus alba, Quercus coccinea,* and *'Pinus rigida* in the Brookhaven Forest from bud-break to abscission. *American Journal of Botany* **61,** 749–753.

Wright, R. F., Roelofs, J. G. M., Bredemeir, M., Blanck, K., Boxman, A. M., Emmett, B. A., Gundersen, P., Hultberg, H., Kjønaas, O. J., Moldan, F., Tieteina, A., van Breeinen, N., and van Dijk, H. E. G. (1995). NITREX: Responses of coniferous forest ecosystems to experimentally changed deposition of nitrogen. *Forest Ecology and Management* **71,** 163–169.

Yanai, R. D. (1991). Soil solution phosphorus dynamics in a whole-tree-harvested northern hardwood forest. *Soil Science Society of America Journal* **55,** 1746–1752.

Yanai, R. D., Fahey, T. J., and Miller, S. L. (1995). Efficiency of nutrient acquisition by fine roots and mycorrhizae. *In* "Resource Physiology of Conifers—Acquisition, Allocation, and Utilization" (W. K. Smith and T. M. Hinckley, eds.), pp. 75–103. Academic Press, San Diego.

Yih, K., Boucher, D. H., Vandermeer, J. H., and Zamora, N. (1991). Recovery of the rain forest of southeastern Nicaragua after destruction by Hurricane Joan. *Biotropica* **23,** 106–113.

Yoder, B. J., and Waring, R. H. (1994). The normalized difference vegetation index of small Douglas-fir canopies with varying chlorophyll concentrations. *Remote Sensing of the Environment* **49,** 81–91.

Yoder, B. J., Ryan, M. G., Waring, R. H., Schoettle, A. W., and Kaufmann, M. R. (1994). Evidence of reduced photosynthetic rates in old trees. *Forest Science* **40,** 513–527.

Youngberg, C. T., and Wollum, A. G. (1976). Nitrogen accretion in developing *Ceanothus velutinu* stands. *Soil Science Society of America Proceedings* **40,** 109–112.

Zagolski, F., Pinel, V., Romier, J., Alcayde, D., Fontanari, J., Gastellu-Etchegorry, J. P., Giordano, G., Marty, G., Mougin, E., and Joffre, R. (1996). Forest canopy chemistry with high spectral resolution remote sensing. *International Journal of Remote Sensing* **17,** 1107–1128.

Zarco-Tejada, P. J., Miller, J. R., Harron, J., Hu, B., Noland, T. L., Goel, N., Mohammed, G. H., and Sampson, P. (2004). Needle chlorophyll content estimation through model inversion using hyperspectral data from boreal conifer forest canopies. *Remote Sensing of Environment* **89,** 189–199.

Zhao, M., Heinsch, F. A., Nemani, R. R., and Running, S. W. (2005). Improvement of the MODIS terrestrial gross and net primary production global data set. *Remote Sensing of Environment* **95,** 164–176.

Zheng, D., Hunt, E. R., Jr., and Running, S. W. (1993). A daily soil temperature model based on air temperature and precipitation for continental applications. *Climatic Research* **2,** 183–191.

Zheng, D., Hunt, E. R., Jr., and Running, S. W. (1996). Comparison of available soil water capacity estimated from topography and soil series information. *Landscape Ecology* **11,** 3–14.

Zhou, L., Tucker, C. J., Kaufmann, R. K., Slayback, D., Shabanov, N. V., and Myneni, R. B. (2001). Variations in northern vegetation activity inferred from satellite data of vegetation index during 1981–1999. *Journal of Geophysical Research* **106,** 20069–20084.

Zhu, A.-X., Band, L. E., Dotton, B., and Nimlos, T. J. (1996). Automated soil inference under fuzzy logic. *Ecological Modelling* **90,** 123–145.

Ziemer, R. R. (1981). Roots and the stability of forested slopes. *International Association of Hydrologic Sciences Publication* **132,** 343–357.

Zimmermann, M. H. (1978). Hydraulic architecture of some diffuse-porous trees. *Canadian Journal of Botany* **56,** 286–295.

Zinke, P. J. (1967). Forest interception studies in the United States. *In* "Forest Hydrology" (W. E. Sopper and W. Lull, eds.), pp. 137–160. Pergamon, Oxford.

Zinke, P. J. (1980). Influence of chronic air pollution on mineral cycling in forests. *USDA Forest Service, Pacific Southwest Forest and Range Experimental Station, General Technical Report* **43,** Berkeley, California.

Zucker, W. V. (1983). Tannins: Does structure determine function? An ecological perspective. *American Naturalist* **121,** 335–365.

INDEX

Printed and bound by CPI Group (UK) Ltd, Croydon, CR0 4YY

09/10/2024

01042641-0001